Linear and Non-linear Numerical Analysis of Foundations

Linear and Non-linear Numerical Analysis of Foundations

Edited by
John W. Bull

CRC Press
Taylor & Francis Group
Boca Raton London New York

CRC Press is an imprint of the
Taylor & Francis Group, an **informa** business
A SPON PRESS BOOK

CRC Press
Taylor & Francis Group
6000 Broken Sound Parkway NW, Suite 300
Boca Raton, FL 33487-2742

First issued in paperback 2019

Typeset in 10/12pt Sabon by Graphicraft Limited, Hong Kong

ISBN13: 978-0-415-42050-1 (hbk)
ISBN13: 978-0-367-38588-0 (pbk)

British Library Cataloguing in Publication Data
A catalogue record for this book is available from the British Library

Library of Congress Cataloging in Publication Data
Bull, John W.
Linear and non linear numerical analysis of foundations / John W. Bull. — 1st ed.
p. cm.
Includes bibliographical references and index.
1. Foundations—Mathematical models. 2. Numerical analysis. I. Title. II. Title: Linear and nonlinear numerical analysis of foundations.
TA775.B85 2009
624.1′5015118—dc22
2008024484

Visit the Taylor & Francis Web site at
http://www.taylorandfrancis.com

and the CRC Press Web site at
http://www.crcpress.com

Contents

List of contributors

Gregor Bachmann, Institut und Versuchsanstalt für Geotechnik, Technische Universität Darmstadt, Germany.

Richard J. Bathurst, GeoEngineering Centre at Queen's-RMC Civil Engineering Department, Royal Military College of Canada, Kingston, Ontario.

John W. Bull, School of Civil Engineering and Geosciences, Newcastle University, Newcastle upon Tyne, UK.

Tadeusz Chmielewski, Technical University of Opole, Opole, Poland.

Deepankar Choudhury, Associate Professor, Department of Civil Engineering, Indian Institute of Technology (IIT) Bombay, Powai, Mumbai, India.

Emilios M. Comodromos, Thessaloniki, Greece.

Ahmed Elgamal, Professor and Chair, Department of Structural Engineering, University of California, San Diego La Jolla, California, USA.

J. S. Goldsworthy, Golder Associates Ltd., Calgary, Alberta, Canada.

Christian Gutberlet, Doktorand, Institut und Versuchsanstalt für Geotechnik, Technische Universität Darmstadt, Germany.

Kianoosh Hatami, Assistant Professor, School of Civil Engineering and Environmental Science, University of Oklahoma, Norman, Oklahoma, USA.

Bingquan Huang, GeoEngineering Centre at Queen's-RMC, Department of Civil Engineering, Queen's University, Kingston, Ontario, Canada.

M. B. Jaksa, University of Adelaide, Australia.

S. S. Jeong, Yonsei University, Korea.

Chai Junrui, College of Civil and Hydroelectric Engineering, China Three Gorges University, Yichang, Hubei Province, P. R. China; College of Hydroelectric Engineering, Xi'an University of Technology, Xi'an, Shaanxi

Province, P. R. China; College of Hydroelectric Engineering, Sichuan University, Chengdu, Sichuan Province, P. R. China.

Rolf Katzenbach, Director of the Institute and Laboratory of Geotechnics Technische Universität Darmstadt, Darmstadt, Germany.

Akiko Kishida, Graduate student, Department of Urban and Environmental Engineering, Graduate School of Engineering, Kyoto University, Kyotodaigaku-Katsura, Nishikyo-ku, Kyoto, Japan.

C. J. Lee, Kangwon National University, Korea.

Shouyi Li, College of Hydroelectric Engineering, Xi'an University of Technology, Xi'an, Shaanxi Province, P. R. China.

Xinan Liu, Department of Mechanical Engineering, University of Maryland, Maryland, USA.

D. S. Liyanapathirana, School of Civil, Mining and Environmental Engineering, University of Wollongong, NSW, Australia.

Richard S. Merifield, Centre for Offshore Foundation Systems, The University of Western Australia, WA, Australia.

C. W. W. Ng, Hong Kong University of Science and Technology, Hong Kong.

H. G. Poulos, School of Civil, Mining and Environmental Engineering University of Wollongong, NSW, Australia.

Izuru Takewaki, Department of Urban and Environmental Engineering, Graduate School of Engineering, Kyoto University, Kyotodaigaku-Katsura, Nishikyo-ku, Kyoto, Japan.

Yanqing Wu, College of Hydroelectric Engineering, Xi'an University of Technology, Xi'an, Shaanxi Province, P. R. China.

Yean-Seng Wu, Engineer, Hydraulic Engineering Department, Sinotech Engineering Consultants, Ltd., Taiwan, ROC.

Haiying Xin, The First Institute of Oceanography, State Oceanic Administration, Qingdao, Shandong, China.

Y. B. Yang, Department of Civil Engineering, National Taiwan University, Taiwan, ROC.

Zhaohui Yang, Department of Structural Engineering, University of California, San Diego La Jolla, California, USA.

Shihua Zhang, Drilling Technology Research Institute of Shengli Oilfield, Dongying, Shandong, China.

Quanan Zheng, Department of Atmospheric and Oceanic Science, University of Maryland, Maryland, USA.

Preface

John W. Bull

The correct understanding, design and analysis of foundations that support structures are fundamental to the safety of those structures. Witness the leaning tower of Pisa, which, if built a short distance from its present location, would have remained upright and have been just another safe structure!

With the introduction of more complex design codes, such as the Eurocodes in Europe, it is becoming increasingly necessary to use linear and non-linear numerical analysis in the design of foundations to model accurately the structure's response to loading.

In order to allow designers, engineers, architects, researchers and clients to understand the advanced numerical techniques used in the analysis and design of foundations, and to guide them into safer, less expensive and longer-lasting structural foundations, a wide range of world experts with knowledge in the latest advances in the design and analysis of foundations has been brought together, and their expertise presented in a clear and logical way.

The chapters in this book provide a review of state-of-the-art techniques for modeling foundations, using linear and non-linear numerical analysis, as they affect a range of infrastructure, civil engineering and structural engineering foundations. The use of these chapters will allow designers, engineers, architects, researchers and clients to understand the advanced numerical techniques used in the analysis and design of foundations, and to guide them into safer, less expensive and longer-lasting structural foundations.

The following topics are covered in the book:

Using probabilistic methods to measure the risk of geotechnical site investigations illustrates that probabilistic methods can be employed successfully to measure the effectiveness of site investigations.

The contribution of numerical analysis to the response prediction of pile foundations considers the necessity of including design procedures that incorporate soil non-linearity and the effects from pile group structural non-linearities.

Uplift capacity of inclined plate ground anchors in soil gives a rigorous numerical study of the ultimate capacity of inclined strip anchors, taking into account the effect of embedment depth, material strength and overburden pressure.

Numerical modeling of geosynthetic reinforced soil walls uses an instrumented full-scale laboratory test wall to identify numerical modeling issues associated with achieving reasonable predictions of key performance features of geosynthetic reinforced soil walls.

Seismic analysis of pile foundations in liquefying soil outlines a dynamic effective-stress-based free-field ground response analysis method and a numerical procedure for the seismic analysis of pile foundations in liquefying soils.

The effect of negative skin friction on piles and pile groups investigates the effects of soil slip at the pile–soil interface on dragload development for single piles and pile groups.

Semi-analytical approach for analyzing ground vibrations caused by trains moving over elevated bridges with pile foundations presents an investigation into ground vibrations induced by trains traveling over a multi-span elevated bridge with pile foundations.

Efficient analysis of buildings with grouped piles for seismic stiffness and strength design shows that a detailed and efficient examination on pile-group effect is necessary in the practical seismic design of buildings from the viewpoint of stiffness and of strength.

Modeling of cyclic mobility and associated lateral ground deformations for earthquake engineering applications focuses on important aspects of soil cyclic mobility and its effects on lateral ground deformations, including liquefaction scenarios.

Bearing capacity of shallow foundations under static and seismic conditions analyses shallow footings in various ground conditions under both static and seismic loading.

Free vibrations of industrial chimneys or communications towers with flexibility of soil shows that soil flexibility under the foundation of a chimney considerably influences the chimney's natural modes and natural periods.

Assessment of settlements of high-rise structures by numerical analysis assesses settlement for both the serviceability limit state and the ultimate limit state of high-rise structures.

Analysis of coupled seepage and stress fields in the rock mass around the Xiaowan arch dam considers the coupled seepage and stress fields in the rock mass around the dam.

Development of bucket foundation technology for operational platforms used in offshore oilfields shows that bucket foundation technology is a reliable, low-cost, environment-friendly technology especially suitable for the construction of oil and gas development platforms in shallow-water areas with thick sea-floor sediment layers.

I am extremely grateful to all of the authors for their diligence in writing their chapter and for giving so generously of their time and knowledge to ensure the high quality of this book. I would like also to express my thanks to my publishers, Taylor & Francis, for their help and guidance in producing this book.

1 Using probabilistic methods to measure the risk of geotechnical site investigations

J. S. Goldsworthy and M. B. Jaksa

1 Introduction

The site investigation phase plays a vital role in foundation design where inadequate characterization of the subsurface conditions may lead to either an under-designed solution, resulting in failure, or an over-designed solution that is not cost-effective. Whether the design is for a building foundation or for a retaining wall for an unstable slope, an investigation of some form is required to predict the soil properties in order to estimate the soil response to applied loading. However, determining whether the scope and type of an investigation is suited to the site and the required design situation is not a straightforward task. Typically, the type and scope of an investigation is determined by a senior geotechnical engineer within the budget and time constraints placed on a project. However, it is rarely known, in other than subjective terms, whether the type and scope of the investigation is adequate or suitable. In order to account for this, as well as for the complex nature of soil behavior under load, geotechnical engineers use higher factors of safety than are otherwise used in other forms of engineering, such as structural engineering.

One of the reasons for the use of high factors of safety is the lack of knowledge regarding the effectiveness of the site investigation performed. Questions such as 'Were sufficient bore holes drilled and/or excavations dug to characterize the stratigraphy appropriately?' and 'Were enough samples taken and tests performed to gain an adequate representation of the ground response to load?' cannot be answered objectively and therefore result in additional redundancies being added to the design. If geotechnical engineers were able to gain a better understanding of the effectiveness of the investigation, the factors of safety used in the design could be selected more appropriately. As such, Jaksa et al. (2003) proposed a framework to measure the effectiveness of geotechnical site investigations using probabilistic methods and risk analysis. The framework was developed into a simulation model and used by Goldsworthy (2006), who presented results that illustrated the risks associated with geotechnical investigations.

The specific goals of this chapter are to illustrate that probabilistic methods can be employed successfully to measure the effectiveness of site investigations. Since a geotechnical engineering system contains several sources of uncertainty, a simulation model such as Monte Carlo simulation provides a tool to incorporate all sources of uncertainty. The results indicate that there are considerable differences between site investigations of different scope, and by incorporating the sources of uncertainty in a probabilistic manner it is possible to draw conclusions regarding more suitable site investigation techniques.

2 Geotechnical site investigations

In all forms of engineering, suitable information and data are required for a successful design. In structural engineering, such information is readily available and typically well defined as materials such as concrete and steel are manufactured to specific quality guidelines. However, geotechnical engineering is very different. Instead of manufactured materials, geotechnical engineers deal with materials that have generally been created by nature. Therefore, the data or information collection about the ground conditions is vital to the accuracy and adequacy of the design.

The purpose of site characterization is to obtain a reasonable representation of the subsurface conditions (Bowles 1997, Lee et al. 1983, Lowe III and Zaccheo 1991). Characterization refers to both reconnaissance and investigation, where the former relates to the review of surrounding geology and the latter involves ground exploration through testing (Baecher and Christian 2003). Lee et al. (1983) further categorized investigative methods into 'areal' and 'local' explorations. Areal explorations are not concerned with obtaining specific soil properties, whereas local explorations are directed toward such properties. Areal explorations may also include drilling and sampling (Lowe III and Zaccheo 1991). Samples obtained from areal exploration techniques are usually undisturbed and are suitable for soil classification or laboratory tests. In-situ geotechnical tests are examples of local explorations and provide results regarding physical and mechanical soil properties (Lee et al. 1983). Sample disturbance is common with local explorations, potentially affecting the accuracy of the resulting properties. Laboratory tests generally provide results that, from a theoretical viewpoint, are better-understood and are sometimes easier to incorporate into design relationships. However, such tests do not always accurately simulate the stresses and in-situ soil conditions (Becker 2001).

2.1 Measuring soil investigation effectiveness

Since soil deposits vary from site to site, it is difficult to measure the effectiveness of one site investigation scheme in comparison with another. The only practical means is to have complete information about the site which can act as a benchmark. However, gaining complete knowledge is beyond current realms of geotechnical testing and is therefore impractical and infeasible.

Independent research has been previously performed to measure the effectiveness of site investigation plans. Whitman (2000) suggested 'search theory', as introduced by Baecher (1979) and Halim and Tang (1993), as a probabilistic method to measure the adequacy of a site investigation. Search theory provides a probabilistic estimate of the ability to locate a flaw or defect in a soil deposit with reasonably known conditions. Parsons and Frost (2002) also used probabilistic methods to measure the 'thoroughness' of site investigations. They used geostatistics and a GIS (geographical information system) to compare the ability of one sampling strategy to another to characterize a known soil. Tsai and Frost (1999) also used thoroughness to examine the quality of site investigations in a similar framework. In relation to guidance with respect to site investigations, Wiesner (1999) suggested that using vertical holes in a 50 m grid is common practice. However, Bowles (1997) suggested that the frequency and arrangement of sampling should be sufficient to investigate the site area that may be affected by the final design. This is very general and provides little quantitative assistance for the practitioner.

On the whole, the geotechnical engineering industry appears to be guided by a combination of experience and national codes. However, like the recommendations of Bowles (1997), national standards are extremely general as they are designed to be applicable for a large proportion of the sites that may be encountered. An example of such recommendations is that 'the investigation shall evaluate the material properties and the volume of ground which will significantly affect the performance of . . . the proposed works' (Australian Standards 1993).

Since site investigation expenditure is a small fraction of the project cost, an increase in scope that reduces the risk of a cost overrun or foundation failure does not have a significant impact on the total cost (Clayton et al. 1982). However, the Site Investigation Steering Group (1993) believed that an improved site investigation may not necessarily incur an additional cost. Instead, they believed that current site investigation practices could be refined to yield designs with a lower risk of failure or cost overrun.

2.2 Using probabilistic methods to measure site investigation effectiveness

Baecher and Christian (2003) suggested that there is definite scope for research measuring the effectiveness of site investigations. In fact, they suggested that many believe there is a solution to the problem of site characterization waiting to be discovered. Furthermore, Clayton (2001) believed that a new risk-based method is required for ground investigations.

Geotechnical engineers typically use two methods to account for the uncertainties in a given system. The more common has been to use a global factor of safety, where either the strength is reduced or the loads are increased to ensure that the design is sufficiently conservative to meet the specified purpose. Phoon et al. (1995) suggested that the factor of safety is typically applied to the material strength or the capacity of the design, as it is the most uncertain component. This philosophy is common in most geotechnical engineering texts (Bowles 1997, Fang 1991, Terzaghi and Peck 1968). The factor of safety is generally determined by the experience and judgment of the geotechnical engineer (Phoon et al. 1995). Furthermore, the use of a single factor of safety does not distinguish between model and parameter uncertainties, and hence does not allow the identification of possible improvements by either increased sampling or alternative testing.

Relatively recently, the geotechnical engineering profession has investigated the use of probabilistic methods to accommodate uncertainty in geotechnical systems. Probabilistic methods have an advantage over typical global factor of safety approaches since (Phoon et al. 1995):

- more cost-effective foundation designs can be targeted;
- any incompatibility between structural and foundation design is minimized;
- the engineer is relieved of assessing the relationship between uncertainties and risk.

However, Phoon et al. (1995) also concluded that closed-form solutions of probabilistic methods are difficult to obtain. Therefore, it is usual that probabilistic methods are solved using either numerical- or simulation-based techniques such as (Baecher and Christian 2003):

- First Order Second Moment (FOSM)
- Second Order Second Moment (SOSM)
- Point Estimation
- Monte Carlo simulation.

As such, Jaksa et al. (2003) proposed that site investigation effectiveness could be examined in a probabilistic manner using Monte Carlo simulation. Goldsworthy (2006) refined the framework in order to quantify the risk of a site investigation based on a resulting foundation design using only serviceability criteria. The general arrangement of this process is encapsulated by Figure 1.1 and is exclusively a numerical modeling activity. In other words, no physical site investigation takes place. This is because comprehensive site investigations, which are required to provide a benchmark for comparison, are prohibitively expensive and impractical.

The process shown in Figure 1.1 is an extension of similar research performed by Fenton and Griffiths (2005), who examined the reliability of

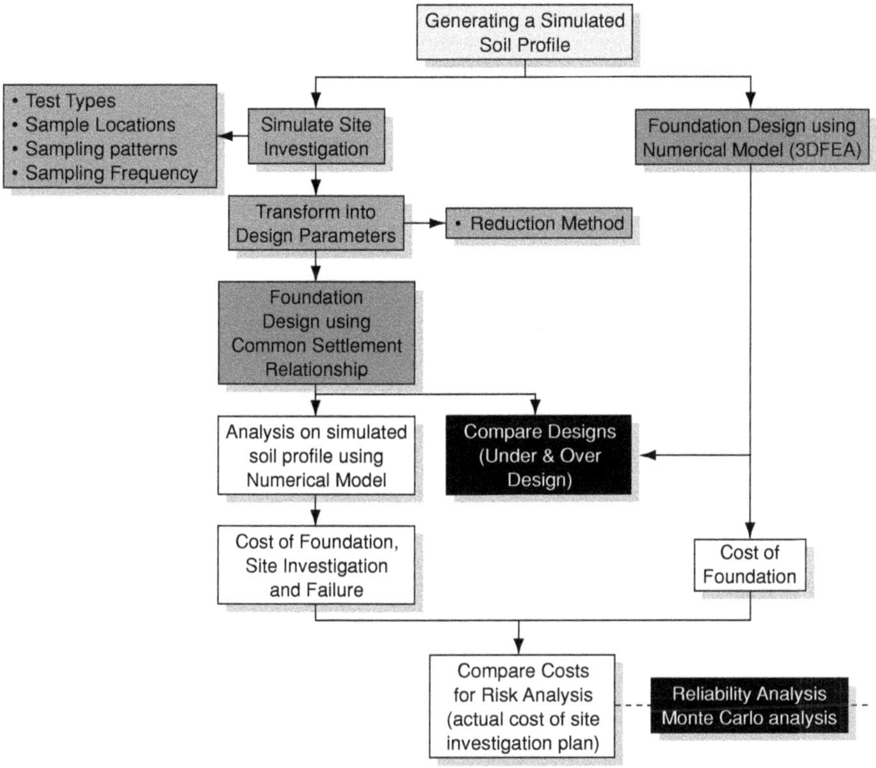

Figure 1.1 Structure of simulation model (after Goldsworthy 2006)

settlement estimates using random field theory to simulate the spatially random nature of soil properties. However, in this case, Goldsworthy (2006) incorporated a design process, additional sources of uncertainty and a costing regime in order to evaluate a measure of financial risk.

The Monte Carlo simulation of the process above incorporates the use of random variables to model various sources of uncertainty. In the example that follows, the Monte Carlo simulation consists of 1000 realizations, where the process illustrated in Figure 1.1 is repeated. The outcome is a distribution of results from which statistical measures, such as the average (an estimation of the mean), standard deviation and coefficient of variation[1] (COV), can be determined.

1 The coefficient of variation is defined as the standard deviation divided by the mean and is a normalized measure of the variability.

3 An example for foundation design

To illustrate the use of probabilistic methods and reliability analysis to measure the risk of a geotechnical site investigation, in the framework described above, a typical foundation design problem is examined. In this case, risk is considered to be financial, where the costs associated with the construction of the foundation and structure are included, as well as the potential costs of a foundation failure resulting from an inappropriate site investigation.

3.1 Design assumptions

The foundation designed in this example is to support a five-story building with a building footprint of 20 m × 20 m with an equal live floor loading of 3 kPa per story and an assumed dead load of 5 kPa per story. The structure is assumed to be supported by nine equi-spaced columns arranged in a grid formation, having a center-to-center spacing of 8 m. Based on tributary slab areas, and proportioning the loads to each column, the corner columns support a load of 960 kN, the center column a load of 1540 kN and the remaining columns loads of 1150 kN.

The nine columns are assumed to be supported by nine individual pad footings that are located at the soil surface (without footing embedment). In this case, the footings are sized to meet only serviceability criteria, and potential failures due to bearing capacity are not considered critical.

The site is assumed to be 50 m × 50 m in plan area, and of a single layer. A statistically homogeneous[2] soil deposit is assumed to extend to a depth of 30 m, where it is underlain by an incompressible stratum. A plan layout of the building footprint and the site boundary is given in Figure 1.2.

3.2 Uncertainties in the foundation design problem

Vanmarcke (1977a, b) suggested that three main sources of uncertainty exist in the estimation of suitable soil properties. These are due to inherent soil variability, statistical uncertainty due to limited sampling, and measurement uncertainties due to associated geotechnical testing errors. In addition, Kulhawy (1992) suggested that sampling error can also be considered statistical uncertainty, and results from limited information about the site can be minimized through additional sampling (Phoon et al. 1995, Vanmarcke 1977a, b). Whitman (2000) adopted a simpler explanation, whereby the uncertainties due to soil variability and random testing errors contribute

2 A soil is considered to be statistically homogeneous if properties vary independently of position in the deposit. Therefore, the mean is independent of location and does not increase nor decrease with depth. For a complete description of statistical homogeneity the reader is referred to Vanmarcke (1977).

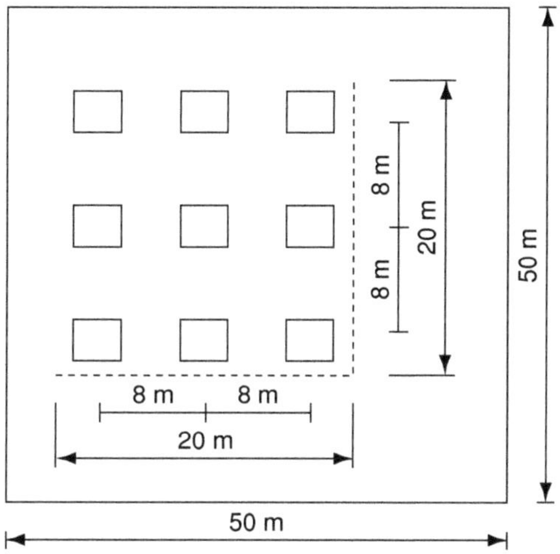

Figure 1.2 Schematic of the 9-pad footing foundation system

to data scatter, while the statistical uncertainty and bias in testing error contribute to systematic errors. Kulhawy and Phoon (2002) have also indicated that soil variability and measurement error have an impact on data scatter.

A slightly different approach to separating the sources of uncertainty has been adopted by Baecher and Christian (2003), who considered the sources to be:

- natural variability
- knowledge uncertainty
- decision model uncertainty

Essentially the first two sources identified by Baecher and Christian (2003) are equivalent to those identified by Kulhawy (1992), where natural and inherent soil variability are the same, and uncertainties due to measurement and transformation model error are equivalent to knowledge uncertainty. Furthermore, Baecher and Christian (2003) categorized knowledge uncertainty into effects dealing with site characterization, model and parameter uncertainty. Site characterization uncertainty accounts for both measurement errors and statistical uncertainty, as described by Filippas et al. (1988) and Kulhawy (1992) respectively, while model and parameter uncertainty is equivalent to transformation model error. The additional source of uncertainty identified by Baecher and Christian (2003), due to decision models, is a function of the decisions an engineer or a client makes regarding conservatism, as well as effects during construction. Such uncertainties are usually due to economic and temporal considerations.

Four sources of uncertainty are considered in this example:

- soil variability
- statistical uncertainty
- measurement error
- transformation model errors

The above sources of uncertainty fall into the categories of natural variability and knowledge uncertainty, as described by Baecher and Christian (2003). Decision model uncertainty is not considered here as modeling engineering judgment and construction effects adds another level of complexity that requires further research and treatment.

The following sections describe how the selected four sources of uncertainty are incorporated in the simulation model to measure the effectiveness of a site investigation for foundation design.

3.2.1 *Inherent variability of soil properties*

Unlike many civil engineering materials, soils are inherently variable, where properties may be significantly different from one location to another. Even when soils are considered reasonably homogeneous, soil properties exhibit considerable variability (Vanmarcke 1977a). This variability is due to the complex and varied physical phenomena experienced during their formation. Variability between soil properties is called spatial variability and has recently been modeled as a random variable (Spry et al. 1988).

Fenton and Griffiths (2003, 2005) used the Local Average Subdivision (LAS) method developed by Fenton (1990) and Fenton and Vanmarcke (1990) to generate three-dimensional random fields that represent the variability of properties within a soil. LAS is well suited to the form of analysis undertaken by Fenton and Griffiths (2003, 2005), because the properties are generated as an average of an element, which can be mapped directly to a finite element mesh. The method generates three-dimensional random fields that conform to a normal distribution and a nominated correlation structure. The distribution is defined by the mean, μ, and the standard deviation, σ, while the correlation structure is defined by the scale of fluctuation, θ. The scale of fluctuation (SOF) can be loosely defined as the distance within which soil properties are considered reasonably correlated (Vanmarcke 1977a, 1983). Fenton (1996) suggested that a simple exponentially decaying, or Markovian, correlation structure best represents the correlation between properties in a soil mass. A three-dimensional exponentially decaying correlation structure is given by:

$$\rho(\tau) = \exp\left(-\sqrt{\left(\frac{2\tau_x}{\theta_x}\right)^2 + \left(\frac{2\tau_y}{\theta_y}\right)^2 + \left(\frac{2\tau_z}{\theta_z}\right)^2}\right) \qquad (1)$$

where $\rho(\tau)$ is the correlation between two properties in the soil mass separated by a lag or distance vector, $\tau = \{\tau_x, \tau_y, \tau_z\}$, and θ_x, θ_y and θ_z are the SOFs in the x, y and z directions respectively.

To generate soil properties using LAS, the 3D soil deposit is discretized into elements of size 0.5 m × 0.5 m × 0.5 m. This leads to an element grid of 100 × 100 × 60, or 600,000 elements in total. Because the shallow footings designed in this example are sized only to meet serviceability criteria, only the elastic soil properties are required. Furthermore, Fenton and Griffiths (2005) suggested that the variability of Poisson's ratio (ν) has little impact on the estimate of footing settlement. As such, a constant Poisson's ratio of 0.3 is assumed, and the only property simulated using probabilistic theory is the elastic modulus.

To model the spatial variability of elastic moduli, a single value is generated for each element and is sampled from a normal distribution of possible values conditioned by a nominated mean and COV. The sampled value is then modified to meet the spatial correlation requirements set out by LAS. Finally, the sample is transformed to a lognormal variant using:

$$X_{\ln} = \exp(\mu_{\ln x} + \sigma_{\ln x} X) \tag{2}$$

where $X_{\ln}$ is the lognormal variant, $\mu_{\ln x}$ and $\sigma_{\ln x}$ are the mean and standard deviation of the lognormal variant respectively, and X is the lognormal variant sampled from the normal distribution. The result is that properties in each element conform to the nominated spatial statistics including mean, COV and SOF in all three directions. Further, since LAS is a top-down discretization method, the spatial statistics of each individual element are also representative of the entire field.

For this example, several combinations of input statistics are used to define different types of soils. The ranges of input statistics are given in Table 1.1. Note the distinction between horizontal and vertical SOF. Using different SOF values allows the investigation of soils with an anisotropic correlation structure. Natural soil deposits typically show greater correlation in the horizontal direction than in the vertical owing to their formation processes, which generally occur in layers (Jaksa et al. 2005).

It is often difficult to visualize the impact of the SOF on soil properties. Therefore, Figure 1.3 presents two different soil deposits, both with identical

Table 1.1 Statistical properties of simulated soil properties for each element

Statistical measures	*Values considered*
Mean (kPa)	30,000
Coefficient of Variation – COV (%)	10, 20, 50, 100
Scale of fluctuation (horizontal) – SOF_h (m)	1, 2, 4, 8, 16, 32
Scale of fluctuation (vertical) – SOF_v (m)	1, 2, 4, 8

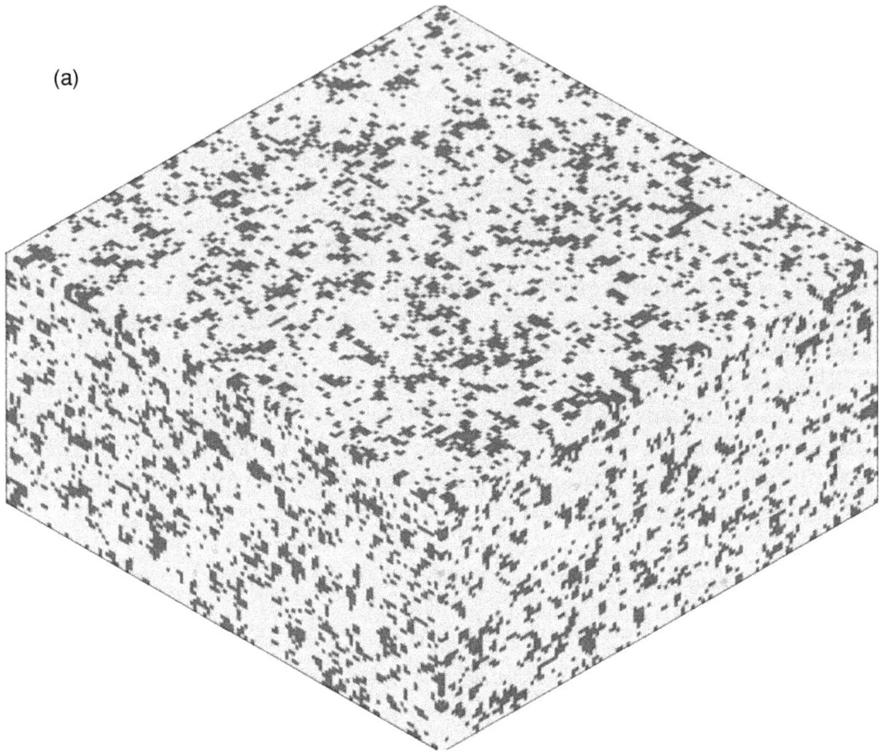

Figure 1.3 Differences between a soil with (*a*) a relatively low SOF and (*b*) a larger SOF

means and COVs, with Figure 1.3(a) illustrating an example of a deposit with a relatively low SOF and Figure 1.3(b) a deposit with a high SOF. Note the regions of similar properties in Figure 1.3(b), which is synonymous with a high SOF. It is clear from Figure 1.3 that a high SOF indicates a more continuously varying deposit.

In the analyses presented below, the type of soil is described by the COV and the SOF in both the horizontal (both planar directions parallel to the surface of the site) and vertical directions. Notation is given as COV(SOF_h, SOF_v).

3.2.2 Statistical uncertainty

The statistical uncertainties associated with a geotechnical model are a result of limited sampling that may not provide an accurate representation of the underlying conditions. Filippas et al. (1988) defined the statistical uncertainty for a set of uncorrelated samples as the variance in the estimate of the mean.

(b)

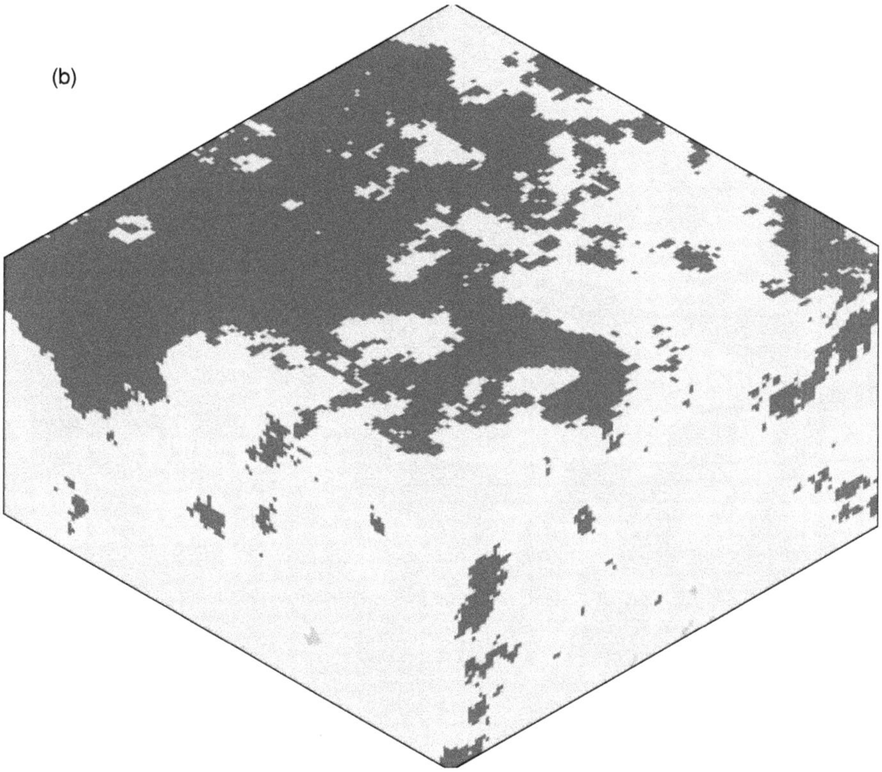

Figure 1.3 (cont'd)

To model the statistical uncertainty of typical site investigation practice, twelve different sampling arrangements are investigated based on two different sampling patterns. The twelve different sampling strategies based on a regular grid pattern are given in Figure 1.4(a), while the twelve strategies based on a stratified random pattern are given in Figure 1.4(b). The least intensive sampling program for both patterns involves a single sampling location, while the most intensive utilizes information from twenty-five different locations.

In addition to modeling the statistical uncertainty of site investigations using different sampling strategies in an areal extent, vertical sampling is also considered. This is because different types of geotechnical testing methods obtain soil information at different depth intervals. For example, a standard penetration test (SPT) is an example of a discrete sampling method, where soil information is usually only retrieved at 1.5 m depth intervals. On the other hand, the cone penetration test (CPT) may acquire information about the soil at much smaller intervals (10 mm to 50 mm). Therefore, the CPT is an example of a continuous sampling method.

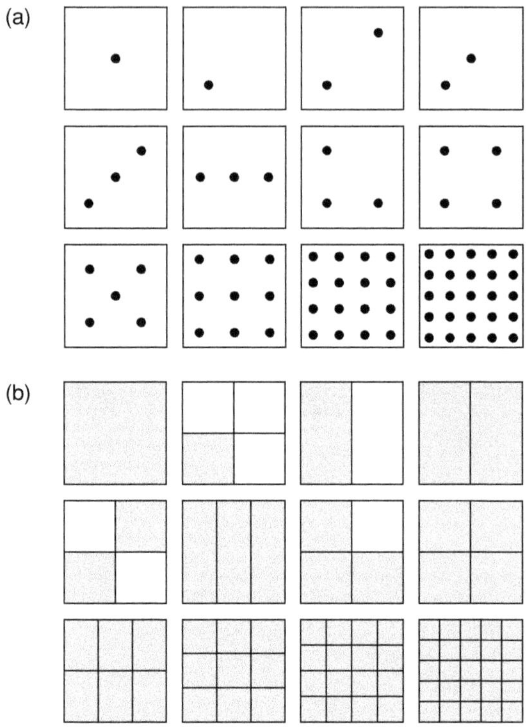

Figure 1.4 Sampling locations for (*a*) the regular grid and (*b*) stratified random patterns (for stratified random pattern, one sampling location is randomly positioned in each shaded region)

The model used in this example accommodates different vertical sampling intervals based on the size of the element used for the simulation of soil properties. In this case, a CPT is assumed to retrieve information from every element in the vertical direction, while the SPT is assumed to obtain soil information from every third element. Two additional test types are also modeled in this example: the triaxial test (TT) and the flat plate dilatometer test (DMT). The DMT is assumed to have the same vertical sampling frequency as the SPT (every third element is used), while the vertical sampling frequency of the TT is assumed to be much coarser because, in typical site investigations, fewer are performed per bore hole. In this case, the TT is assumed to utilize soil information from only two elements in each bore hole, where the bore hole is assumed to be the depth of the soil layer (30 m). The vertical sampling intervals of each test method are summarized in Table 1.2.

Another procedure that has a direct impact on the statistical uncertainty of the soil property estimate is the manner in which soil properties obtained

Table 1.2 Assumed vertical sampling frequency for test methods investigated

Test type	*Vertical sampling frequency*	
	Element	*Depth interval (m)*
Standard penetration test (SPT)	Every third	1.5
Cone penetration test (CPT)	Every	0.5
Triaxial test (TT)	Every third	1.5
Flat plate dilatometer (DMT)	2 per bore hole	15

at the same depth from different sampling locations are combined. The most common method is to estimate the mean of the values by calculating the average. However, there are several alternatives to combining the results from various sampling locations. Five such methods are considered in this example:

- standard arithmetic average (SA)
- geometric average (GA)
- harmonic average (HA)
- inverse distance weighted (ID)
- 1st quartile threshold (1Q)

Goldsworthy et al. (2005) considered the method used to combine values from multiple sampling locations as the 'reduction technique' and identified that each method has a very different impact on the effectiveness of a site investigation. Such results are shown later. For completeness, the formulation of each reduction technique is given in Appendix A.

3.2.3 Measurement error

Measurement errors arise from the inability of geotechnical tests to estimate accurately the soil properties being tested. Sources of measurement error can be separated into two categories: random and systematic (Filippas et al. 1988). Random testing effects are inherent to the test type but cannot be attributed to the spatial variability of soil properties. Their effects are generally considered to have zero mean and influence the results of the soil properties equally above and below the mean (Baecher 1979, Snedecor and Cochran 1980). Filippas et al. (1988) suggested that the best way to evaluate random testing effects is by undertaking several tests under essentially identical conditions. Systematic errors consistently under- or overestimate the property and are generally due to operator and procedural effects and inadequacies with the equipment (Jaksa et al. 1997). Lumb (1974) considered such errors as a bias.

Table 1.3 Assumed COV values representing measurement errors

Test type	*Measurement error COV*	
	Bias	*Random*
Standard penetration test (SPT)	20%	40%
Cone penetration test (CPT)	15%	20%
Triaxial test (TT)	20%	20%
Flat plate dilatometer (DMT)	15%	15%

For this example, uncertainties due to measurement error are separated into bias and random effects, and assigned a different degree of uncertainty for different testing methods, as is the case in reality. Each source of uncertainty is characterized by a unit-mean lognormal variable with a nominated COV value. The COV corresponding to each test method examined in this example (SPT, CPT, TT and DMT) is given in Table 1.3. The COV values given in Table 1.3 are consistent with relative errors between test types as described by Orchant et al. (1988) and Phoon and Kulhawy (1999a).

The measurement errors for bias and random effects are included in a two-step process. Consider a test result as shown in Figure 1.5(a) with properties taken directly from the random field representation of the soil. The set of properties have a sample mean, as given by the broken line in Figure 1.5(a). The bias error is determined by multiplying the sample mean with a unit-mean lognormal variable with a COV representative of the type of test, as given in Table 1.3. In the example shown in Figure 1.5, the difference between the biased mean and the sample mean is shown in Figure 1.5(b). The difference between the biased mean and the sample mean is then added to each sample property, to yield a biased sample, as shown in Figure 1.5(c). The bias amount is recalculated for each test location.

The random error is included by multiplying each biased sample value by a unit-mean lognormal variable with a COV representative of the test type as given in Table 1.3. This changes each sample value by a different amount and yields a resultant sample as shown in Figure 1.5(d).

The process of adding both bias and random errors to the elastic modulus value sampled directly from the random field representation of the soil profile can be represented by:

$$E_r^* = (m_b m_r) E_f \tag{3}$$

where E_r^* is the resultant elastic modulus value with both bias and random measurement errors included, m_b and m_r are unit mean lognormal variables representing bias and random measurement error respectively, with COVs as given in Table 1.3, and E_f is the elastic modulus value sampled directly from the random field representation of the soil profile.

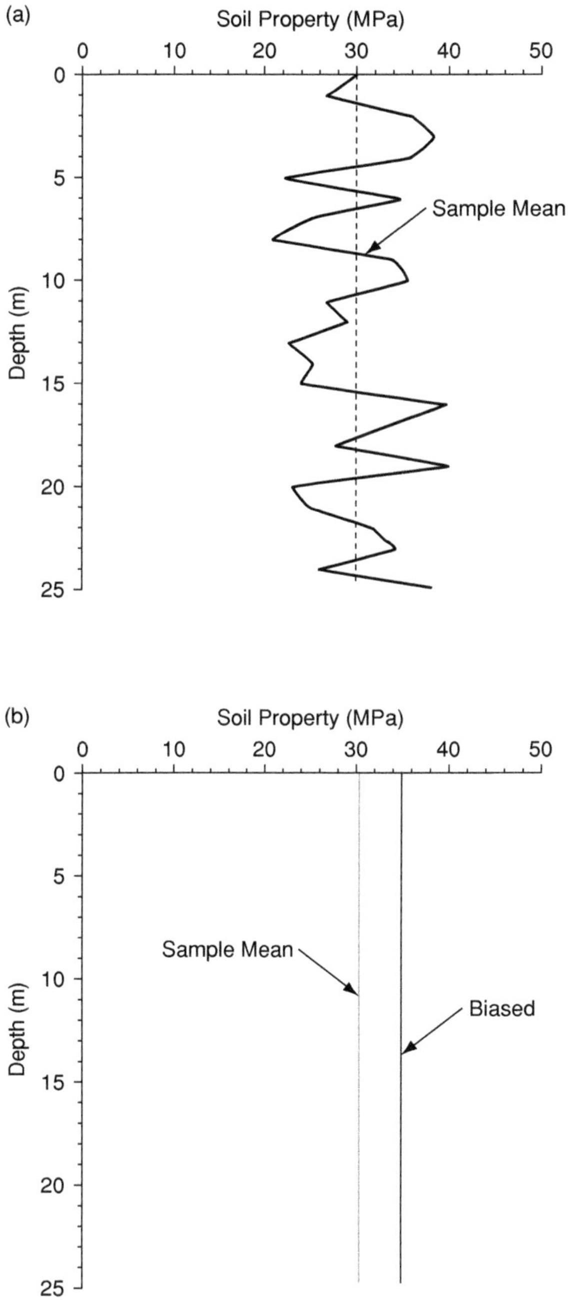

Figure 1.5 Process of attributing test uncertainties

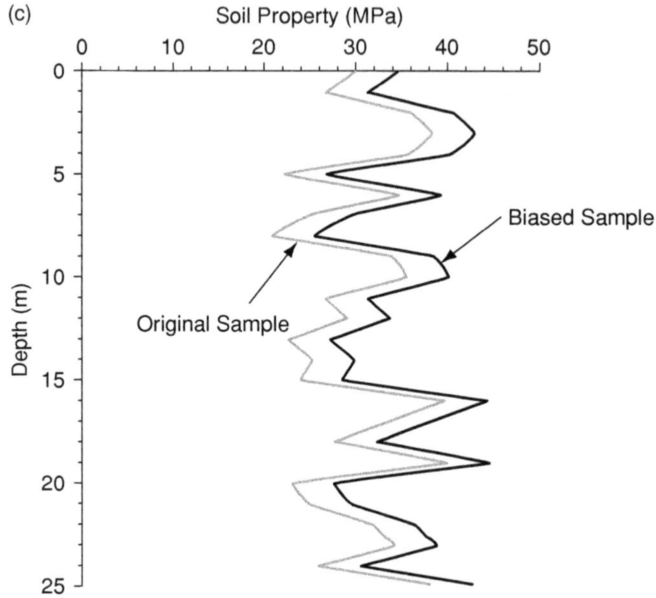

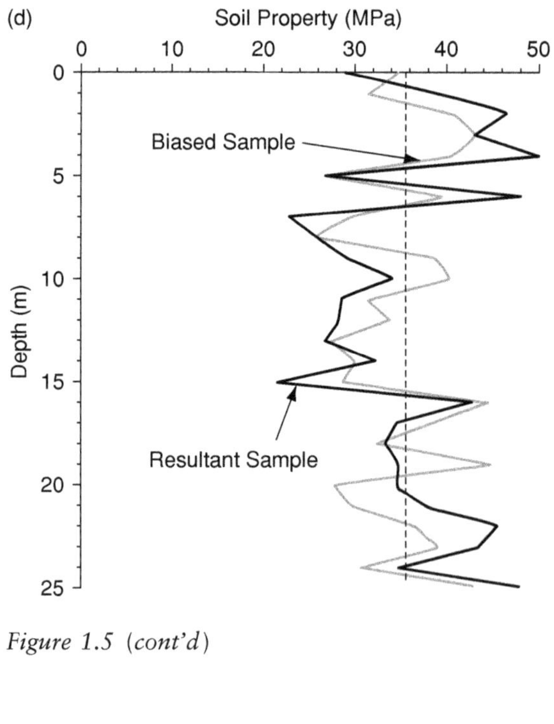

Figure 1.5 (*cont'd*)

3.2.4 *Transformation model uncertainty*

Transformation model errors result from the fact that common geotechnical tests do not always provide applicable soil properties that are useful for design relationships (Phoon and Kulhawy 1999b). Rather, the raw test data are processed using a transformation model into a suitable design parameter. Such models are often obtained empirically through back-substitution or calibration. Accordingly, a degree of uncertainty is added to the estimation of the design parameter. Phoon and Kulhawy (1999b) further stated that uncertainty still exists if the transformation is based on a theoretical relationship because of idealizations and simplifications in the theory. Therefore, it is important to consider the uncertainties due to transformation model error. Similar to the uncertainties for the measurement error described above, the transformation model errors are simulated using a unit mean lognormal variable with a nominated COV that is different for the various test methods considered in this example. The COV values for each test method are summarized in Table 1.4.

The transformation model error is added to the model by multiplying the resultant elastic modulus value as given by E_r^* in Equation (3) with a unit mean lognormal variable with a COV representative of the test type, as given in Table 1.4. This process is expressed by:

$$E_r = (tm)E_r^* \tag{4}$$

where E_r is the resultant elastic modulus including both measurement and transformation model errors, tm is a unit mean lognormal variable representing transformation model errors with a COV as given in Table 1.4, and E_r^* is the resultant elastic modulus including measurement errors, as given in Equation (3).

Transformation model errors are applied consistently to all test locations within an investigation program. However, a different transformation model error is applied for different investigation programs, or the same investigation program on a different soil.

3.3 *Foundation design process*

In a serviceability limit state design, footings are sized to ensure that the maximum and the differential settlements do not exceed specified limits. In the analyses that follow, the maximum settlement limit is set to 25 mm, while the differential settlement limit is set to 0.0025 m/m.

To begin the design process, an initial minimum footing size of 0.5 m × 0.5 m is adopted for all footings in the foundation system. If the settlement of any footing, or the differential settlement between any two footings, exceeds the limits indicated above, the footing size is increased by 0.1 m in one of the plan directions. In the case of the differential settlement exceeding

Table 1.4 Assumed COV values representing transformation model errors

Test type	*Transformation model COV*
Standard penetration test (SPT)	25%
Cone penetration test (CPT)	15%
Triaxial test (TT)	0%
Flat plate dilatometer (DMT)	10%

the limit, the size of the footing with the largest predicted settlement is increased. This process is repeated until the predicted settlement of all footings is less than the maximum settlement limit, and the predicted differential settlement between every pair of footings is less than the differential settlement limit. If a footing size is required to be increased more than once, the plan dimensions of the footing are increased in alternate directions for subsequent iterations. For example, the plan size of the footing will be increased from 0.5 m × 0.5 m to 0.5 m × 0.6 m then 0.6 m × 0.6 m, 0.6 m × 0.7 m, and so on.

In this example, footing settlements used for design purposes are estimated with the Schmertmann prediction technique incorporating the '2B-0.6' strain influence distribution (Schmertmann 1970). Although others (Holtz 1991, Small 2001) believe this technique is applicable only for cohesionless soils, Bowles (1997) suggested that it provides a good alternative settlement estimate for all soil types. Goldsworthy (2006) investigated the risk of geotechnical site investigations based on the use of other foundation design techniques.

The effect of adjacent footings is also considered, by predicting the settlement of one footing due to another footing. This process, which is based on the principle of superposition, is detailed in Appendix B. It is expected, however, that the settlement of a footing due to adjacent foundations will have minimal impact on the results, because the minimum footing spacing is 8 m. This was confirmed by Goldsworthy (2006). However, for completeness, the effects of adjacent footing are still considered here.

3.4 Calculating financial risk

Since financial risk, in this example, is considered to be a function of the costs, including consequential costs, associated with performing site investigations of varying scope, costs are assigned to:

- the site investigation phase
- the construction of the foundation
- the superstructure
- rehabilitation works required in the event of a foundation failure

The sum of these yields the total cost of the design, C_{TOT}, as given by:

$$C_{TOT} = C_{SI} + C_{Con} + C_{Rhb} \quad (5)$$

where C_{SI} is the cost of the site investigation, C_{Con} is the cost of the construction of both the foundation and the superstructure, and C_{Rhb} is the cost associated with rehabilitation works due to a foundation failure. In the event that the site investigation yields a foundation design that is much larger than required, the cost associated with the rehabilitation, C_{Rhb}, will be zero, but the cost associated with construction will be somewhat higher. Conversely, if the site investigation yields a foundation design that is smaller than required, the construction cost will be low but the cost associated with rehabilitation will be greater than zero.

The costs associated with the site investigation are based on industry rates in South Australia (Jaksa 2004) and are expressed in Australian dollars ($AUD). The adopted rates are dependent on the type of test, as given in Table 1.5. Rates for the SPT, CPT and DMT also include the cost of hiring a drilling rig, while the costs associated with the TT are based on a consolidated undrained test and include extra costs associated with drilling and sampling from the site. Costs associated with the SPT, CPT and DMT are based on a full depth test of 30 m, while the costs associated with the TT are based on two test samples per sampling location.

Costs associated with the construction of the foundation are based on a cubic metre rate of $510, adopted from Rawlinsons (2004). This rate includes: excavation; supply and placement of steel reinforcing; supply, placement and finishing of concrete; and backfilling. The volumetric size of each footing is determined using the plan area designed to meet the serviceability criteria and a thickness estimated using the Australian Standard for beam and punching shear, as described by Warner et al. (1998).

Construction costs of the superstructure are based on a five-story, fully serviced office building. Costs for such a building, given by Rawlinsons (2004), also include the construction cost of the foundation. However, because the foundation cost is variable in this form of analysis, a nominal percentage of 1.8 percent (determined by estimating the proportion of cost associated with the construction of the foundation alone) is removed from the

Table 1.5 Rates adopted for different test types

Test type	*Cost ($AUD/sample location)*
Standard penetration test (SPT)	2900[1]
Cone penetration test (CPT)	3300[2]
Triaxial test (TT)	2650[3]
Flat plate dilatometer (DMT)	3600[1]

1 Based on vertical test rate of 1.5 m
2 Based on vertical test rate of 0.5 m
3 Based on 2 tests per sample location

Table 1.6 Rehabilitation costs

Category	*Description*	*Cost ($AUD/m²/story)*
Minor	Patching, repainting and minor plumbing repairs	410
Major	Significant patching, structural retrofitting, major plumbing repairs and foundation underpinning	2035 + 1730/m² footing

Rawlinsons (2004) rate, to yield a superstructure construction cost rate of $11,650 per square metre of plan area.

Rehabilitation costs are incorporated based on the severity of the foundation failure. For example, if the foundation design is found to be highly inadequate and the actual settlement of the foundation system is large, the rehabilitation cost is also large. On the other hand, if the foundation failure is found to be minimal, the rehabilitation cost is also minimal. For the extreme case, when the foundation failure is so large that the building could not be used for its intended purpose, the rehabilitation cost is based on demolishing and rebuilding the structure.

Costs associated with a minor and major degree of rehabilitation are given in Table 1.6. The minor rehabilitation costs are based on minor repair works to the superstructure including crack-filling, repainting, and repair works to plumbing. The major rehabilitation works are based on major repairs works including repointing of walls and foundation underpinning. Rates for the minor and major works are adopted from Rawlinsons (2004).

The severity of a foundation failure is determined by analyzing the foundation design (originally based on the information gained in a site investigation) on a soil considering each and every elastic modulus. This is considered a 'complete knowledge analysis'. The original framework, proposed by Jaksa et al. (2003), envisaged that this analysis would be conducted using a three-dimensional finite element analysis (3D FEA) that accommodates the spatial variability of the soil. However, the computational time and restrictions on the design regarding footing discretization render the use of 3D FEA inappropriate. Therefore, the authors have used an alternative settlement prediction technique, based on the Schmertmann method, to analyze the foundation utilizing complete knowledge of the soil. This method involves averaging the elastic moduli within a specific influence region, to yield a characteristic value that suitably represents the stiffness of the soil under the footing. The optimal influence region size was determined by calibrating settlement estimates with those using 3D FEA.

To assign rehabilitation costs, based on the severity of a foundation failure, both settlement and differential settlement limits requiring minor and major works and complete reconstruction are nominated. These limits

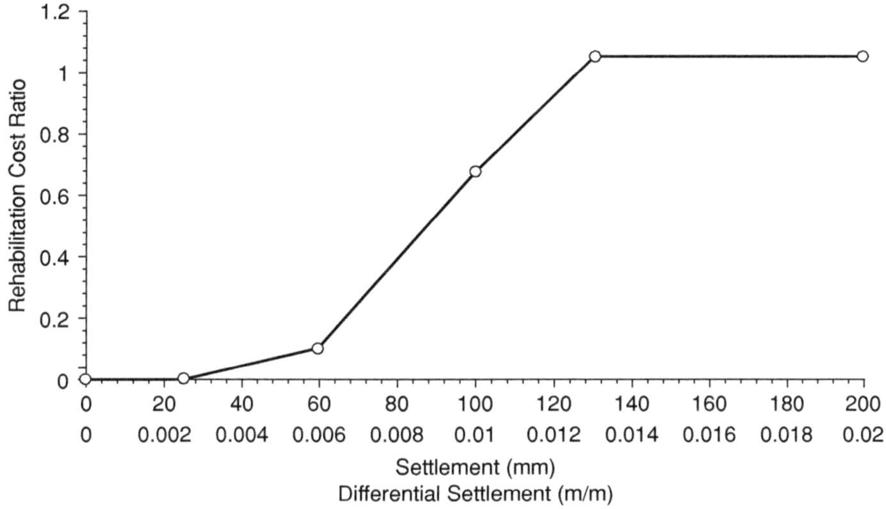

Figure 1.6 Relationship between footing settlement and rehabilitation cost ratio

are based on research by Day (1999) and Boone (2004). The rehabilitation cost associated with a foundation settlement that occurs between these limits is based on linear interpolation. Such interpolation yields a relationship between the cost of rehabilitation and the settlement of the foundation, as shown in Figure 1.6. Goldsworthy (2006) has found that the magnitude of the settlement limits has little effect on the analyses that follow.

A rehabilitation cost ratio is assigned to each foundation design based on the maximum settlement and maximum differential settlement in the foundation system. For example, if two of the footings in the system are found to settle greater than the 25 mm threshold, the rehabilitation cost ratio is assigned in relation to the largest settlement. In other words, the rehabilitation cost ratio is not additive and has a maximum of 1.1.

The rehabilitation cost, C_{Rhb}, is calculated by multiplying the rehabilitation cost ratio by the construction cost, C_{Con}. For the maximum rehabilitation cost ratio of 1.1, the rehabilitation cost equals 110 percent of the original cost of the structure. This accounts for the demolition of the previous structure and reconstruction of a new structure.

3.5 Model results

3.5.1 Design area

Before examining the results which express the impacts of the site investigation scope on the financial risk of a project, it is beneficial to examine the trends resulting from the simulation model. First, since the simulation model designs a footing to meet specified design criteria, it is possible to examine the influence of additional sampling on the design area. Figure 1.7 illustrates the average total footing area of the design foundation based on a site investigation with CPT soundings arranged in a regular grid pattern (RG) and combined using the Standard Arithmetic Average (SA). The results in Figure 1.7(a) show how the soil variability influences the results, while Figure 1.7(b) indicates the effect of soil SOF.

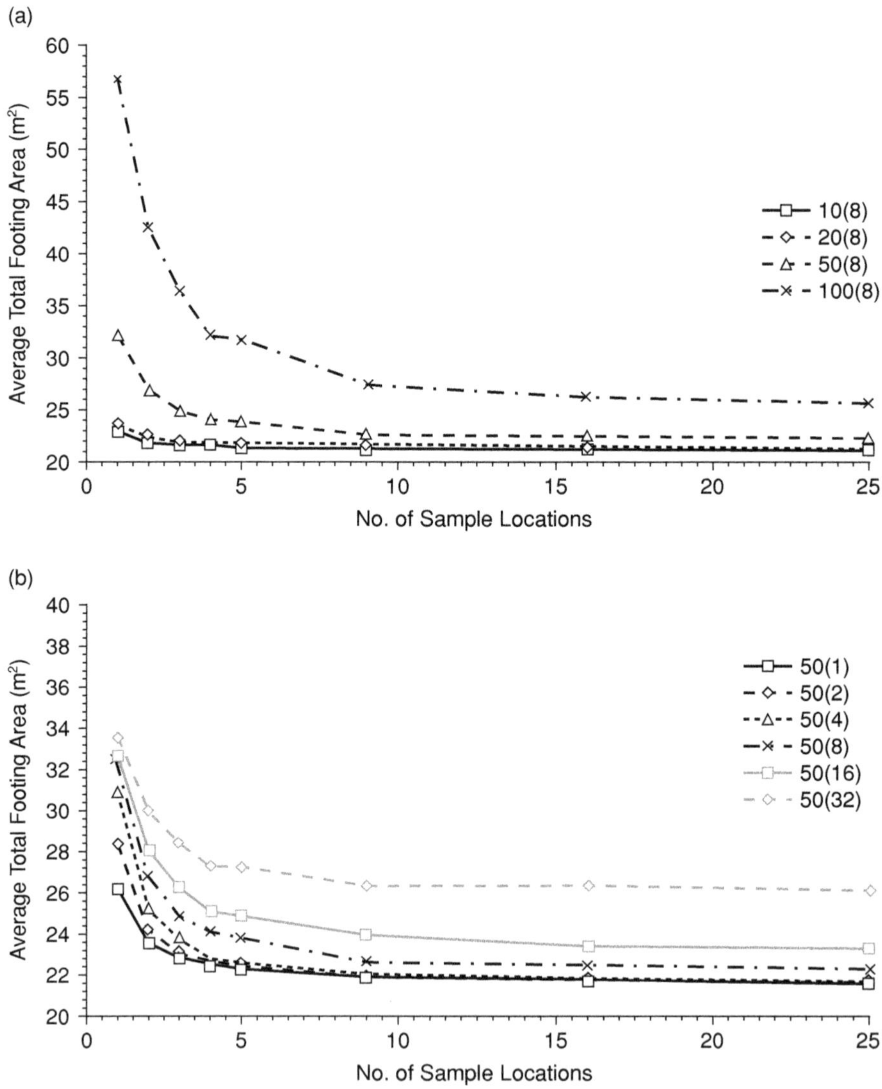

Figure 1.7 Average total design area for a footing design based on a site investigation consisting of CPT soundings arranged in an RG pattern and combined using the SA

It should be noted that the average total footing area is based on the sum of the areas from all nine footings. The total footing area is calculated for each Monte Carlo realization and then averaged over the suite of 1000 realizations to estimate the mean. The results, in Figure 1.7, clearly show that the total footing area reduces considerably as the number of sample locations increases. In a practical sense, this suggests that, if more information is used in the design, on average, a smaller footing will result. It is important to remember at this stage that the previous conclusion is based on an average and will not occur for every case.

The results in Figure 1.7 allow conclusions to be drawn regarding how the spatial statistics of the soil deposit influence the impact of additional sampling. It is clear from Figure 1.7(a) that additional sampling is very beneficial in highly variable soils. However, it is also interesting to note that the average design area tends to a minimum, once a certain degree of testing is performed. In the case of the soil with high variability (COV of 100 percent), the minimum average total design area is not achieved until approximately sixteen sample locations are used; whereas the minimum appears to be achieved once five sampling locations are included in the analysis for soils with a COV of 10 percent and 20 percent. Further, the minimum average total design area is greater for soils with a higher COV. This suggests that, on average, a larger footing design is required for highly variable soils.

In contrast to the results in Figure 1.7(a), the reduction of average design error for increased sampling does not appear to be heavily influenced by the SOF of the soil. This is evident in Figure 1.7(b) by the fact that the average design error reduces at a similar rate for the soil conditions examined. However, the results in Figure 1.7(b) do show that designs on a soil with a higher SOF, on average, require a larger area.

A further examination of the impact of increased sampling on the design area is made by plotting the COV of total design area against increased sampling. Such results are presented in Figures 1.8(a) and (b) for soils with varying COV and SOF respectively.

It is clear from the results in Figure 1.8 that the COV of total footing area is reduced considerably as additional samples are included in the analysis. In practical terms this means that, when more sampling locations are considered, there is increased confidence in the design. It is also interesting to note that the COV is a function of the average or sample mean, which was plotted in Figure 1.7 for the same conditions. Since the average design area also reduces for increased sampling, it is expected that the variability of design area is actually reduced considerably more than that shown in Figure 1.8. Therefore, it appears that increased sampling has a greater impact on the variability of the design area than the average. It is also probable that, since the lognormal distribution is used exclusively in the analysis, there are second-order effects in the average, where the variance has a large influence.

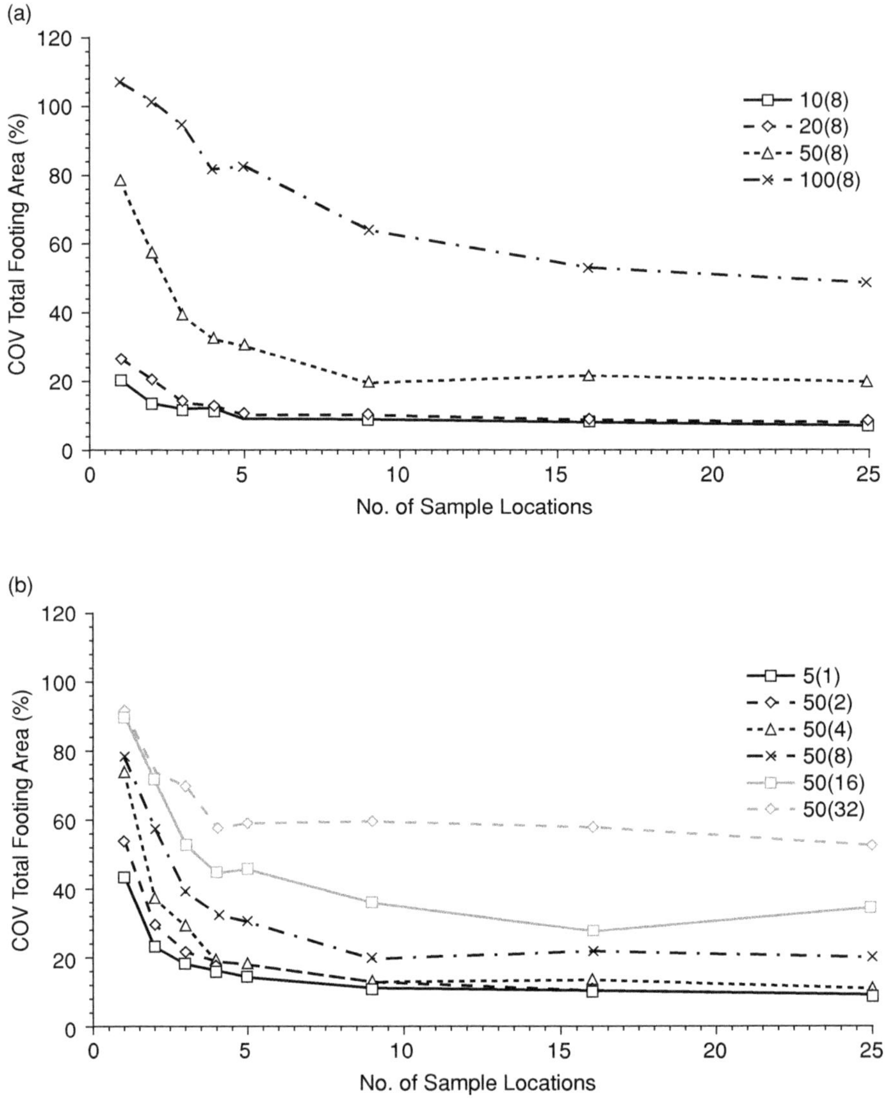

Figure 1.8 COV of total design area for a footing design based on a site investigation consisting of CPT soundings arranged in an RG pattern and combined using the SA with varying (*a*) COV and (*b*) SOF

The results presented in Figure 1.7(b) indicated that the soil SOF had little influence on the impact of additional sampling. However, such results were based on a soil with an anisotropic correlation structure. Using LAS and post-conditioning of the elastic modulus field, it was possible to generate a soil deposit with an anisotropic correlation structure, where the SOF

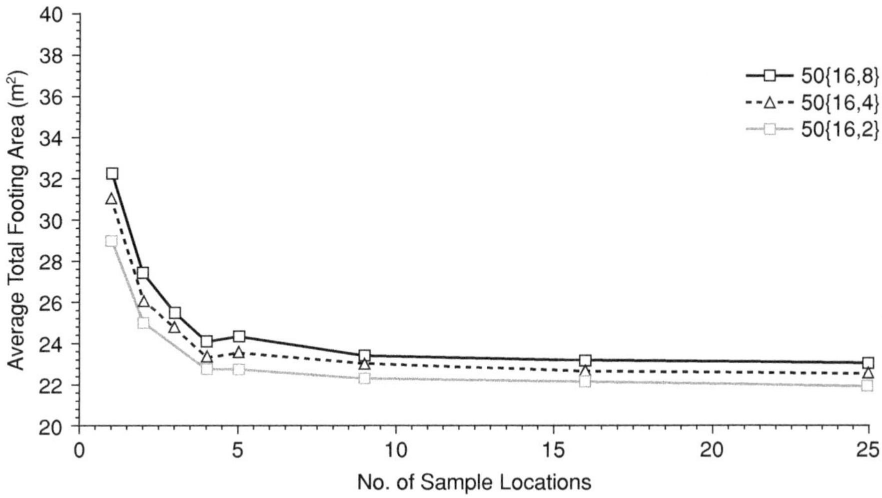

Figure 1.9 Average total design area for a footing design based on a site investigation consisting of CPT soundings arranged in an RG pattern and combined using the SA for a soil with an anisotropic correlation structure

in the vertical direction is different from that in the horizontal or planar direction. The reason for such analysis is to identify whether soils that are horizontally layered, or weakly correlated in the vertical direction, have an impact on the effectiveness of a site investigation. The results presented in Figure 1.9 are based on the same site investigation strategy as those presented earlier (Figures 1.7 and 1.8), but this time the horizontal SOF is held constant at 16 m, while the vertical SOF is varied between 2 and 8 m.

The results in Figure 1.9 indicate that soils with an anisotropic correlation structure have little impact on the benefits of additional sampling. In fact, it appears that the results are offset and the trend is identical. Therefore, horizontal layering has little to no effect on planning site investigations. Consequently, the remaining results given for this example deal only with soils with an isotropic correlation structure.

3.5.2 Construction and rehabilitation cost

Similarly to assessing the impact of additional sampling on the designed footing area, there is benefit in examining foundation construction and rehabilitation costs separately. Such results are given in Figure 1.10 for different soil types and a site investigation that consists of CPTs arranged in a regular grid (RG) pattern and uses the standard arithmetic average (SA) to combine results from multiple sampling locations.

It should be noted that most of the results presented in this section illustrate average costs, where the costs associated with the foundation

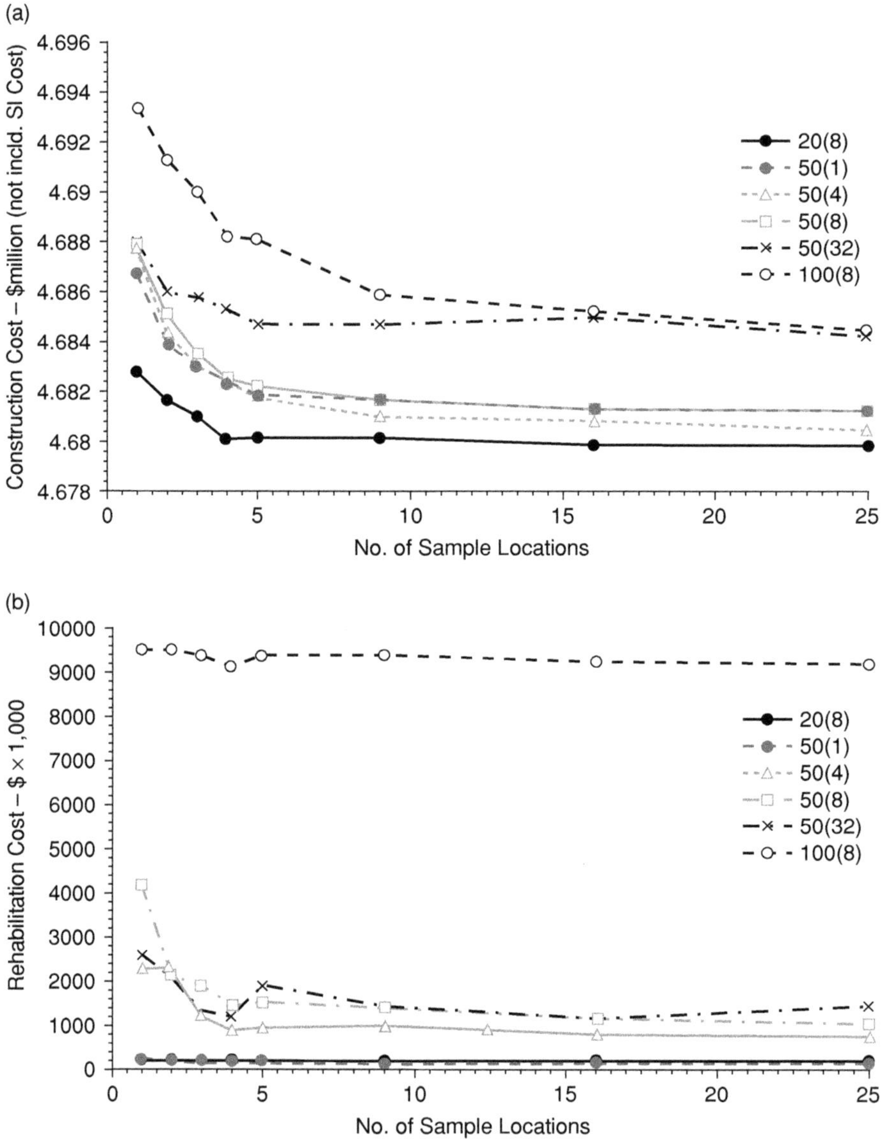

Figure 1.10 Average costs for foundation (*a*) construction and (*b*) rehabilitation based on a site investigation consisting of CPT samples arranged in an RG pattern and reduced using the SA

construction and rehabilitation, and total costs, are averaged over a suite of 1000 Monte Carlo realizations. Although this does not constitute a probabilistic result, there is greater practical benefit in exploring the following results as they provide an indication of the expected costs of the project.

Goldsworthy (2006) presented probabilistic results for similar analyses that demonstrate trends in the mean, standard deviation and COV.

The results presented in Figure 1.10 indicate that, in general, both the construction and rehabilitation costs decrease as the amount dedicated to a site investigation increases. Therefore, additional information regarding a site results in a smaller construction cost and less potential for having to repair the foundation. One obvious outlier shown in Figure 1.10(b) is the result based on the soil with a COV of 100 percent and a SOF of 8 m. In this case, there appears to be little benefit in performing additional testing. However, the results in Figure 1.10(a) suggest that additional testing reduces the construction cost. As such, there is still benefit in having additional information. A similar result was shown in Figure 1.7 where the average total design area was seen to reduce for additional sampling. Of course, the construction cost of the foundation design is tied directly to the total footing area.

A likely cause of the high rehabilitation cost and little evidence of a downward trend shown for soil 100(8) in Figure 1.10(b) is the degree of variability in the soil properties. Although additional testing should result in additional information, and therefore a better representation of the site conditions, a soil with a COV of 100 percent is very variable, and even a sampling program consisting of twenty-five sampling locations may not be sufficient to account for the variability. Further, since the costs given in Figure 1.10 are based on an average of the 1000 Monte Carlo realizations, there is a possibility that the costs obtained from each realization are highly variable. Goldsworthy (2006) observed that the resulting costs closely followed a lognormal distribution. Therefore, based on the form of a lognormal distribution, if the variability increases, the mean must also increase. This results in a high average rehabilitation cost, as shown in Figure 1.10(b).

3.5.3 Total cost/financial risk

By combining the costs associated with the construction of the foundation and any possible rehabilitation, the total cost of the project is estimated and a measure of the financial risk is obtained. Such results are given in Figure 1.11 for site investigations with the same components as those which were used to obtain the results shown in Figure 1.10.

In this case, the total cost or financial risk of a design is measured against site investigation cost expressed in terms of a percentage of the construction cost. This allows direct comparisons with research undertaken by Jaksa (2000), who suggested that it is commonplace that site investigation expenditure can be as low as 0.0025 percent of the total project cost.

The results in Figure 1.11 closely resemble those in Figure 1.10 for the rehabilitation costs. This infers that costs associated with rehabilitation have a greater impact on the total cost of the design and, therefore, the financial risk of the project. This is clearly reasonable, since a foundation

Figure 1.11 Expected total cost based on a site investigation consisting of CPT samples arranged in an RG pattern and reduced using the SA

failure affects the entire building, whereas a foundation over-design only impacts the foundation. The greatest reduction in total cost appears to occur for a 50(8) soil. Therefore, the remaining results in this section are based on a soil with a COV of 50 percent and a SOF of 8 m. Results for additional soil types have been further investigated and discussed by Goldsworthy (2006).

The two different sampling patterns were investigated to determine whether the arrangement of sampling locations would have an influence on the benefits of additional sampling. The results are provided in Figure 1.12 and suggest, for the example given, that there is little difference between using a regular grid (RG) and stratified random (SR) sampling pattern.

Different reduction techniques were also investigated to gauge their impact on the benefits of additional sampling. Such results are given in Figure 1.13 and provide interesting conclusions. First, there are considerable differences between the reduction techniques, especially the inverse distance (ID) technique, which shows an upward trend once the site investigation expenditure approaches and exceeds 1 percent of the construction cost. A detailed treatment of the use of the ID technique in this application was made by Goldsworthy et al. (2005). The conclusions of such research suggested that the ID technique is not preferred, as increased sampling does not necessarily reduce the variability of the resulting design. Therefore, the increased confidence in the design size, as shown in Figure 1.8, does not reduce the rehabilitation cost and, therefore, the total cost.

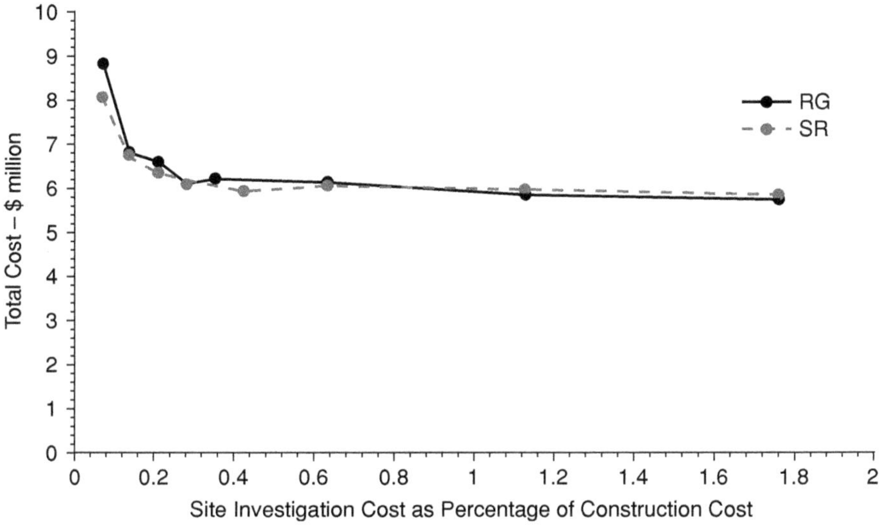

Figure 1.12 Expected total cost based on a site investigation consisting of CPT samples and reduced using the SA in a soil with a COV of 50% and SOF of 8 m

Figure 1.13 Expected total cost based on a site investigation consisting of CPT samples and arranged in an RG pattern on a soil with a COV of 50% and SOF of 8 m

The results in Figure 1.13 suggest that the 1Q and HA are the preferable methods. Both these techniques are slightly more conservative than the other methods examined. Therefore, a more conservative design is obtained, which protects against possible foundation failure, and since rehabilitation costs were identified to be the greatest influence on the total cost a smaller rehabilitation costs leads to a lower average total cost.

Four different geotechnical test types were also included in the analysis of this example. The results to date have dealt exclusively with CPT soundings. However, the SPT, DMT and TT were also examined to gauge their influence on the impact of additional sampling. Results are presented in Figure 1.14 for a soil with a COV of 50 percent and SOF of 8 m, where sample locations are arranged in an RG pattern and combined using the SA.

Based on the results in Figure 1.14, it appears that the CPT yields the lowest total cost for most sampling efforts. Therefore, this test type is recommended for the conditions examined. However, it should be noted that the performance of the different test types is heavily dependent on the assumed uncertainties and the vertical sampling rate. Therefore, the little difference between test types may be a manifestation of the simulation model, and more research in this field is recommended. Further, it is interesting to note that the SPT shows an increase in total cost after a site investigation expenditure of approximately 0.2 percent of the construction cost. Goldsworthy (2006) concluded that this phenomenon was caused by the

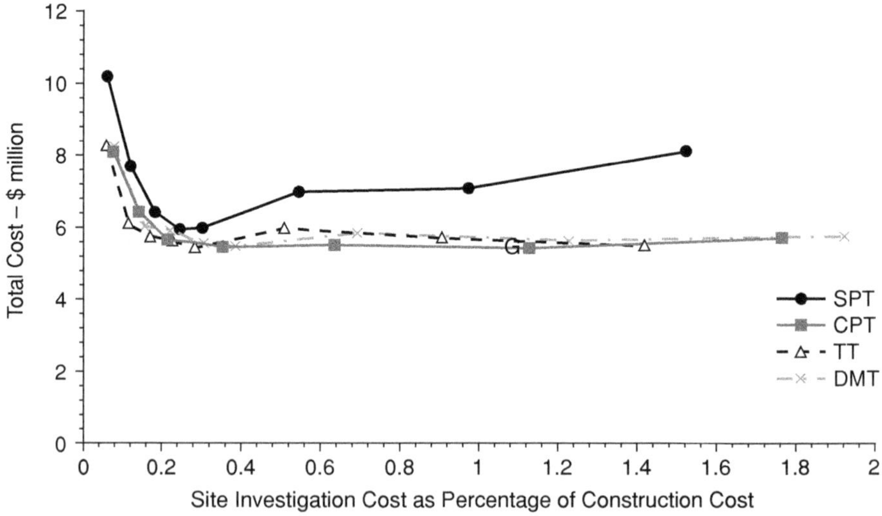

Figure 1.14 Expected total cost based on a site investigation consisting of samples arranged in an RG pattern and reduced using the SA on a soil with a COV of 50% and SOF of 8 m

high transformation model error assumed for the SPT (Table 1.4), which is not reduced by additional sampling. Therefore, increased sampling does not have a significant impact on the variability of the design, nor the rehabilitation costs. In practical terms, large quantities of SPTs may be redundant and provide no benefit to the design.

In all the results presented, the average total cost of the design is shown to reduce considerably as the site investigation expenditure increases. For example, when using the 1Q reduction technique, an RG pattern and the CPT, the average total cost of the design reduces from $AUD 8 million to less than $AUD 6 million, for an increase in site investigation expenditure of 0.05 percent to 0.4 percent (Figure 1.11). An increase of 0.35 percent in site investigation expenditure relates to an increase of approximately $AUD 13,000. In a benefit–cost ratio, where the benefits are the savings in expected total cost and the costs are the increase in site investigation expenditure, this extra testing relates to a benefit equal to 170 times the cost. However, it should also be noted that the expected construction cost, based on a site investigation expenditure of 0.4 percent, is nearly $AUD 40,000 higher than an expenditure of 0.05 percent (Figure 1.10[a]).

4 Conclusions and recommendations

As site investigations play a vital role in any geotechnical engineering design, it is important that such investigations are adequately planned to characterize the subsurface conditions. However, because no two soil sites are the same, and many uncertainties exist in a geotechnical system, it is difficult to prescribe minimum guidelines for site investigations. Further, it is not easy to compare objectively two site investigations with each other, since a quantitative manner to measure the effectiveness of a site investigation does not exist. However, probabilistic methods have been shown to provide some insight into the effectiveness of one site investigation strategy compared with another.

Probabilistic and, in this case, simulation methods provide a rational framework in which to treat the uncertainties that exist in a geotechnical system. By modeling the natural variability of soil properties using a numerical technique and incorporating uncertainties due to sampling, measurement and transformation model error, a representative analysis was performed.

The example presented in this chapter has further identified how, by using probabilistic methods to develop trends and relationships, the results of complex analyses can be used to illustrate simple conclusions. It is anticipated that designers will make use of results of a similar ilk to those presented in this chapter to identify potential risk savings at the site investigation phase of a project. Such results identify the increased risk of conducting site investigations of relatively small scope and, therefore, the additional benefit of increasing the scope of such an investigation.

5 References

Australian Standards (1993) *AS1726 Geotechnical Site Investigations.*

Baecher, G. B. (1979) Analysing exploration strategies, in *Site Characterisation and Exploration*, Evanston, Ill.: Northwestern University, pp. 220–46.

Baecher, G. B. and Christian, J. T. (2003) *Reliability and Statistics in Geotechnical Engineering*, Chichester: John Wiley.

Becker, D. E. (2001) Site characterization, in *Geotechnical and Geoenvironmental Engineering Handbook*, ed. R. Rowe, Boston, Mass.: Kluwer Academic Publishers, pp. 69–106.

Boone, S. J. (2004) Evaluating damage to buildings from settlement and construction-induced movement, *The International Conference on Structural and Foundation Failures*, Singapore, pp. 612–23.

Bowles, J. E. (1997) *Foundation Analysis and Design*, Singapore: McGraw-Hill.

Clayton, C. R. I. (2001) Managing geotechnical risk: time for change?, *Geotechnical Engineering*, 149 (1): 3–11.

Clayton, C. R. I., Simons, N. E. and Matthews, M. C. (1982) *Site Investigation*, London: Granada.

Day, R. (1999) *Forensic Geotechnical and Foundation Engineering*, New York: McGraw-Hill.

Fang, H. Y. (1991) *Foundation Engineering Handbook*, New York: Van Nostrand Reinhold.

Fenton, G. A. (1990) Simulation and analysis of random fields, PhD thesis, Princeton University.

Fenton, G. A. (1996) *Probabilistic Methods in Geotechnical Engineering*, Logan, Utah.

Fenton, G. A. and Griffiths, D. V. (2003) Bearing-capacity prediction of spatially random c-ϕ soils, *Canadian Geotechnical Journal*, 40 (1): 54–65.

Fenton, G. A. and Griffiths, D. V. (2005) Three-dimensional probabilistic foundation settlement, *Journal of Geotechnical and Geoenvironmental Engineering*, 131 (2): 232–9.

Fenton, G. A. and Vanmarcke, E. H. (1990) Simulation of random fields via local average subdivision, *Journal of Engineering Mechanics*, 116 (8): 1733–49.

Filippas, O. B., Kulhawy, F. H. and Grigoriu, M. D. (1988) Reliability-based foundation design for transmission line structures: uncertainties in soil property measurement, *EL-5507*, Vol. 3, Cornell University/Electric Power Research Institute, Palo Alto, Calif.

Goldsworthy, J. S. (2006) Quantifying the risk of geotechnical site investigations, PhD thesis, University of Adelaide.

Goldsworthy, J. S., Jaksa, M. B., Kaggwa, W. S., Fenton, G. A., Griffiths, D. V. and Poulos, H. G. (2005) Reliability of Site Investigations Using Different reduction techniques for foundation design, *9th International Conference on Structural Safety and Reliability*, Rome, pp. 901–8 (on CD).

Halim, I. S. and Tang, W. H. (1993) Site exploration strategy for geologic anomaly characterisation, *Journal of Geotechnical Engineering*, 119 (2): 195–213.

Harr, M. E. (1977) *Mechanics of Particulate Media – a Probabilistic Approach*, New York: McGraw-Hill.

Holtz, R. D. (1991) Stress distribution and settlement of shallow foundations, in *Foundation Engineering Handbook*, ed. H. Y. Fang, New York: Van Nostrand Reinhold, pp. 166–223.

Jaksa, M. B. (2000) Geotechnical risk and inadequate site investigations: a case study, *Australian Geomechanics*, 35 (2): 39–46.

Jaksa, M. B. (2004) Geotechnical Engineering and Design III, University of Adelaide.

Jaksa, M. B., Brooker, P. I. and Kaggwa, W. S. (1997) Inaccuracies associated with estimating random measurement errors, *Journal of Geotechnical and Geoenvironmental Engineering*, 123 (5): 393–401.

Jaksa, M. B., Goldsworthy, J. S., Fenton, G. A., Kaggwa, W. S., Griffiths, D. V., Kuo, Y. L. and Poulos, H. G. (2005) Towards reliable and effective site investigations, *Géotechnique*, 55 (2): 109–21.

Jaksa, M. B., Kaggwa, W. S., Fenton, G. A. and Poulos, H. G. (2003) A framework for quantifying the reliability of geotechnical investigations, *9th International Conference on the Application of Statistics and Probability in Civil Engineering*, San Francisco, Calif., pp. 1285–91.

Kulhawy, F. H. (1992) On evaluation of static soil properties, in *Stability and Performance of Slopes and Embankments II (GSP 31)*, ed. R. B. Seed and R. W. Boulanger, New York: ASCE, pp. 95–115.

Kulhawy, F. H. and Phoon, K. K. (2002) Observations on geotechnical reliability-based design development in North America, *International Workshop on Foundation Design Codes and Soil Investigations in View of International Harmonisation and Performance Based Designs*, pp. 31–48.

Lee, I. K., White, W. and Ingles, O. G. (1983) *Geotechnical Engineering*, Marshfield, Mass.: Pitman.

Lowe III, J. and Zaccheo P. F. (1991) Subsurface exploration and sampling, in *Foundation Engineering Handbook*, ed. H. Fang, New York: Chapman Hall, pp. 1–71.

Lumb, P. (1974) Application of statistics in soil mechanics, in *Soil Mechanics – New Horizons*, ed. I. Lee, London: Butterworth, pp. 44–111.

Orchant, C. J., Kulhawy, F. H. and Trautmann, C. H. (1988) Reliability-based foundation design for transmission line structures: critical evaluation of in-situ test methods, *EL-5507*, Vol. 2, Cornell University/Electric Power Research Institute, Palo Alto, Calif.

Parsons, R. L. and Frost, J. D. (2002) Evaluating site investigation quality using GIS and geostatistics, *Journal of Geotechnical and Geoenvironmental Engineering*, 128 (6): 451–61.

Perloff, W. H. (1975) Pressure distribution and settlement, in *Foundation Engineering Handbook*, ed. H. Winterkorn and H. Fang, New York: Van Nostrand Reinhold, pp. 148–96.

Phoon, K. K. and Kulhawy, F. H. (1999a) Characterization of geotechnical variability, *Canadian Geotechnical Journal*, 36 (4): 612–24.

Phoon, K. K. and Kulhawy, F. H. (1999b) Evaluation of geotechnical property variability, *Canadian Geotechnical Journal*, 36 (4) 625–39.

Phoon, K. K., Kulhawy, F. H. and Grigoriu, M. D. (1995) Reliability based design of foundations for transmission line structures, PhD thesis, Cornell University.

Rawlinsons (2004) *Australian Construction Handbook 2004*, Perth: Rawlhouse Publishing.

Schmertmann, J. H. (1970) Static cone to compute static settlement over sand, *Journal of the Soil Mechanics and Foundations Division*, 96 (SM3): 1011–43.

Site Investigation Steering Group (1993) *Site Investigation in Construction*, London: Thomas Telford.

Small, J. C. (2001) Shallow foundations, in *Geotechnical and Geoenvironmental Engineering Handbook*, ed. R. Rowe, Boston, Mass.: Kluwer Academic Publishers, pp. 223–60.

Snedecor, G. W. and Cochran, W. G. (1980) *Statistical Methods*, Ames, Iowa: The Iowa State University Press.

Spry, M. J., Kulhawy, F. H. and Grigoriu, M. D. (1988) Reliability-based foundation design for transmission line structures: geotechnical site characterization strategy, *EL-5507*, Vol. 1, Cornell University/Electric Power Research Institute, Palo Alto, Calif.

Terzaghi, K. and Peck, R. B. (1968) *Soil Mechanics in Engineering Practice*, New York: John Wiley.

Tsai, Y. C. and Frost, J. D. (1999) Using geographic information system and knowledge base system technology for real-time planning of site characterization activities, *Canadian Geotechnical Journal*, 36 (2): 300–12.

Vanmarcke, E. H. (1977a) Probabilistic modeling of soil profiles, *Journal of the Geotechnical Engineering Division*, 103 (GT11): 1227–46.

Vanmarcke, E. H. (1977b) Reliability of earth slopes, *Journal of the Geotechnical Engineering Division*, 103 (GT11): 1247–65.

Vanmarcke, E. H. (1983) *Random Fields: Analysis and Synthesis*, Cambridge, Mass.: MIT Press.

Warner, R. F., Rangan, B. V., Hall, A. S. and Faulkes, K. A. (1998) *Concrete Structures*, Melbourne: Addison Wesley Longman.

Whitman, R. V. (2000) Organising and evaluating uncertainty in geotechnical engineering, *Journal of Geotechnical and Geoenvironmental Engineering*, 126 (7): 583–93.

Wiesner, T. (1999) Site characterisation, *Australian Geomechanics*, 34 (4): 41–57.

Appendix A. Formulation of reduction techniques

The reduction techniques used to estimate the characteristic value from a series of test results are given by:

$$x_{SA} = \frac{1}{n}\sum_{i=1}^{n} x_i$$

$$x_{GA} = \left(\prod_{i=1}^{n} x_i\right)^{1/n}$$

$$\frac{1}{x_{HA}} = \frac{1}{n}\sum_{i=1}^{n}\frac{1}{x_i}$$

$$x_{ID} = \sum_{i=1}^{n}\frac{d_i}{d_{tot}}x_i$$

$$x_{1Q} = \frac{x_{\max} - x_{\min}}{4}$$

where x_i is the result from the ith sample location, n is the total number of sample locations, d_i is the distance between the ith sample location and the footing, d_{tot} is the total distance between all sample locations and the footing, x_{max} is the maximum result and x_{min} is the minimum result.

Appendix B. Method to account for settlement from an adjacent footing

Procedure to estimate the settlement of Footing 1 due to an adjacent Footing 2, based on superposition:

(1) Determine the distance between the center of Footing 1 and the farthest point of Footing 2, r_o;

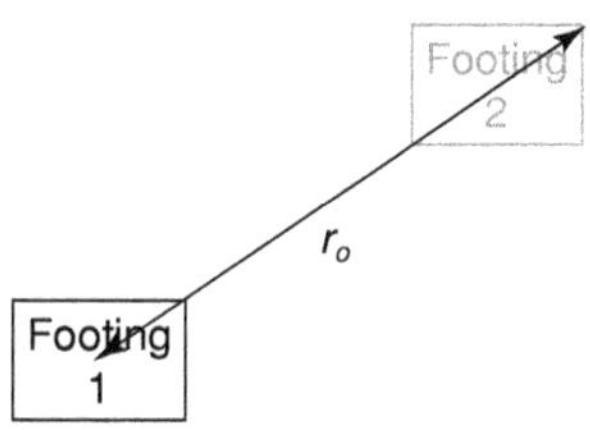

(2) Determine the distance between the center of Footing 1 and the closest point of Footing 2, r_i;

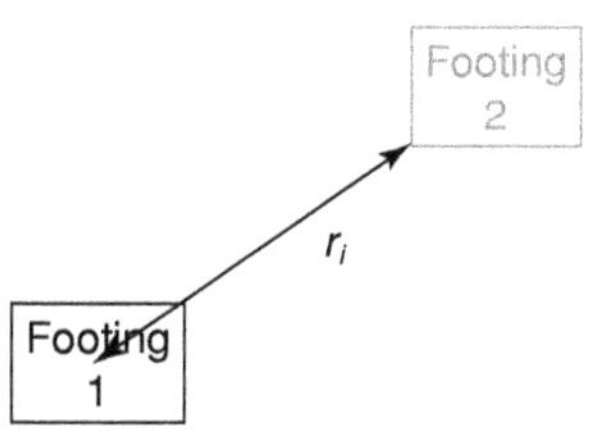

(3) Estimate the settlement of a rigid annulus, δ_{ann}, with an external radius equal to r_o, an internal radius equal to r_i and an applied pressure equal to the pressure on Footing 2. Using theory of elasticity (Perloff 1975) with influence values from Harr (1977);

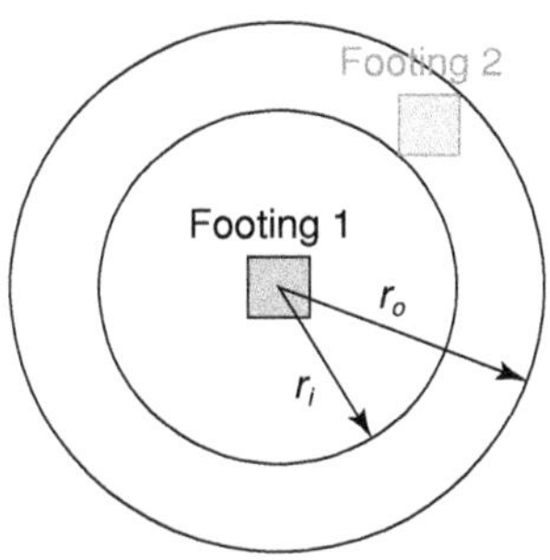

(4) Estimate the proportion of annulus area occupied by Footing 2, by estimating the angle α, made with two lines from the center of Footing 1 to the mid-point of the two closest sides of Footing 2;

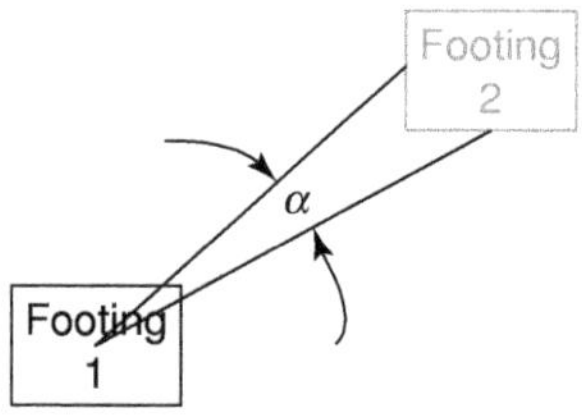

(5) Calculate the settlement of Footing 1 due to Footing 2, $\delta_{1/2}$, by proportioning the settlement of the rigid annulus, δ_{ann}, by the angle α, using:

$$\delta_{1/2} = \frac{\alpha}{360}\delta_{\text{ann}}$$

2 The contribution of numerical analysis to the response prediction of pile foundations

Emilios M. Comodromos

1 Introduction

The response of a pile group under axial and lateral loading is considered to be a factor which considerably affects the behavior of the superstructure. The suitability of the conventional load capacity approach to predict the actual behavior of pile foundations is considered questionable despite the fact that it is still widely used in practice. Present codes and regulations, essentially based on this approach, act as a restraint rather than as a stimulus and need some revision, (Mandolini et al. [1]). Continuous advances in technology, methods of analysis and design approaches have been the subject of intensive research during the past three decades. Nevertheless, many questions regarding the interaction between soil, piles and superstructure remain for the scientific community to resolve.

Exact closed-form solutions are available for a rather limited set of conditions and fundamental assumptions simplifying the soil profile to an elastic halfspace. Empirical relationships gained through the last few decades of research were proposed to estimate the reduction factors on both the bearing capacity and the stiffness of a pile group due to the interaction between the piles in a group. Moreover, specific values for these factors have been proposed, in tabular or graphical forms, resulting mainly from simplified analyses based on elastic continuum theory and the principle of superposition. Simplified numerical analyses based on load-transfer methods to estimate the response of a single pile, and to assess the interaction between the piles, have also been used.

Depending on the level of sophistication, Poulos [2] classified the procedures of analysis and design of single piles and pile groups into three main categories. Categories 1 and 2 are principally simplified approaches based on empiricism, involving the use of simple computational methods that rarely need the use of simplified computer codes. Category 3, on the other hand, is applied when a higher level of accuracy is required in the estimation of the bearing capacity and the response of a single pile or a pile group. Category 3 is further subdivided into three categories of which 3C is the most sophisticated, accurate and consequently the most computationally demanding. The computational demands of Category 3C could be regarded as a considerable restraint; however, the rapid advances in numerical methods and the development of increasingly powerful computers have made the use of 3C more applicable and effective. Nevertheless, it should be noted that the successfulness in the prediction of a pile foundation response is highly dependent on the values for the subsoil strength and deformations parameters as well as on the interface parameters. Bearing in mind that the stress path developing around a pile does not correspond to conventional laboratory tests, pile load tests play an important role in value engineering and in the geotechnical and the structural optimization [2]. A full-scale test may contribute to the elimination of practically all the uncertainties arising from the soil behavior and interface parameters which govern the mechanism of pile–soil interaction. Furthermore a thorough back-analysis of a pile test could provide appropriate design values for the aforementioned parameters, while by back-figuring from the results of pile load tests further information regarding the shear strength mobilization could be attained. This is the foremost reason that pile design regulations require that the pile design calculations should be related to results from pile load tests [3, 4].

Assuming a pile load test could satisfactorily predict the response of a similar single pile, the response of a pile group remains the crucial point for the design of a foundation and the effect on the superstructure. Thus, whenever the design of a superstructure significantly affects the response of the pile foundation (where settlements, displacements or rotations provoke stress concentrations in the superstructure), a three-dimensional (3D) non-linear analysis in conjunction with appropriate pile load tests could give the necessary data regarding the foundation response and allow an accurate, economical and safe design.

Sections 2 and 3 are devoted to the vertical and the horizontal loading of single piles and pile groups respectively. More specifically, a general introduction is given for these two types of loading, followed by a brief description of the conventional methods and procedures in estimating the bearing capacity and/or simplified load-displacement relationships. The advantages of 3D non-linear analysis are then discussed and some instructive applications are presented, for both single piles and pile groups under vertical and lateral loading.

Extensive numerical experiments in 3D non-linear analysis made possible a precise correlation of the response of single piles and pile groups, and within this framework suitable relationships with the ability to predict the response of pile groups based on that of a single pile are given in Sections 2.2.2 and 3.3.2 for vertical and horizontal loading respectively. The distribution of a load applied to the cap of a pile group is considered also as a key issue for the design of pile foundations. This particular subject is discussed in Sections 2.2.3 and 3.3.3 for piles groups under vertical and lateral loading.

Finally, Section 4 deals with the application of the experience gained through using 3D non-linear analysis to the design of complex structures under the concept of soil–structure interaction. The proposed relationships in conjunction with predefined comparative mode of characteristic piles of a group and the technique of substructuring could be employed to simplify complex problems. An iterative procedure of the above method is described, reducing the complexity and the solution time for intricate soil–structures interaction problems. The application of this iterative procedure allows for an accurate solution and an optimum design of the superstructure and pile foundation, as the foundation effects can be estimated from the piles' response and be exactly introduced in the analysis.

2 Piles under vertical loading

2.1 Introduction

The earliest processes for single pile design subject to vertical loading were conventional capacity-based approaches integrating the shaft and tip resistance using simplified formulae. They are well known as 'geotechnical methods' since they are focused mainly on geotechnical data, while their computational requirements are minimized to simple hand-calculations. The majority of these methods are still widely used in practice and, together with empirical approaches, led to the development of the existing codes and regulations in many countries. Most of the European codes were created within this framework, in which soil–pile interface and tip shear strength are the determinant factors in estimating the pile-bearing capacity.

Interface shear strength is derived from the surrounding soil shear strength parameters by applying a reduction factor, which is governed mostly by the installation method and the soil shear strength. The German code, DIN 4014, is a representative example of such codes, enabling the estimation of the bearing capacity and the allowable load of a single pile [3]. Moreover, it could be used to determine the response of the shaft and tip resistance as well as the overall resistance with pile head settlement at characteristic levels of settlements as given in Figure 2.1. The French code, Fascicule No. 62, follows similar principles but is more detailed in classifying the soil categories and proposed interface reductions factors [5]. Instead of providing the load-settlement response, the French code makes available the shaft (t-z) and the tip (q-z_b) stiffness in a form of multilinear relationships in simplified one-dimensional numerical algorithms and defines the response of a single pile under vertical loading. Both codes make use of results from laboratory and in-situ testing.

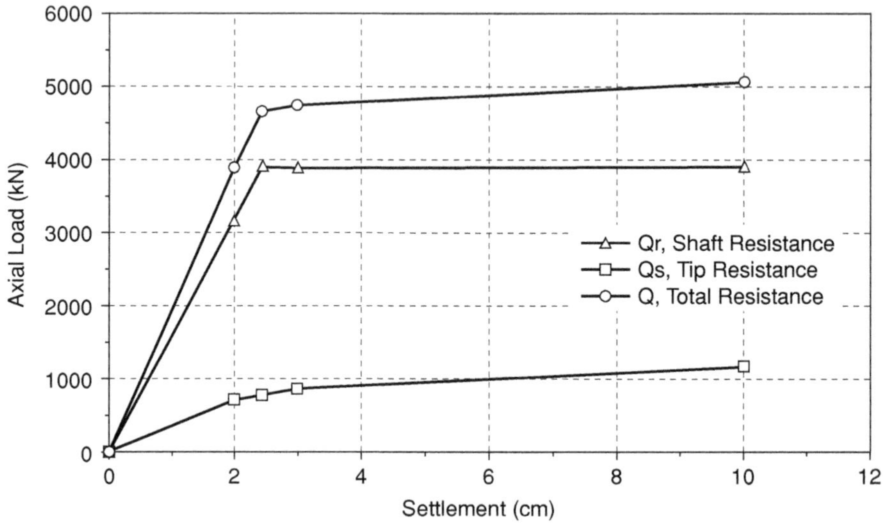

Figure 2.1 Typical load-settlement curves estimated using DIN 4014

2.2 *Single pile under vertical loading*

Pile–soil interaction is generally considered as a three-dimensional problem. However, in the particular case of a pile in a homogeneous or even stratified soil profile under vertical loading, a 2D axisymmetric approach could be applied. Numerical simulation of a pile using 1D elements and an idealization of soil response by discrete springs with linear or multi-linear branches, well known as the '*t-z*' approach, represents further simplifications of the issue. Owing to the limited number of degrees of freedom, the required computational time is practically negligible, rendering the method most attractive. Many computer codes have been especially written based on the '*t-z*' approach [6–8], while the method can be simply implemented into finite element or finite difference general codes. However, it should be noted that an accurate evaluation of the '*t-z*' relationships, even in the case where adequate geotechnical data are available, is always questionable and in most cases empirically derived. This could be considered as the main drawback of the method.

Three-dimensional non-linear analyses, on the other hand, may be very computationally demanding, depending upon the size of the problem. The continuous development of more powerful computers renders possible a three-dimensional analysis. Sophisticated constitutive models for simulating soil behavior are also available, and their use may improve solution precision provided that adequate geotechnical data are introduced. Bearing in mind that conventional laboratory tests are not able to reproduce stress path development around a pile, full-scale tests are performed. Precise values for both shear strength and deformation parameters could be back-calculated from a full-scale test in conjunction with a 3D analysis. The pile load test setup affects the response of the tested pile depending upon the pile load test method. The most widely used method in testing piles under vertical loading is that making use of reaction piles or ground anchors. However, in most cases it leads to a significant over-estimation of the pile head stiffness [9–11]. More specifically, Comodromos et al. carried out a 3D numerical simulation of a static pile load test consisting of a large tested pile and four reaction piles in a spacing of 4.0-D, as shown in the layout given in Figure 2.2 [10]. A significant over-estimation of head stiffness was derived attaining the order of 200 percent, which was mainly due to the earlier mobilization of the shear strength of the soil between the pile under test and the reaction piles. Eventually, a simultaneous downward movement of the tested pile and upward movement of the reaction piles produced a higher level of shear strain; and, as a result, the shear strength mobilization is developed earlier than in the corresponding case of a single pile. The level of over-estimation significantly depends upon the pile test layout, the soil behavior and the level of loading. It should be emphasized that, contrary to the observed significant effect on the stiffness of the tested pile, the impact on the bearing capacity of the pile was found negligible.

Figure 2.3 shows the load-settlement curve of the test loading, together with the numerically established response of the tested pile using 3D non-linear analysis. To establish the single pile response, the same simulation process was then applied without activating the reaction piles. The single pile response is shown in Figure 2.3. A comparison of the behavior of the tested pile and the single pile validates the widely accepted approach that the existence of tension piles affects the response of the tested pile, rendering it stiffer. Owing to the development of uplift forces to the surrounding soil, the tension piles provoke the earlier mobilization of shaft resistance of the tested pile, as can be verified from Figure 2.4. It can be seen that a full shaft mobilization occurs at the settlement level of less than 1 percent for the tested pile, while in the case of the single pile in the same conditions the level of full shaft mobilization attains the value of 2 percent, rendering the single pile 50 percent less stiff than the tested pile. Further loading stimulates gradual soil yielding, and the effect of interaction gradually decreases until the resistance–settlement relationship of both cases is equalized. Figure 2.5 illustrates the settlement of the tested pile and the reaction piles corresponding to an applied load of 12 MN. It can be also seen in Figure 2.6 that, for the same load of 12 MN, the single pile exhibits almost double the settlement. Figure 2.7 reveals the distribution of the pile axial forces along the depth for the single pile and the tested pile. For the same load the tested pile always manifests higher shaft resistance, for the reasons given above. This is valid until a total plastification of the soil around the pile occurs (application of loads higher than 15 MN)

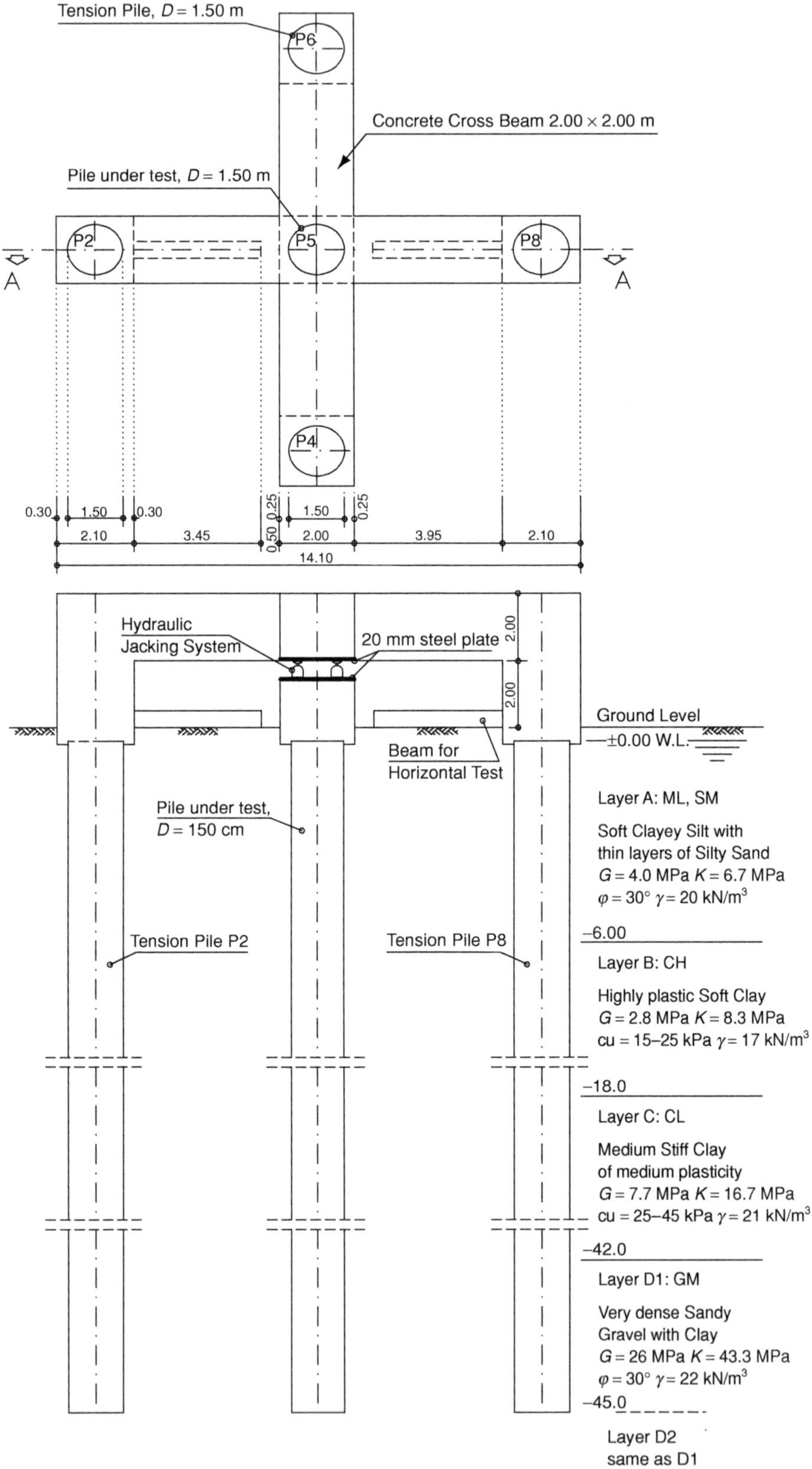

Figure 2.2 Pile load arrangement and design soil profile (Comodromos et al. [10])

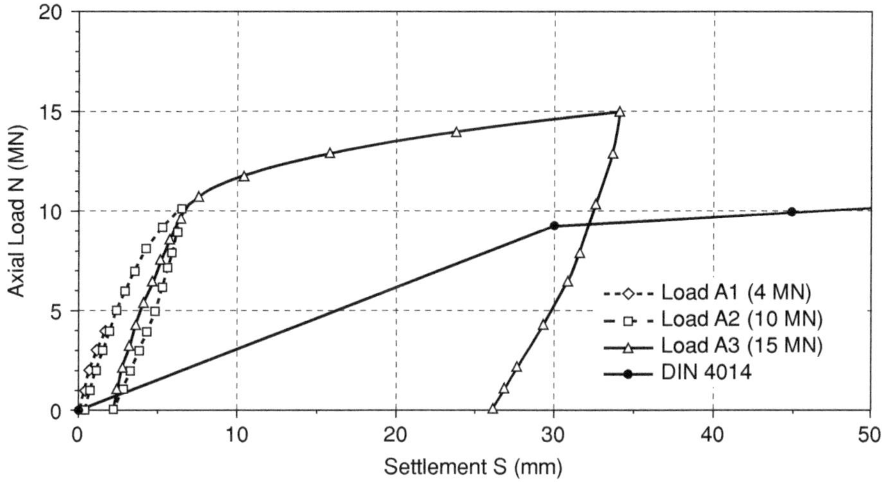

Figure 2.3 Pile load test and DIN 4014 load-settlement curves (Comodromos et al. [10])

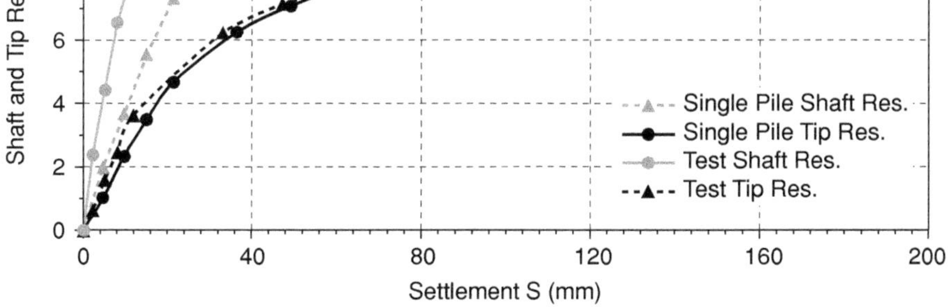

Figure 2.4 Shaft and tip resistance vs settlement relationship for test pile and the single pile (Comodromos et al. [10])

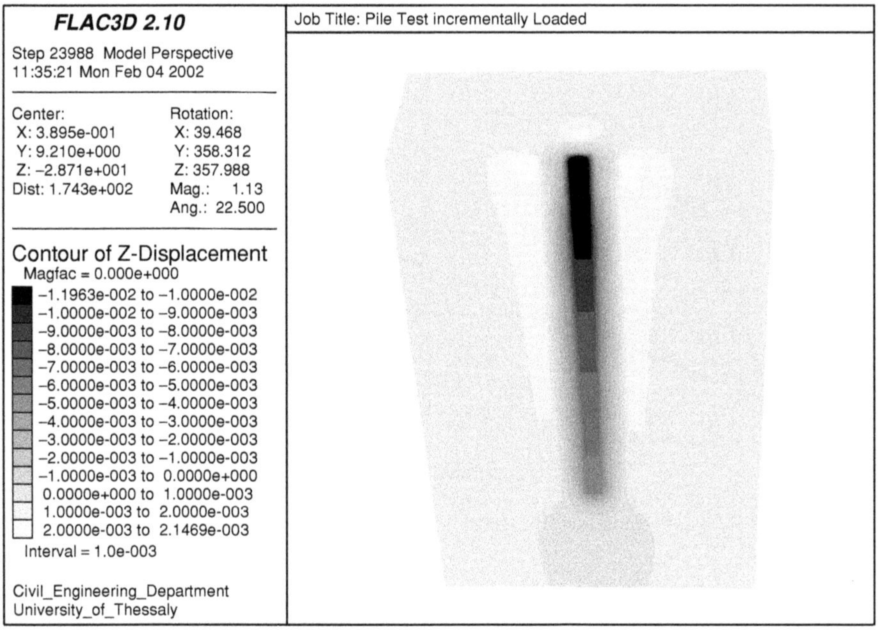

Figure 2.5 Vertical displacement field around the test pile for N = 12MN (Comodromos et al. [10])

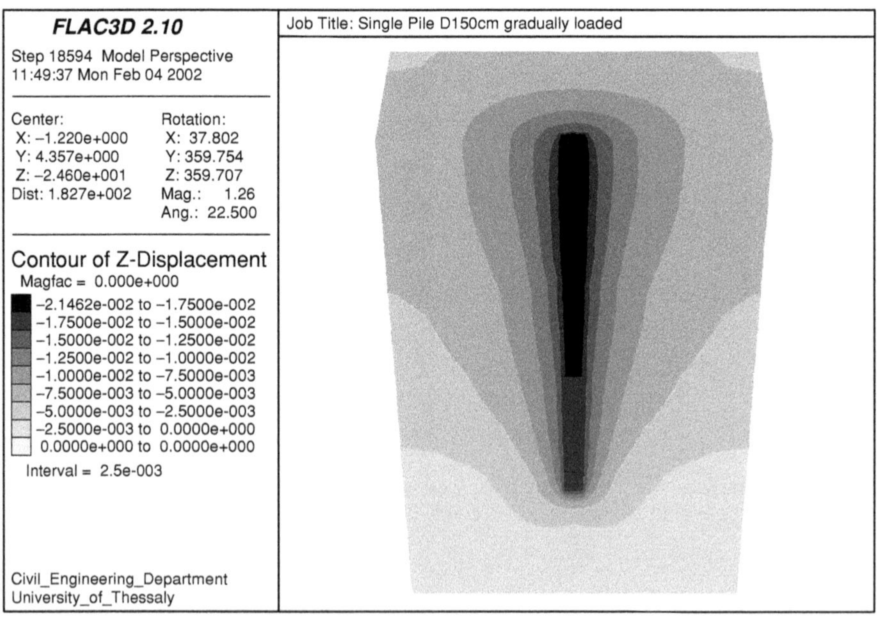

Figure 2.6 Vertical displacement field around the single pile for N = 12MN (Comodromos et al. [10])

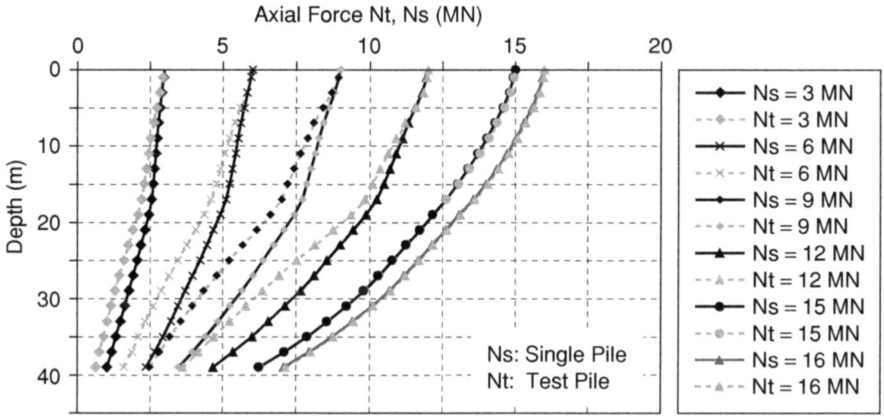

Figure 2.7 Axial force vs depth relationship for test pile and single pile (Comodromos et al. [10])

rendering the effect of the interaction of tested pile negligible and therefore both the tested pile and the single pile exhibit the same axial force distribution.

2.3 Pile group under vertical loading

2.3.1 Response of pile groups

As stated previously, while the interaction effect among the piles in a group under axial loading remains a topic of interest for the research community, the response of a pile group under axial loading is considered one factor among those which most affect the behavior of the superstructure. Simplified approaches to this 3D problem were published in the 1960s, focusing mainly on the development of theoretical or even simplified numerical methods. Randolph and Wroth [12] and Chow [13] proposed simplified analytical methods relating the settlement of the soil around the pile caused by the shaft shear stress to the radial distance from the pile. Lee [14] used the load-transfer *t-z* method to estimate the response of a single pile and the solution of Mindlin [15] to assess the interaction between the piles, while Randolph and Wroth [16], Randolph [17] and Horikoshi and Randolph [18] used the notion of an equivalent pier to simplify and solve the problem of a pile group. Most of the above methods involve soil profile simplifications and other idealizations rendering them computer cost-effective, with the drawback, however, of limited accuracy in many cases. The boundary element method has been used by Poulos [2], Butterfield and Banerjee [19], Poulos and Davis [20] and Mandolini and Viggiani [21] to estimate the effect of soil–pile interaction and, as a result, they proposed specific values for bearing capacity and stiffness reduction factors in tabular or graphical form. Elastic continuum analysis and the principle of superposition are the main simplifications of the above approach. The incorporation of non-linearities arising from soil behavior or even the behavior of the interface along the soil–pile contact was not feasible until powerful 3D numerical methods such as the finite element method (FEM) and the finite difference method (FDM), in conjunction with a variety of constitutive laws, were available. Such analyses are advantageous, leading to a better understanding of the behavior of single piles and pile groups.

Based on a 3D finite element analysis, Katzenbach and Moormann [22] demonstrated a remarkable interaction between the piles in a group with a 3.0-D spacing that was still significant even when the spacing increased to 6.0-D. Moreover, using 3D non-linear analysis, Comodromos et al. [10] and Comodromos [23] demonstrated that the pile group's bearing capacity efficiency factor did not deviate significantly from unity. In contrast, it was revealed that the interaction considerably affected the group stiffness efficiency factor, which depends not only on the pile arrangement but also on the settlement level.

2.3.2 Response prediction

Despite the fact that effective computer codes and very powerful computers are now available, making a 3D non-linear analysis feasible, a pile group analysis remains very computationally demanding. To overcome this drawback, Comodromos [23] proposed a simplified relationship in order to predict the response of pile groups, provided that the response of a single pile is known using Equation 1.

$$\begin{aligned} S_{mG} &= R_a S_{mLs} \\ S_{nG} &= R_a S_{ns} \end{aligned} \tag{1}$$

in which

S_{nG}: pile group normalized settlement to pile diameter
S_{ns}: single pile normalized settlement to pile diameter
S_{mG}: pile group settlement
S_{mLs}: single pile settlement

Illustrative responses of a single pile and a pile group corresponding to piles in stiff clay and the variation of factor R_a with settlement level are given in Figure 2.8 [24]. The application of this relationship was limited to the commonly applied pile spacing of 3.0-D and to soil profiles similar to the one used in the analysis. Based on the results of an extensive parametric numerical 3D non-linear analysis on pile groups, response carried out by Comodromos and Bareka [25], the originally proposed relationship by Comodromos [23] for predicting the response of pile groups for the widely

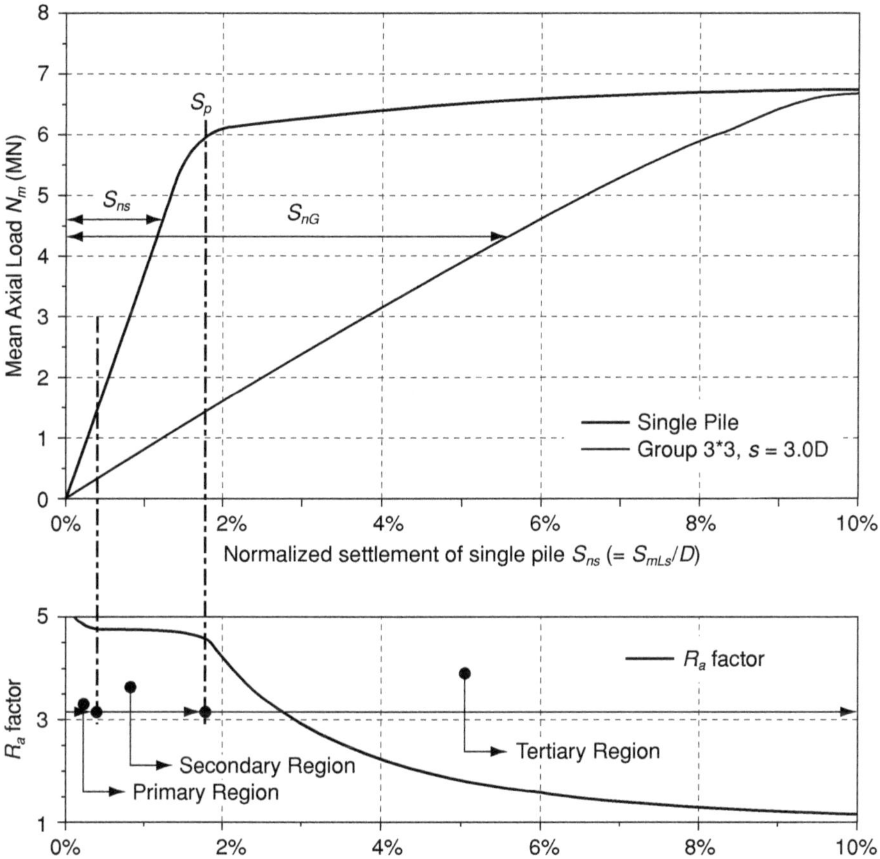

Figure 2.8 Variation of settlement amplification factor R_a with settlement level

applied spacing of 3.0D was modified in order to extend its validity to all commonly adopted pile spacings. The proposed relationship for estimating the settlement amplification factor R_a for spacings varying from 2.0D to 5.0D is given in Equation 2,

$$R_a = A[S_{ns}^B(1.23N_R)^C + S_{ns}^E e^{0.54N_R}]\ln(1.25 + \frac{5}{d}) \tag{2}$$

in which

$$N_R = \frac{n}{n_r + n_c} \tag{3}$$

where

R_a: settlement amplification factor,
S_{ns}: normalized settlement of the single pile to the pile diameter D,
n: total number of piles in the group,
n_r, n_c: number of rows and columns in the pile group respectively
d: normalized axial spacing to pile diameter D,
A, B, C, E: constants determined by numerical process

To determine the most suitable values for parameters A, B, C and E a curve-fining procedure was applied similar to that given by Comodromos and Pitilakis [26]. More specifically, the notion of the stiffness mean error and the potential energy error were considered [24], and common values for parameters A, B, C and E were defined, in order to achieve an acceptable level of error for all analysed cases. Appropriate values for the parameters A, B, C and D were estimated based on the numerical experiments carried out in Reference 27. More specifically, from all 148 analysed cases (78 cases in clayey soils, 60 in sandy soils and 10 in multi-layered soils), the most appropriate values for the parameters above were found to be $A = 0.8$, $B = 0.07$, $C = 1.9$ and $E = -0.08$. Figures 2.9–11

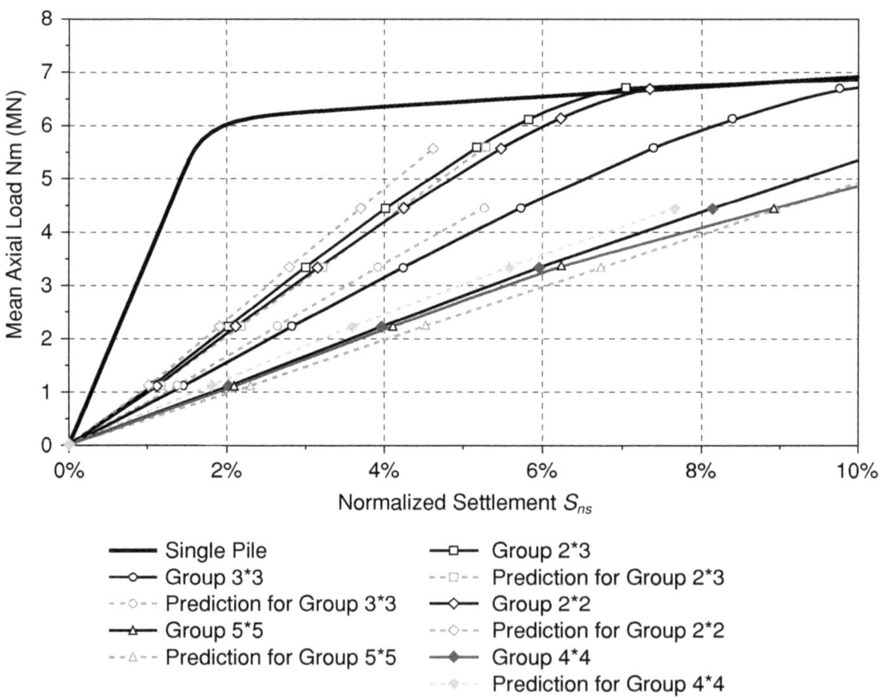

Figure 2.9 Comparison between numerically established load-settlement curves using FLAC3D and those predicted by equations 1, 2 and 3, for soil type C3, relative length L/D = 25 and relative spacing d = 3.0D (Bareka [25])

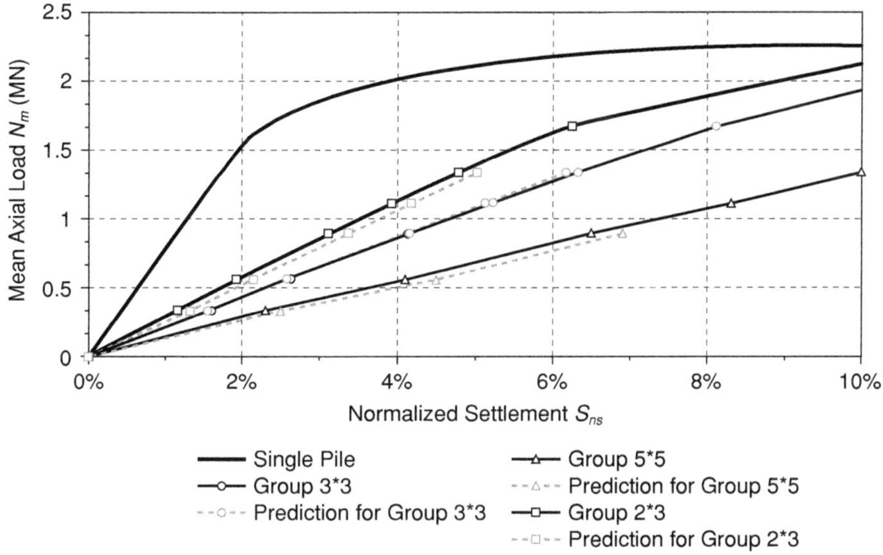

Figure 2.10 Comparison between numerically established load-settlement curves using FLAC3D and those predicted by equations 1, 2 and 3, for soil type C1, relative length L/D = 25 and relative spacing d = 4.0D (Bareka [25])

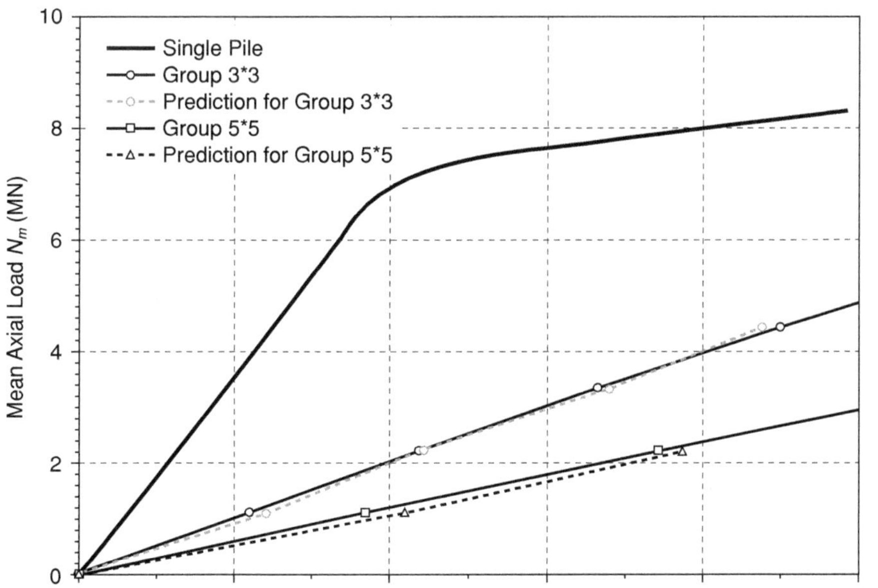

Figure 2.11 Comparison between numerically established load-settlement curves using FLAC3D and those predicted by equations 1, 2 and 3, for soil type C2, relative length L/D = 50 and relative spacing d = 4.0D (Bareka [25])

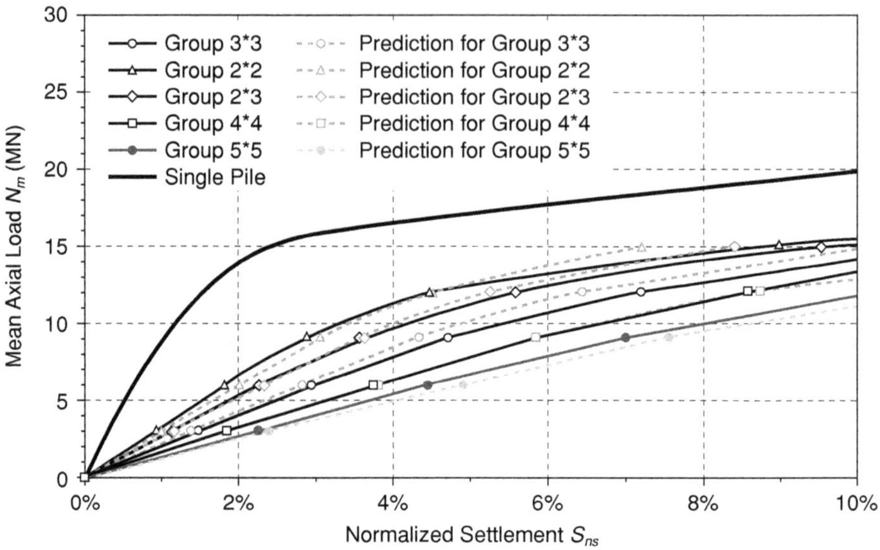

Figure 2.12 Comparison between numerically established load-settlement curves using FLAC3D (Comodromos [23]) and those predicted by equations 1, 2 and 3, for various fixed pile groups configurations with relative spacing d = 3.0D

illustrate the load-settlement curves calculated using the FD code FLAC3D [28] given by the continuous lines, while the dashed lines with the same markers stand for predicted curves using the proposed relationship and the above values for the parameters A, B, C and E. The accuracy provided by the predicted curves could be considered satisfactory for the design of most superstructures. Similar results have been derived from the application of the proposed relationship for all cases. The application of this new relationship provided a satisfactory prediction for the specific site and soil profile examined by Comodromos [23], as presented in Figure 2.12.

Based on Comodromos and Bareka's [24] observations, the stiffness and the potential energy mean error prediction remains less than the 15 percent. An error of that level is considered as acceptable, with no significant impact on the design of the superstructure, contrary to the approaches where the effect of the interaction between the piles of a group is not taken into account. Moreover, for a given mean load N_m (defined as the total load of a group divided by the number of piles in the group), the secant stiffness of a single pile K_s and the secant stiffness of a group K_G are related through Equation 4:

$$K_G = \frac{K_S}{R_a} n \tag{4}$$

where n denotes the number of piles in the group.

2.3.3 Load distribution in characteristic piles

Previous research on this subject demonstrated that, in the case of the fixed head pile, for the same settlement, the piles within the group carry different proportions of the applied load [2, 20, 23]. Figure 2.13 shows the response of the characteristic piles of a 3 × 3 fixed-head pile group, i.e. the *corner* pile P1, the *perimetric* pile P2 located at the external middle position and that of the *central* pile P3. As anticipated, for the same settlement, the central pile carries the least load, presenting the minimum stiffness; while the external piles carry the most load, presenting the maximum stiffness. It is worthwhile noticing that at a certain level of settlement all the piles behave the same. This is the level where the surrounding soil has yielded considerably and therefore the influence of the interaction is considered negligible. To achieve this level, a larger load is needed as spacing increases, while bigger normalized settlement level corresponds to group layout with smaller spacings. Figure 2.14 illustrates the variation of the response of the piles with the level of settlements in a more revealing way. Comparing Figures 2.14 and 2.15, it could be concluded that, as the number of piles in a group increases, the influence of the interaction increases accordingly.

It should be noted, however, that the load distribution among the piles of the group given above corresponds to a fixed-head pile group, i.e. a pile group with a pile head of infinite rigidity. In the case of pile cap thickness of the same order as pile diameter, the load distribution highly depends on the type (concentrated or distributed) and the level of loading. Figure 2.16 demonstrates the variation of the normalized load carried by the characteristic piles of a 3 × 3 pile group in medium stiff clayey soil as a variation of the pile cap thickness. It can be seen that when the group's allowable load is applied as a concentrated load at the centre of the pile cap, the central load carries almost 140 percent of the mean load for a relatively flexible pile cap. On the contrary, when a relatively rigid pile cap is considered, the load resisted is less than 60 percent of the mean load. As anticipated, the corner pile demonstrates inverse behavior. Finally, it should be noted that the effect of the load distribution to the piles making up the group becomes less important when the pile group load is uniformly distributed on the pile cap; see Figure 2.17.

3 Piles under horizontal loading

3.1 Introduction

The response of laterally loaded pile foundations is also significantly important in the design of structures. As presented for vertically loaded pile foundations, in many cases the criterion for the design of piles to resist lateral loads is not the ultimate lateral capacity but deflection. In the case of bridges or other structures founded on piles, a few centimetres of

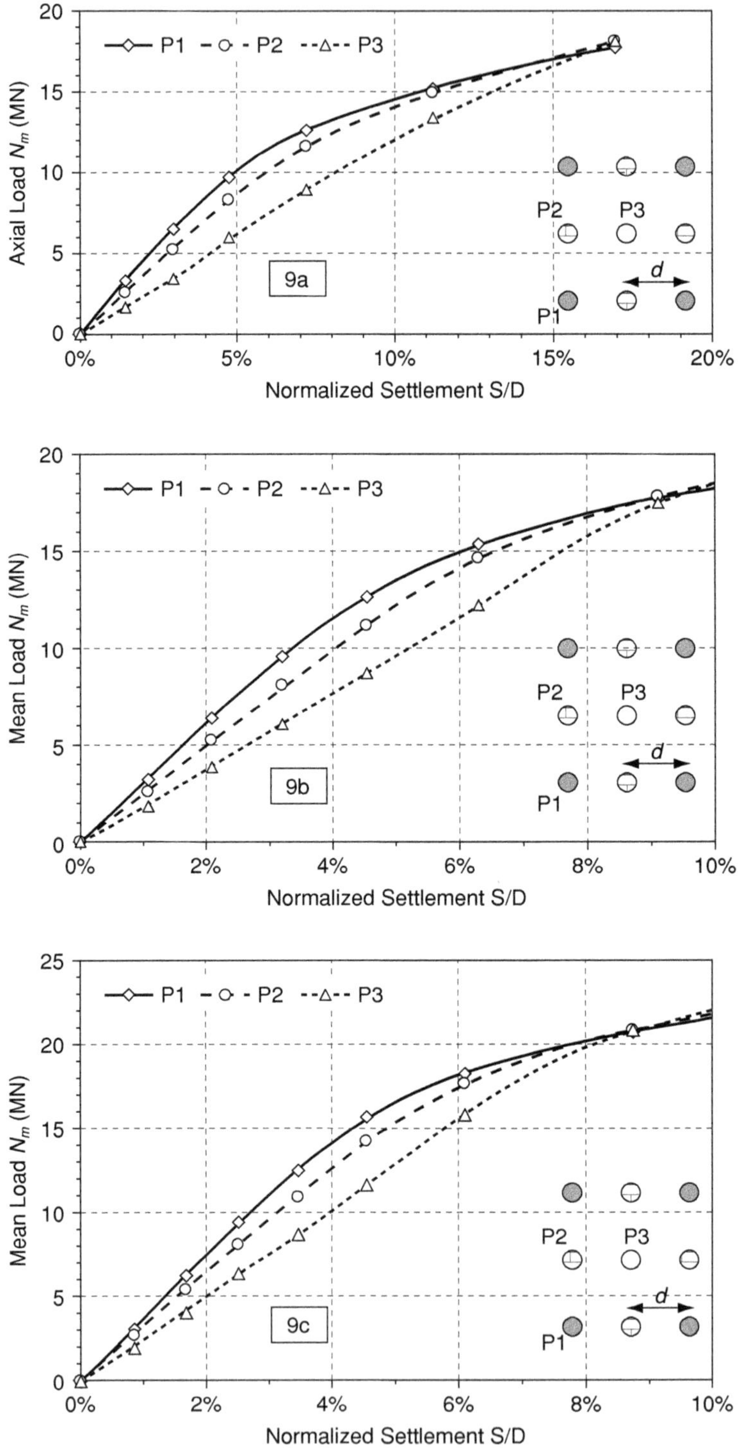

Figure 2.13 Load-settlement response of piles P1, P2 and P3 in a 3 × 3 layout with a spacing of 3D (9a), 4.5D (9b) and 6D (9c) (Comodromos [23])

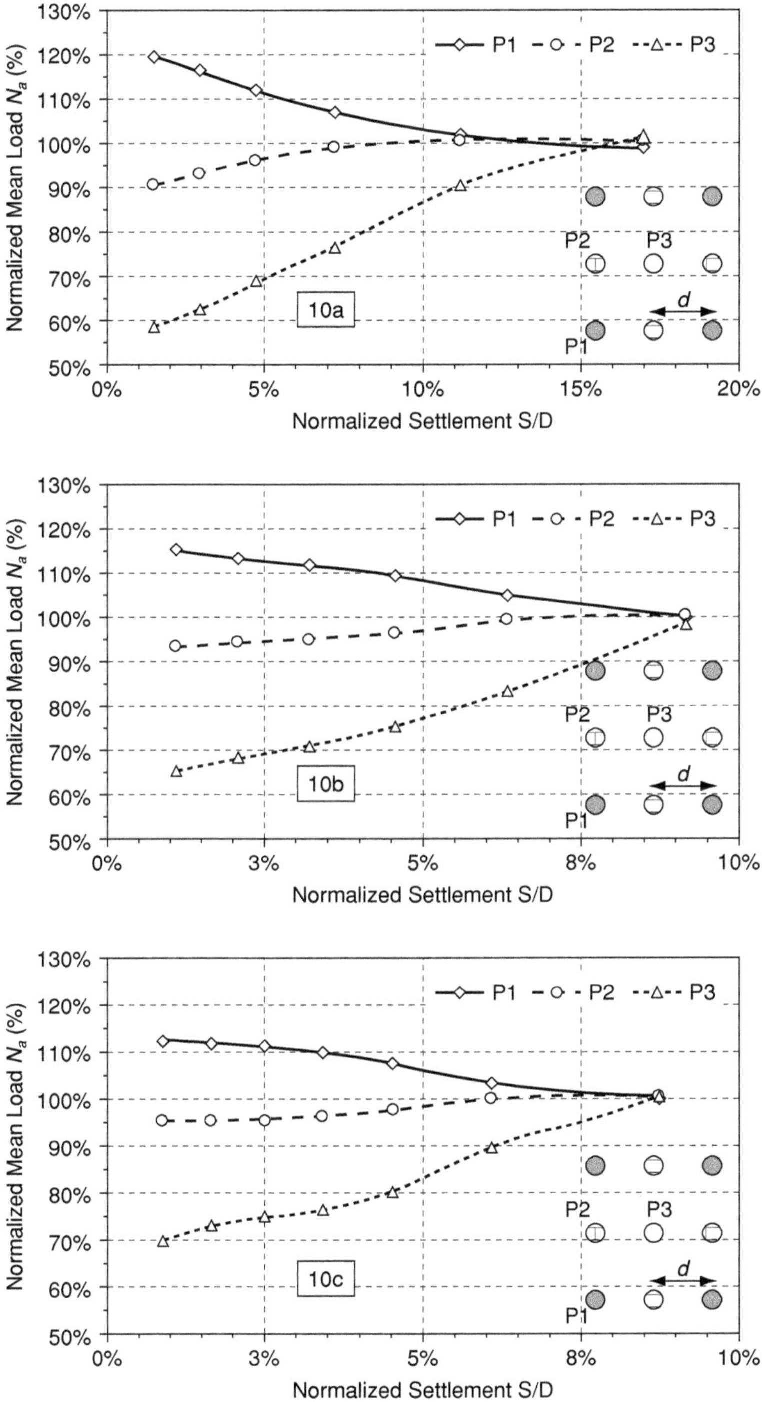

Figure 2.14 Variation of normalized load with normalized settlement for piles P1, P2 and P3 in a 3 × 3 layout with a spacing of 3D (10a), 4.5D (10b) and 6D (10c) (Comodromos [23])

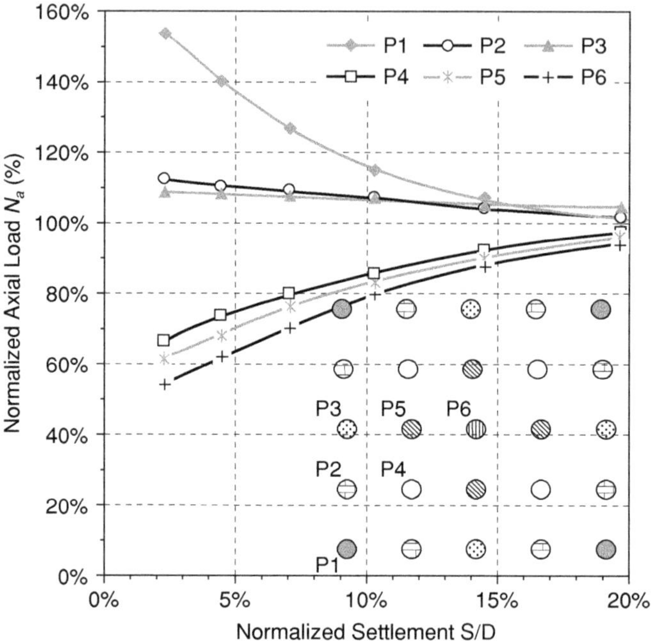

Figure 2.15 Variation of normalized load with normalized settlement for piles P1 to P6 in a 5 × 5 layout with a spacing of 3D (Comodromos [23])

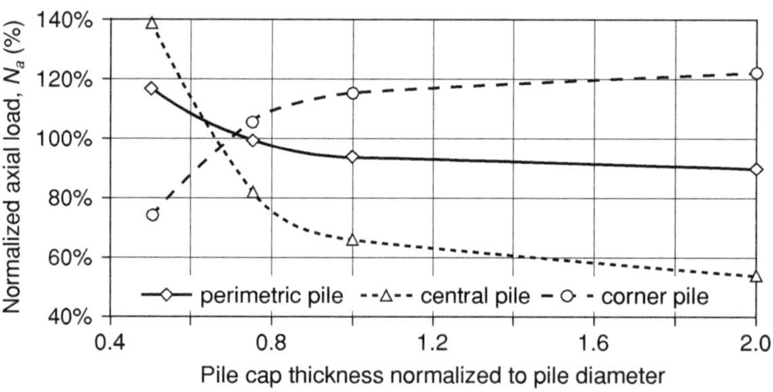

Figure 2.16 Effect of pile cap thickness to the load carried by the characteristic piles; case of concentrated applied load

displacement could cause significant stress development in these structures. The load-deflection curve of a single free-head pile can be determined from the results of pile load tests. However, it should be noted that the response of a single free-head pile under lateral loading significantly differs from that

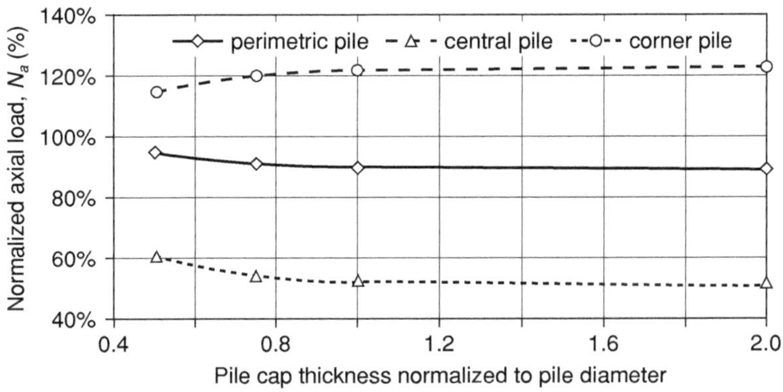

Figure 2.17 Effect of pile cap thickness to the load carried by the characteristic piles; case of distributed applied load

of a free or fixed-head pile group, depending upon the particular group configuration. On the other hand, full-scale pile group tests for determining the response of a pile group are very rare owing to the extremely high costs involved.

Simplified numerical approaches were proposed in order to take into account non-linearities arising from the soil–pile interaction. The well-known '*p-y*' method, proposed by Reese [29], is considered as the most representative and effective among the simplified numerical methods in determining the response of single piles under lateral loading. Although the method is reliable for evaluating the response of a single pile under horizontal load, it is questionable for assessing the response of pile groups. It is, however, commonly accepted that, for the same mean load, the piles of a pile group exhibit significantly greater deflection than an identical single pile. This behavior is attributed to the resisting soil zones behind the piles overlapping. Clearly the effect of the overlapping becomes evident as the spacing between piles decreases and when the soil around the piles yields, forming a large yielded zone surrounding the pile group.

Despite the fact that, in order to analyze the problem, a 3D non-linear numerical analysis is required, such an application is rarely utilized owing to the complexity in simulating the non-linearities of the interaction between the soil and the piles. In addition, such a procedure is very computationally demanding when compared with simplified numerical methods. Nevertheless, it is the most powerful tool for pile group response evaluation under horizontal or other types of loading, since it is able to predict both stiffness and ultimate resistance reduction factors, particularly in the case of sensitive soils undergoing plastification even at a low load level. Moreover, in most cases of pile lateral loading, the effect of pile–soil separation develops near the top and behind the pile owing to the soil having a limited ability to sustain tension. The depth to which separation may extend depends upon pile and soil stiffnesses as well as the level of loading. When incorporating interface elements between soil and piles in a 3D non-linear analysis, the development of separation should be taken into account according to a predefined criterion.

The next section deals with the response of a single pile followed by a paragraph devoted to the response of pile groups.

3.2 Single pile under horizontal loading

The equation of an elastic beam supported on an elastic foundation given by Hetenyi [30] can be used to analyze the behavior of a single pile under lateral loading. This simplified approach led to the well-known fourth-order differential beam-bending Equation 5, which is able to estimate the response of a laterally loaded single pile for both the free-head and the fixed-head conditions:

$$EI\frac{d^4y}{dx^4} + P_x\frac{d^2y}{dx^2} + E_s y = 0 \tag{5}$$

where

EI: flexural stiffness of the pile,
P_x: axial load of the pile,
E_s: soil modulus,
y: deflection and
x: length along the pile.

Equation 5 becomes increasingly complex if we take into account the fact that the third term, corresponding to the lateral soil reaction p, is not linear, since the soil modulus varies with the deflection level. It could, however, be approximated by a family of curves giving soil resistance as a function of deflection and depth below the ground surface, well known as the '*p-y*' method. The simplicity of the method in conjunction with the well-defined procedures for establishing the '*p-y*' curves (Matlock [31], Reese et al. [32], Reese and Welch [33]) made the method the most widely used. The method can be easily implemented in a simplified FE or FD code [34, 35], where a pile is subdivided into a certain number of structural beam elements, while soil resistance is introduced at nodal points via multi-linear springs resulting from '*p-y*' curves. Pile loads (lateral force, bending moment or even a load–moment combination) and various constraints (lateral displacement, rotation or a combination) are usually applied on the pile head. The pile head can be a free or a fixed head. A solution is achieved by applying an iterative scheme when a predefined criterion of convergence has been achieved.

Estimation of *p-y* curves in hard Clay

(After Reese et al. *[33])*

INPUT DATA	
Shear Stress Su (kPa):	120.00
Pile Diameter D (m)	1.00
Depth from soil surface x (m)	6.00
Buoyant Unit Weight of Soil γ' (kN/m^3)	11.00
Strain ε_{50} Soft Caly: 0.020, Medium Stiff: 0.010, Stiff: 0.005	0.005
Loading Conditions (static/cyclic)	static
Number of cycles N (cyclic loading)	7

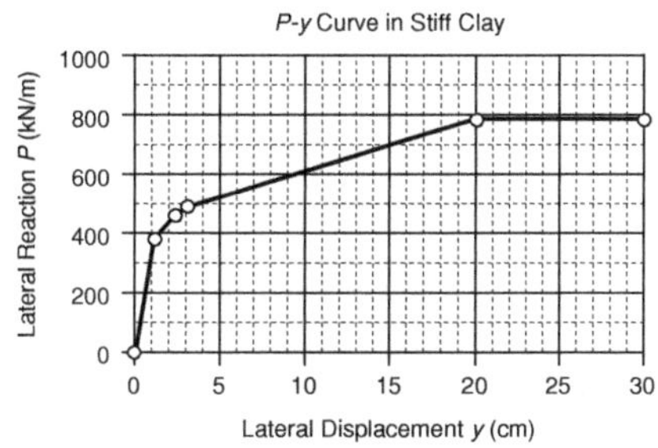

OUTPUT	
Limit Lateral Resistance P_{ult} (kN/m)	*786.00*
Lateral Displacement P = $P_{ult}/2\ y_{50}$ (m)	*0.013*

EQUATIONS

– Static Conditions

<u>*Calculation of Limit Lateral Resistance:*</u>

P_{ult} = min ($P_{ut} \cdot P_{ud}$) OTTOU:
P_{ut} = (3 + (γ'/S_u)*x + (0.5/b)*x) * S_u * b και P_{ud} = 9 * S_u * b
Y': Buoyant Unit Weigth of Soil Y' (kN/m^3)
S_u: Shear strength (kPa)
D: Pile Diameter (m)
x: Depth from soil surface (m)

<u>*P-y Curve (2 segments):*</u>

α) curve segment: $p = 0.5^*P_{ult}{}^*(y/y_{50})^{1/4}$. $y_{50} = 2.5^*\varepsilon_{50}{}^*$b for $y \leq 16^*y_{50}$ and
β) linear segment $p = P_{ult}$

–Cyclic Conditions

<u>*P-y Curve:*</u>

α) curve segment p-y_c : $p = 0.5^*p_{ult}{}^*(y/y_{50})^{1/4}$ $\gamma\iota\alpha\ y \leq 16^*y_{50}$
$y_c = y_s + y_{50}{}^*C^*\log N$, $C = 9.6^*(p/p_u)^4$
y_s: lateral displacement under static conditions
y_c: lateral displacement under cyclic conditions for N cycles
β) horizontal part $p = p_{ult}$

RESULTS				
a/a	*P* (t/m)	*y* (cm)	K (t/m^3)	Remarks
1	0.0	0.00		–
2	384.9	1.15	334.7	–
3	457.7	2.30	199.0	–
4	492.2	3.08	160.1	–
5	786.0	20.00	39.3	$p = P_{ult}$ $y = 16^*y_{50}$
6	786.0	30.00	26.2	P_{ult}

Figure 2.18 Calculation process for *p-y* curves in hard clays according to Reese et al. [33]

Figures 2.18 and 2.19 present simple EXCEL worksheets in order to determine the '*p-y*' curves for clayey and sandy soils based on the observations of Reese et al. [32, 33]. The main limitation of the method is that the '*p-y*' curves are independent of one another and therefore the continuous nature of the soil along the pile length is not explicitly modeled. To overcome this shortcoming, a shear coupling between the springs was proposed by Georgiadis [36]. Many attempts have been carried out to extend the application of the method to the analysis of pile groups by introducing pile–soil–pile interaction factors or displacements multipliers. However, the application of the method to pile groups remains questionable as the method does not incorporate the effect of plastic zones overlapping around the individual piles nor does it incorporate the pile–soil separation developing at a certain level of displacement.

The '*p-y*' method could also be used to back-calculate values for soil parameters when pile test data under horizontal loading are available. Figure 2.20

Estimation of *p-y* curves in sandy soils

(After Reese et al. *[32])*

INPUT DATA	
Internal Friction φ (°)	34.00
Pile Diameter D (m)	1.00
Depth from soil surface x (m)	3.00
Buoyant Unit Weight of Soil Y' (kN/m³)	11.00
Critical depth $P_{ct} = P_{ucd}\ X_t$ (m)	15.90
Coefficient k	35000
Loading Conditions (static/cyclic)	Cyclic

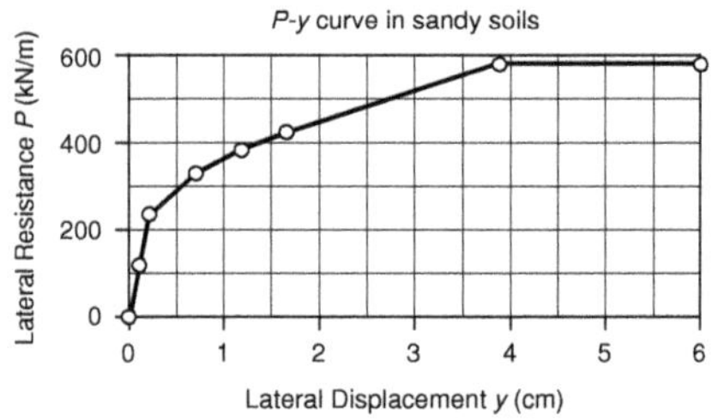

OUTPUT	
$\alpha = \varphi/2$	17.0
$\beta = 45 + \varphi/2$	62.00
$K_o = 1 - \sin\varphi$	0.44
$K_a = \tan^2(45 - \varphi/2)$	0.28
Limit Lateral Resistance p_c (kN/m)	603.43
Dimensionless Coefficient A*	0.96
Dimensionless Conefficient B*	0.70
Lateral Displacement for $p = p_u\ y_u$ (m)	0.0375

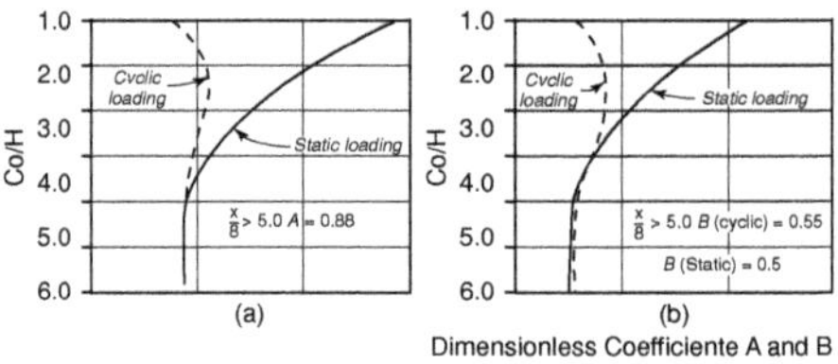

EQUATIONS

Limit lateral Resistance of sand calculated as:

Limit lateral resistance for depth less than critical ($x < x_t$):

$$P_{ct} = \gamma H\left[\frac{K_o H \tan\varphi \sin\beta}{\tan\beta - \varphi)\cos\alpha} + \frac{\tan\beta}{(\tan\beta - \varphi)}(b + H\tan\beta\tan\alpha) + K_o H \tan\beta(\tan\varphi\sin\beta - \tan\alpha) - K_a b\right]$$

Limit lateral resistance for depth > than critical ($x \geq x_t$).

$$p_{cd} = K_a b\gamma H(\tan^8\beta - 1) + K_o b\gamma H\tan\varphi\tan^4\beta$$

P-y curve (4 different segments):

α) Linear part: $p = K^* x^* y$, $y \leq y_k$ where $y_k = (C/kx)^{n(n-1)}$

$C = p_m / y_m^{1/n}$

$p_m = B'^* p_c$ | $m = (p_u - p_m)/(y_u - y_m)$

$y_m = b/60$ $y_u = 3b/80$ | $q = p_m/(m^* y_m)$

loose sand: $k = 5500$ kN/m³, medium dense sand: $k = 17000$ kN/m³, dense sand $k = 35000$

β) Hyperbofic part: $p = C^* y^{1/n}$, $y_k \leq y \leq y_m$

γ) Linear part with slope m for $y_m \leq y \leq y_u$

δ) Horizontal part $p_u = A'^* pc$, $y \geq y_u$

Results				
1	0.00	0.00		–
2	117.92	0.11	1,050.0	–
3	235.85	0.22	1,050.0	$y = y_k$
4	329.62	0.71	467.3	–
5	383.75	1.19	323.6	–
6	423.91	1.87	254.3	y_m
7	578.93	3.75	154.4	y_u
8	578.93	5.63	102.9	–

Figure 2.19 Calculation process for p–y curves in sandy soils according to Reese et al. [32]

shows the response of such a test corresponding to the pile geometry and soil profile presented in Figure 2.21. A detailed description of the site, the pile test configuration and the soil conditions at which the free-head pile test was carried out is given by Comodromos [37], while a brief description is given below. Figure 2.20 shows the pile load test arrangement and the soil profile.

Comodromos and Pitilakis [26] carried out a series of parametric analyses of fixed-head pile groups on the above soil profile using the FD code FLAC^{3D}. Figure 2.22 illustrates the finite difference grid of a 4 × 4 pile group, consisting of 19,800 brick elements and 18,903 nodes. Pile elements were activated or not activated as soil material according to the specific group

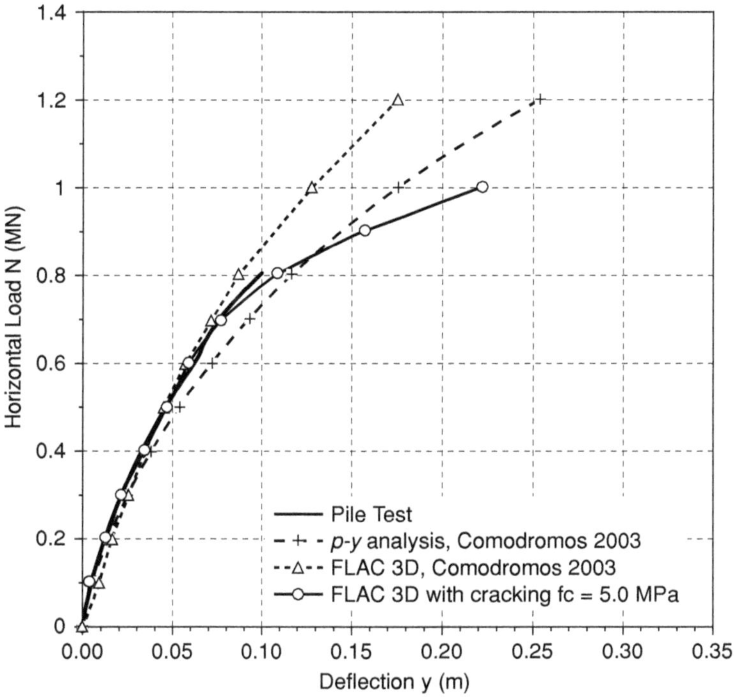

Figure 2.20 Load deflection curves for pile test, '*p-y*' analysis and 3D analysis with and without cracking

layout. In their analysis the elastic perfectly plastic Mohr–Coulomb constitutive model was used, in conjunction with a non-associated flow rule, to simulate the non-linear elasto-plastic behavior of the soil, while interface elements were introduced to allow pile separation from the surrounding soil. Separation occurs near the top and behind the pile generally no deeper than 20 percent of pile length, depending on the pile and the soil stiffness. Together with the local yield at the top of the soil, where large compressive stresses are developed in front of the soil, separation is considered as the main reason for the non-linear behavior.

The constitutive model of the interface elements in FLAC^{3D} is defined by a linear Coulomb shear-strength criterion that limits the shear force acting at an interface node, a dilation angle that causes an increase in effective normal force on the target face after the shear strength limit is reached and at a tensile strength limit. Figure 2.23 illustrates the components of the constitutive model acting at an interface node. The interface elements are allowed to separate if tension develops across the interface and exceeds the tensile limit of the interface. Once a gap is formed at the pile–soil interface, the shear and normal forces are set to zero.

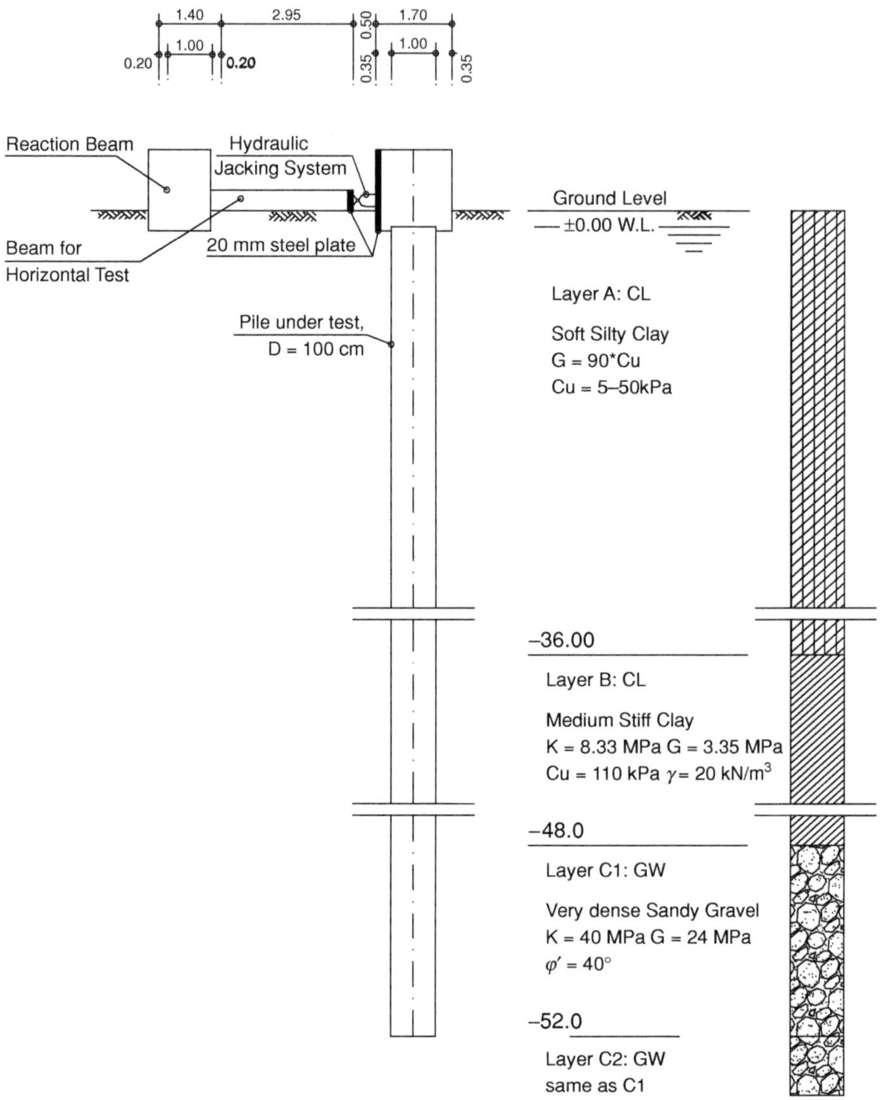

Figure 2.21 Soil profile and design parameters (Comodromos and Pitilakis [26])

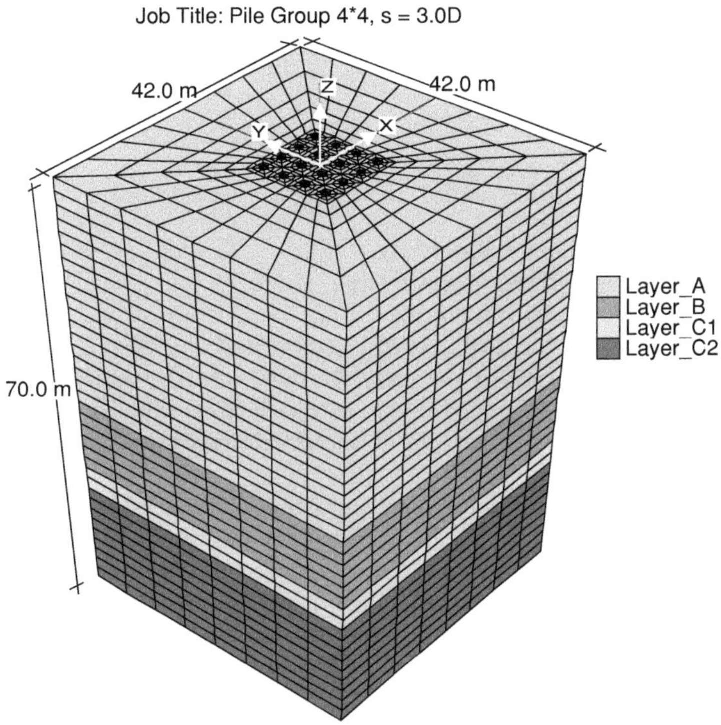

Figure 2.22 Finite difference grid for a 4 × 4 pile group (Comodromos and Pitilakis [26])

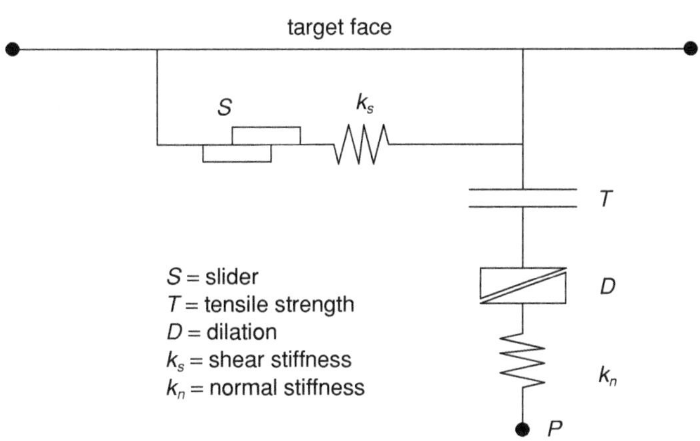

Figure 2.23 Components of the interface constitutive model in FLAC3D

The normal and shear forces are determined by the following equations,

$$F_n^{(t+\Delta t)} = k_n u_n A + \sigma_n A \tag{6}$$

$$F_{si}^{(t+\Delta t)} = F_{si}^{(t)} + k_s \Delta u_{si}^{(t+0.5\Delta t)} A + \sigma_{si} A \tag{7}$$

where:

F_n, F_{si}: normal and shear force respectively
k_n, k_s: normal and shear stiffness respectively
A: area associated with an interface node,
Δu_{si}: incremental relative shear displacement vector, and
u_n: absolute normal penetration of the interface node into the target face
σ_n: the additional normal stress added due to interface stress initialization
σ_{si}: the additional shear stress vector due to interface stress initialization.

Piles consisted of class C30/35 concrete, and Comodromos and Pitilakis [26] considered their behavior as linearly elastic. The modulus of elasticity was estimated to be 42 GPa, including the stiffening caused by the existence of steel reinforcement bars. A reduction in the moment of inertia I of the order of 50 percent for the upper part of the pile was applied, to incorporate the fact that the load applied on the tested pile was extreme enough to produce concrete cracking. In the present work, further 3D non-linear analysis has been carried out in order to take into account pile cracking in a more precise manner. To achieve this target, concrete has been modeled using the Mohr–Coulomb constitutive law with the tension cut-off option when exceeding the tension yield strength of the concrete. Compression and tension strength and the bond resistance between the concrete and the steel bars were taken as 30, 5 and 5 MPa respectively. Steel bars were also simulated using truss elements and an elastic-perfectly plastic constitutive law. Young's modulus for the Grade S550 steel bars was taken as 200 GPa, while the yielding strength was 550 MPa. A very close view of the mesh at the pile top (represented with grey shadow) at the plane $y = 0$ is given in Figure 2.24, in which the steel bars are also shown. This type of analysis is extremely computer-demanding in both mesh preparation and time required to achieve the solution.

The predicted load-deflection relationships are given in Figure 2.19, together with the prediction from the '*p-y*' analysis and the response of the pile test. Predicted pile response using the FE code NFEAG [38], in which the '*p-y*' approach was implemented, is in close agreement with the response of the tested pile. It should be noted that the difference amplifies as the displacement increases. At this point, non-linearity effects resulting from soil–pile separation and concrete cracking need to be considered. FLAC^{3D} analysis by Comodromos and Pitilakis [26] provides a deflection prediction which is closer to the pile test response. However, in a similar

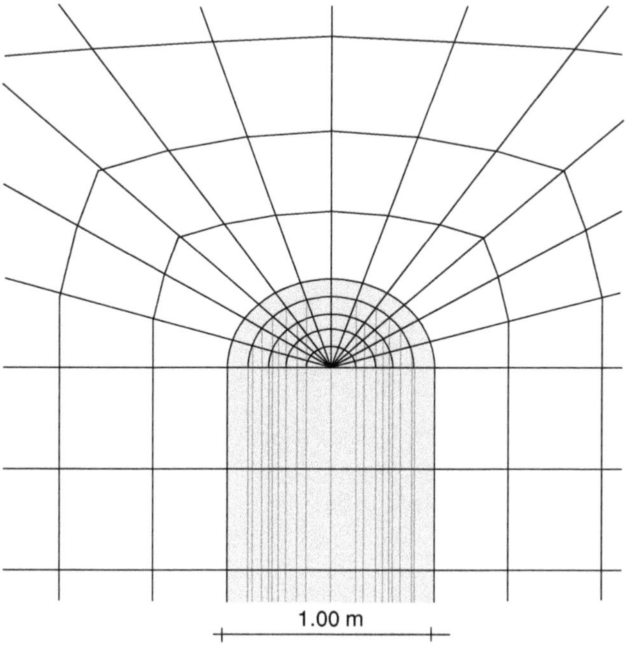

Figure 2.24 Mesh close view; pile elements and steel bars

manner to the previous analysis, the prediction becomes less satisfactory as deflection increases attributed to the effect of cracking, which at high deflection levels becomes important and was not taken into account. It should be stated that, with respect to practical design considerations, both predictions of the above analyses can be considered quite satisfactory. When incorporating the effect of cracking into the FLAC^{3D} analysis the prediction remains very close to the pile test curve, even for high levels of deflection at which cracking was encountered. According to the results of the analysis, the effect of cracking starts at an applied lateral load of 0.4 MN. The effect becomes significant when the load exceeds 0.8 MN. Further loading of the pile produces pile cracking and yielding, resulting in significant deflections for even a small increment of applied load. Furthermore, it could be seen that both the '*p-y*' analysis and the 3D analysis continue to provide resistance as the effect of cracking was not appropriately incorporated in these types of analyses. Based on the results of current research, it can be said that the effect of cracking could be a significant factor in the response of a free-head pile when it is significantly loaded. However, in the case of fixed-head piles, the effect occurs when the applied load exceeds 10 percent of the allowable vertical load. Moreover, for this level of lateral loading, if a significant vertical load acts simultaneously on the pile head, the effect is almost negligible.

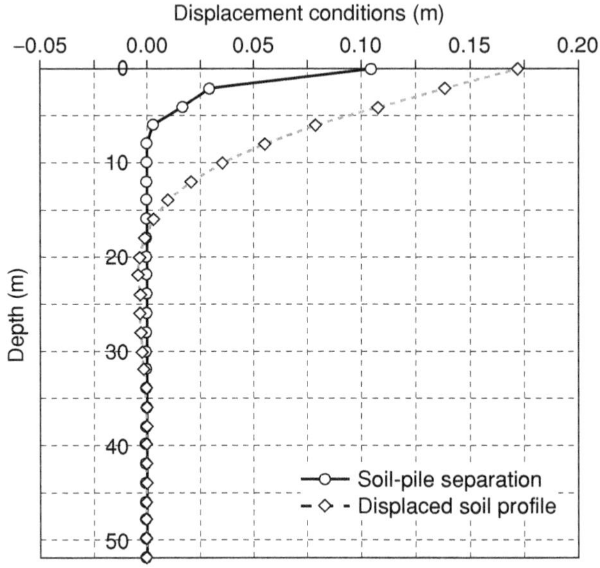

Figure 2.25 Soil–pile separation along the pile for the maximum load of 1.2 MN (Comodromos and Pitilakis [26])

Figure 2.25 shows the soil–pile separation along the pile for the maximum load of 1.2 MN predicted by 3D FD analysis. The separation is equal to 10.5 cm at the top of the pile and 0.3 cm at a depth of 6 m, while the displacement at the head of the pile is 17.5 cm. According to the results, the mobilized area is larger in front of the pile than behind it, mainly owing to the existence of the interface surface around the pile, which allows soil–pile separation when the soil tensile strength is reached. In the given soil profile, where the clayey soil at the surface allows separation, the interface was considered as necessary to avoid any hanging of soil elements to those of the pile.

When cracking does not occur, pile bending moment at a given depth can be estimated from the displacement field at its center using Equation 8. Curvature of the pile is obtained by numerical differentiation.

$$M = -EI\frac{d^2y}{dx^2} \tag{8}$$

When cracking occurs, pile rigidity decreases and the neutral axis moves. In this case, Equation 9 is used to estimate the pile curvature at a given location.

$$\varphi = \frac{(\varepsilon_t - \varepsilon c)}{ds} \tag{9}$$

where

ε_t: tensile strain of the steel bar in the middle of the pile in the direction of loading
ε_c: compressive strain of the steel bar in the middle of the pile in the direction of loading
ds: distance between the above opposed bars.

The pile bending moment is calculated as the sum of the moments resisted by concrete and that carried by the steel bars according to Equations 10 and 11 respectively.

$$M_c = -E_c I_{unc} \varphi \quad (10)$$

$$M_s = E_s \varphi \sum_{1}^{n} (I_{sn} + A_{sn} x_n^2) \quad (11)$$

$$M = M_c + M_s \quad (12)$$

where

M_c: bending moment undertaken by the uncracked part of the concrete,
M_s: bending moment carried out by steel bars,
E_c: concrete Young's modulus,
E_s: steel Young's modulus,
I_{unc}: moment of inertia of the uncracked concrete area,
φ: curvature at the neutral axis,
A_{sn}: section of a given steel bar,
X_n: distance of a given steel bar form the neutral axis
n: number of steel bar reinforcement.

Comparing the three methods, the following conclusion can be drawn:

(1) The '*p-y*' approach is a simple, effective and in most cases satisfactory method of determining the response of a single pile under lateral loading.
(2) Three-dimensional analysis may be required to account for effects of pile–soil separation and when stress and/or the kinematic field of the surrounding soil is needed.
(3) Three-dimensional analyses, when incorporating an accurate effect of pile cracking, render the problem very computationally demanding; and, to the knowledge of the author, no such a detailed analysis has been carried out. According to preliminary results of ongoing research, it can be confidently stated that the effect of cracking remains insignificant for fixed-head piles under combined vertical and lateral loading, when the level of lateral loading remains less than 20 percent or even

10 percent of the vertical load. The effect becomes considerable, in the case of free-head piles, when the applied load produces high values of pile curvature, which generally occurs as the deflection reaches the critical value of 10 percent of the pile diameter.

3.3 Pile group under horizontal loading

3.3.1 Response of pile groups

The effect of the interaction between the piles of a group under lateral loading was described in 1972 by Oteo [39] and since then has been verified and in some cases quantified by many researchers. According to Prakash and Sharma [40], the lateral group efficiency n_L, defined by Equation 13, may reach 40 percent, depending upon the number of piles in and the layout of the group.

$$n_L = \frac{\text{ultimate lateral load capacity of a group}}{n * \text{ultimate lateral load capacity of single pile}} \tag{13}$$

In order to take into account the effect of pile groups on the stiffness, both the Canadian Foundation Engineering Manual [41] and the Foundations and Earth Structures Design Manual 7.2 [42] recommend the application of a reduction factor R, to the lateral subgrade modulus. More specifically, a reduction factor of $R = 0.3$ is proposed for groups with a pile spacing of 3.0-D. The value is linearly increased for spacings up to 8.0-D, at which no further change is considered, $R = 1.0$.

As previously mentioned, the load-deflection curve could be the determining factor for the design of a project, and therefore the group stiffness reduction factor caused by a lateral load is of greater importance than the group efficiency factor. Poulos [2, 43] introduced four different kinds of interaction and reduction factors for piles under lateral load, depending on the loading at the pile head and the type of deformation. Moreover, based on the elastic continuum approach, Randolph [44] proposed a relationship for estimating the interaction factors in fixed-head piles, demonstrating that the interaction under lateral loading decreases much more rapidly with spacing between piles than for axial loading. Wakai et al. [45] used 3D elasto-plastic finite element analysis to estimate the effect of soil–pile interaction within model tests for free or fixed-head pile groups. In that analysis, thin frictional elements were inserted between the pile and the soil to consider slippage at the pile–soil interface. It should, however, be noted that similarly to the reasoning explained for single piles, in many cases where the pile–soil interaction is governed by non-linearities arising from the soil separation behind the pile and the yield of soil in front of the pile, a 3D non-linear analysis, including interface elements around the piles, is considered more accurate in providing the pile group response.

Comodromos [37] utilized 3D FD analysis to evaluate the response of free-head pile groups. A similar procedure was applied by Comodromos and Pitilakis [26] to evaluate the response of fixed-head pile groups of various layouts. A group of nine piles fixed in a pile cap in a 3 × 3 arrangement was initially considered, while other layouts of 3 × 2, 3 × 1 and a 4 × 4 were also examined. Pile spacing was taken equal to 2, 3 and 6 times the pile diameter. The case of a fixed-head single pile was also considered since its response is to be used to compare with the group responses. The loading sequence included the initial step, at which the initial stress field was established, followed by incremental loading. Figures 2.26–28 illustrate the displacement contours in the direction of loading at the plane $y = 0$ for the case of the 3 × 3 layout with spacing of 2, 3 and 6 diameters. The displacement contours correspond to a mean load of 0.8 MN at the pile cap. The level of pile–soil–pile interaction could be seen qualitatively by the unification of the displacement contours. When the spacing is too small (as in Figure 2.28), a common displacement is observed at the soil surface between the piles, while from a certain load level the resisting zones behind the piles overlap. When these zones become plastic, the lateral load capacity becomes the load capacity of an equivalent single pile containing the piles rather than the summation of the lateral load capacity of the piles. A comparison between Figures 2.26 and 2.28 demonstrates that, as the spacing

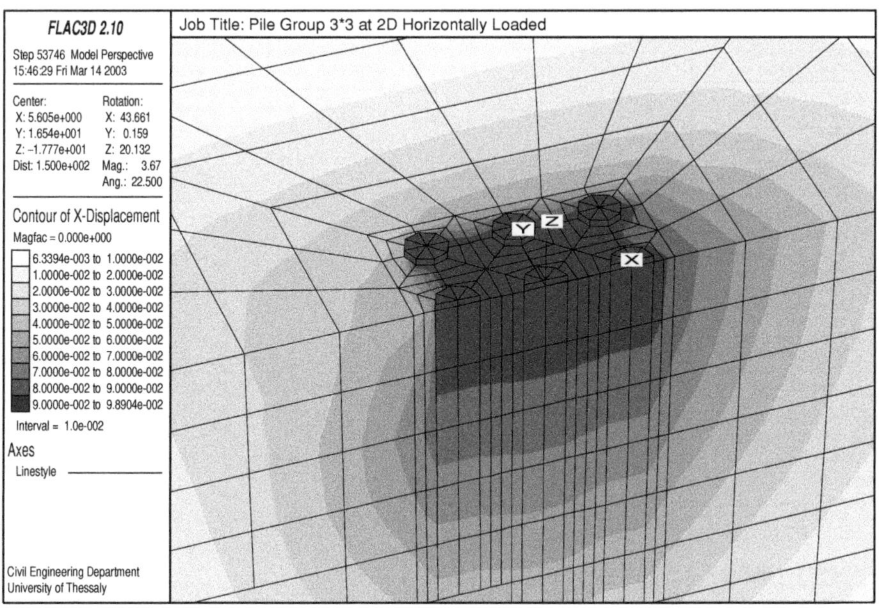

Figure 2.26 Displacement contours along the direction of loading for the case of a 3 × 3 layout with a spacing of 2.0D (Comodromos and Pitilakis [26])

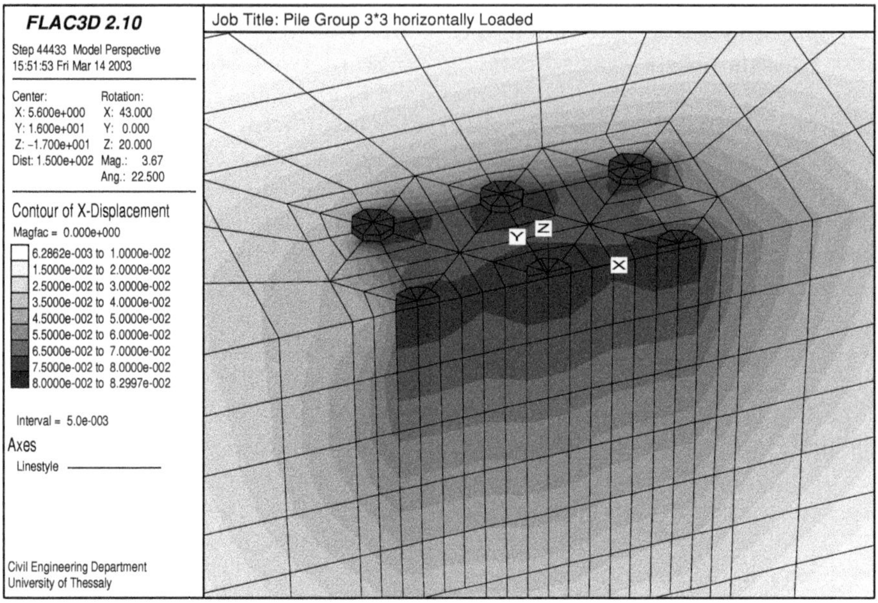

Figure 2.27 Displacement contours along the direction of loading for the case of a 3 × 3 layout with a spacing of 3.0D (Comodromos and Pitilakis [26])

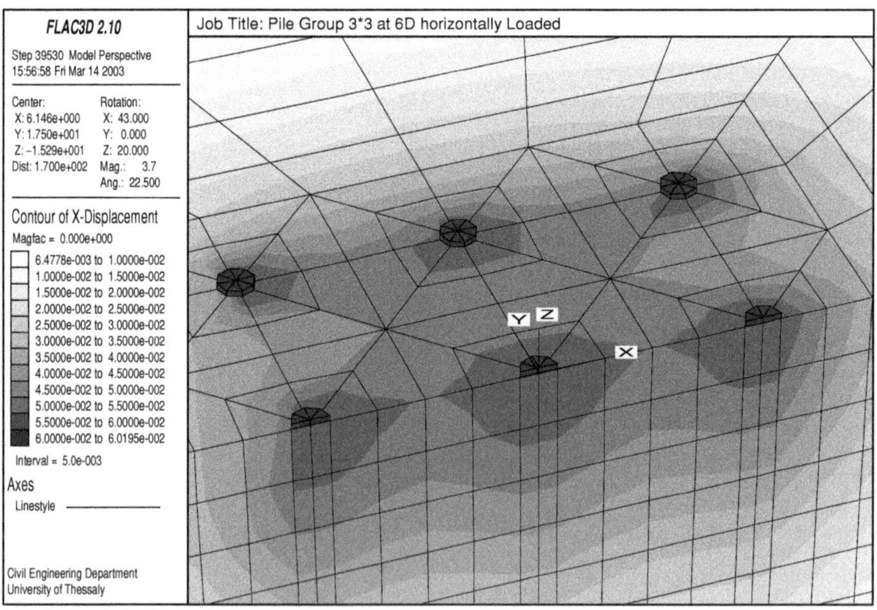

Figure 2.28 Displacement contours along the direction of loading for the case of a 3 × 3 layout with a spacing of 6.0D (Comodromos and Pitilakis [26])

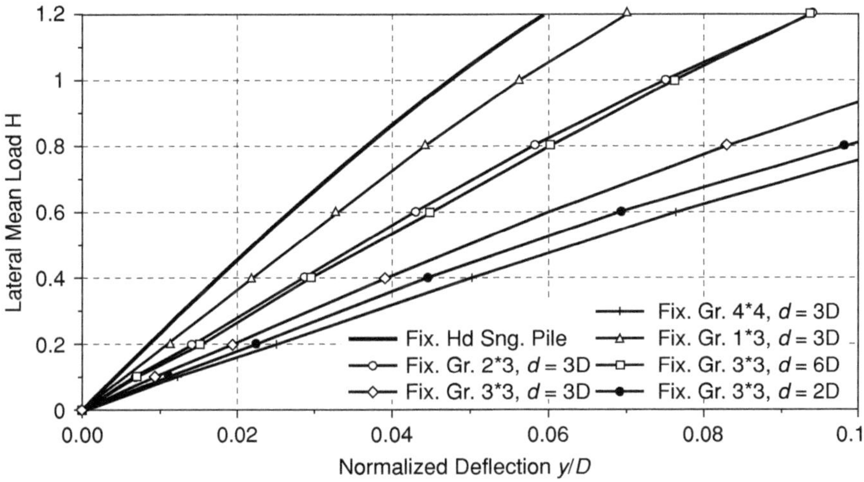

Figure 2.29 Numerically established lateral load-deflection curves for the fixed-head single pile, and various configurations of fixed-head groups (Comodromos and Pitilakis [26])

increases, the effect of overlapping between the resisting zones becomes less significant.

A detailed comparison of the results demonstrated that spacing significantly affects the load-deflection curves, while the number of rows and the total number of piles also play an important but less significant role. Figure 2.29 illustrates the load-deflection curves at the top of the pile for various pile groups together with that of the fixed-head single pile. It is concluded that the stiffer group is the group consisting of three piles in a row in the direction of loading (layout 3 × 1) with a spacing of 3.0-D, followed by the 3 × 2 having the same spacing. When examining the groups in a 3 × 3 layout, it could be verified that, as the spacing decreases, the stiffness of the group declines. Finally, the load-deflection curve of the 4 × 4 group with a spacing of 3.0-D shows the lowest stiffness, indicating that the number of piles affects the response of the group. Despite the variation of the load-deflection curve of each group, it could be concluded that all curves demonstrate a form similar to that of the fixed-head single pile.

A comparison between the deflection of the single pile and that of the pile group under the same mean load provides the stiffness efficiency factor defined by the following equation.

$$R_G = \frac{y_{mLs}}{y_{mG}} \tag{14}$$

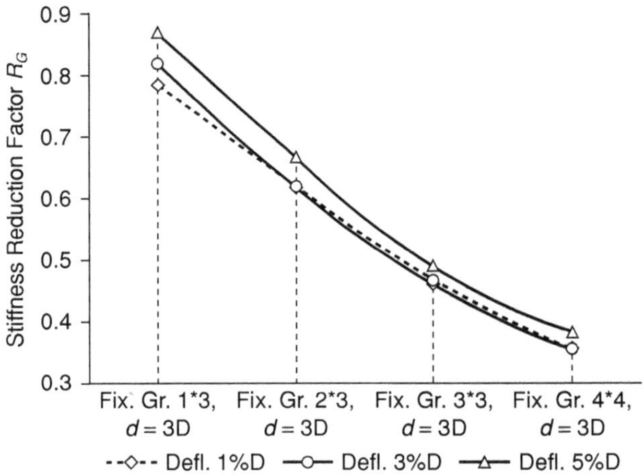

Figure 2.30 Variation of stiffness reduction factor with group size for a deflection of 1%, 3% and 5%D at the head of a fixed-head pile group (Comodromos and Pitilakis [26])

in which y_{mG} and y_{mLs} stand for the deflection at the head of the pile group and the single pile under the same horizontal mean load H_m respectively. The stiffness of a pile group for a given mean load H_m is then calculated using Equation 15.

$$K_G = R_G * K_S * n_p \tag{15}$$

in which K_S is the stiffness of the single pile for a given load. K_G denotes the stiffness of the pile group for the same load, and n_p is the number of piles in the group. Figures 2.30 and 2.31 illustrate the variation of the stiffness reduction factor with row numbers and spacings respectively. It can be seen that the reduction, as defined by Equation 14, may attain the level of 40 percent for groups with multiple rows. For the commonly adopted 3 × 3 pile group with spacing of 3.0-D, the predicted stiffness reduction factor R_G for a range of normalized settlements of 1–5%-D varies from 0.45 to 0.48. These values are in close agreement with the average *p*-multipliers of 0.47 proposed by Peterson and Rollins [46]. The effect becomes less significant in the case of a single row group, where the reduction factor was calculated to be in the order of 0.80. Figure 2.31 illustrates the effect of pile spacing related to the response of a pile group. More specifically, for a 3 × 3 pile group in a spacing of 6.0-D, the effect varies from 0.57 to 0.65 for normalized deflection of 1 and 5 percent respectively. The corresponding values, when spacing decreases to 2.0-D, are dramatically reduced to 0.39 and 0.42 respectively, demonstrating the significant impact of spacing. It should also be emphasized about the importance of the number of

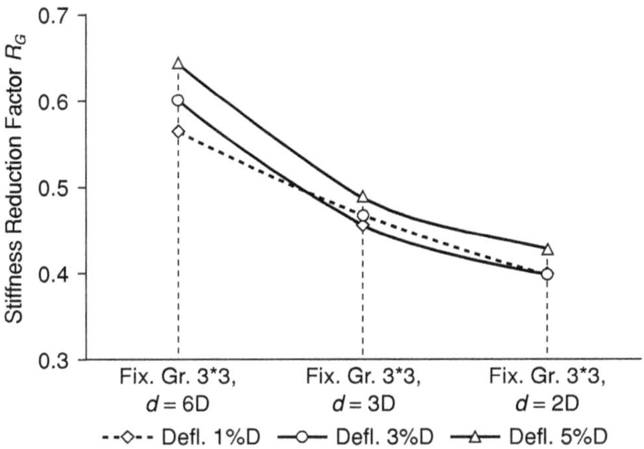

Figure 2.31 Variation of stiffness reduction factor with spacing, for a deflection of 1%, 3% and 5%D at the head of a fixed-head pile group (Comodromos and Pitilakis [26])

piles in a group. Figure 2.31 demonstrates that, for the same spacing, the greater the number of piles in a group the greater the stiffness reduction.

3.3.2 Response prediction

Based on the fact that the load-deflection curves of each group have a similar form to that of the single pile, it is essential to derive a relationship giving the ability to define the load-deflection curve of a given pile group using that of a single pile. The latter can be established using 3D analysis and pile test or even an accurate '*p-y*' analysis. It is evident that such a relationship would be eventually affected by the load-deflection curve of a single pile in a given soil profile, by the spacing and the number of columns and rows in the pile group, and the total number of piles. Within this framework, Comodromos and Pitilakis [26], based on the various numerical experiments and on curve refining procedure, defined a precise form of relationship to predict a pile group response founded on that of the single pile. Equation 16 was proposed to estimate the variable amplification factor and to allow for the prediction of the response of pile groups with a rigid cap. The most important variables determining the deflection amplification factor for groups are the deflection of the single pile, the spacing between the piles, the number of rows and columns in a pile group and the total number of piles which are included in Equation 16 (Figures 2.39–40). The deflection is profoundly non-linear, and for this reason at least three components were needed in Equation 16, in which the deflection of the single pile is introduced with a different weighting factor (α, β and γ).

$$R_a = \frac{\left(\frac{3}{n_x}\right)^{0.2} \frac{y_D^\alpha}{d} + \frac{1.1\sqrt{d}}{d} \ln(n_x + n_y) y_D^\beta \log\left[\frac{(n_x^4 + n_y)^{0.8}}{y_D}\right] \exp(\frac{1}{d})}{(0.7 y_D^\gamma d^3)^{0.03}} \tag{16}$$

where

R_a: deflection amplification factor,
y_D: normalized deflection of the fixed-head single pile defined as y/D,
d: relative pile spacing defined as s/D,
n_x, n_y: number of piles in the direction of and perpendicular to the loading respectively,
α, β, γ: parameters to be determined by the curve-fining procedure.

Using the deflection amplification factor from Equation 16 for a given mean horizontal load, Equation 17 provides the normalized group deflection y_G.

$$y_G = R_a y_d \tag{17}$$

The most suitable values for α, β and γ were automatically defined by the curve-fining procedure as $\alpha = 0.8$, $\beta = 0.2$, and $\gamma = 0.1$. In Figure 2.32, the bold lines represent the pile group load-deflection curves calculated using FLAC^{3D}, while the dashed line corresponds to the predicted curves using Equations 16 and 17. The calculated and predicted curves demonstrate notable agreement.

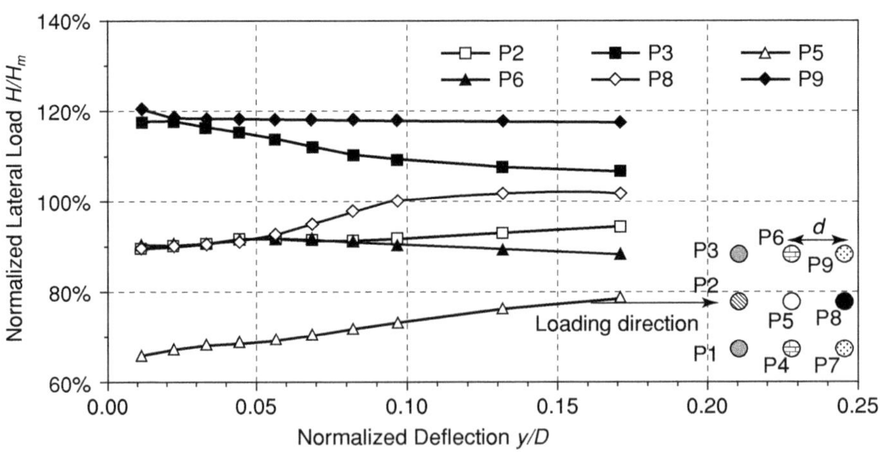

Figure 2.32 Variation of normalized load with normalized deflection for piles P2, P3, P5, P6, P8 and P9 in a 3 × 3 layout with a spacing of 2.0D (Comodromos and Pitilakis [26])

The validity of Equation 16 has also been verified using the experimental results given by Wakai et al. [45] for a fixed-head single pile and fixed-head pile groups. The model consists of aluminium piles with an outside diameter of 50 mm and a pile length of 1500 mm in a sandy soil (Onahama sand) arranged in a 3 × 3 layout with a spacing of 2.5-D. Figure 2.33 illustrates the response of the fixed-head single pile and that of the fixed-head pile group. In the same figure, the prediction by Wakai et al. [45] resulting from a 3D non-linear analysis is shown together with the prediction using Equation 16, in which the determined values $\alpha = 0.8$, $\beta = 0.2$, and $\gamma = 0.1$ were used. It can be seen that this equation provides a prediction sufficiently close to both the measured curve and to the curve provided by the 3D analysis of Wakai et al. [45].

The verification of the methodology for full or large-scale fixed-head pile groups was not feasible as no measurements for such tests subjected to lateral loading were available. However, in order to examine the validity to comparable conditions, the data from a pile group with moment-free connection [47] and a free-head pile group [48] were used. Brown et al. [47] carried out tests on a large-scale pile group subjected to lateral loading. Their model consists of closed-end steel piles with an outside diameter of 273 mm and a pile length of 13.1 m. The piles were driven into a pre-consolidated clay formation arranged in a 3 × 3 layout with a spacing of 3.0-D. Equal deflection level was applied to all piles using a loading frame with moment-free connections. Figure 2.33 illustrates the response of the single pile and that of the moment-free pile group. The prediction provided

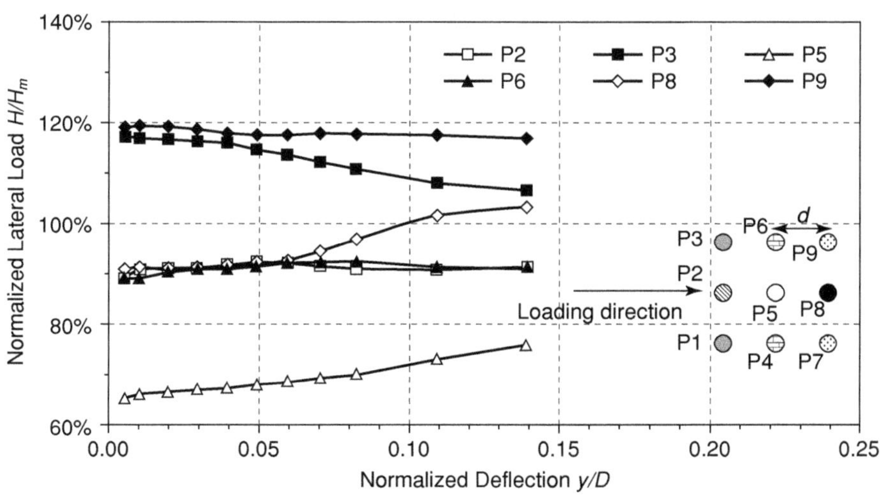

Figure 2.33 Variation of normalized load with normalized deflection for piles P2, P3, P5, P6, P8 and P9 in a 3 × 3 layout with a spacing of 3.0D (Comodromos and Pitilakis [26])

by Equation 16, using the previously determined values $\alpha = 0.8$, $\beta = 0.2$, and $\gamma = 0.1$ is again satisfactorily close to the measured deflection values.

Rollins et al. [48] performed a lateral loading test on a large-scale free-head pile group. Their model consists of closed-end steel piles with an outside diameter of 324 mm and a pile length of 9.1 m. The piles were driven into a composite soil profile consisting of gravel fill at the top while layers of clayey, silty and sandy soil were encountered down to the depth of 11.0 m. The pile arrangement was in a 3 × 3 layout with a spacing of 3.0-D. Load was applied to each pile using different load cells which were connected to a common loading frame. Thus, for the same central frame, each pile was able to carry different loads and deflections (free-head pile) in accordance with the soil-resistance distribution along each pile. Figure 2.41 illustrates the response of the free-head single pile and that of the free-head pile group after averaging the group loads and deflections. The prediction using Equation 16 and values of $\alpha = 0.8$, $\beta = 0.2$, and $\gamma = 0.1$ is presented by the line with the triangular markers. The experimental relationship conforms to the prediction curve based on the proposed equation. Should the need arise for an accurate prediction for this free-head pile group, the application of the curve improvement subroutine given in Comodromos and Pitilakis [26] suggests the use of $\alpha = 1.05$, $\beta = 0.25$, and $\gamma = 0.10$, for which the prediction shown by the line with circular markers is very close to the measured points.

In conclusion, owing to the observed shape similarity between the response of a single pile, a group consisting of piles of the same size, in the same soil profile and under lateral loading, it was possible to propose a relationship with the ability of predicting the response of a given pile group based on that of a single pile. The relationship was originally proposed by Comodromos [37] for free-head piles, and its validity was checked with numerical experiments. A similar procedure has been followed by Comodromos and Pitilakis [26] for fixed-head piles. Several pile groups have been considered, and a curve fitting procedure was formulated to determine precise values for the parameters affecting the prediction. The values $\alpha = 0.8$, $\beta = 0.2$, and $\gamma = 0.1$ were proven to be appropriate for the fixed-head piles just as in the case of free-head piles. Furthermore, the validity of the proposed relationships was examined for different soil profiles (sandy and clayey soils) as well as for a small experimental-size and a large-scale single pile and pile group. It was concluded that the relationship predicts successfully the response of a pile group in completely different soil profiles, for pile sizes other than for those originally used and for different methods of construction. However it would be unwise for it to be used in every soil profile. The results were extremely encouraging even for large-scale pile groups with different boundary conditions (moment-free, free and fixed-head piles). Nevertheless, the applicability of the proposed formulae to a different soil profile must be verified or the proposed equations readjusted by numerical analyses preferably in conjunction with *in-situ* test results.

For all important projects, a 3D non-linear analysis of a single pile and a pile group is recommended. Based on the numerical results parameters α, β and γ, Equation 16 may need to be re-adjusted, using the curve fining procedure given in Reference 26. The readjusted relationship could then be applied to define the response of any pile group for the same soil profile (it usually remains invariable within the limits of common projects) and the same pile geometry.

Finally, it should be noted that the proposed methodology is valid for monotonic loading and for more complex cases, such as cyclic or dynamic loading, with the use of more advanced constitutive models to simulate the soil among the piles, where warranted.

3.3.3 Load distribution in characteristic piles

In order to investigate the effects of load distribution to the piles within a pile group, the responses of the piles in a 3 × 3 layout were examined. As anticipated, the central pile carries the lowest load for the same deflection, presenting the minimum stiffness, while the two corner piles on the direction of loading (P7 and P9) carry the largest load, presenting the maximum stiffness. Figures 2.34–36 illustrate the normalized load undertaken by the piles of the group as a function of the normalized deflection. The central pile P5 initially carries the 65 percent, 65 percent and 69 percent of the mean load for spacings of 2.0-D, 3.0-D and 6.0-D respectively. These percentages gradually increase to 78 percent, 75 percent and 75 percent

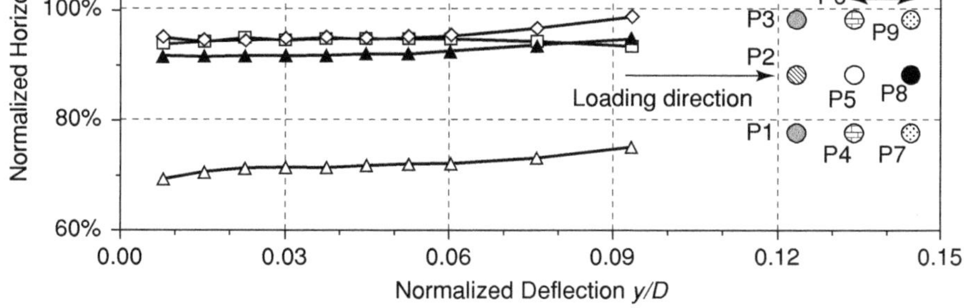

Figure 2.34 Variation of normalized load with normalized deflection for piles P2, P3, P5, P6, P8 and P9 in a 3 × 3 layout with a spacing of 6.0D (Comodromos and Pitilakis [26])

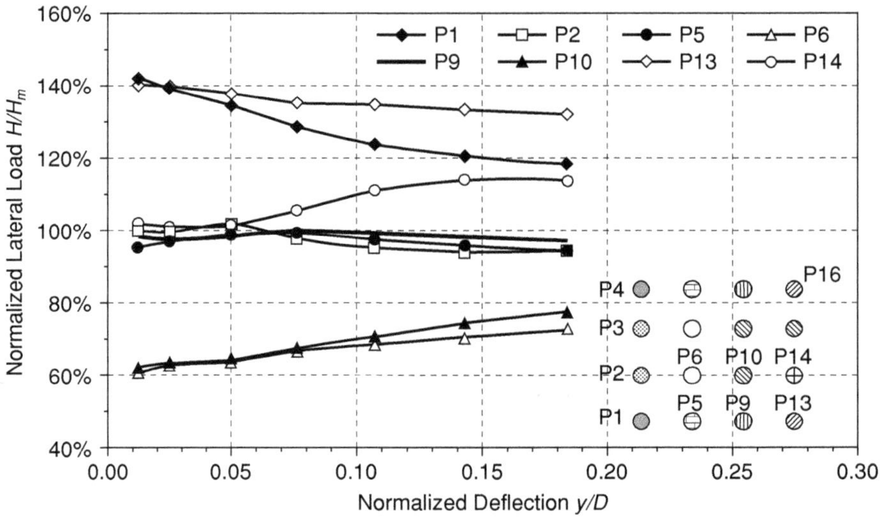

Figure 2.35 Variation of normalized lateral load with normalized deflection of piles P1, P2, P5, P6, P6, P9, P10, P13 and P14 in a 4 × 4 group with an axial distance of 3.0D (Comodromos and Pitilakis [26])

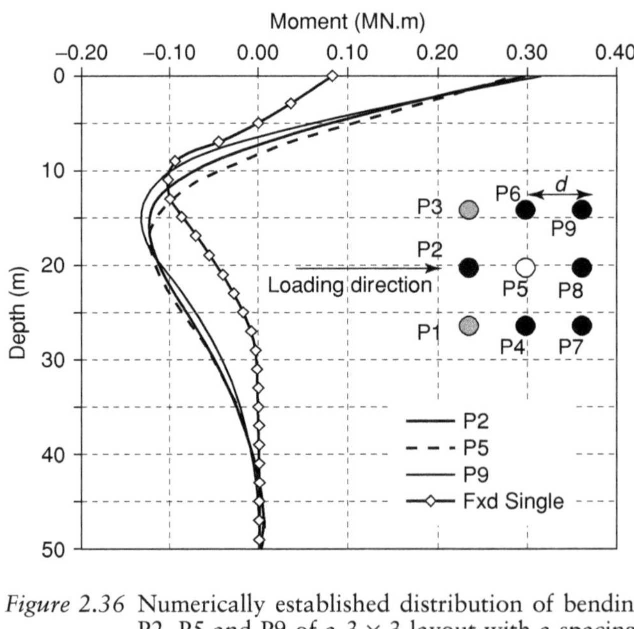

Figure 2.36 Numerically established distribution of bending moments along piles P2, P5 and P9 of a 3 × 3 layout with a spacing of 3.0D, compared to the predicted curve of an identical fixed-head single pile, for a mean lateral load of 0.4MN (Comodromos and Pitilakis [26])

respectively when the deflection level becomes greater than 10 percent of the pile diameter. On the other hand, pile P9 initially carries 120 percent, 120 percent and 115 percent of the mean load. This percentage gradually decreases with deflection level, becoming 117 percent, 116 percent or 112 percent when the deflection increases to 10 percent of the pile diameter. The loads transferred to the other piles of the group remain within the limits of these two characteristic piles. It can be observed that the load carried by the piles of the layout with a spacing of 6.0-D remains invariant no matter what the deflection levels and that the response of the piles is almost linear (Figure 2.36). The results are in agreement with experimental results [46] demonstrating that the leading pile row carries the higher loads. It should be noted that the external piles, which are less affected by the overlapping effect of resisting zones, demonstrate higher resistance. The higher load is always transmitted to the corner pile in the leading row. The corner pile of the trail row initially presents the same resistance, which gradually reduces to a value slightly greater than that of the pile in the middle of the leading row. The central pile always demonstrates significantly less resistance since it is greatly affected by the overlapping of resisting zones created by the surrounding piles.

Comparable results are shown in Figure 2.37 for the 4 × 4 group with a spacing of 3.0-D. As anticipated, in this case, the effect of interaction is

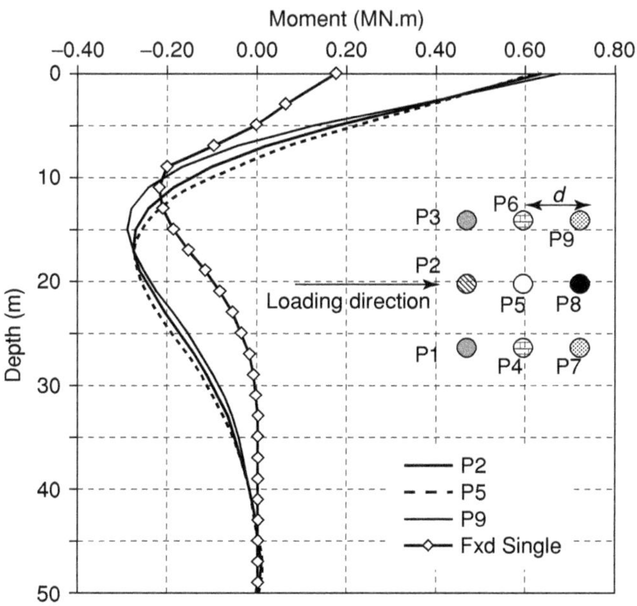

Figure 2.37 Numerically established distribution of bending moments along piles P2, P5 and P9 of a 3 × 3 layout with a spacing of 3.0D, compared to the predicted curve of an identical fixed-head single pile, for a mean lateral load of 0.8MN (Comodromos and Pitilakis [26])

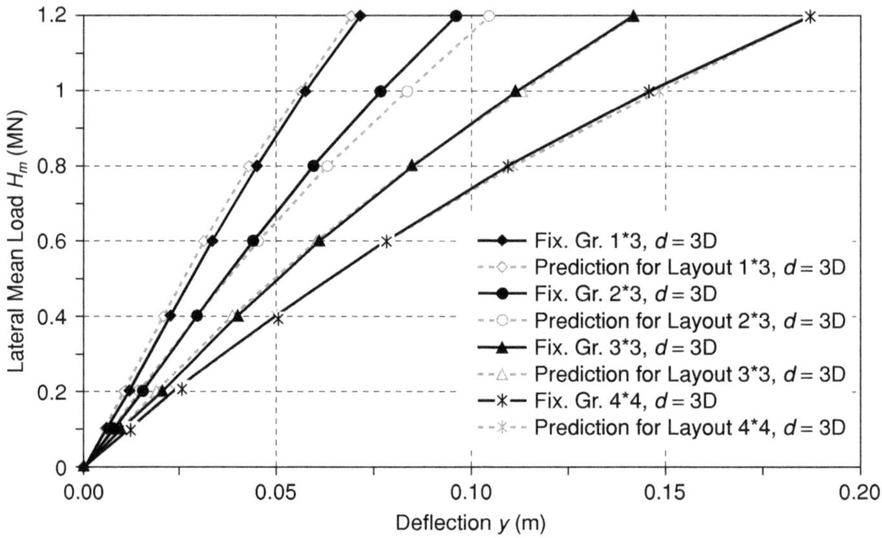

Figure 2.38 Comparison between numerically established load-settlement curves using FLAC3D and those predicted by Equations 11 and 12, for various fixed-head pile groups configurations with spacing s = 3.0D (Comodromos and Pitilakis [26])

higher. The central pile P10 initially carries 60 percent of the mean load and the corner pile, P13, 140 percent. These percentages gradually change with the deflection level, becoming 134 percent and 70 percent when the deflection attains a value of 10 percent of the pile diameter.

Figure 2.38 illustrates the bending moment in piles P2, P5 and P9 corresponding to an applied mean load of 0.4 MN for a 3 × 3 group with a spacing of 3.0-D. Pile P2 is in the middle of the rear row, P5 is the central pile and P9 is at the corner of the front row. It may be noticed that the differences between bending moment of these piles are less than 10 percent, despite the fact that the load carried by the corner piles is almost double the load of the central pile. This can be attributed to the resistance of soil zones in front of the piles carrying higher loads, since the effect of the interaction at these zones is negligible. The predicted bending moment curve for an identical fixed-head single pile is essentially different. At the head of the pile, the bending moment predicted for the single pile is three times less than the values predicted for the piles of the 3 × 3 group. The difference of the predicted values for the maximum bending moment along the piles between the single pile and the piles of the group, on the other hand, is not exceeding 20 percent. It may be noticed, however, that, while the bending moment of the single pile approaches zero at the mid-depth of the pile, the piles of the group are subjected to bending moments for

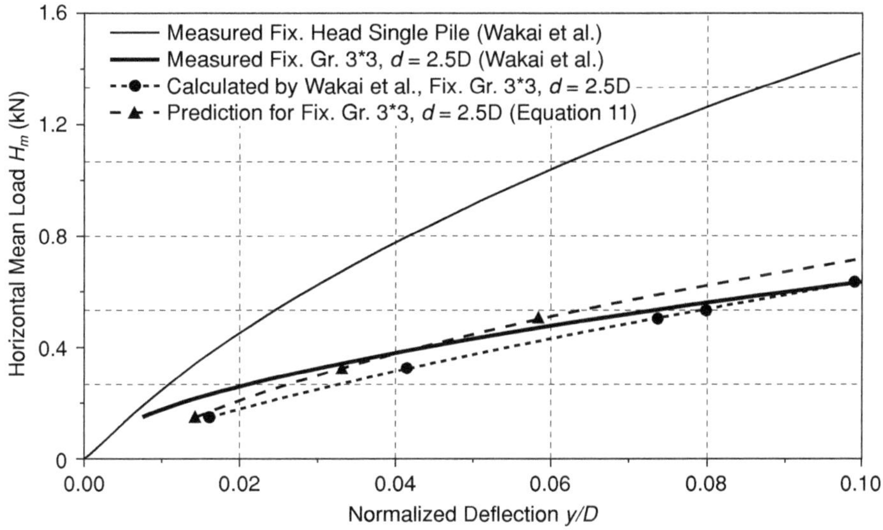

Figure 2.39 Comparison between measured, calculated (Wakai et al.) and predicted by Equation 11 load-deflection curve (Comodromos and Pitilakis [26])

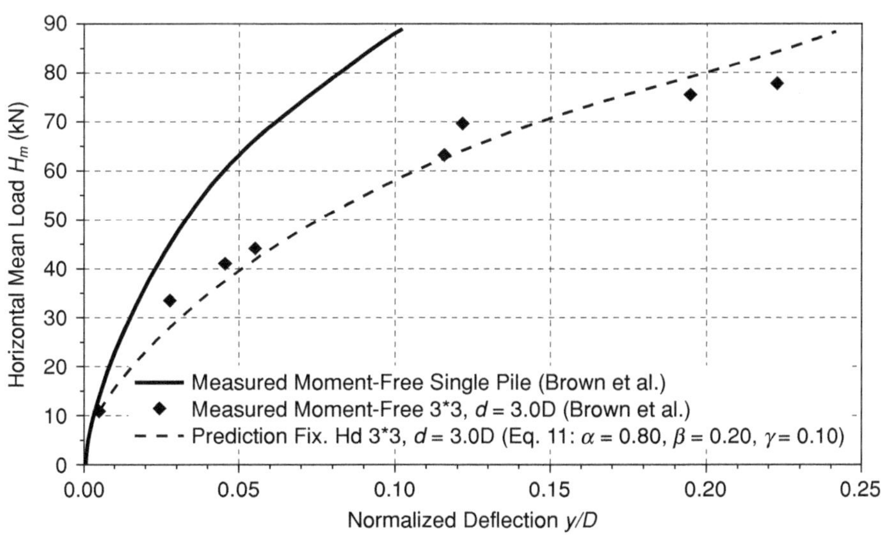

Figure 2.40 Comparison between measured load-deflection curve (Brown et al.) and prediction by Equation 16 (Comodromos and Pitilakis [26])

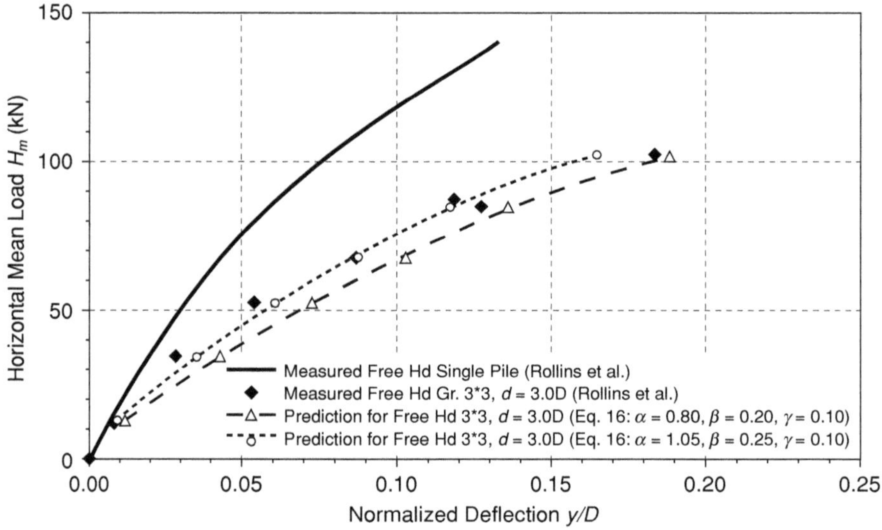

Figure 2.41 Comparison between measured load-deflection curve [48] and prediction by Equation 16 (Comodromos and Pitilakis [26])

significantly greater depth. The same conclusions can be drawn when the applied mean load is increased to 0.8 MN, as illustrated in Figure 2.36.

4 Application to Design Process

A straightforward analysis of a superstructure based on a pile foundation can be achieved only by assuming linear elasticity in both the foundation and in the superstructure. This conventional design procedure cannot be applied when soil non-linearity and effects from pile group response must be considered. In this case, it is usually necessary to adopt some powerful numerical tools having the ability to model both the soil and the structural non-linearities. This procedure has not yet been incorporated into design practice owing to its complexity and time demands. Instead, in the case of very important structures, an effective iterative procedure could be gradually applied to re-adjust the stiffness of pile foundations within the notion of the substructuring technique. The application of this approach allows the economical analysis of very large finite element or finite difference systems. According to the use of substructuring, it is possible to couple a non-linear analysis of a complex system by using, independently, a non-linear superstructure analysis code and a non-linear foundation analysis code. More specifically, by firstly iterating the superstructure analysis code the loads acting on the foundation can be determined. Having calculated the response of the foundation following the guidelines given in the previous paragraphs, the equivalent linear foundation stiffness, corresponding to the loads, calculated from the superstructure analysis can be estimated. These linear springs are then introduced into the second step of the superstructure analysis, and, if necessary, the foundation stiffness is re-adjusted. As a typical example, the proposed procedure for the design of a bridge is as follows:

(1) Consider the bridge piers to be fixed in the soil and solve for the superstructure. Define the loads acting on pile caps.
(2) Determine the response of the single pile under vertical and lateral loading for the existing soil conditions using 3D non-linear analysis. In case of the non-availability of a 3D FE or FD non-linear code, a '*t-z*' and a '*p-y*' analysis could be applied to determine the pile foundation response under vertical and lateral loading respectively. The level of accuracy may be reduced when applying this alternative scheme, particularly in the vertical response owing to uncertainties regarding the '*t-z*' functions.

Define the pile group layout, for every pier, able to sustain loads from the superstructure analysis. To achieve this target, Equations 2, 3, 4, 16 and 17 may be used in conjunction with the previously defined response of the single pile to estimate the response of pile groups.

Compute the components of the linear 6×6 stiffness matrix replacing the pile foundation, defined as the ratio of forces (or moments) over the corresponding estimated displacement (or rotation). Estimate, approximately, the rotational stiffness by considering the pile cap as a rigid body and the axial pile forces acting as a couple to resist to rotation of the pile cap.

(3) Reanalyse the superstructure using linear springs for simulating the foundation resistance. The loads acting on the pile caps can now be estimated to a good approximation. Re-adjust the values and repeat step 2 only in the case of important differences between the assumed and the calculated stiffnesses.
(4) An acceptable level of coupling can be achieved within two-to-four iterations, and a designer can proceed to a separate design of the substructures. Carry out a 3D non-linear analysis of the pile groups of the piers under the precise value of acting load from the superstructure analysis. The process is not so time-demanding since it is limited to a particular loading, contrary to a scheme of parametric analysis in which an incremental loading for various pile groups is carried out.

When a 3D computer code is not available an alternative, simpler but less accurate and straightforward process can be applied. In that case the pile cap is modeled using plate elements, while the piles are simulated by linear springs acting at the top of them. Loads from the superstructure analysis are applied at the location of pier columns. Numerical solution of the above model provides displacements, stress and moment distributions within the pile cap, while axial and lateral loads acting on each particular pile are determined from the corresponding springs.

The above procedure can be modified according to the level of the required accuracy and the capabilities of the particular FE or FD code available for the analysis of the superstructure and the pile foundation.

References

Bareka, S. V. Contribution to the response prediction of vertically loaded pile groups, PhD thesis, University of Thessaly, Volos, 2007.

Brown, D. A., Reese, L. C. and O'Neill, M. W. Cyclic lateral loading of a large-scale pile group, *J. Geotech. Engng*, ASCE, 1987, 113 (11): 1326–43.

Butterfield, R. and Banerjee, P. K. The elastic analysis of compressible piles and pile groups, *Géotechnique*, 1971, 21 (1): 43–60.

Canadian foundation engineering manual. *Canadian Geotechnical Society*, Vancouver, 1985.

Chow, Y. K. Analysis of vertically loaded pile groups, *Int. J. Numer. Anal. Methods Geomech*, 1986, 10 (1): 59–72.

Comodromos, E. M., Anagnostopoulos, C. T. and Georgiadis, M. K. Numerical assessment of axial pile group response based on load test, *Comput. Geotech*, 2003, 30 (6): 505–15.

Comodromos, E. M. Response evaluation of axially loaded fixed-head pile groups using 3D nonlinear analysis, *Soils Found.*, 2004, 44 (2): 31–9.

Comodromos, E. M. Response prediction of horizontally loaded pile groups, *Geotechnical Engineering Journal*, 2003, 34 (2): 123–33.

Comodromos, E. M. and Bareka, S. V. On the response prediction of axially loaded fixed-head pile groups in clayey soils (submitted for publication in *Int. J. Numer. Anal. Methods Geomech.* 2008).

Comodromos, E. M. and Bareka, S. V. Response prediction of pile groups in clays using 3D nonlinear analysis, *Proc. 16th Int. Conf. Soil Mech. Geotech. Engng*, Osaka, 2005, pp. 2095–8.

Comodromos, E. M. and Pitilakas, K. D. Response evaluation of horizontally loaded fixed-head pile groups using 3D nonlinear analysis, *Int. J. Numer. Anal. Methods Geomech.*, 2005, 29 (6): 597–625.

Comodromos, E., Hatzigogos, T. and Pitilakis, K. Multi-stage finite element algorithm for simulating excavations in elastoplastic soils. *Computer and Structures*, 1993, 46 (2): 289–98.

DIN 4014. Bored piles: construction procedure, design and bearing behaviour, *German code*, Berlin, 1990.

Eurocode 7: Geotechnical design – Part 1: *General rules together with the United Kingdom National Application Document*. DD ENV 1997-1, London, 1995.

Foundations and earth structures, Design manual 7.2. NAVFAC DM-7.2, Department of the Navy, Naval Facilities Engineering Command, 1982.

Georgiadis, M. Development of *p-y* curves for layered soils, in *Geotechnical Practice in Offshore Engineering*, ed. S. G. Wright, New York: American Society of Civil Engineers, 1983, pp. 536–45.

Hetenyi, M. *Beams on Elastic Foundations*, Ann Arbor, Mich.: University of Michigan Press, 1946.

Horikoshi, K. and Randolph, M. Estimation of overall settlement of piled raft, *Soils Found.*, 1999, 39 (2): 59–68.

Itasca Consulting Group: *FLAC*^{3D}, Fast Lagrangian analysis of continua user's and theory manuals Ver. 3.0, Minneapolis, Minn., 2005.

Katzenbach, R. and Moormann, C. Design of axially loaded piles and pile groups – German practice, *Design of Axially Loaded Piles – European Practice*, ed. F. de Cock and C. Legrand, Rotterdam: Balkema, 1997, pp. 177–201.

Kitiyotom, P., Matsumoto, T. and Kanefusa, N. Influence of reaction piles on the behaviour of a test pile in static load testing, *Can. Geotech. J.*, 2004, 41: 408–20.

Kraft, L. M., Ray, R. P. and Kagava, T. Theoretical *t-z* curves, *J. Geotech. Engng*, ASCE, 1981, 119 (6): 984–97.

Lee, C. Y. Pile group settlement analysis by hybrid layer approach, *J. Geotech. Engng*, ASCE, 1993, 119 (6): 984–97.

Mandolini, A. and Viggiani. Settlement of piled foundations, *Géotechnique*, 1997, 47 (3): 791–816.

Mandolini, A., Russo, G. and Viggiani. Pile foundations: experimental investigations, analysis and design, *Proc. 16th Int. Conf. Soil Mech. Geotech. Engng*, Osaka 2005, pp. 177–213.

Matlock, H. Correlations for design of laterally loaded piles in soft clay, *Proceedings of 2nd Offshore Technology Conference, Houston*, 1970, pp. 577–94.

MELT: Règles techniques de conception et de calcul des fondations des ouvrages de génie civil. CCTG, Fascicule No 62-Titre V. Ministère de l'Equipement du Logement et des Transports, Paris, 1993.

Mindlin, R. D. Force at a point in the interior of a semi-infinite solid, *Physics*, 1936, 7: 192–202.

Oteo, C. S. Displacement of a vertical pile group subjected to lateral loads, *Proceedings of 5th European Conference of Soil Mechanics and Foundation Engineering, Madrid*, 1972, pp. 397–405.

Peterson, K. T. and Rollins, K. M. Static and dynamic lateral load testing of a full-scale pile group in clay, Res. Rep. CEG 96-02, Department of Civil Engineering, Brigham Young University, 1996.

Poulos, H. G. Behaviour of laterally loaded piles: I – single pile and II – pile group, *Journal of Soil Mechanics and Foundation Division*, 1971, 97: 711–51.

Poulos, H. G. Pile behaviour – theory and application, *Géotechnique*, 1989, 39 (3): 366–415.

Poulos, H. G. Pile testing – from the designer's viewpoint, *Stamnatic Loading Test '98*, 2000, Rotterdam: Balkema, pp. 3–21.

Poulos, H. G. and Davis, E. H. *Pile Foundation Analysis and Design*, New York: John Wiley, 1980.

Prakash, S. and Sharma, D. *Pile Foundations in Engineering Practice*. New York: John Wiley, 1990.

Randolph, M. F. Design method for pile group and piled raft, *Proc. 13th Int. Conf. Soil Mech. Found. Engng*, New Delhi, 1994, 5: 61–82.

Randolph, M. F. *Ratz – Load transfer analysis of axially loaded piles.* Research report No. 86003, University of Western Australia, 1986.

Randolph, M. F. The response of flexible piles to lateral loading, *Géotechnique*, 1981, 31 (2): 247–59.

Randolph, M. F. and Wroth, C. P. An analysis of the vertical deformation of pile groups, *Géotechnique*, 1979, 29 (4): 423–39.

Randolph, M. F. and Wroth, C. P. Analysis of deformation of vertically loaded piles, *J. Geotech. Engng*, ASCE, 1978, 104 (12): 1465–88.

Reese, L. C. Laterally loaded piles: program documentation, *J. Geotechn. Engnng Div.*, 1977, 103: 287–305.

Reese, L. C. and Wang, S. T. Computer program LPILE Plus Ver. 3.0 – a program for the analysis of piles and drilled shafts under lateral loads, *Technical Manual*, Ensoft Inc., 1997.

Reese, L. C., Cox, W. R. and Koop, F. D. Analysis of laterally loaded piles in sand, *Proceedings of 6th Offshore Technology Conference, Houston*, 1974, pp. 473–83.

Reese, L. C. and Welch, R. C. Lateral loadings of deep foundations in stiff clay, *Journal of Geotechnical Engineering Division*, 1975, 101: 633–49.

Rollins, K. M., Peterson, K. T. and Weaver, T. J. Lateral load behaviour of full-scale group in clay, *J. Geotechn. and Geoenv. Engng*, ASCE, 1998, 124 (6): 468–78.

Smith, I. M. and Griffiths, D. V. *Programming the Finite Element Method*, 2nd edn, New York: John Wiley, 1988.

Wakai, A., Gose, S. and Ugai, K. 3-D Elasto-plastic finite element analyses of pile foundations subjected to lateral loading, *Soils and Foundations*, 1999, 39 (1): 97–111.

Wang, S. T. and Reese, L. C. COM624 – laterally loaded pile analysis program for the microcomputer, FHWASA-91-048: Computer program documentation, 1993.

3 Uplift capacity of inclined plate ground anchors in soil

Richard S. Merifield

1 Introduction

The design of many engineering structures requires foundation systems to resist vertical uplift or horizontal pull-out forces. In such cases, an attractive and economic design solution may be achieved through the use of tension members. These members, which are referred to as soil anchors, are typically fixed to the structure and embedded in the ground to sufficient depth that they can resist pull-out forces with safety. Soil or 'ground' anchors are a lightweight foundation system designed and constructed specifically to resist any uplifting force or overturning moment placed on a structure.

As the range of applications for soil anchors continues to increase to include support for substantially larger and more elaborate structures, greater demands are placed on anchor design and performance. One such recent application includes providing mooring support to floating systems for offshore oil and gas facilities. Unfortunately, research into the behavior of plate anchors has not kept up with overall performance demand and is very limited in comparison to other foundation systems. In particular, the numerical study of soil anchor behavior has attracted limited attention.

Current understanding regarding the behavior of buried foundations, and anchor plates in particular, is somewhat unsatisfactory. The complex nature of anchor behavior, and the sheer number of variables that influence soil uplift capacity, has meant that there are many conflicting theories reported in the literature. Most currently proposed theories have significant underlying assumptions based on experimental observations regarding the likely failure mode of anchors. Unfortunately, it would appear that these assumptions are responsible for the general lack of overall agreement on soil uplift theory. The advantage of using rigorous numerical methods to study anchor behavior is that a good indication of the likely failure mechanism can be obtained without any assumptions being made in advance.

To date, most anchor studies have been concerned with either the vertical or the horizontal uplift problem. In many instances, anchors are placed at inclined orientations depending on the type of application and loading (e.g. transmission tower foundations). However, the important effect of anchor inclination has received very little attention by researchers. This research applies numerical limit analysis and displacement finite element analysis to evaluate the stability of inclined strip anchors in undrained clay and sand. Results are presented in the familiar form of break-out factors based on various anchor geometries.

2 Background

Typically, soil anchors are used to transmit tensile forces from a structure to the soil. Their strength is obtained through the sheer strength and dead weight of the surrounding soil. The types of soil anchors used in civil engineering practice vary considerably; however, in general, anchors can be divided into four basic categories:

- 'deadman' anchors/plate anchors
- screw anchors
- grout injected anchors
- anchor piles

The method of load transfer from the anchor to the surrounding soil provides the distinction between these various forms of anchorage. Load can be transferred to the soil through direct bearing (plate anchors, screw anchors), shaft friction (grout injected anchors), or a combination of both direct bearing and shaft friction (anchor piles). In this research, anchors that obtain some capacity through shaft friction are not considered, and discussion will be limited to those anchors that obtain their strength through direct bearing. These anchors will be referred to as plate anchors.

The range of applications for plate anchors has been widely reviewed in the literature (e.g. Das 1990) and includes the following:

- foundations for transmission towers, utility poles and marine moorings (Figure 3.1[a]). Transmission tower foundations;
- tieback support for retaining structures (Figure 3.1[b]);
- break-out support for submerged pipelines and other structures subject to uplift pressures (Figure 3.1[c]).

As the range of applications for anchors expands to include the support of more elaborate and substantially larger structures, a greater understanding of their behavior is required.

Anchors are typically constructed from steel or concrete and may be circular (including helical), square or rectangular in shape (Figure 3.2). They may be placed horizontally, vertically, or at an inclined position depending on the load orientation or type of structure requiring support. A general layout of the problem to be analyzed is shown in Figure 3.3.

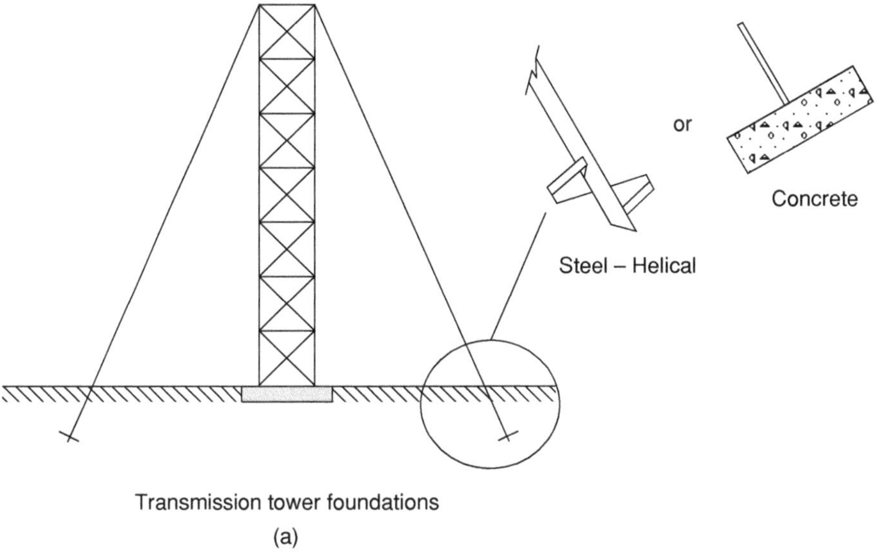

Figure 3.1 Applications for soil anchors

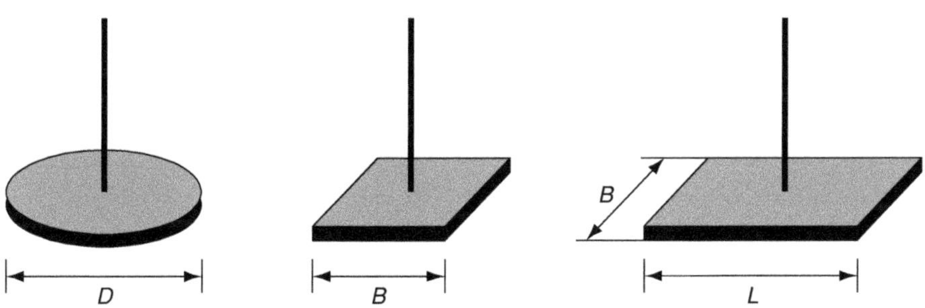

Figure 3.2 Anchor shapes

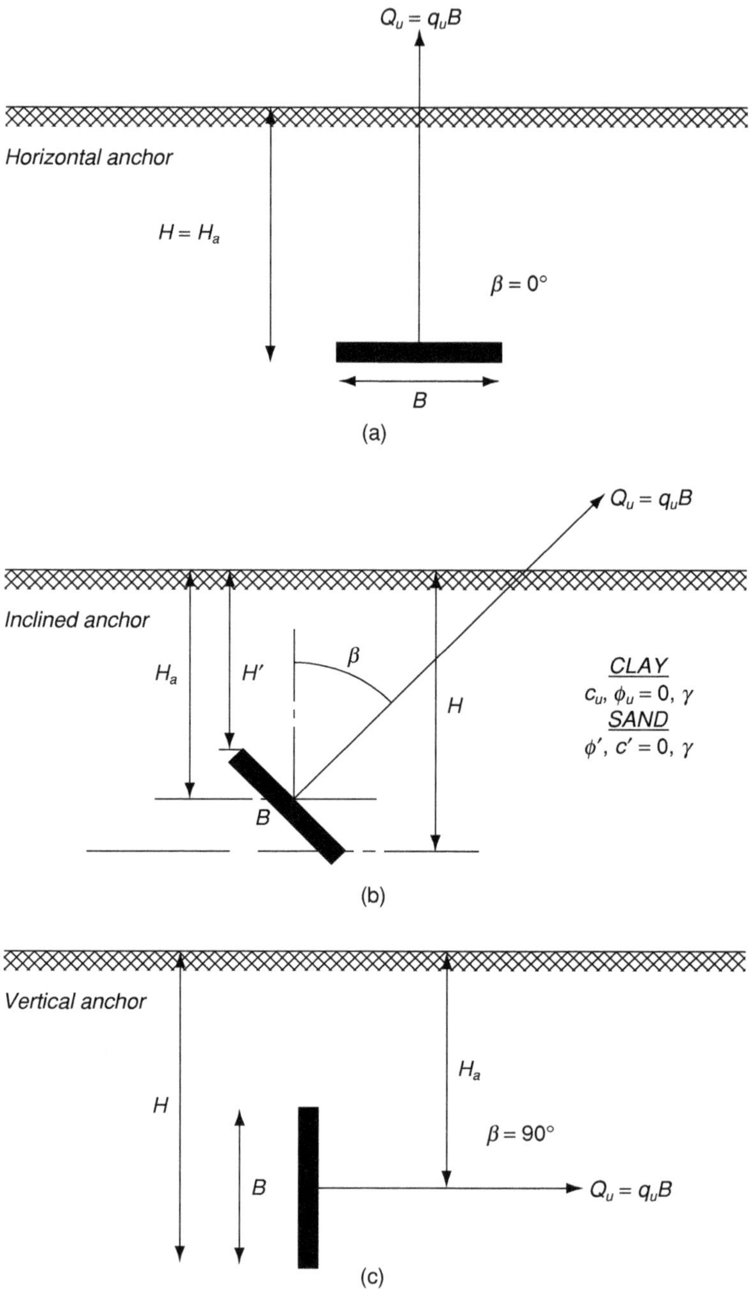

Figure 3.3 Problem definition

Plate anchors can be installed by excavating the ground to the required depth, placing the anchor, and then backfilling with soil. For example, when used as a support for retaining structures, anchors are installed in excavated trenches and connected to tie rods which may be driven or placed through augered holes. This type of anchor is the subject of interest in this research.

To date, most anchor studies have been concerned with either the vertical or the horizontal pull-out problem (Merifield et al. 2001, Merifield and Sloan 2006). In many instances, anchors are placed at inclined orientations depending on the type of application and loading (e.g. transmission tower foundations). However, the important effect of anchor inclination has received very little attention by researchers.

A thorough investigation of the effect of anchor inclination was performed by Merifield et al. (2001) for anchors in clay. Many of these results have been presented below in Section 4 for completeness.

The purpose of this research is to take full advantage of the ability of recent numerical formulations of the limit theorems to bracket the actual collapse load of inclined anchors accurately from above and below. The lower and upper bounds are computed respectively, using the numerical techniques developed by Lyamin and Sloan (2002a, b) and Sloan and Kleeman (1995). In addition, the displacement finite element formulation presented by Abbo (1997) and Abbo and Sloan (2000) has also been used for comparison purposes. This research software, named SNAC (Solid Nonlinear Analysis Code), was developed with the aim of reducing the complexity of elasto-plastic analysis by using advanced solution algorithms with automatic error control.

3 Numerical modeling details

3.1 Displacement finite element 'SNAC' analysis (DFEA)

The use of the finite element method is now widespread amongst researchers and practitioners. Theoretically, the finite element technique can deal with complicated loadings, excavation and deposition sequences, geometries of arbitrary shape, anisotropy, layered deposits and complex stress–strain relationships. However, the method has not been used widely when estimating the capacity of soil anchors.

The finite element formulation used in this research is that presented by Abbo (1997) and Abbo and Sloan (2000), named SNAC (Solid Nonlinear Analysis Code). The resulting formulation greatly enhances the ability of the finite element technique to predict collapse loads accurately, and avoids many of the locking problems discussed by Toh and Sloan (1980) and Sloan and Randolph (1982). A brief discussion of SNAC's more distinctive features can be found in Merifield (2002).

The accuracy of the finite element method depends not only on the size and distribution of elements, but also on the type of element and its corresponding displacement approximation. Sloan and Randolph (1982) proved that the condition of zero volume change generates a large number of independent constraints on the deformation of that element. It is therefore necessary to ensure that the number of degrees of freedom within each element exceeds the number of constraints imposed by the undrained material behavior. It was concluded that, while most element types are suitable in plane strain, only the cubic strain triangle is satisfactory for axisymmetric conditions. Similar arguments also apply for drained conditions, where the Mohr–Coulomb flow rule stipulates that the ratio of the principal plastic strain rates is a constant.

For the displacement finite element (SNAC) analyses, between 200 and 400 15-noded triangular elements (1500–4000 nodes) were used depending on the problem geometry. Checks were made to ensure the overall SNAC mesh dimensions were adequate to contain the zones of plastic shearing and the observed displacement fields. It is anticipated that such a high-order element will provide a good estimate of the anchor collapse load.

To determine the collapse load of an inclined anchor, displacement defined analyses were performed where the anchor was considered as being perfectly rigid. That is, a uniform prescribed displacement was applied to those nodes representing the anchor. The total displacement was applied over a number of substeps, and the nodal forces along the anchor were summed to compute the equivalent force.

The soil has been modeled as an elasto-plastic Mohr–Coulomb material. The material parameters used are the soil undrained shear strength c_u, soil friction angle ϕ', soil weight γ, Poisson's ratio ν and Young's modulus E.

3.2 Numerical limit analysis

Another approach for analyzing the stability of geotechnical structures is to use the lower and upper bound limit theorems developed by Drucker et al. (1952). These theorems are based on defining the statically admissible stress fields and the kinematically admissible velocity fields respectively and can be used to bracket the exact ultimate load from below and above. The limit theorems are most powerful when both types of solution can bracket the true limit load to within a few percent, which can be sufficient for use in the design practice. However, to bracket the true collapse load this accurately is seldom possible using analytical upper and lower bound formulations. The large number of applications of geotechnical stability analyses using limit theorems are found to be based on the upper bound theorem alone (e.g. see Chen 1975, Michalowski 1995). It is usually simpler to postulate a good kinematically admissible failure mechanism than it is to construct a good statically admissible stress field. This is one reason why significantly fewer lower-bound solutions for stability problems exist in the literature.

A numerical breakthrough of the bounding theorems in geotechnical stability application is attributed to Sloan (1988, 1989) and Sloan and Kleeman

(1995), who introduced finite element and linear programming (LP) formulations based on an active set algorithm that permits large two-dimensional problems to be solved efficiently on a standard personal computer. Despite the great success of Sloan's linear programming solutions for two-dimensional problems, these approaches are unsuitable for performing general three-dimensional stability analysis. The potential of Sloan's approach was increased dramatically when Lyamin and Sloan (2002a, b) proposed a new formulation using linear finite elements and non-linear programming (NLP) in which the yield criteria are employed in their original non-linear form. Lyamin and Sloan's NLP optimization procedure has been proven to be many times faster than Sloan's LP algorithm, and accurate bounding estimates for large stability problems can always be efficiently computed using a standard personal computer.

Estimates of the pull-out capacity of inclined anchors herein have been obtained by using the procedures developed by Lyamin and Sloan (2002a, b), and Sloan and Kleeman (1995). Full details of the formulations can be found in the relevant references and will not be discussed here.

The soil has been modeled as a rigid, perfectly plastic Mohr–Coulomb material. The material parameters used are the soil undrained shear strength c_u, soil friction angle ϕ' and soil weight γ.

3.2.1 Mesh details

Previous numerical studies (Merifield and Sloan 2006, using the formulations of Lyamin and Sloan 2002a, b) have provided several important guidelines for mesh generation. These, and a number of other studies, indicated that successful mesh generation typically involves ensuring that:

(*a*) the overall mesh dimensions are adequate to contain the computed stress field (lower bound) or velocity/plastic field (upper bound);
(*b*) there is an adequate concentration of elements within critical regions.

When generating upper and lower bound meshes there are several points worth considering. First, a greater concentration of elements should be provided in areas where high stress gradients (lower bound) or high velocity gradients (upper bound) are likely to occur. For the problem of a soil anchor, these regions exist at the anchor edges. Second, in areas where there is a significant change in principal stress direction (lower bound), appropriate-shaped meshes should be used. Such principal stress rotations are, for example, observed at the edges of footings and anchors in cohesionless soil. Finally, where possible, elements with severely distorted geometries should be avoided.

In accordance with the above discussion, the final finite element mesh arrangements (both upper and lower bound) were selected only after considerable refinements had been made. The process of mesh optimization followed an iterative procedure, and the final selected mesh characteristics were those that were found either to minimize the upper bound or maximize the lower bound solution. This will have the desirable effect of reducing the error bounds between both solutions, hence bracketing the actual collapse load more closely.

A typical upper bound mesh for the problem of an inclined anchor, along with the applied velocity boundary conditions and mesh dimensions, is shown in Figure 3.4. The distribution of elements for each upper bound mesh was typically uniform over the whole domain. This provided excellent results and, although the regions around the anchor edges indicated large velocity gradients, a significant increase in element concentrations in this region was not required.

A typical lower bound mesh for the problem of an inclined anchor, along with the applied stress boundary conditions, is shown in Figure 3.5.

The overall mesh dimensions required to contain the statically admissible stress field for both the vertical and the horizontal anchor cases were found to be similar to those required for the corresponding upper bound solutions.

To allow the underside of the anchor to separate from the soil (immediate breakaway) in a lower bound analysis, the stress discontinuity just below the anchor is removed. The shear stress and normal stress are set equal to zero behind the anchor (Figure 3.5). This effectively creates a 'free surface' below the anchor.

For the upper bound case, the addition of a line element to model the anchor (Figure 3.4.T) creates a series of velocity discontinuities between the anchor and the soil. To allow immediate breakaway for a purely cohesive soil, the discontinuities between the soil and rear/underside of the anchor are removed. The same discontinuities are not removed when analyzing the cohesionless soil case as they provide the interface that is needed correctly to model active collapse should it occur.

3.2.2 Limit analysis collapse load determination

An upper bound solution is obtained by prescribing a certain set of velocity boundary conditions, which depend on the type of problem being analyzed. As an example, for an inclined anchor problem a unit velocity is prescribed to the nodes along the line element that represents the anchor. The unit velocity will have both horizontal and vertical components which are related to the anchor orientation or inclination angle β. After the corresponding optimization problem is solved for the imposed boundary conditions, the collapse load is found by equating the internal dissipated power to the power expended by the external forces.

A lower bound solution for the anchor problem is obtained by maximizing the integral of the compressive stress along the soil anchor interface. The task is to find a statically admissible stress field which maximizes the collapse load over the area of the anchor.

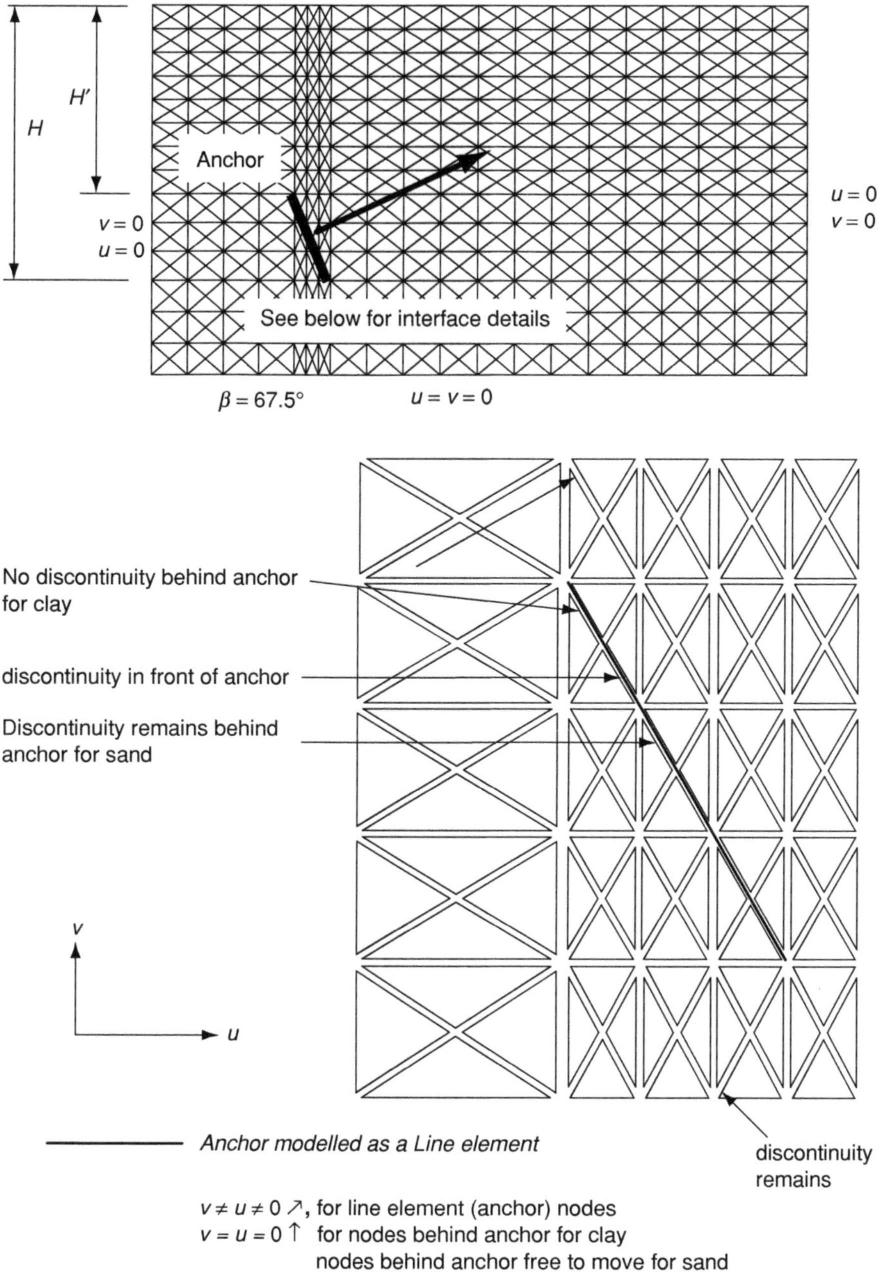

Figure 3.4 Example upper bound mesh details $\beta = 67.5$

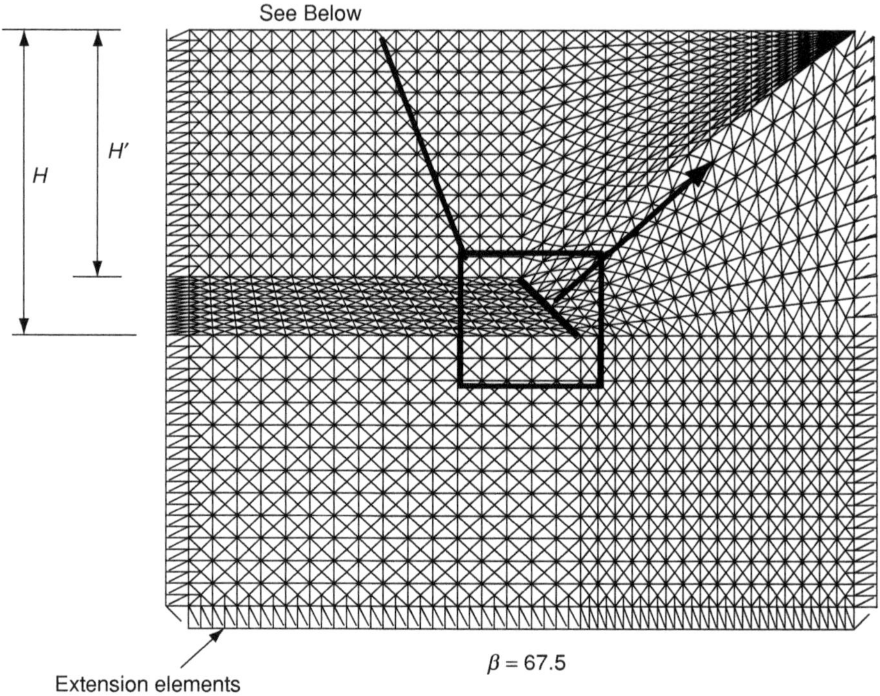

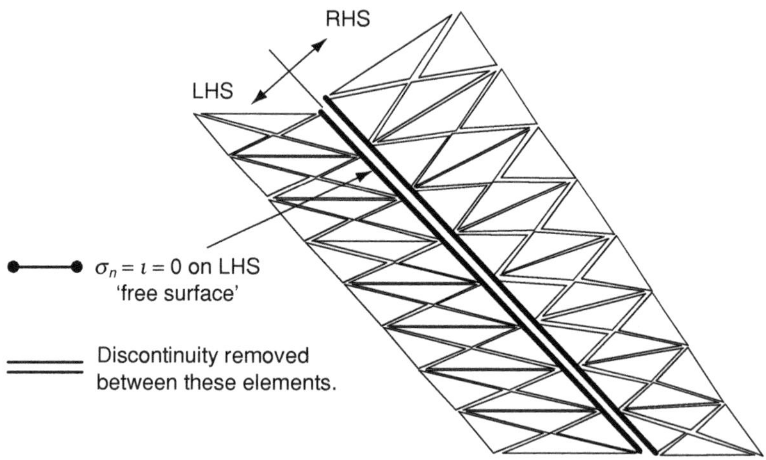

Figure 3.5 Example lower bound mesh details $\beta = 45$

4 Capacity of inclined anchors in clay

The problem geometry to be considered is shown in Figure 3.3(b). An inclined anchor will be defined as an anchor placed at an angle β to the vertical (Figure 3.3[b]). A horizontal anchor is one where $\beta = 0°$ (Figure 3.3[a]), while a vertical anchor is one where $\beta = 90°$ (Figure 3.3[c]). The direction of pull-out is perpendicular to the anchor face, and the depths $H'H_a$ and H are respectively the depths to the top, middle and bottom of the anchor from the soil surface. The capacity of anchors inclined at $\beta = 22.5°$, 45° and 67.5° will be investigated.

After Rowe and Davis (1982), the analysis of anchor behavior can be divided into two distinct categories, namely those of 'immediate breakaway' and of 'no breakaway'. In the immediate breakaway case it is assumed that the soil–anchor interface cannot sustain tension, so that, upon loading, the vertical stress immediately below the anchor reduces to zero and the anchor is no longer in contact with the underlying soil. This represents the case where there is no adhesion or suction between the soil and the anchor. In the 'no breakaway' case the opposite is assumed, with the soil–anchor interface sustaining adequate tension to ensure that the anchor remains in contact with the soil at all times. This models the case where an adhesion or suction exists between the anchor and the soil. In reality, it is likely that the true breakaway state will fall somewhere between the extremities of the 'immediate breakaway' and 'no breakaway' cases.

The suction force developed between the anchor and the soil is likely to be a function of several variables including the embedment depth, soil permeability, undrained shear strength and loading rate. As such, the actual magnitude of any adhesion or suction force is highly uncertain and therefore should not be relied upon in the routine design of anchors. For this reason, the anchor analyses presented in this research are performed for the 'immediate breakaway' case only. This will result in conservative estimates of the actual pull-out resistance.

The ultimate anchor pull-out capacity of horizontal and vertical anchors in purely cohesive soil is usually expressed as a function of the undrained shear strength in the following form (Merifield et al. 2001):

$$q_u = \frac{Q_u}{A} = c_u N_c \tag{1}$$

where for a homogeneous soil profile

$$N_c = \left(\frac{q_u}{c_u}\right)_{\gamma \neq 0} = N_{co} + \frac{\gamma H_a}{c_u} \tag{2}$$

and the term N_{co} is defined as

$$N_{co} = \left(\frac{q_u}{c_u}\right)_{\gamma=0} \tag{3}$$

In the above, c_u is the undrained soil strength and N_c is known as the anchor break-out factor. Note that $H_a = H$ for horizontal anchors (Figure 3.3[a]) and $H_a = H - B/2$ for vertical anchors (Figure 3[c]).

Implicit in (1) is the assumption that the effects of soil unit weight and cohesion are independent of each other and may be superimposed. It was shown by Merifield et al. (2001) that this assumption generally provides a good approximation to the behavior of anchors in purely cohesive undrained clay.

For an inclined anchor in purely cohesive soil, the ultimate capacity will be given by (1) where

$$N_c = N_{co\beta} + \frac{\gamma H_a}{c_u} \tag{4}$$

and a new break-out factor $N_{co\beta}$ is introduced which has a value somewhere between the break-out factors N_{co} given in (3) for vertical and horizontal anchors. Only the homogeneous case is considered.

It should be noted that the break-out factor N_c given in (4) does not continue to increase indefinitely, but reaches a limiting value which marks the transition between shallow and deep anchor behavior. This process is explained in greater depth by Merifield et al. (2001) and Rowe (1978). The limiting value of the break-out factor is defined as N_{c*} for a homogeneous soil profile (Merifield et al. 2001).

4.1 Results and discussion

The computed upper and lower bound estimates of the anchor break-out factor $N_{co\beta}$ (Equation [4]) for homogeneous soils with no soil weight are shown graphically in Figures 3.6 and 3.7. Sufficiently small error bounds were achieved, with the true value of the anchor break-out factor typically being bracketed to within ±7%. The greatest variation between the bounds solutions occurs at small embedment ratios ($H_a/B \leq 2$) where the error bounds grow to a maximum of ±10%.

Also shown in Figures 3.6 and 3.7 are the SNAC results. These results plot close to the upper bound solution and are typically within ±5%.

The variation of break-out factor with angle of inclination is clearly presented in Figure 3.8. In this figure, the break-out factor is presented as a ratio of the break-out factor for an inclined anchor to that of a vertical anchor. This ratio is defined as the inclination factor i according to $H_a/B = 2,3,4,5,7,10$

$$i = \frac{N_{co\beta}}{N_{co?90}} \tag{5}$$

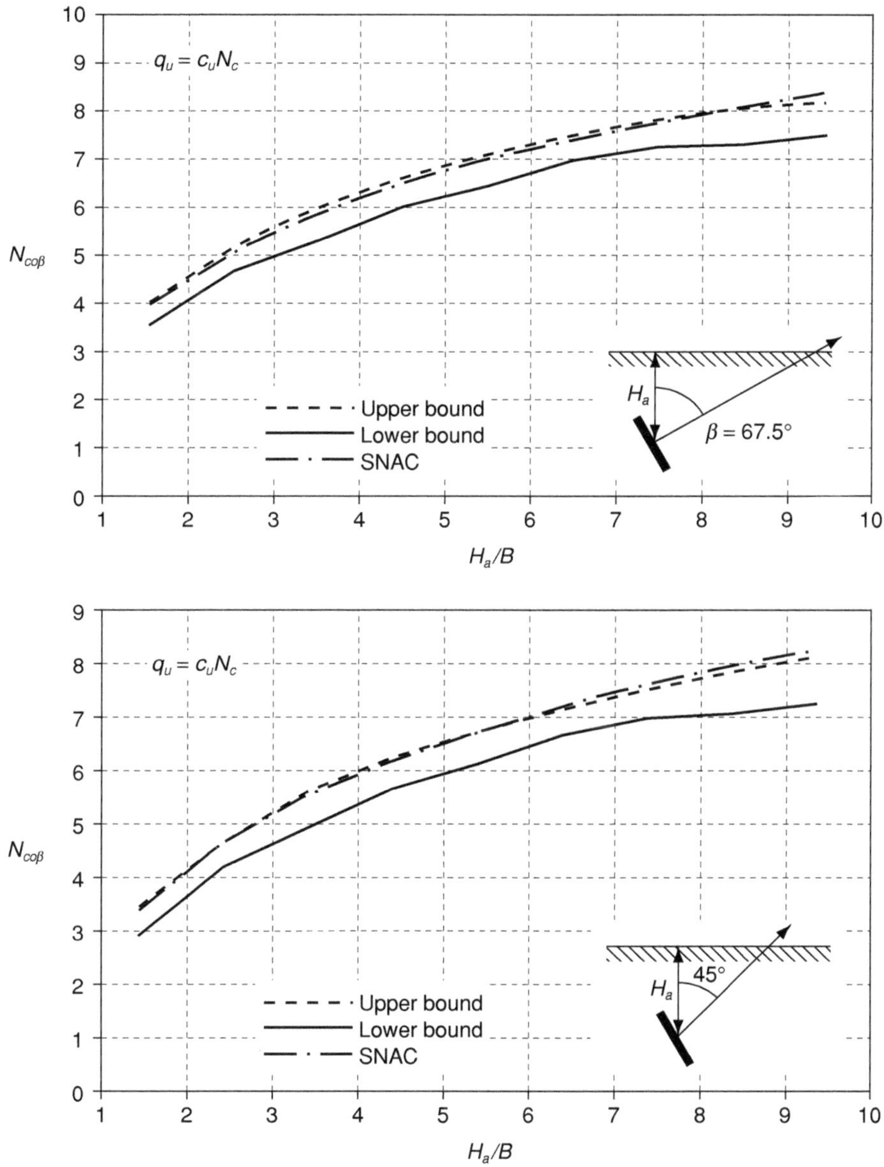

Figure 3.6 Break-out factors for inclined anchors in purely cohesive weightless soil

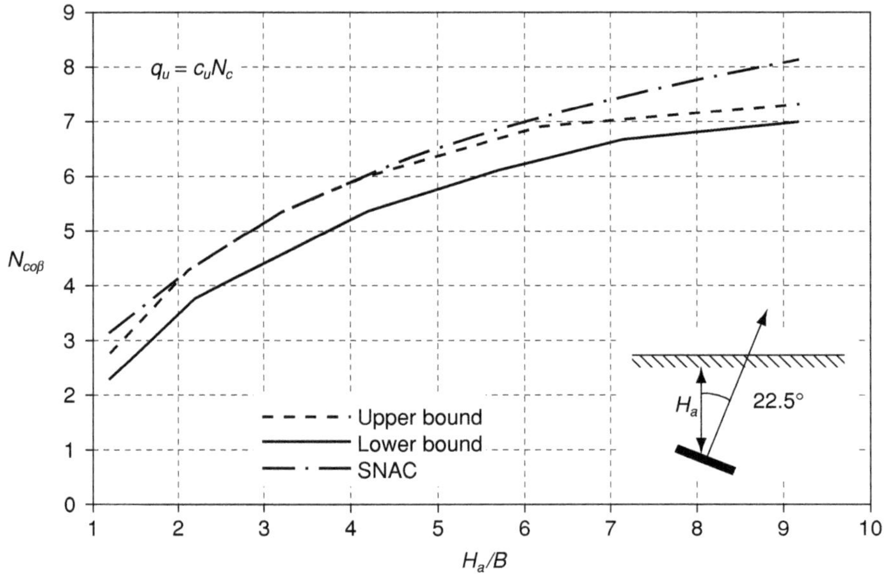

Figure 3.7 Break-out factors for inclined anchors in purely cohesive weightless soil

where i is the inclination factor, $N_{co\beta}$ is the break-out factor for an inclined anchor at an embedment ratio of H_a/B (Figure 3.6 or Figure 3.7), and $N_{co?90}$ is the break-out factor for a vertical anchor at the same embedment ratio H_a/B given by

$$N_{co90} = N_{co(\beta=90, H/B=H_a/B+0.5)}$$

The value of the break-out factor N_{co} can, with sufficient accuracy, be approximated by the following equations (Merifield et al. 2001):

$$N_{co90} = N_{co} = 2.46 \ln\left(2\frac{H}{B}\right) + 0.89 \text{ Lower Bound} \tag{6}$$

Equation (6) predicts the break-out factor to within ±2% of the numerical solutions.

The inclination factor can be seen to increase in a non-linear manner with increasing inclination from $\beta = 0°$ to $\beta = 90°$. This observation is consistent with the laboratory study of Das and Puri (1989). Figure 3.8 also suggests that there is very little difference between the capacity of a horizontal anchor ($\beta = 0°$) and anchors inclined at $\beta \leq 22.5°$. The greatest rate of increase in anchor capacity appears to occur once $\beta \geq 30°$.

The failure mechanisms observed for inclined anchors are illustrated by the upper bound velocity diagrams and SNAC displacement plots in

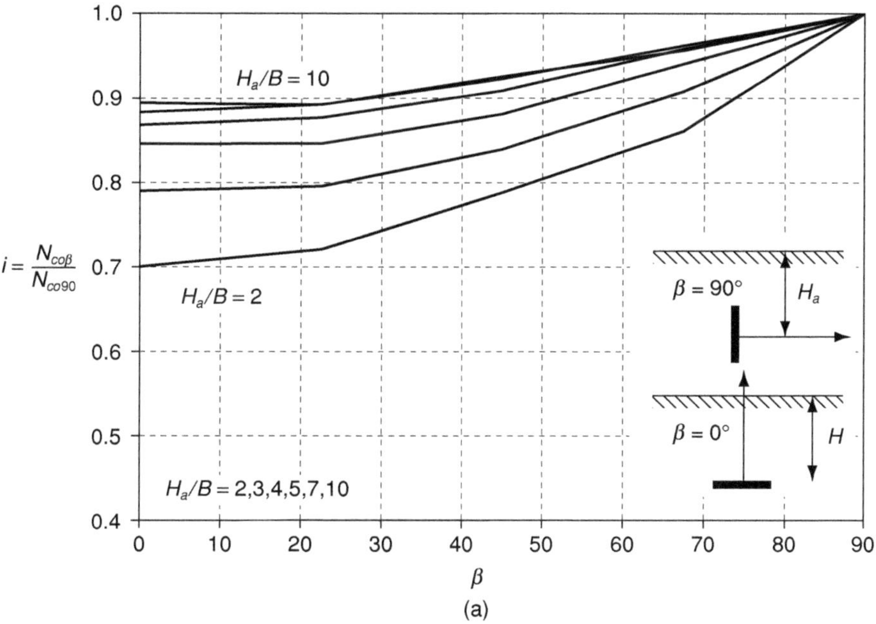

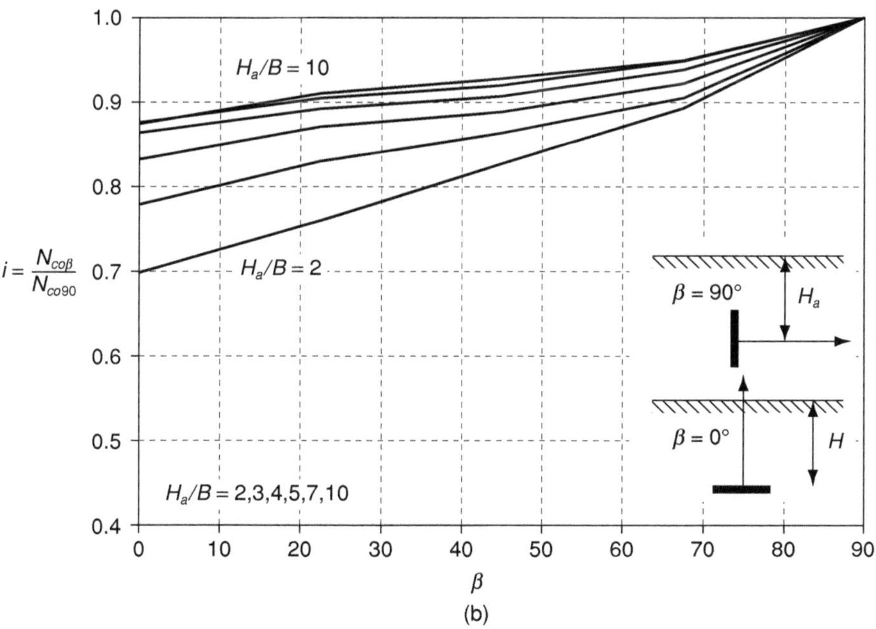

Figure 3.8 Inclination factors for anchors in purely cohesive weightless soil (*a*) lower bound, (*b*) SNAC

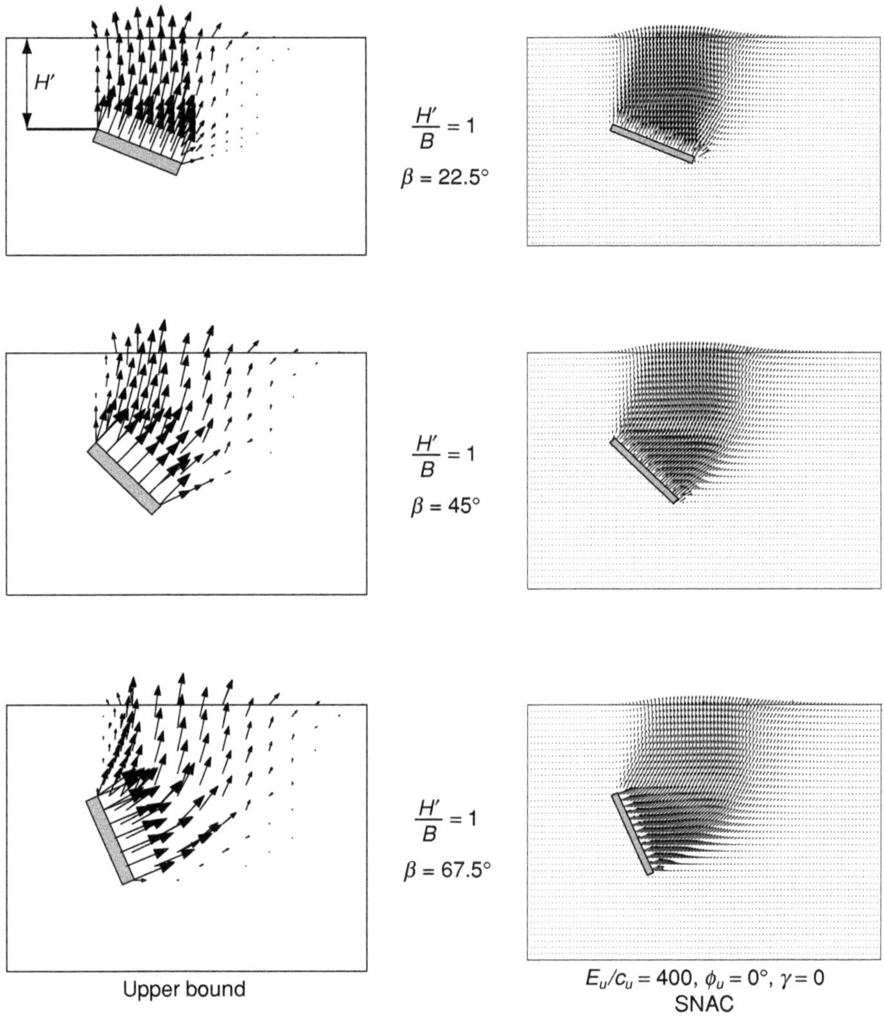

Figure 3.9 Failure modes for inclined anchors in purely cohesive weightless soil

Figures 3.9 and 3.10. As expected, the vector and displacement fields obtained from both types of analyses are very similar. A direct comparison is shown for anchors at H'/B = 1 in Figure 3.9.H'.

The lateral extent of surface deformation increases with increasing embedment depth and inclination angle. This is consistent with the findings for both the horizontal and the vertical anchor cases (Merifield et al. 2001). As expected, the actual magnitude of the surface deformations decreases with the embedment ratio and, at H'/B = 10, these are predicted to be negligible.

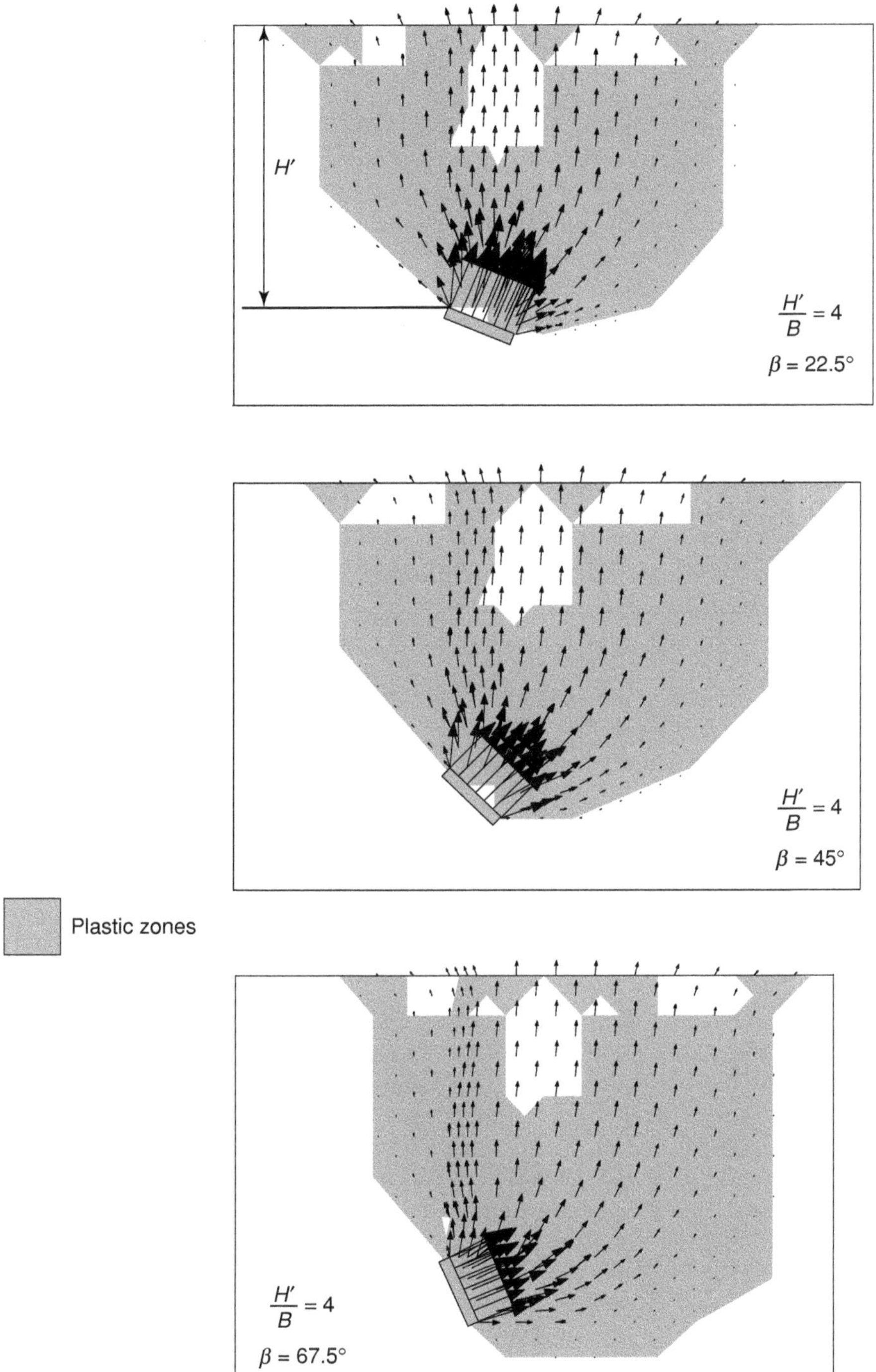

Figure 3.10 Upper bound failure modes and zones of plastic yielding for inclined anchors in purely cohesive weightless soil (upper bound)

Localized elastic zones were observed near the soil surface at most embedment ratios and inclination angles. Several of these zones are shown in Figure 3.10 for anchors at $H'/B = 4$. In addition, very little plastic shearing was observed below the bottom edge of anchors inclined at $\beta <$ 45°.22.5°.

A limited number of results for the capacity of inclined square and strip anchors can be found in the works of Meyerhof (1973). The study of Das and Puri (1989) appears to be the most significant attempt to quantify the capacity of inclined anchors. In their tests, the capacity of shallow square anchors embedded in compacted clay with an average undrained shear strength of 42.1 kPa was investigated. Pull-out tests were conducted on anchors at inclinations ranging between 0° (horizontal) and 90° (vertical) for embedment ratios (H/B) of up to four. A simple empirical relationship was suggested for predicting the capacity of square anchors at any orientation which compared reasonably well with the laboratory observations. Das and Puri (1989) also concluded that anchors with aspect ratios (L/B) of 5 or greater would, for all practical purposes, behave as a strip anchor.

The relationship proposed by Das and Puri (1989) is of the form

$$N_{co\beta} = N_{co(\beta=0°)} + [N_{co(\beta=90°)} - N_{co(\beta=0°)}\left(\frac{\beta°}{90}\right)^2 \tag{7}$$

where N_{co} is obtained at the same value of N_a for each inclination angle β. The value of $N_{co\beta=0}$ is the break-out factor for a horizontal anchor and can, with sufficient accuracy, be approximated by the following expression (Merifield et al. 2001):

$$N_{co(\beta=0)} = N_{co} = 2.56\ln\left(2\frac{H}{B}\right) \text{ Lower Bound} \tag{8}$$

Unfortunately, the tests of Das and Puri (1989) were limited to square anchors, and their results cannot be compared directly to those presented here. None the less, out of curiosity, Equation (7) has been used to estimate the break-out factors for strip anchors, and a comparison between these estimates and the results from the current study are shown in Figure 3.11. The limit analysis and SNAC results (90 points) for inclination angles of 22.5°, 45°, 67.5° and embedment depths of H_d/B of 1 to 10 are shown in estimated value Equation (7).

Figure 3.11 indicates that, although the empirical equation of Das and Puri (1989) was specifically proposed for inclined square anchors, it also provides a reasonable estimate for the capacity of inclined strip anchors. Equation (7) plots almost central to the data and, on average, the estimated values are within ±5% of the actual values. This is considered an adequate level of accuracy for design purposes. The discrepancy between the predicted and the actual break-out factors tends to be marginally larger for

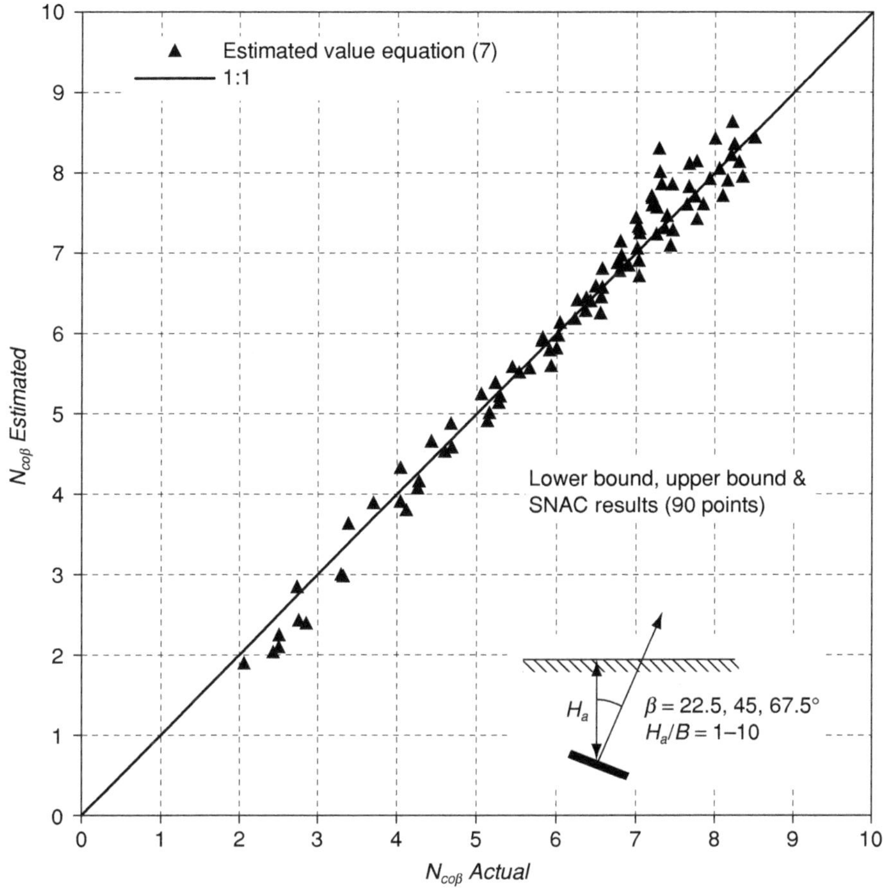

Figure 3.11 Comparison of break-out factors for inclined strip anchors in purely cohesive weightless soil

smaller embedment ratios ($H/B \leq 2$) where the predicted value is expected to be slightly conservative. It is therefore concluded that the empirical relation given by Equation (7) may be used to estimate the capacity of inclined strip anchors.

4.2 Effect of overburden pressure

The numerical results discussed above are limited to soil with no unit weight, and therefore the effect of soil weight (overburden) needs to be investigated. If our assumption of superposition is valid, then it would be expected that the ultimate anchor capacity, as given by Equations (1) and (2), would increase linearly with the dimensionless overburden pressure $\gamma H_a/c_u$. The results from further lower bound analyses that include cohesion and soil weight, shown in Figure 3.12(a), confirm that this is indeed the case. This conclusion is in agreement with the observations of Merifield et al. (2001) and Rowe (1978).

The error due to superposition can be expressed in the following form:

$$F_s = \frac{q_{actual}}{q_{predicted}} \tag{9}$$

and is shown in Figure 3.12(b). This figure indicates that the superposition error is likely to be insignificant.

Figure 3.12(a) indicates that the ultimate anchor capacity increases linearly with overburden pressure up to a limiting value. This limiting value reflects the transition of the failure mode from being a non-local one to a local one. At a given embedment depth the anchor failure mode may be non-localized or localized, depending on the dimensionless overburden ratio $\gamma H_a/c_u$. For shallow anchors exhibiting non-localized failure, the mode of failure is independent of the overburden pressure.

For deep anchors, the average limiting values of the break-out factor N_{c*} for all values of β were found to be 10.8 (lower bound) and 11.96 (upper bound). These values sit somewhere between the values obtained by Merifield et al. (2001) for deep horizontal and vertical anchors, and compare well with the analytical solutions of Rowe (1978), who found lower and upper bounds of 10.28 and 11.42 for the horizontal anchor case. Intuitively, the value of N_{c*} should, for a comparable mesh density, be independent of β. Unfortunately, successfully bracketing the collapse load for a deep anchor (N_{c*}) proved difficult and very much mesh dependent. For deep anchors, the form of the velocity field at collapse is essentially independent of the overburden pressure.

4.3 Suggested procedure for estimating uplift capacity

(1) Determine representative values of the material parameters c_u *and* γ.
(2) Knowing the anchor size B and embedment depth H_a calculate the embedment ratio H_a/B and overburden ratio $\gamma H_a/c_u$.
(3) Calculate N_{co90} using Equation (6) with $H/B = H_a/B + 0.5$.
(4) (i) For an anchor at $\beta = 22.5°$, 45° or 67.5°, estimate the break-out factor N_{co} using Figure 3.6 or Figure 3.7, depending on the anchor orientation.
    (ii) For anchors at other orientations, estimate the anchor inclination factor i using Figure 3.8(a) and the value of H_a/B obtained in (3). Then calculate N_{co} as per Equation (5). A value of N_{co} could also be estimated from Equation (7) using Equations (6) and (8).
(5) Adopt $N_{c*} = 10.8$.

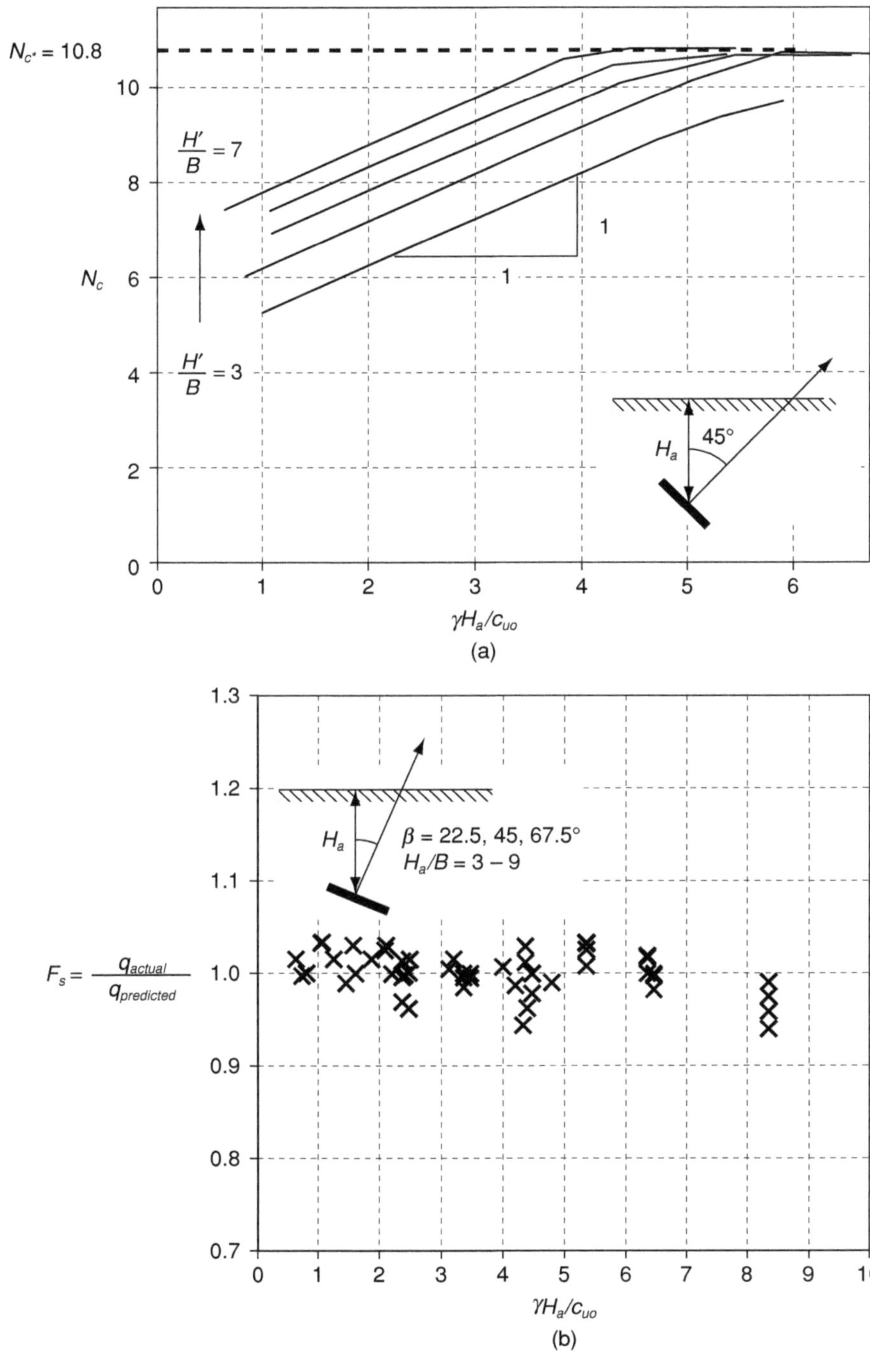

Figure 3.12 Effect of overburden pressure for inclined anchors in purely cohesive soil – lower bound

(6) (i) Calculate the break-out factor N_c using Equation (2).
(ii) If $N_c \geq N_{c*}$, then the anchor is a deep anchor. The ultimate pull-out capacity is given by Equation (1) where $N_c = N_{c*} = 10.8$.
(iii) If $N_c \leq N_{c*}$, then the anchor is a shallow anchor. The ultimate pull-out capacity is given by Equation (1) where N_c is the value obtained in 6 (i).

4.3.1 Example

We now illustrate how to use the results presented to determine the ultimate pull-out capacity of an inclined anchor in clay.

Problem: A plate anchor of width 0.25 m is to be embedded at $H_a = 2$ m at an orientation of 45°. Determine the ultimate pull-out capacity given that the clay has a shear strength $c_u = 100$ kPa and unit weight $\gamma = 17$ kN/m^3.

The systematic procedures given above will now be used to determine the ultimate anchor capacity.

(1) Given $c_u = 100$ kPa and $\gamma = 17$ kN/m^3.
(2) The embedment ratio can be calculated as $H_a/B = 2/0.25 = 8$
The dimensionless parameter $\gamma H_a/c_u = (17 \times 2)/100 = 0.34$
(3) $N_{co90} = 2.46 \ln(2H/B) + 0.89 = 2.46 \ln(28 + 0.5) + 0.89 = 7.86$
(4) (i) From Figure 3.6, $N_{co} = 7.1$ (lower bound)
(5) Adopt $N_{c*} = 10.8$.
(6) (i) From Equation (2), $N_c = 7.1 + 0.34 = 7.44$
(ii) $N_c < N_{c*}$ and therefore the anchor is 'shallow' and using Equation (1)
$q_u = c_u N_c = 100 \times 7.44 =$ **744** kPa
$Q_u = 744 \times (0.25) =$ **186** kN per m run

5 Capacity of inclined anchors in sand

In this section, the effect of anchor inclination on the pull-out capacity of anchors in cohesionless soil will be examined. The problem geometry to be considered is shown in Figure 3.3(b). Where possible, the results will then be compared with existing theoretical- and experimental-based solutions.

For numerical convenience, the ultimate anchor capacity q_u will be presented in a form analogous to Terzaghi's equation which is used to analyse surface footings, namely:

$$q_u = \gamma H N_\gamma \tag{10}$$

where N_γ is referred to as the anchor break-out factor.

For the case of a rough inclined anchor, the ultimate anchor pull-out capacity in cohesionless soil will be expressed as a function of the soil unit weight and embedment depth according to

$$q_u = \gamma H_a N_{\gamma\beta} \tag{11}$$

where H_a is the vertical depth to the center-line of the anchor, and $N_{\gamma\beta}$ is the new anchor break-out factor for inclined anchors. The capacity of anchors inclined at $\beta = 22.5°$, $45°$, and $67.5°$ will be investigated. The capacity of vertical ($\beta = 90°$) or horizontal anchors ($\beta = 0°$) have been presented previously by Merifield and Sloan (2006).

5.1 Upper bound mechanisms for anchor plates in sand

By deriving several analytical solutions for the ultimate capacity of anchors, a useful check can be made on the estimates obtained from the numerical finite element schemes.

Both the upper and lower bound techniques lend themselves to hand calculations that can be used to determine the collapse load for geotechnical problems. These calculations not only provide the engineer with a physical feeling for how the soil mass may fail, but also provide a useful check on the collapse load obtained from numerical finite element schemes. Generally speaking, the upper bound theorem is applied more frequently than the lower bound theorem when analyzing soil behavior. The main reason for this is that, in many cases, it is difficult to construct a statically admissible stress field that extends to infinity. In contrast, it is usually easier to construct a good kinematically admissible upper bound failure mechanism.

This section illustrates the use of the upper bound theorem to determine the collapse load of inclined anchors in cohesionless soil.

The upper bound method has been used to predict the collapse load of anchor plates in cohesionless soil by several authors, including Murray and Geddes (1989), Regenass and Soubra (1995) and Kumar (1999). Based on several assumed failure modes, the break-out factors for anchors at any orientation can be obtained by equating the internal power dissipation to the power expended by the external loads, and then using a suitable minimization algorithm.

The simplest form of failure mechanism for an anchor inclined at an angle ω to the horizontal consists of straight slip surfaces emanating from the anchor edges, as shown in Figure 3.13.

Also shown in Figure 3.13 are the velocity vectors which, because of an associated flow rule, are inclined at an angle ϕ' to the velocity discontinuities. In order to be a valid collapse mechanism, the velocity field must be continuous and also meet the requirements of compatibility. These requirements dictate the allowable form and general arrangement of the velocity discontinuities. For the simple mode of failure shown, the mechanism can be characterized by the angular parameters α and ω. Noting here that $\beta = 90 - \alpha$ in the context of the definitions for anchor inclination shown in Figure 3.3.

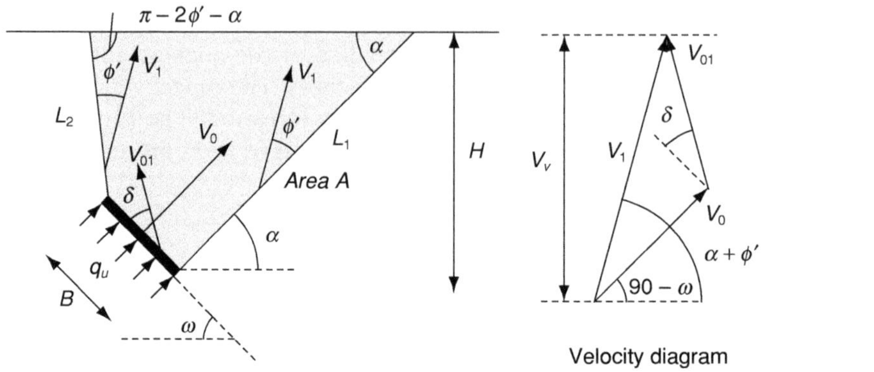

Figure 3.13 Simple upper bound mechanism for anchors in cohesionless soil – Mechanism 1

An upper bound to the exact collapse load is found by equating the power dissipated internally to the power expended by the external loads. During an increment of movement, power may be dissipated internally by plastic yielding of the soil mass, as well as by sliding along the velocity discontinuities. The work expended by the external loads during an increment of movement consists of the work done by the applied load as well as the work done by the soil self-weight. For a rigid block mechanism like that shown, each block moves with a constant velocity, and therefore no internal energy is dissipated by plastic yielding of the soil mass. Also, for an associated cohesionless material, the internal work done by the stresses on the slip surfaces is zero. Therefore, the total internal power dissipated by a rigid block mechanism in a cohesionless soil is zero.

Referring to Figure 3.13, the rate of external work can be written as

$$W_{ext} = q_u \cdot B \cdot V_0 - \gamma \cdot A \cdot V_v$$

and equating this with the internal work $W_{int} = 0$ yields

$$W_{int} = 0 = q_u \cdot B \cdot V_0 - \gamma \cdot A \cdot V_v \tag{12}$$

For a unit anchor width B, (12) may be expressed in terms of the break-out factor $N_{\gamma\beta}$ (Equation [11]), according to

$$N_{\gamma\beta} = \frac{q_u}{\gamma H} = \frac{AV_v}{HV_0} \tag{13}$$

where

$$\frac{V_v}{V_0} = \frac{\sin(\alpha + \phi')\sin(\pi/2 + \delta)}{\sin(\pi - \alpha - \phi' - \omega - \delta)}$$

$$A = \frac{H^2}{2\tan\alpha} + \frac{B\cos\omega(2H - B\sin\omega)}{2} + \frac{(H - B\sin\omega)^2\ \tan(\alpha + 2\phi' - \pi/2)}{2}$$

The soil–anchor interface roughness may also be incorporated into the solution by allowing dilation, so that the velocity V_{01} is inclined at an angle δ to the anchor interface. The angle δ may be varied between 0° and ϕ', where these limits constitute a perfectly smooth and rough anchor respectively. For the particular case of a horizontal anchor with $\omega = 0°$, no velocity jump occurs across the soil–anchor interface for this mechanism, and the solution is valid for both smooth and rough anchors.

The solution to (13) can be found using a suitable minimization routine.

More elaborate failure mechanisms may be adopted to obtain better (smaller) upper bound estimates, but these require more work to develop and verify. One such upper bound failure mechanism is shown in Figure 3.14.

Mechanism 2 has been used previously by Murray and Geddes (1989) to determine the break-out factors of anchors. The break–out factors obtained by this mechanism are used for comparison purposes.

Mechanism 2 consists of a radial shear zone sandwiched between two rigid blocks. The radial shear zone is limited by a log spiral slip surface. This type of mechanism has been used extensively by Chen (1975) to study both active and passive earth pressure problems. The mechanism is characterized by two angular parameters ρ and υ. Following Murray and Geddes (1989), the anchor break-out factor is given by

$$N_r = -\frac{W_1 + W_2 + W_3}{\gamma BHV_1} \cdot \frac{\cos\delta}{\cos\rho + \delta}$$

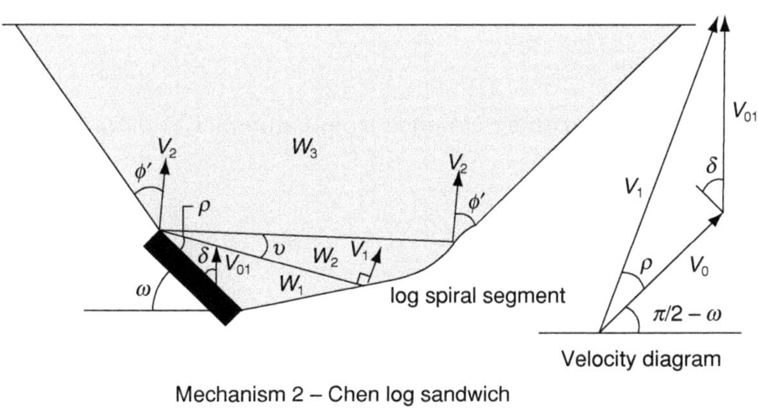

Figure 3.14 Upper bound failure mechanism for anchors in cohesionless soil

where δ again denotes the angle of friction at the anchor–soil interface and W_1, W_2 and W_3 denote the power expended by the soil unit weight in the three regions of the failure mechanism.

5.2 Results and discussion

The finite element limit analysis and SNAC estimates of the break-out factor $N_{\gamma\beta}$ are shown in Figures 3.15 and 3.16. For the limit analysis results, sufficiently small error bounds were achieved, with the true value of the anchor break-out factor typically being bracketed to within ±10%. The greatest variation between the bounds solutions occurs at an inclination of $\beta = 67.5°$ with $\phi' = 40°$. In this case the error bounds increase to around 30% (±15%). It may be possible to reduce these bounds through a greater level of mesh refinement, but this is not pursued here.

The displacement finite element results generally plot close to the upper bound solution for $\phi' \leq 30°$, but for larger friction angles they tend to provide a better (lower) estimate of the anchor capacity. The only exception to this observation occurs when $\beta = 67.5°$ and $\phi' = 40°$ (Figure 3.16). For this single case, the displacement finite element solutions exceed the upper bounds by up to 25 percent, which is somewhat surprising. As further mesh refinement was unable to reduce these error bounds, additional investigation of this case is warranted.

The ratio of the break-out factor for an inclined anchor to that of a vertical anchor can be expressed as

$$i = \frac{N_{\gamma\beta}}{N_{\gamma?90}}$$

where i is defined as the inclination factor, $N_{\gamma\beta}$ is the break-out factor for an inclined anchor at an embedment ratio of H_a/B, and $N_{\gamma?90}$ is the break-out factor for a vertical anchor at the same embedment ratio H_a/B given by

$$N_{\gamma 90} = N_{\gamma(\beta=90,H/B=H_a/B+0.5)}$$

The break-out factor $N_{\gamma 90}$ can be obtained from Figure 3.17 (lower bound) at an embedment

$$N_{\gamma 90} = e^{\left[0.645 \ln \frac{H}{B} + 3.07 \tan\phi - 0.764\right]} \tag{14}$$

Figure 3.17 has been taken from Merifield and Sloan (2006).

Values of the inclination factor based on the finite element lower bound and SNAC results are shown in Figure 3.18 and Figure 3.19(a). A line of best fit has been drawn through the data obtained from both methods of analysis. The inclination factor i increases in a non-linear manner with

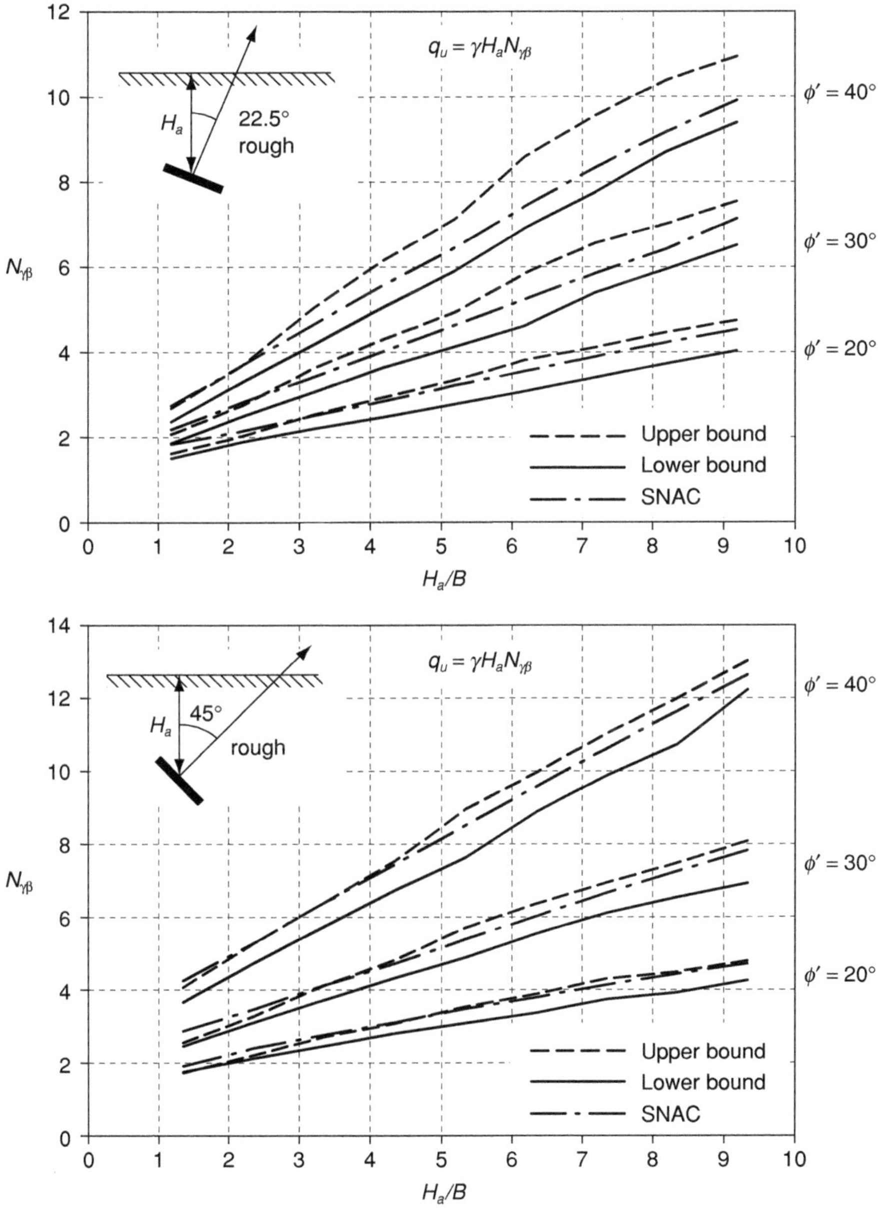

Figure 3.15 Break-out factor for inclined rough anchors in cohesionless soil

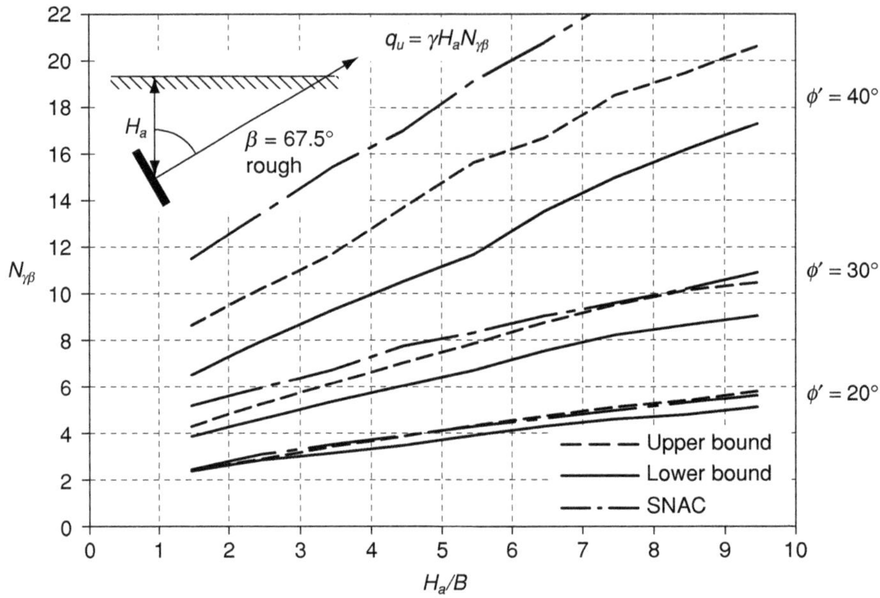

Figure 3.16 Break-out factor for inclined rough anchors in cohesionless soil

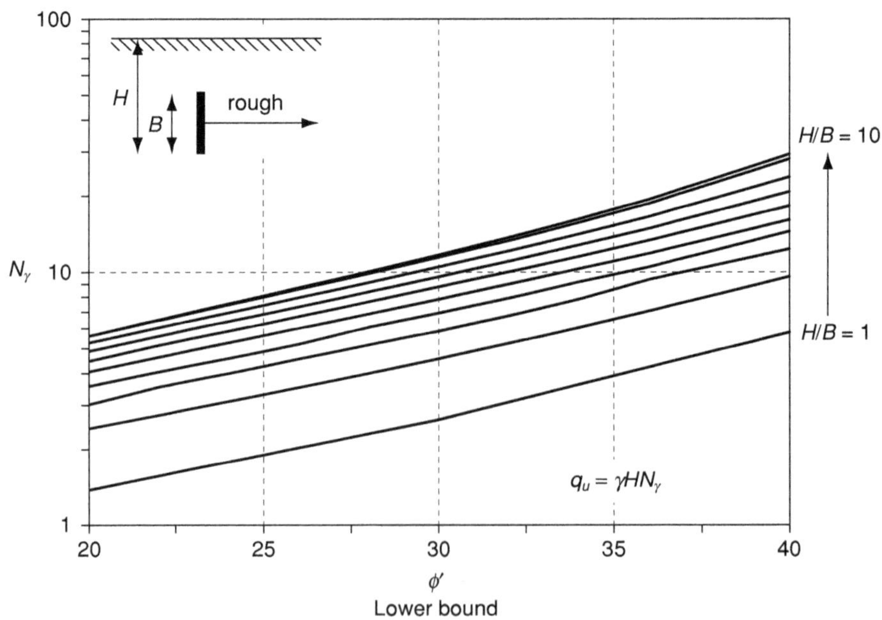

Figure 3.17 Break-out factors for rough vertical strip anchors in cohesionless soil

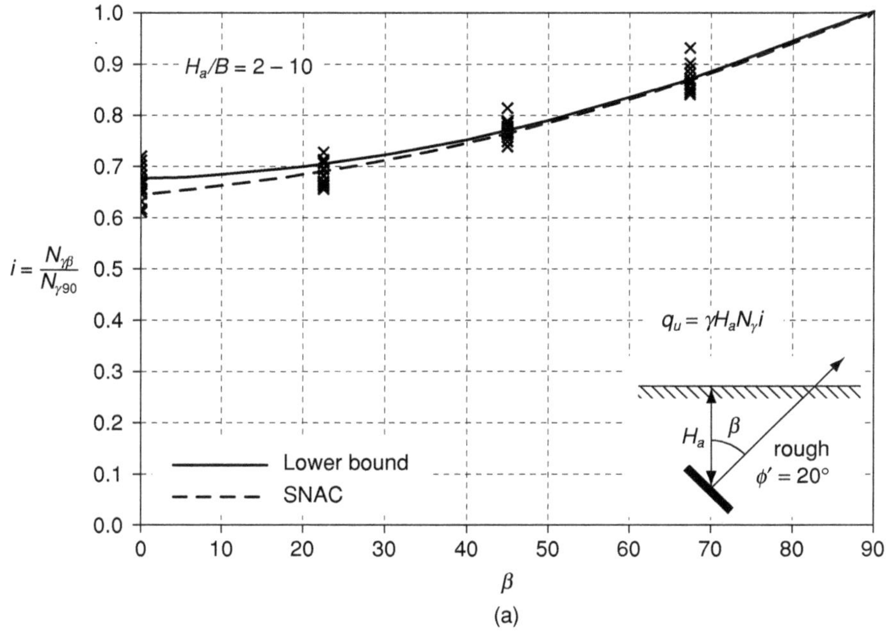

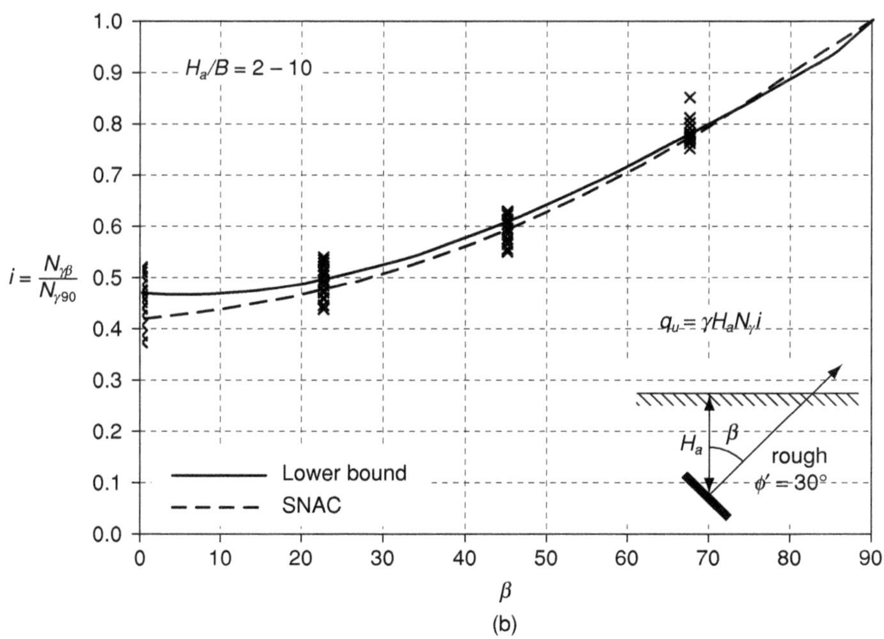

Figure 3.18 Inclination factors for inclined rough anchors in cohesionless soil – lower bound and SNAC

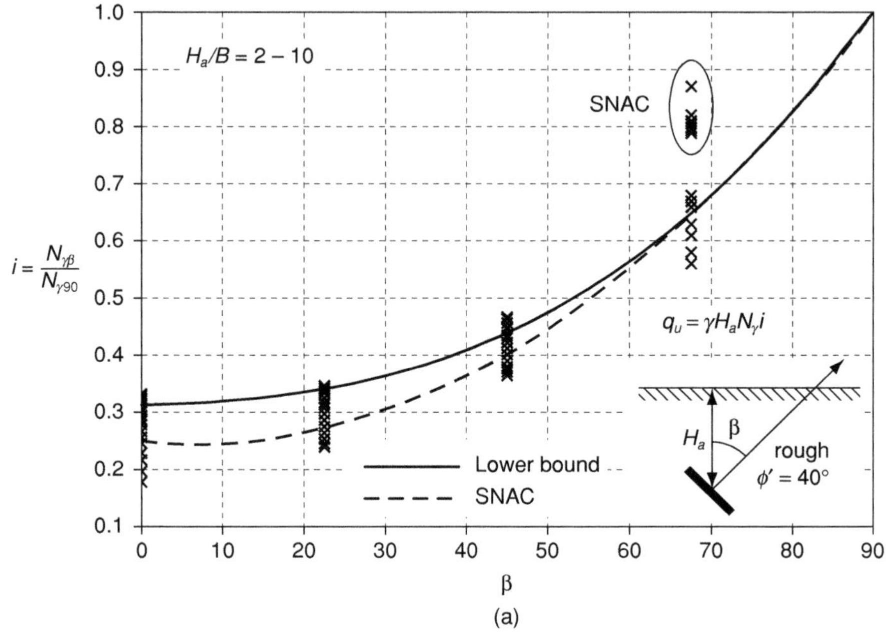

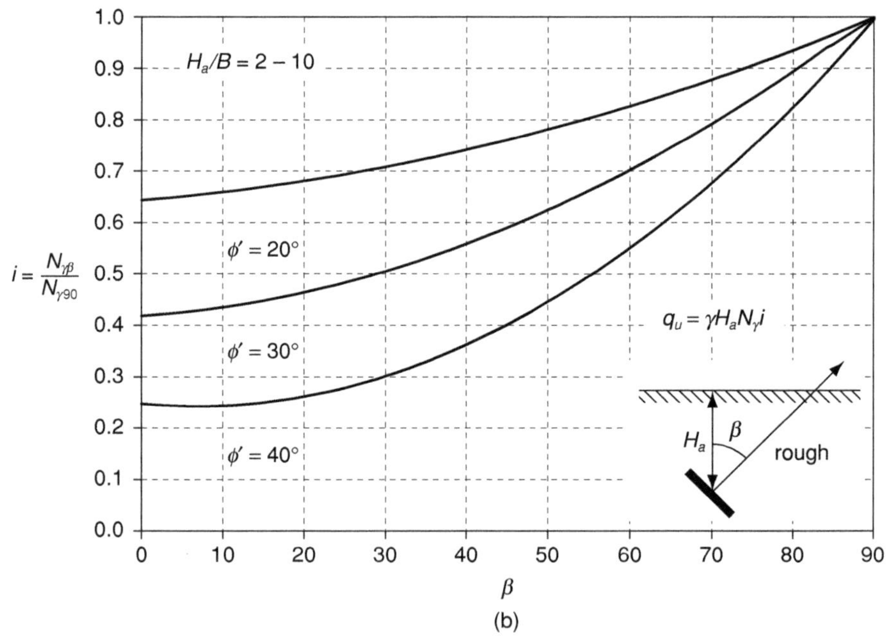

Figure 3.19 Inclination factors for inclined rough anchors in cohesionless soil: (*a*) lower bound & SNAC, (*b*) proposed design chart

increasing angle of inclination β before reaching unity at $\beta = 90°$ (vertical). A similar observation was made for inclined anchors in purely cohesive soil (Figure 3.8).

A proposed design chart is presented in Figure 3.19(b), where the ultimate capacity of an inclined anchor is given by

$$q_u = \gamma H_a N_\gamma i \tag{15}$$

The types of failure modes observed for inclined anchors are illustrated by the upper bound velocity diagrams and SNAC displacement plots in Figures 3.20–4. In general, when $\beta \geq \phi'$ failure is characterized by the development

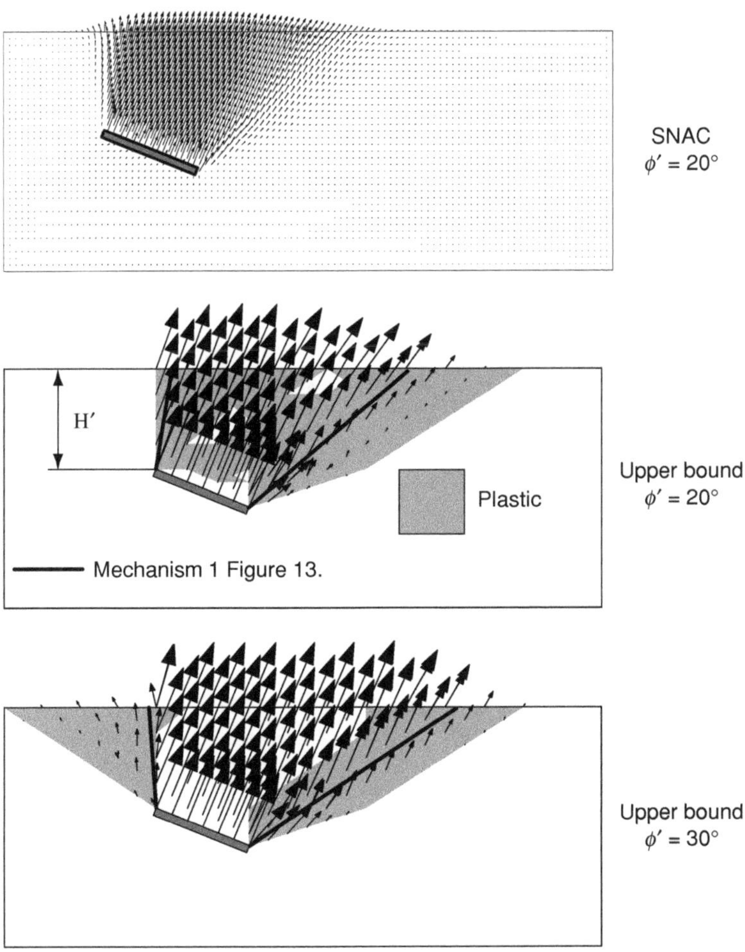

Figure 3.20 Upper bound and SNAC failure modes for inclined rough anchors in cohesionless soil with $H'/B = 1$ and $\beta = 22.5°$

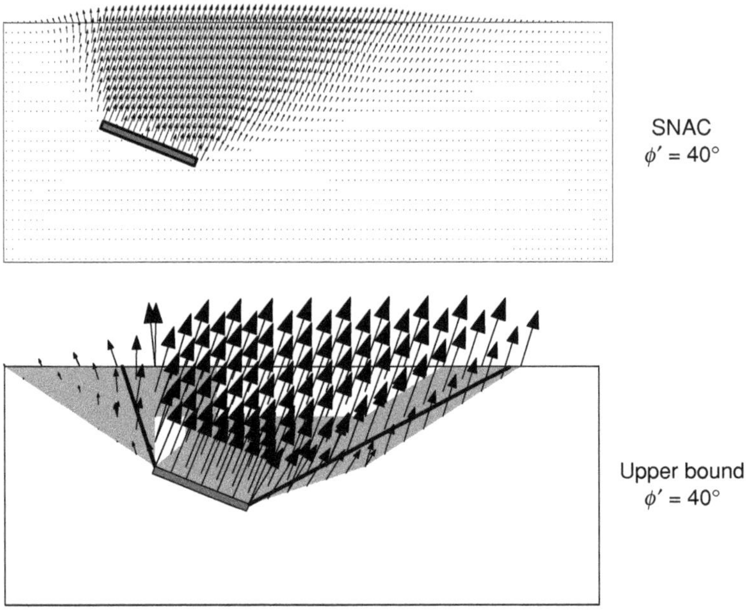

Figure 3.20 (*cont'd*)

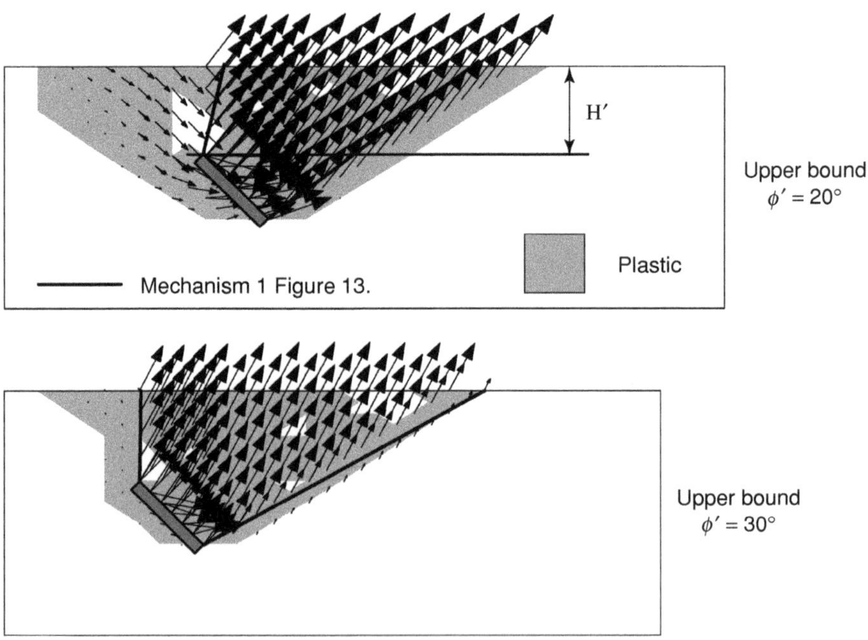

Figure 3.21 Upper bound and SNAC failure modes for inclined rough anchors in cohesionless soil with $H'/B = 1$ and $\beta = 45°$

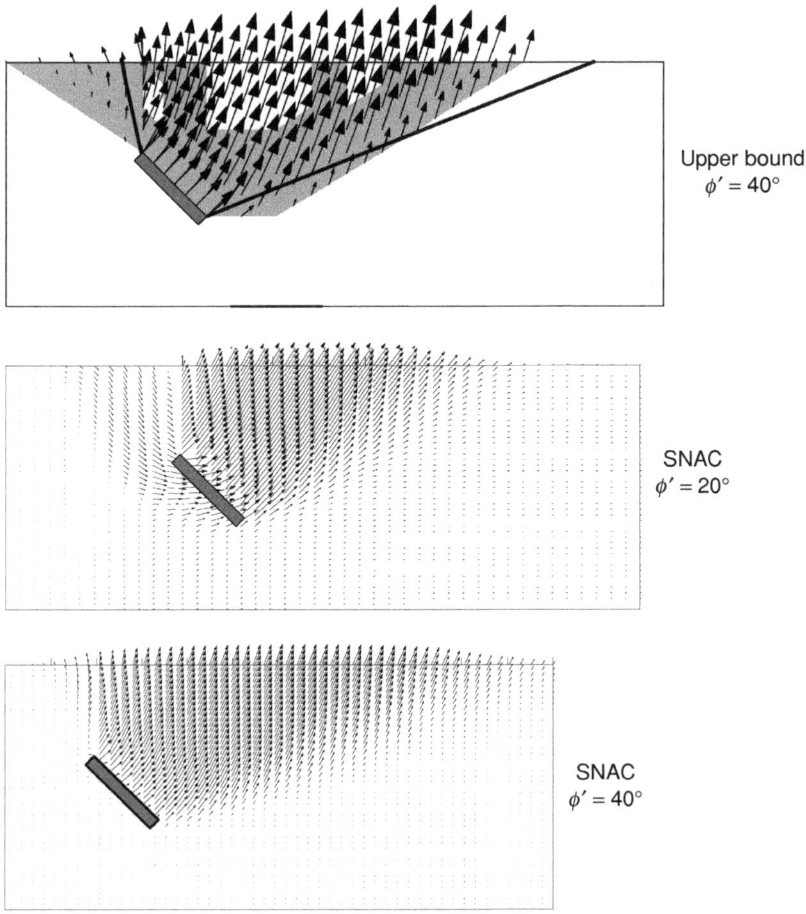

Figure 3.21 (*cont'd*)

of an active failure zone immediately behind the anchor, and an extensive passive failure zone in front of and above the anchor. This type of collapse mechanism is consistent with that observed for vertical anchors ($\beta = 90°$) by Merifield (2002).

The anchor break-out factors from Mechanism 1 (Figure 3.13) are compared to the break-out factors from the finite element upper bound method in Figures 3.25 and 3.26. These figures suggest that this mechanism provides a reasonable estimate of the break-out factor when $\beta \leq 45°$. This may be explained by comparing the predicted collapse mechanisms. For shallow anchors the mode of failure predicted by Mechanism 1 compares reasonably well with the mode of failure predicted by the upper bound

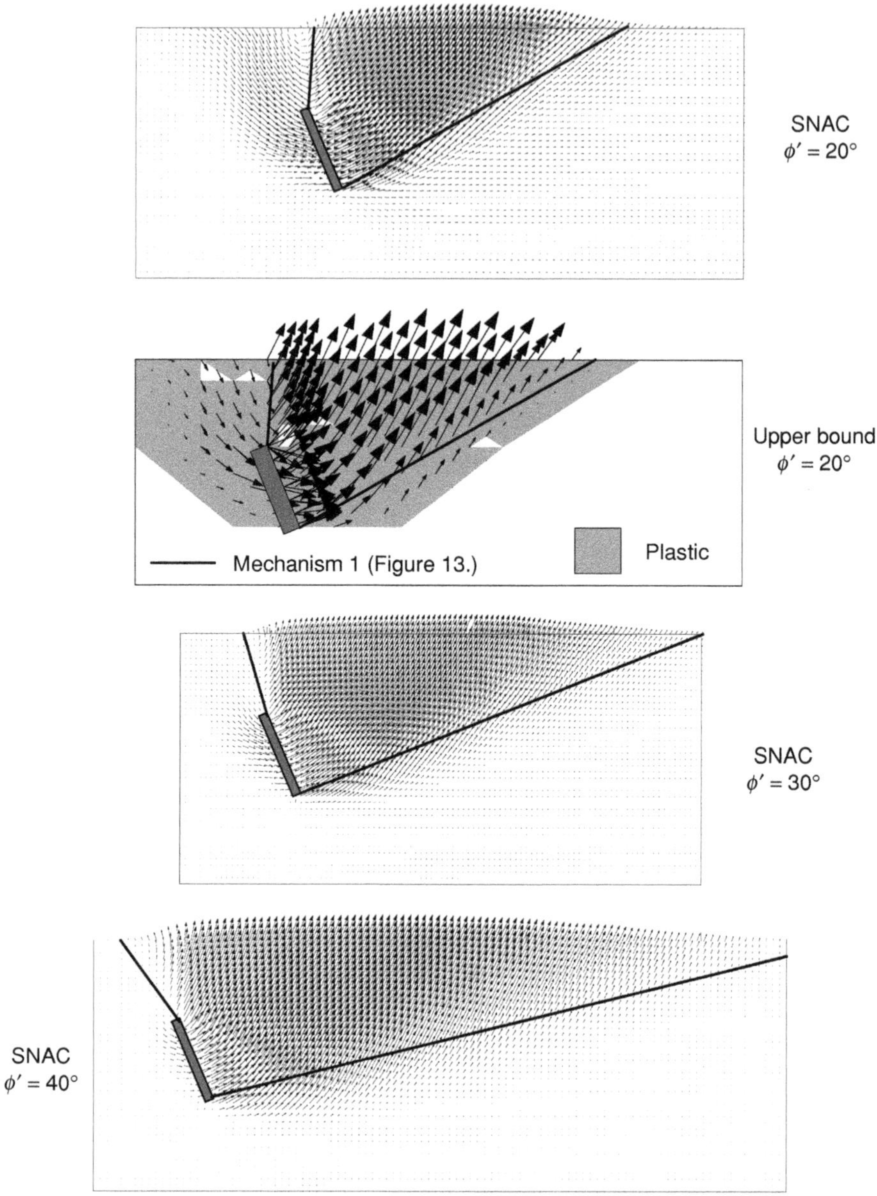

Figure 3.22 Upper bound and SNAC failure modes for inclined rough anchors in cohesionless soil with $H'/B = 1$ and $\beta = 67.5°$

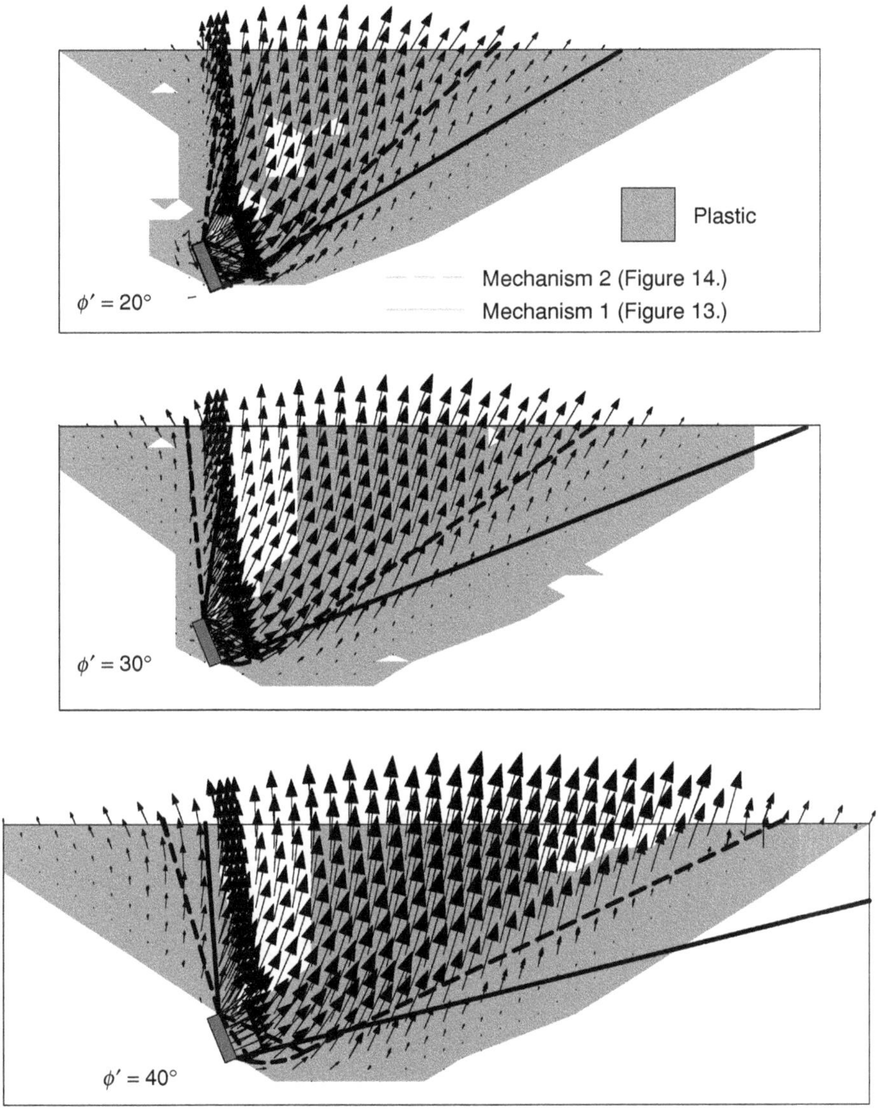

Figure 3.23 Upper bound and SNAC failure modes for inclined rough anchors in cohesionless soil with $H'/B = 4$ and $\beta = 67.5°$

method (see Figures 3.20, 3.21 and 3.24). It should noted, however, that this mechanism is unable to account for any active pressure on the back of the anchor. Thus, for inclination angles greater than 45°, Mechanism 1 can significantly over-estimate the break-out factor, particularly for high friction angles and embedment ratios. This can be seen for the case of $H/B = 4$ and $\beta = 67.5°$ in Figure 3.23.

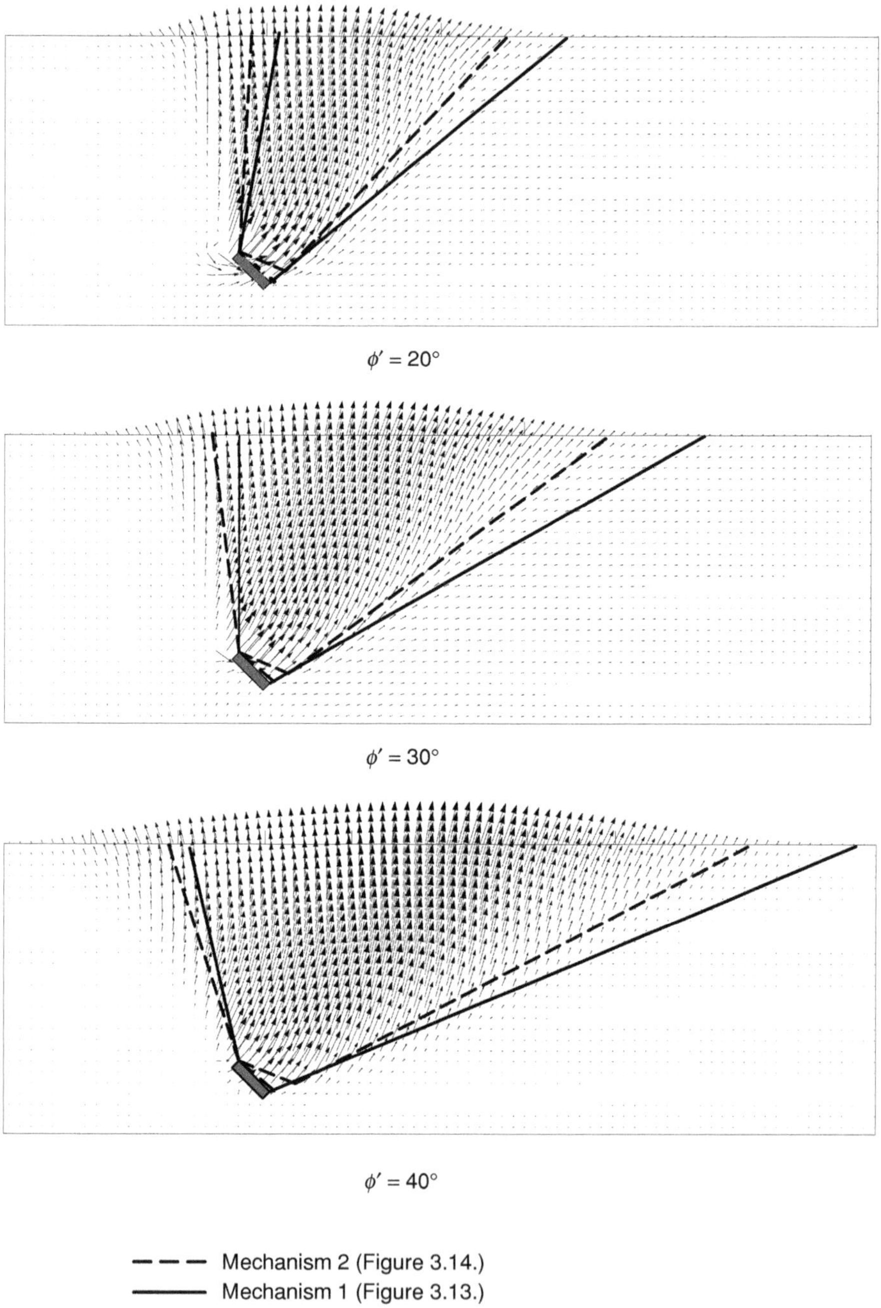

Figure 3.24 Upper bound and SNAC failure modes for inclined rough anchors in cohesionless soil with $H'/B = 5$ and $\beta = 45°$

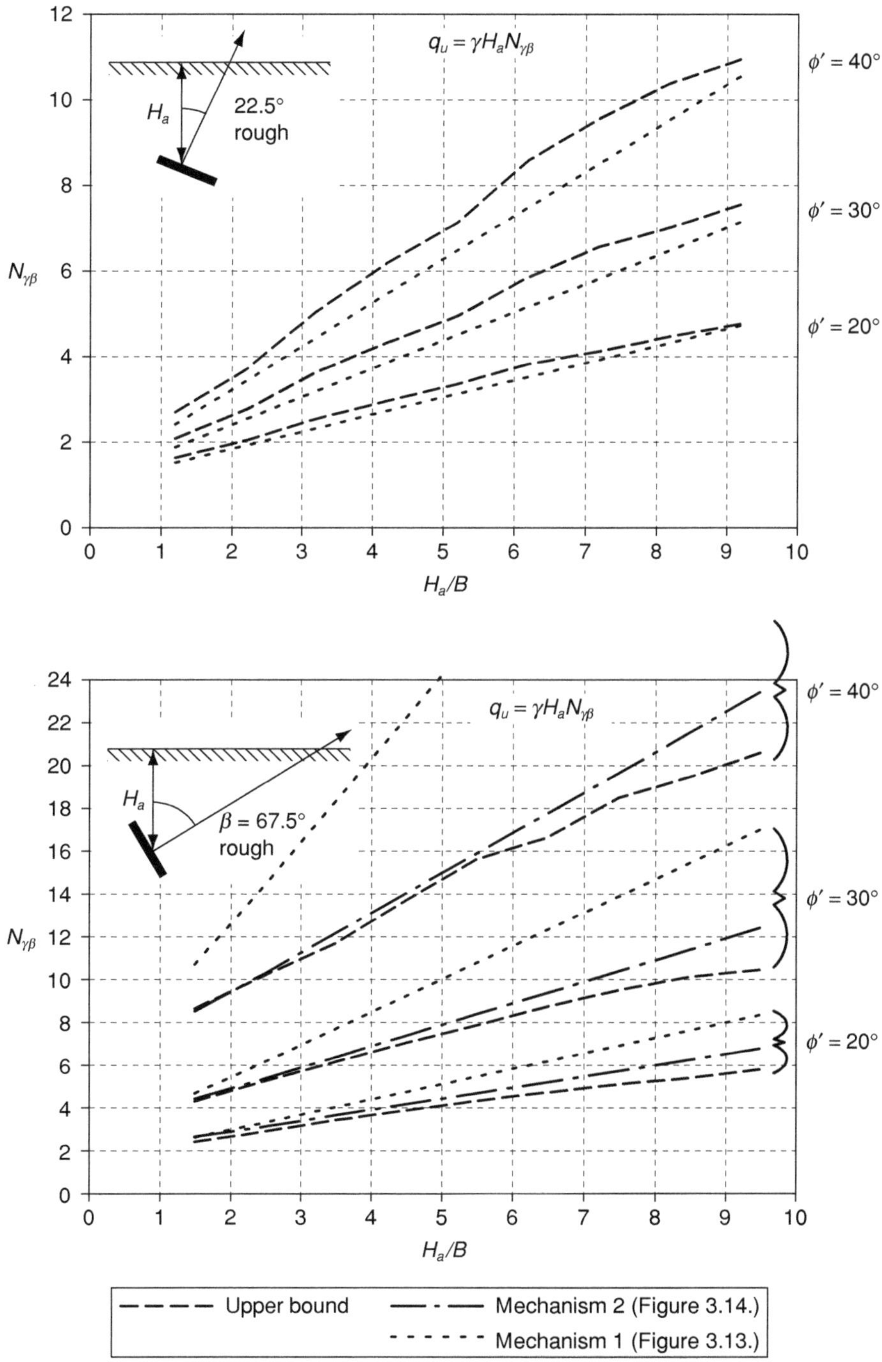

Figure 3.25 Comparison of theoretical break-out factors for inclined anchors in cohesionless soil

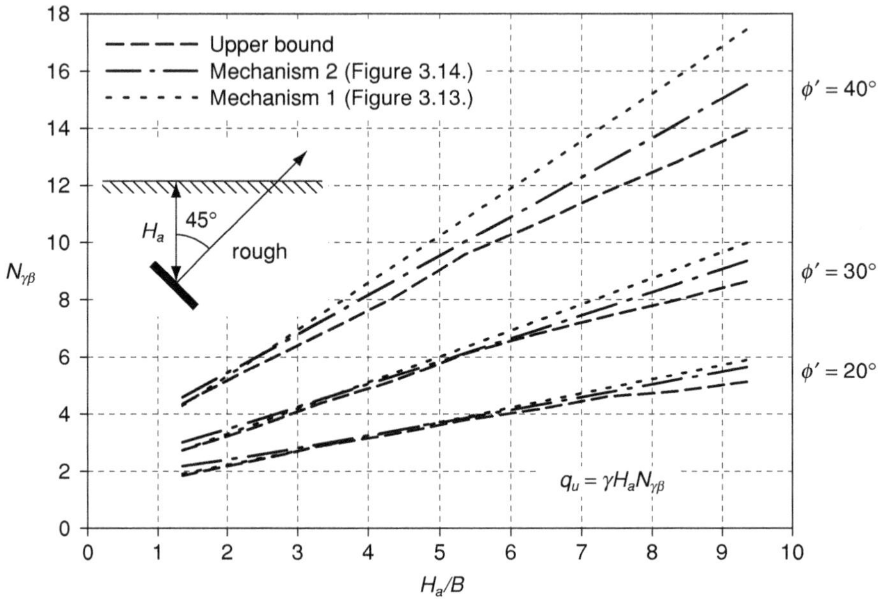

Figure 3.26 Comparison of theoretical break-out factors for inclined anchors in cohesionless soil

The more complex Mechanism 2, shown in Figure 3.14, provides better estimates of the break-out factor over the range of problems analyzed. This fact is illustrated in Figures 3.25 and 3.26 where the break-out factors are typically within 10 percent of the numerical upper bound estimates. For anchors inclined at $\beta \leq 22.5°$, this mechanism collapses such that the estimate of the break-out factor is identical to that obtained from Mechanism 1.

Existing numerical and laboratory studies that address the inclined capacity of anchors in cohesionless soil are scarce. The literature on inclined anchors includes papers by Meyerhof (1973), Das and Seeley (1975a, b), Hanna et al. (1988), Frydman and Shamam (1989), Murray and Geddes (1989) and Singh and Basudhar (1992).

Meyerhof (1973) extended the theory of Meyerhof and Adams (1968) to include inclined anchors. The theory used by Meyerhof (1973) is based on active and passive earth pressure theory. The break-out factor proposed by Meyerhof is estimated by the following relationship (Das 1990)

$$N_\gamma = \frac{1}{2}K_b\left(\frac{H_a}{B}\right)\left(1 + \frac{1}{2\frac{H_a}{B}}\sin\beta\right)^2 + \left(1 + \frac{1}{2\frac{H_a}{B}}\sin\beta\right)\cos^2\beta \tag{16}$$

where K_b is defined as an uplift coefficient that is obtained from the earth pressure coefficients for an inclined wall (Caquot and Kerisel 1949).

Equation (16) has been utilized to estimate the break-out factors for anchors inclined at 45° and 67.5° for embedment ratios ranging from one to ten. These estimates are compared to the lower bound results in Figure 3.27. In general, the predictions of Meyerhof (1973) agree well with the lower bound predictions, particularly for $\beta = 45°$ where the estimates are slightly conservative. For the case of $\beta = 67.5°$, the solutions of Meyerhof are conservative when $H/B \leq 5$ or $\phi' > 30°$.

Hanna et al. (1988) developed an analytical method for predicting the capacity of inclined anchors where β varies from 0° to 60°. The theory behind their procedure is similar to that of Meyerhof's (1968, 1973), and combines laboratory observations along with the earth pressure tables of Caquot and Kerisel (1949). The break-out factors obtained from this method are compared to the finite element lower bound in Figure 3.28. This figure indicates that the method proposed by Hanna et al. significantly over-predicts the anchor break-out factor in all cases. This conclusion was also made by Das (1990).

An extensive laboratory testing program to determine the capacity of inclined anchors was carried out by Murray and Geddes (1989). Results obtained from this study, along with the results of several field tests conducted by Frydman and Shamam (1989), are compared to the numerical limit analysis results in Figure 3.29. Direct comparisons are again difficult because information regarding the soil dilation angle was not provided by the authors. Notwithstanding the difference in interface friction δ and aspect ratio L/B for these laboratory and field studies, there is reasonable agreement between the theoretical and experimental findings. The most significant discrepancy in the comparison appears in the results of Murray and Geddes (1989) for $\beta = 45°$ at small embedment ratios. This is most likely due to the difference in anchor interface roughness between the laboratory study of Murray and Geddes ($\delta = 10.6°$) and the finite element lower bound ($\delta = \phi' = 40°$), as this factor is always significant for vertical anchors at small embedment ratios (Merifield et al. 2006).

The results and discussion presented above indicate that the pull-out capacity of inclined anchors in cohesionless soil is a non-linear function of the inclination and soil friction angle. Although Figure 3.19(b) is useful for design purposes, a single parametric equation that can be used for a whole range of anchor inclination and soil friction angles is also desirable.

A simple empirical relationship for estimating the ultimate capacity of shallow inclined anchors embedded in cohesionless soil was proposed by Maiah et al. (1986), which is of the form

$$N_{\gamma\beta} = N_{\gamma\beta=0} + [N_{\gamma\beta=90} - N_{\gamma\beta=0}]\left(\frac{\beta}{90}\right)^n \tag{17}$$

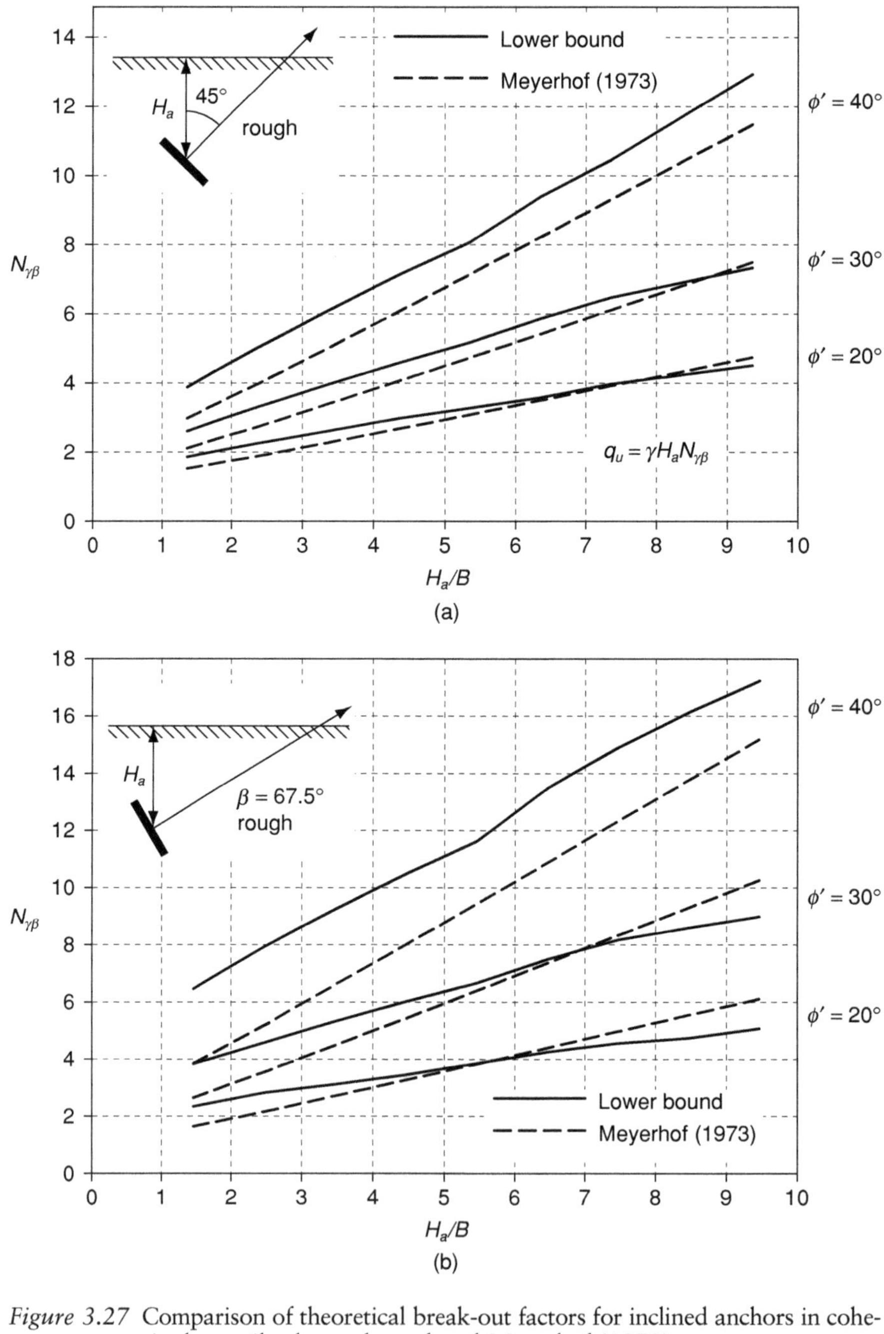

Figure 3.27 Comparison of theoretical break-out factors for inclined anchors in cohesionless soil – lower bound and Meyerhof (1973)

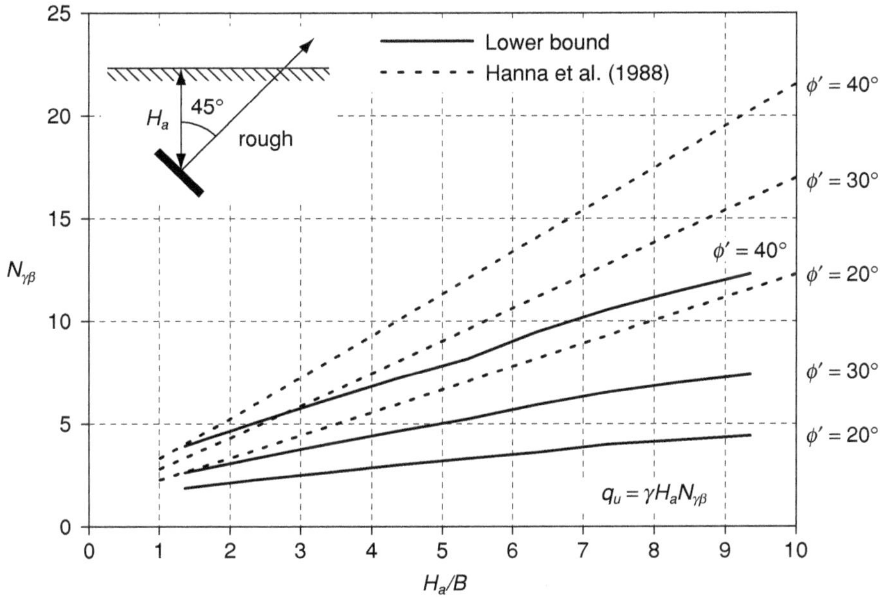

Figure 3.28 Comparison of theoretical break-out factors for inclined anchors in cohesionless soil – lower bound and Hanna et al. (1988)

By substituting the break-out factors obtained for the vertical and horizontal anchor cases into (17), a best-fit estimate of the break-out factor for an anchor inclined at any angle β can be obtained. The limit analysis and SNAC results (180 data points) have been used in conjunction with Equation (17) to estimate the break-out factors for anchors inclined at angles of 22.5°, 45°, 67.5° over a range of embedment ratios from H_a/B of 1 to 10. To obtain the best fit, a value of $n = 2.25$ has been adopted. These estimates are compared to the actual results from the current study in Figure 3.30. The lower bounds predicted by (17) are within ±5% of the actual values, while the SNAC results are within ±7%. The greatest discrepancy between the actual and predicted estimates occurs at small embedment ratios where the estimated values tend to be slightly conservative. None the less, it can be concluded that Equation (17) provides a reasonable estimate for the capacity of inclined strip anchors.

5.3 Suggested procedure for estimating uplift capacity

(1) Determine representative values of the material parameters ϕ' and γ.
(2) Knowing the anchor size B and embedment depth H_a calculate the embedment ratio H_a/B.
(3) Calculate $N_{\gamma 90}$ from Figure 3.17 or by using Equation (14) with $H/B = H_a/B + 0.5$.

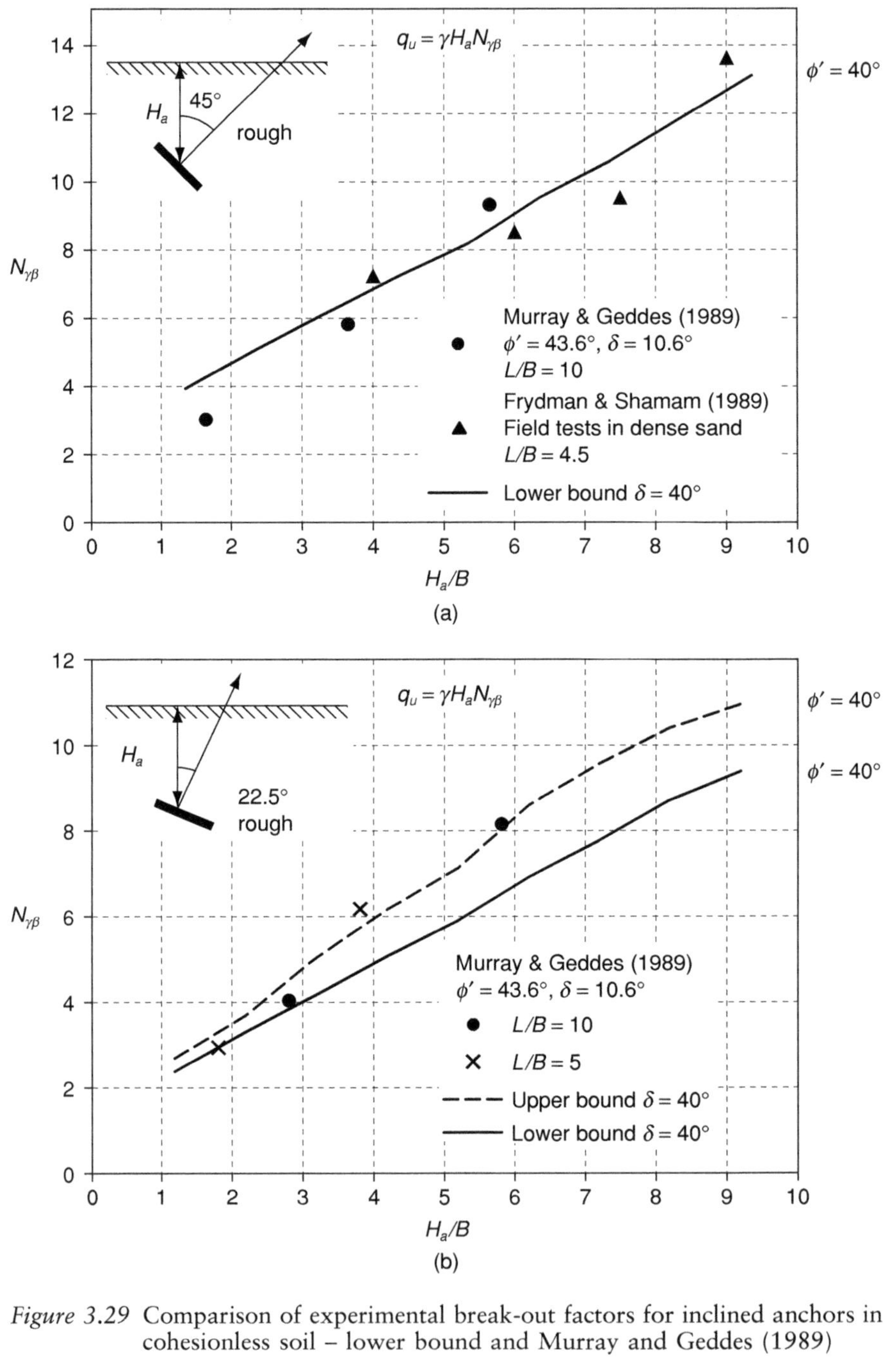

Figure 3.29 Comparison of experimental break-out factors for inclined anchors in cohesionless soil – lower bound and Murray and Geddes (1989)

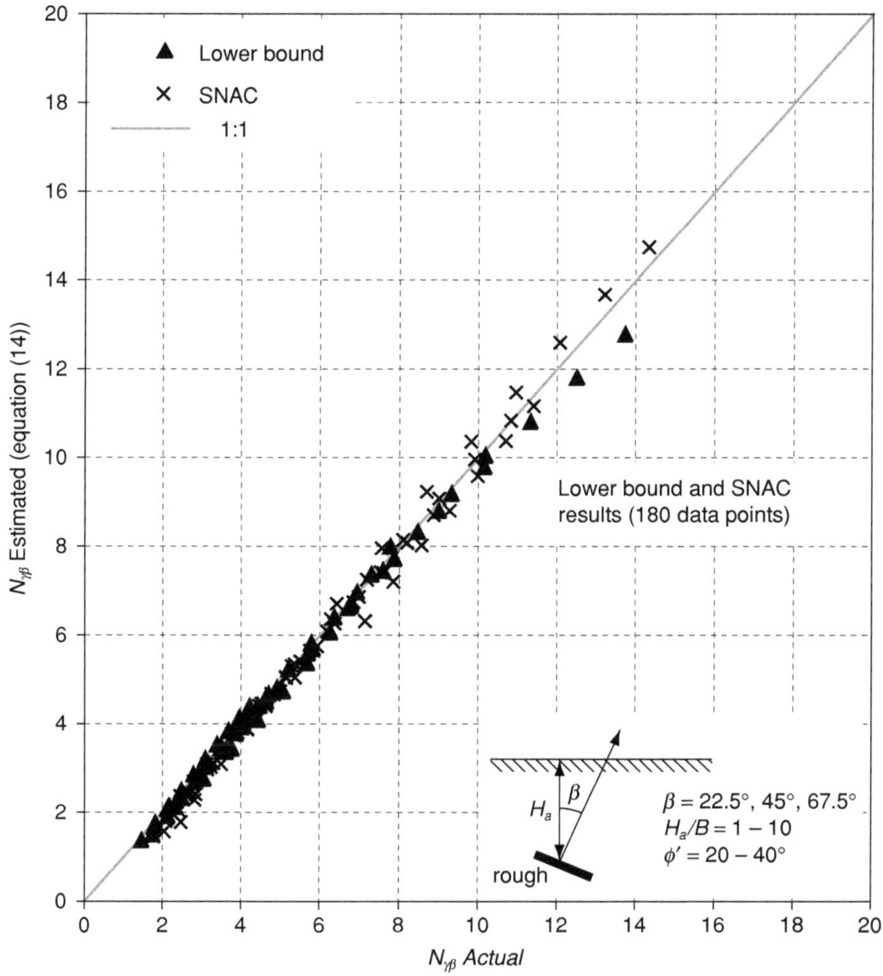

Figure 3.30 Empirical relationship for the break-out factors for inclined strip anchors in cohesionless soil

(4) For an anchor at $\beta = 22.5°$, 45° or 67.5°, estimate the inclination factor i using Figure 3.19(b), depending on the anchor orientation.
(5) The ultimate pull-out capacity is given by Equation (15).

5.3.1 Example

We now illustrate how to use the results presented to determine the ultimate pull-out capacity of an inclined anchor in clay.

Problem: A plate anchor of width 0.5 m is to be embedded at $H_a = 1.5$ m at an orientation of 45°. Determine the ultimate pull-out capacity given that the clay has a friction angle of $\phi' = 30°$ and unit weight $\gamma = 20$ kN/m^3.

The systematic procedures given above will now be used to determine the ultimate anchor capacity.

(1) Given $\phi' = 30°$ and $\gamma = 20$ kN/m^3.
(2) The embedment ratio can be calculated as $H_a/B = 1.5/0.5 = 3$
(3) From Figure 3.17 using $H/B = H_a/B = 1.5/0.5 = 3$, $N_{\gamma 90} \approx 6$

Using Equation (14) gives $N_{\gamma 90} = 5.6$

(4) From Figure 3.19(b), the inclination factor $i \approx 0.6$
(5) The ultimate pull-out capacity is given by Equation (15)
$q_u = \gamma H_a N_\gamma i = 20 \times 1.5 \times 6 \times 0.6 = \mathbf{108}$ kPa
$Q_u = 108 \times (0.5) = \mathbf{54}$ **kN** per m run

6 Conclusions

A rigorous numerical study into the ultimate capacity of inclined strip anchors has been presented. Consideration has been given to the effect of embedment depth, material strength and overburden pressure. The results have been presented as break-out factors in both chart and best-fit equation form to facilitate their use in solving practical design problems.

Using the lower and upper bound limit theorems, small error bounds were achieved on the true value of the break-out factor for inclined anchors in clay or sand. Displacement finite element results (SNAC) were found to compare favorably with the numerical bounds solutions, and plot close to or just above the upper bound solutions over the range of embedment ratios considered.

The effect of anchor inclination on the pull-out capacity of anchors in clay has been investigated. A simple empirical equation has been proposed which, on average, provides collapse load estimates within ±5% of the actual values.

The effect of anchor inclination on the pull-out capacity in cohesionless soil was also investigated. The numerical results obtained showed reasonable agreement with existing theoretical predictions and laboratory results. A simple empirical relationship for estimating the ultimate capacity of shallow inclined anchors embedded in cohesionless soil was developed. This can be used for routine design purposes.

Acknowledgments

I gratefully acknowledge the contributions made by Professor Scott Sloan and Associate Professor Andrei Lyamin when undertaking the above work, which was performed with the financial assistance received from the Australian Research Council Discovery Projects Grants Scheme. I am also thankful for the support provided by the School of Engineering at the University of Newcastle, Australia.

References

Abbo, A. J. (1997) Finite element algorithms for elastoplasticity and consolidation, PhD thesis, Department of Civil, Surveying and Environmental Engineering, University of Newcastle, NSW.

Abbo, A. J. and Sloan, S. W. (2000) SNAC users manual Version 2.0.

Caquot, A. and Kerisel, L. (1949) *Traité de Mécanique des Sols*, Paris: Gauthier-Villars.

Chen, W. F. (1975) *Limit Analysis and Soil Plasticity*, Amsterdam: Elsevier.

Das, B. M. (1990) *Earth Anchors*, Amsterdam: Elsevier.

Das, B. M. and Puri, V. K. (1989) Holding capacity of inclined square plate anchors in clay, *Soils and Found.*, Japan, 29 (3): 138–44.

Das, B. M. and Seeley, G. R. (1975a) Passive resistance of inclined anchors in sand, *J. of Geotech. Eng.*, ASCE, 101 (3): 353–6.

Das, B. M. and Seeley, G. R. (1975b) Inclined load resistance of anchors in sand, *J. of Geotech. Eng.*, ASCE, 101 (9): 995–8.

Drucker, D. C., Greenberg, H. J. and Prager, W. (1952) Extended limit design theorems for continuous media, *Quarterly Journal of Applied Mathematics*, 9: 381–9.

Frydman, S. and Shamam, I. (1989) Pullout capacity of slab anchors in sand, *Can. Geotech. J.*, 26: 385–400.

Hanna, A. M., Das, B. M. and Foriero, A. (1988) Behaviour of shallow inclined plate anchors in sand, *Special Topics in Foundations*, Geotech. Spec. Tech. Pub. No. 16, ASCE, pp. 54–72.

Kumar, J. (1999) Kinematic slices approach for uplift analysis of strip foundations, *Int. J. Num. Methods in Geomechanics*, 23: 1159–70.

Lyamin, A. V. and Sloan, S. W. (2002a) Lower bound limit analysis using non-linear programming, *International Journal for Numerical Methods in Engineering*, 55: 573–611.

Lyamin, A. V. and Sloan, S. W. (2002b) Upper bound limit analysis using linear finite elements and non-linear programming, *International Journal for Numerical and Analytical Methods in Geomechanics*, 26: 181–216.

Maiah, A. A., Das, B. M. and Picornell, M. (1986) Ultimate resistance of shallow inclined strip anchor plate in sand, *Proceedings Southeastern Conference on Theoretical and Applied Mechanics* (SECAM XIII), Columbia, SC, 2: 503–9.

Merifield, R. S. (2002) Numerical modelling of soil anchors, PhD thesis, University of Newcastle, Australia.

Merifield, R. S., Lyamin, A. V. and Sloan, S. W. (2005) The stability of inclined plate anchors in purely cohesive soil, *Journal of Geotechnical and Geoenvironmental Engineering*, ASCE, 131 (6): 792–9.

Merifield, R. S. and Sloan, S. W. (2006) The ultimate pullout capacity of anchors in frictional soils, *Canadian Geotechnical Journal*, NRC Research Press, 43 (8): 852–68.

Merifield, R. S., Sloan, S. W. and Yu, H. S. (2001) Stability of plate anchors in undrained clay, *Géotechnique*, 51 (2): 141–53.

Meyerhof, G. G. (1973) Uplift resistance of inclined anchors and piles, *Proceedings, 8th International Conference on Soil Mechanics and Foundation Engineering, Moscow*, 2: 167–72.

Meyerhof, G. G. and Adams, J. I. (1968) The ultimate uplift capacity of foundations, *Canadian Geotechnical Journal*, 5 (4): 225–44.

Michalowski, R. L. (1995) Slope stability analysis : a kinematical approach, *Géotechnique*, 45 (2): 283–93.

Murray, E. J. and Geddes, J. D. (1989) Resistance of passive inclined anchors in cohesionless medium, *Géotechnique*, 39 (3): 417–31.

Regenass, P. and Soubra, A. H. (1995) *Anchors in Theory and Practice*, Rotterdam: Balkema, pp. 117–23.

Rowe, R. K. (1978) Soil structure interaction analysis and its application to the prediction of anchor behaviour, PhD thesis, University of Sydney, Australia.

Rowe, R. K. and Davis, E. H. (1982) The behaviour of anchor plates in clay, *Géotechnique*, 32 (1): 9–23.

Singh, D. N. and Basudhar, P. K. (1992) A note on the optimal lower bound pull-out capacity of inclined strip anchors in sand, *Can. Geotech. J.*, 29: 870–3.

Sloan, S. W. (1988) Lower bound limit analysis using finite elements and linear programming, *International Journal for Numerical and Analytical Methods in Geomechanics*, 12: 61–7.

Sloan, S. W. (1989) Upper bound limit analysis using finite elements and linear programming, *International Journal for Numerical and Analytical Methods in Geomechanics*, 13: 263–82.

Sloan, S. W. and Kleeman, P. W. (1995) Upper bound limit analysis using discontinuous velocity fields, *Computer Methods in Applied Mechanics and Engineering*, 127: 293–314.

Sloan, S. W. and Randolph, M. F. (1982) Numerical prediction of collapse loads using finite element methods, *International Journal for Numerical and Analytical Methods in Geomechanics*, 6 (1): 47–76.

Toh, C. T. and Sloan, S. W. (1980) Finite element analysis of isotropic and anisotropic cohesive soils with a view to correctly predicting impending collapse, *International Journal for Numerical and Analytical Methods in Geomechanics*, 4: 1–23.

4 Numerical modeling of geosynthetic reinforced soil walls

Richard J. Bathurst, Bingquan Huang and Kianoosh Hatami

1 Introduction

Geosynthetic reinforced soil walls are now a mature technology for the solution of earth retaining wall problems. In the USA they have been demonstrated to be 50 percent of the cost of traditional concrete gravity structures. Conventional Rankine and Coulomb earth pressure theories and slope stability methods have been adopted in modified form to carry out the analysis and design of geosynthetic reinforced soil walls. These methods are limit equilibrium approaches that assume simplified failure mechanisms and employ either global factors of safety (AASHTO 2002, Elias et al. 2001, NCMA 1997) or partial factors (BS8006 1995, AASHTO 2007) to design against serviceability failure and/or collapse (i.e. limit states design). However, the operational performance of these structures (i.e. under working stress conditions) is controlled by deformation limits that cannot be accounted for explicitly using limit equilibrium-based methods. In addition, reinforced soil walls are complex systems typically involving a structural facing (such as concrete panels or stacked modular concrete blocks), soil backfill and horizontal layers of polymeric reinforcement. Limit equilibrium and semi-empirical analysis approaches require simplifying assumptions regarding the mechanical properties of polymeric reinforcement products and the interaction between components of reinforced soil wall systems. As a result, these approaches provide very limited insight into the fundamental behavior of reinforced soil walls since failure mechanisms are assumed a priori.

Review and analysis of a large number of instrumented full-scale walls has shown that current limit equilibrium based analysis methods overestimate reinforcement loads under operational conditions (on average) by a factor of 2 to 3 (Allen et al. 2003, Bathurst et al. 2005, Bathurst et al. 2008). As the emphasis shifts to the prediction of geosynthetic reinforced soil walls under operational (serviceability) conditions, the demand for improved and more accurate design models for these systems increases. Furthermore, the need to develop calibrated serviceability limit states design models for design engineers requires data that are difficult to obtain owing to the limited number of monitored field structures.

Numerical methods hold promise as a design and research tool to investigate the entire response of reinforced soil retaining walls. An advantage of many modern computer codes is that advanced constitutive models for polymeric materials and soils can also be implemented in the analyses. Furthermore, numerical models can be used to carry out parametric analyses of the influence of wall geometry, facing type and mechanical properties of the constituent materials on wall behavior. Calibrated numerical models can also be used to extend the database of carefully instrumented field or laboratory-scale structures and hence contribute to the development of rational design methods that are, nevertheless, based on conventional concepts of earth pressure theory.

Numerical modeling of geosynthetic reinforced soil walls has been reported in the literature. Examples are Karpurapu and Bathurst (1995), Rowe and Ho (1997), Ling and Leshchinsky (2003), Leshchinsky and Vulova (2001) and Yoo and Song (2006) amongst many others. However, very few cases exist in the literature in which an attempt has been made to verify the results of numerical models against carefully constructed, instrumented and monitored full-scale walls. Examples where this has been done are Ling (2003), Hatami and Bathurst (2005, 2006) and Guler et al. (2007). Useful reviews of geosynthetic reinforced soil wall numerical modeling efforts can be found in the papers by Bathurst and Hatami (2001), Ling (2003) and Hatami and Bathurst (2005).

This chapter describes the numerical modeling details of one full-scale physical geosynthetic reinforced soil wall from a series of structures constructed at the Royal Military College of Canada (RMC). Similar numerical modeling of other walls in the RMC test program has been carried out by the authors. However, owing to the chapter word limit, the focus is on the first wall in this series. Nevertheless, recommendations that are applicable to modeling of the other walls in the RMC test series appear in the sections to follow. Details of the numerical model and constitutive modeling of the component materials are described. Numerical predictions are compared to measured response features of the physical test to assess the accuracy of the numerical approach. Next, example parametric analyses are carried out using the verified numerical code to investigate the influence of wall toe stiffness and facing block interface stiffness on a theoretical wall of 6 m in height. The lessons learned here have applications to other types of geosynthetic reinforced soil walls and are of value to modelers who wish to explore the mechanical behavior of these systems, generate data to fill in gaps in performance data from the limited number of monitored structures reported in the literature or carry out parametric analyses.

2 RMC physical models

The results of numerical models described in this chapter are compared to measurements taken from the control (reference) physical model test in a series of eleven full-scale reinforced soil walls constructed at the RMC (e.g.

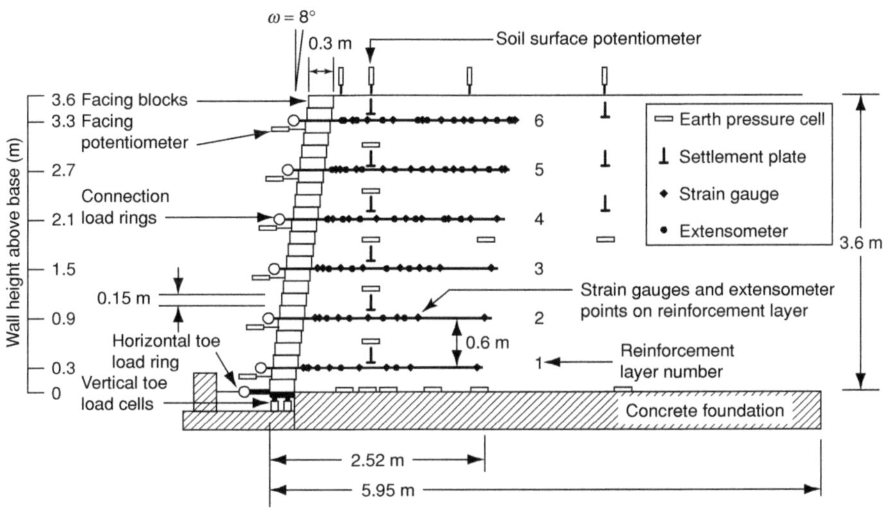

Figure 4.1 Cross-section of the RMC instrumented wall

Bathurst et al. 2000, 2006). The walls were nominally the same, with one component changed from the control wall considered in this chapter. The walls varied with respect to facing type, facing batter, reinforcement type and reinforcement spacing. A cross-section view of the control wall is shown in Figure 4.1. The wall was 3.6 m in height and was constructed with dry-stacked solid masonry blocks 300 mm wide (toe to heel), 200 mm long and 150 mm high, arranged in a staggered pattern when viewed from the front of the structure. Each block was cast with a continuous concrete key in the top and a matching groove in the bottom to assist with wall alignment and to provide shear transfer along the height of the wall. The backfill soil was seated on a rigid concrete foundation and the facing column on a rigid instrumented footing. The backfill was uniformly graded sand and was reinforced with six layers of integral drawn biaxial polypropylene geogrid. The wall was constructed in 150 mm lifts matching the height of each facing unit course. The soil was compacted using a lightweight vibratory plate compactor (Hatami and Bathurst 2006). The incremental construction and a small amount of interface shear compliance between blocks resulted in a moving datum as wall height increased. Following construction, the backfill surface was uniformly surcharged in increments up to a maximum pressure of 130 kPa.

3 Numerical modelling

The numerical simulations were performed using the 2D finite difference computer program FLAC (Itasca 2005). Figure 4.2 shows the FLAC numerical grid used in this study. The construction process was modeled using

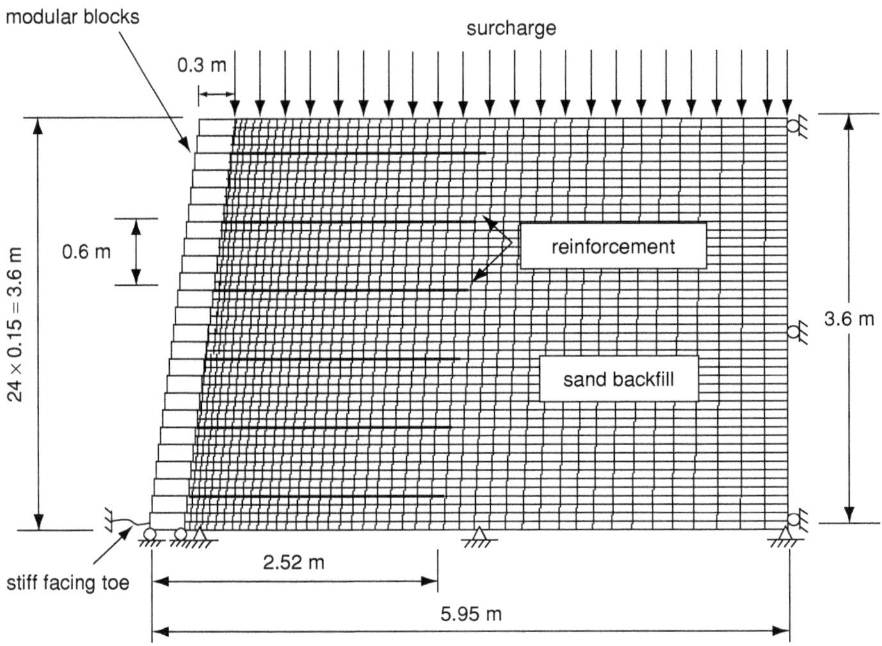

Figure 4.2 Numerical FLAC grid

sequential bottom-up numerical grid increments 0.15 m thick. The influence of dynamic compaction loads was simulated by a transient uniform pressure applied to each soil lift. The magnitude of the transient uniform surcharge pressure was taken as 8 kPa for this wall. However, some of the RMC walls were constructed using a heavier vibratory rammer, and for these structures the transient load was increased to 16 kPa (Hatami and Bathurst 2005, 2006). The influence of the compaction method on the stiffness of the soil was captured by varying the magnitude of the selected constitutive model parameters as described later. Computations were carried out in large-strain mode to ensure sufficient accuracy in the event of large wall deformations or reinforcement strains and to accommodate the moving local datum as each row of facing units and soil layer was placed during construction simulation. The ratio of the wall height to width of the numerical grid was 0.65 matching the physical model and thus capturing the influence (if any) of the proximity of the far-end boundary on model behavior.

3.1 Soil

Design codes typically recommend that the backfill soil be granular so that it is free-draining. Furthermore, these materials are more likely to have greater strength, stiffness and better constructibility than soils with high fines content. For this reason, numerical modeling efforts by the authors have focused on walls constructed with granular fills. A number of possibilities exist with respect to the choice of the constitutive model for the soil. In order of complexity the authors have examined the following types: (*a*) linear elastic-plastic with Mohr–Coulomb failure criterion, (*b*) non-linear elastic (hyperbolic) models or variants (e.g. Duncan et al. 1980), and (*c*) Lade's single hardening model (Kim and Lade 1988, Lade and Kim 1988a, b).

3.1.1 Linear elastic-plastic model (Mohr–Coulomb)

Soil linear elastic behavior is expressed by the generalized Hooke's law with constant Young's modulus ($E = 48$ or 96 MPa depending on the compaction method) and constant Poisson's ratio (assumed as $\nu = 0.3$). Limitations of the Mohr–Coulomb model for frictional soils are: (i) the influence of confining pressure and strain level on soil stiffness is not captured; and (ii) shear-softening behavior is not included. In addition, some judgment is required to select the value of E from the initial portion of measured stress–strain curves. The elastic modulus was adjusted from laboratory tests using the slope of the line plotting surcharge pressure versus vertical strain in the 3.6 m high soil column located equidistance from the back of the facing column and the back of the test facility. The larger modulus value corresponds to walls constructed with heavier compaction equipment.

The yield (failure) criterion chosen is the Mohr–Coulomb failure criterion for cohesive-frictional materials and expressed in terms of principal stresses as:

$$\sigma_1 = \sigma_3 N_\varphi + 2c\sqrt{N_\varphi} \tag{1}$$

where $N_\varphi = (1 + \sin\phi)/(1 - \sin\phi)$. For the RMC walls, the Mohr–Coulomb parameters are $c = 0.2$ and $\phi = 44$ degrees. The peak friction angle was computed from the results of plane strain testing of the RMC sand at principal stress levels estimated at mid-elevation in the wall backfill. The plane strain friction angle was also in agreement with the peak plane strain friction angle computed from direct shear tests and triaxial test results using the empirical relationships proposed by Bolton (1986) and Lade and Lee (1976). The small value of cohesion was introduced to prevent premature yielding (numerical instability) of soil zones that were subject to temporary low confining stress during simulated construction.

A non-associated flow rule is adopted in the Mohr–Coulomb model and the shear potential function defined as:

$$\sigma_1 = \sigma_3 N_\psi \tag{2}$$

where $N_\psi = (1 + \sin\psi)/(1 - \sin\psi)$ and ψ is the dilation angle. The value of the dilation angle was computed from the results of the direct shear box tests.

Table 4.1 Backfill soil properties

Model parameters	*Value*
Duncan–Chang (hyperbolic) model	
K_e (elastic modulus number)	800 (1968)
K_{ur} (unloading–reloading modulus number)	960 (2362)
n (elastic modulus exponent)	0.5
R_f (failure ratio)	0.86
ν_t (tangent Poisson's ratio)	0–0.49
B_i/p_a (initial bulk modulus number)	110 (270)
ε_u (asymptotic volumetric strain value)	0.012
Mohr–Coulomb model	
E (Young's modulus) (MPa)	48 (96)
ν (Poisson's ratio)	0.3
ψ (dilation angle) (degrees)	11
Lade's model	
M, λ, ν (elastic properties)	550 (1360), 0.25, 0.3
m, η_1, a (failure criterion)	0.107, 36.0, $200/p_a$
Ψ_2, μ (plastic potential)	−3.65, 2.425
h, α (yield criterion)	0.432, 0.34
C, p (hardening/softening law)	0.145 (0.073) × 10^{-3}, 1.22
Common parameters (as applicable)	
γ (unit weight) (kN/m^3)	16.8
ϕ (friction angle) (degrees)	44
c (cohesion) (kPa)	0.2
p_a (atmospheric pressure) (kPa)	101

Note: Values in parentheses are for soil compacted using heavier compaction equipment

3.1.2 Hyperbolic model

The compacted backfill sand was assumed as an isotropic, homogeneous, non-linear elastic material using the Duncan–Chang hyperbolic model. The elastic tangent modulus is expressed as:

$$E_t = \left[1 - \frac{R_f(1-\sin\phi)(\sigma_1-\sigma_3)}{2c\cos\phi + 2\sigma_3\sin\phi}\right]^2 K_e p_a \left(\frac{\sigma_3}{p_a}\right)^n \tag{3}$$

where: σ_1 = major principal stress, σ_3 = minor principal stress, p_a = atmospheric pressure and other parameters are defined in Table 4.1. The original Duncan–Chang model was developed for axisymmetric (triaxial) loading conditions, and therefore the bulk modulus is a function of only σ_3. However, under plane strain conditions that typically apply to reinforced soil walls, $\sigma_2 > \sigma_3$. Hatami and Bathurst (2005) have shown that the Duncan–Chang parameters back-fitted from triaxial tests on the RMC sand under-estimated the stiffness of the same soil when tested in a plane strain test apparatus owing to higher confining pressure. The bulk modulus formulation proposed by Boscardin et al. (1990) can be used to capture better the effect of the confining pressure on the bulk stiffness, hence:

$$B_t = B_i\left[1 + \frac{\sigma_m}{B_i \varepsilon_u}\right]^2 \tag{4}$$

where: σ_m = mean pressure = $(\sigma_1 + \sigma_2 + \sigma_3)/3$; B_i and ε_u are material properties that are determined as the intercept and the inverse of the slope from a plot of $\sigma_m/\varepsilon_{\text{vol}}$ versus σ_m in an isotropic compression test where ε_{vol} is volumetric strain. In the user-defined model in this study, the B value was restricted to the following range:

$$\frac{E_t}{(1 - 2\nu_{t\min})} \leq B \leq \frac{E_t}{(1 - 2\nu_{t\max})} \tag{5}$$

where $\nu_{t\max} = 0.49$ and $\nu_{t\min} = 0$.

The parameters used in this modified Duncan–Chang model were taken from Boscardin et al. (1990), with some minor adjustments to match a high-quality SW (Unified Soil Classification System designation) sand material compacted to 95 percent of the Standard Proctor. In the simulation of other RMC tests with heavier compaction, values for K_e and K_{ur} were increased by a factor of 2.5 to reflect greater soil stiffness due to higher compaction effort (see Table 4.1). Parameter K_{ur} is the unload–reload modulus number in the Duncan–Chang model and represents the soil inelastic behavior. A ratio $K_{ur}/K_e = 1.2$ was used to estimate K_{ur} for a dense sand as recommended by Duncan et al. (1980).

3.1.3 Lade's model

Lade's model is an elasto-plastic work-hardening and work-softening constitutive model with a single yield surface developed for frictional geo-materials (Kim and Lade 1988, Lade and Kim 1988a, b). Lade's model was implemented within the FLAC code using a user-defined subroutine written in C++. Only a brief description of the model is described here for completeness.

The elastic soil behavior was simulated by the generalized Hooke's law using an elastic model developed by Lade and Nelson (1987). Poisson's ratio is assumed to be constant, and the Young's modulus (E) is calculated from:

$$E = Mp_a\left[\left(\frac{I_1}{p_a}\right)^2 + 6\left(\frac{1+\nu}{1-2\nu}\right)\frac{J_2'}{p_a^2}\right]^\lambda \tag{6}$$

A symmetric bullet-shape failure surface in principal stress space is defined by the following failure criterion:

$$\left(\frac{I_1^3}{I_3} - 27\right)\left(\frac{I_1}{p_a}\right)^m = \eta_1 \tag{7}$$

Once the stress path reaches the failure surface strain-softening may take place. In plasticity theory, the flow rule can be expressed as:

$$d\varepsilon^p = d\lambda_p \frac{\partial g_p}{\partial \sigma} \tag{8}$$

Lade's model adopts a non-associated flow rule, and the plastic potential function (g_p) is given by:

$$g_p = \left(\psi_1 \frac{I_1^3}{I_3} - \frac{I_1^2}{I_2} + \psi_2\right)\left(\frac{I_1}{p_a}\right)^\mu \tag{9}$$

I_1, I_2, I_3 and J_2' are invariants of the stress tensor. The yield surfaces are defined as contours of constant total plastic work. The yield criterion is a function of the stress tensor and plastic work, and can be expressed as:

$$f_p'(\sigma) = f_p''(W_p) \tag{10}$$

where

$$f_p'(\sigma) = \left(\psi_1 \frac{I_1^3}{I_3} - \frac{I_1^2}{I_2}\right)\left(\frac{I_1}{p_a}\right)^h e^q \tag{11}$$

and

$$f_p''(W_p) = \left(\frac{1}{D}\right)^{1/\rho}\left(\frac{W_p}{p_a}\right)^{1/\rho} \quad \text{(for hardening)} \tag{12a}$$

$$f_p''(W_p) = Ae^{-BW_p/p_a} \quad \text{(for softening)} \tag{12b}$$

Under isotropic compression conditions, the plastic work W_p is related to the hardening parameter C in Lade's model by:

$$W_p = Cp_a\left(\frac{I_1}{p_a}\right)^p \tag{13}$$

here, I_1 is three times the isotropic pressure. The influence of greater compaction on the soil stiffness was accounted for by reducing the parameter C in Lade's model by 50 percent and increasing parameter M by the same factor as the stiffness modulus values in the Duncan–Chang model (factor of 2.5).

3.2 Polymeric reinforcement

Geosynthetic reinforcement materials are manufactured from polyolefins (high-density polyethylene or polypropylene) or polyester. These materials are visco-elastic-plastic materials and hence offer challenges in numerical modeling. In practical terms these materials have properties that are strain level-dependent and load rate-dependent. A large number of constitutive models have been proposed. Some are very complex, such as the non-linear three-component model described by Kongkitkul and Tatsuoka (2007). A summary of earlier models can be found in the paper by Bathurst and Kaliakin (2005). For the simulation of reinforced soil walls subjected to constant or monotonically increasing load, such as the RMC walls, two models have been proposed by the authors and co-workers in which axial load (T) is related to strain (ε) by a secant axial stiffness value (J) that is time-level and strain-level dependent, hence:

$$T(\varepsilon, t) = J(\varepsilon, t) \times \varepsilon \tag{14}$$

where t is time. Load and stiffness values are described in units of force per unit width of material. The characterization of the load–strain-time behavior of polymeric reinforcement products can be carried out using data from constant load (creep) tests or from a series of constant rate-of-strain tests (Walters et al. 2002). As an example, the creep data for the polypropylene geogrid material used in the RMC walls is shown in Figure 4.3a. Data points from each load curve at the same load duration can be used to create the isochronous load–strain curves illustrated in Figure 4.3b.

Hatami and Bathurst (2006) proposed a hyperbolic function to model the load–strain-time behavior of polymeric reinforcement materials under constant load:

$$T(\varepsilon, t) = \left(\frac{1}{\dfrac{1}{J_o(t)} + \dfrac{\eta(t)}{T_f(t)}\varepsilon} \right) \varepsilon \tag{15}$$

where: $J_o(t)$ is the initial tangent stiffness, $\eta(t)$ is a dimensionless scaling function and $T_f(t)$ is the stress-rupture function for the reinforcement. The values of initial tangent stiffness are taken from the origin of the isochronous curves and plotted against time in Figure 4.3c. The data show that the initial stiffness value decreases at a decreasing rate with log time and can be reasonably approximated by a constant value after about 1000

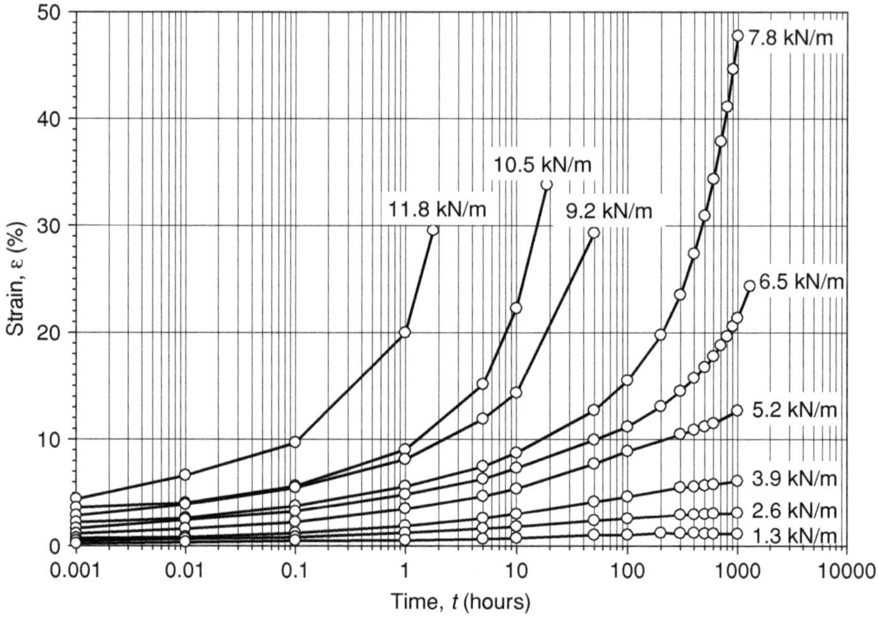

Figure 4.3a Constant load (creep) curves

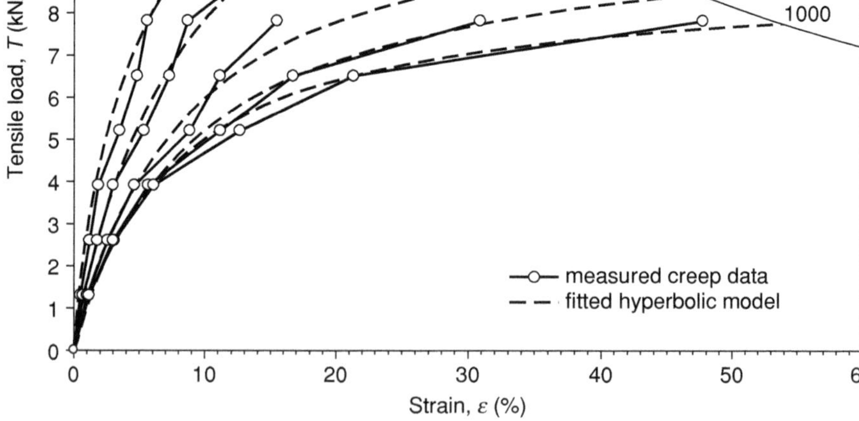

Figure 4.3b Isochronous load–strain curves

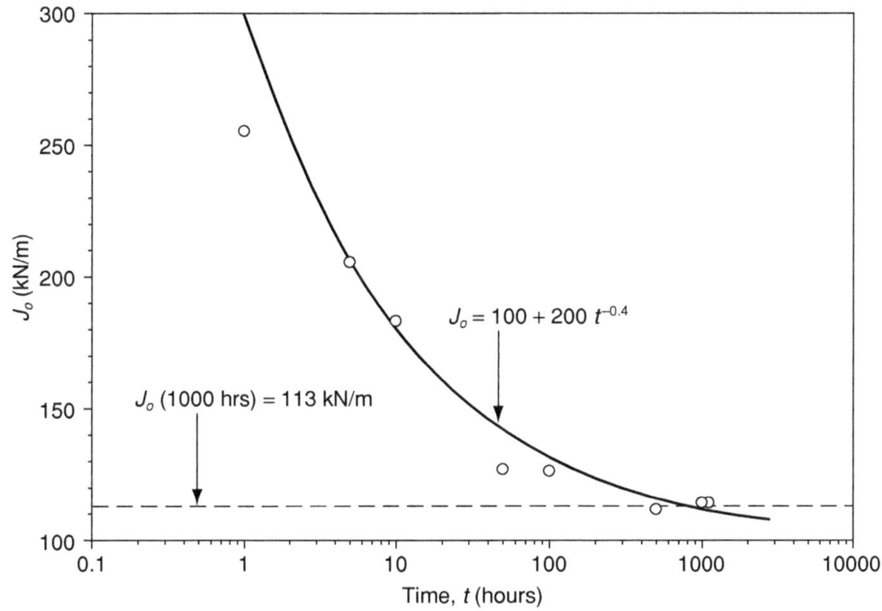

Figure 4.3c Initial stiffness parameter

hours. The measured stress-rupture data from the constant load tests that terminated in rupture are approximated by a log-linear function as illustrated in Figure 4.3d. This function, together with the plots generated using Equation 15, is used to generate the cut-off curve (T_{rupture}) in Figure 4.3b. Equation 15 and the rupture curve fully define the general constitutive model for polymeric reinforcement materials. The scaling factor term η, which is also time dependent, is computed using constant load data from Figure 4.3a and from Equation 15 after expressions for $J_o(t)$ and $\eta(t)$ are defined. In the limit of linear elastic-plastic reinforcement (e.g. metallic reinforcement), Equation 15 reduces to $T = J_o \times \varepsilon$, and T_{rupture} is a constant. Polyester reinforcement materials are less prone to creep. Hatami and Bathurst (2006) showed that, for a polyester geogrid used in one of the RMC walls, the mechanical behavior of the material was linear elastic under monotonic loading to strain levels of practical interest, and hence could be modeled using the linear elastic form of Equation 15.

In most numerical codes the tangent modulus of the planar reinforcement is required. In geosynthetics practice, stiffness in units of force over length is used since cross-section area and thickness of these materials varies and is hence problematic. In most numerical codes this value must be converted to a modulus for an equivalent solid material with constant area and thickness (i.e. $E = J/d$ where d is thickness).

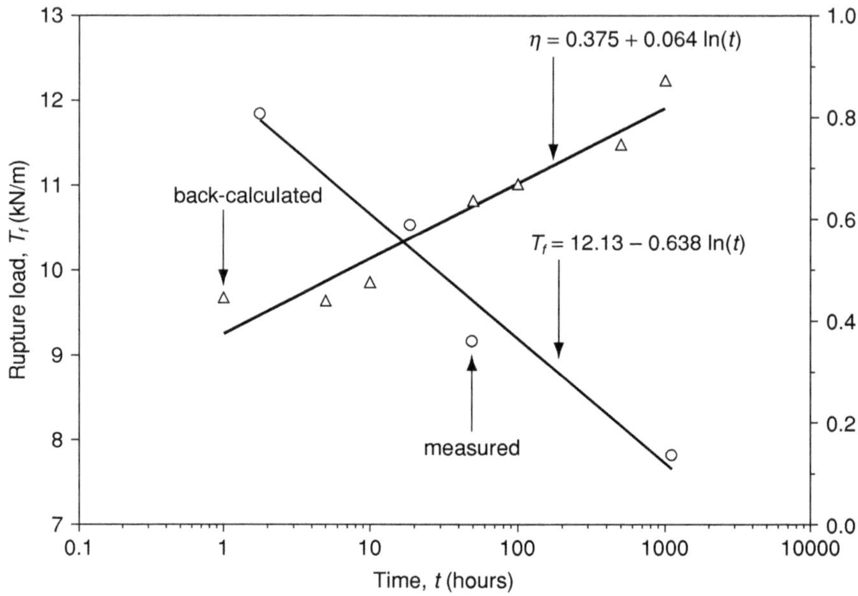

Figure 4.3d Rupture load and scaling factor
Figure 4.3 Constant load (creep) data for polypropylene geogrid and fitting of hyperbolic model parameters

In program FLAC, reinforcement layers can be simulated using cable elements. The properties of the elements can be continuously updated using the equivalent tangent stiffness function $J_t(\varepsilon, t)$ (after conversion to equivalent modulus) (Hatami and Bathurst 2006):

$$J_t(\varepsilon, t) = \frac{1}{J_o(t)\left(\dfrac{1}{J_o(t)} + \dfrac{\eta(t)}{T_f(t)}\varepsilon\right)^2} \tag{16}$$

The plots in Figures 4.3b, c and d show that, as time increases, values of the parameters used to describe the hyperbolic model can be assumed to approach constant values. Walters et al. (2002) interpreted in-isolation (and in-soil) constant load, constant rate-of-strain and relaxation test data for different polymeric reinforcement products and concluded that the long-term stiffness of these materials was, for practical purposes, the same regardless of test method, and constant with time greater than 1000 hours. Since most walls are constructed over times greater than 1000 hours, this conclusion leads to a simplifying approximation for reinforcement stiffness values if the objective of numerical modeling is the long-term behavior of

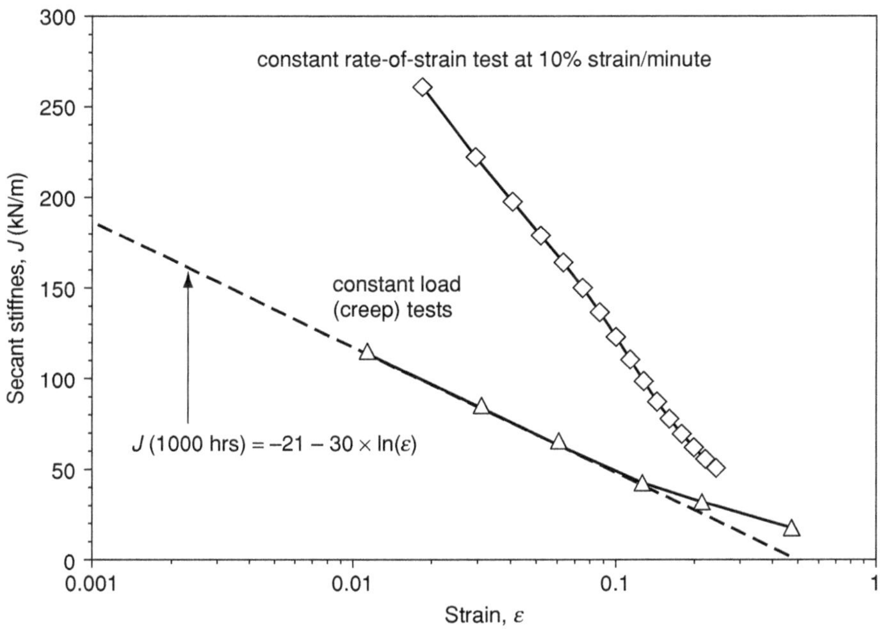

Figure 4.4 1000-hour secant stiffness values from constant load and conventional constant rate-of-strain test

these systems under operational (serviceability) conditions. The data from the constant load tests in Figure 4.3a at 1000 hours can be used to compute quickly the 1000-hour secant stiffness values as shown in Figure 4.4 using a log-linear expression of the form:

$$J(\varepsilon, t) = A(t) + B(t) \ln \varepsilon \tag{17}$$

In this example, the fitted parameters for $t = 1000$ hours and ε in the range of 1–10 percent are $A(t) = -21$ kN/m and $B(t) = -30$ kN/m. Equations 17 and 14 have been used to estimate reinforcement loads in instrumented walls as a check against the accuracy of current and proposed working stress design methods that use closed-form solutions for internal stability design of geosynthetic reinforced soil walls (e.g. Bathurst et al. 2008). If the simplified approach is used in numerical codes, the tangent stiffness value is expressed as:

$$J_t(\varepsilon, t) = A(t) + B(t)(1 + \ln \varepsilon) \tag{18}$$

It should be noted that it is difficult to measure strains less than 1 percent using conventional creep testing. The data in Figure 4.4 for $\varepsilon < 1\%$ have been extrapolated back only to 0.1%. In practice, the authors have capped the maximum stiffness value using Equation 17 to the value computed at 0.1% strain.

Also shown in the plot are the secant stiffness values computed from the results of a conventional in-isolation constant rate-of-strain test. This test was carried out at 10 percent strain per minute in accordance with the ASTM D 4595 protocol. Because the mechanical properties of this material are load-rate dependent, the material appears much stiffer. However, the rate of loading is well above rates of loading expected in the field. Using stiffness values from conventional tensile tests on polymeric materials (particularly polyolefins) will result in excessive estimates of reinforcement stiffness. If the objective of numerical modeling is design, this can lead to potentially unsafe structures.

The reinforcement layers (FLAC cable elements) were assumed to interact with the backfill soil using the FLAC grout utility. A large bond strength value was selected along the reinforcement–backfill interface to prevent slip and thus keep the interpretation of the numerical results as simple as possible. This assumption is probably sufficiently accurate for geogrid products in combination with stiff, well-compacted sand backfill (the case for the RMC walls). For other reinforcement materials (e.g. some geotextiles) and/or more cohesive backfills, provision for interface shear deformation may be warranted. Unfortunately, quantitative determination of interface stiffness values from independent laboratory tests is problematic at the present time. The value of K_b shown in Table 4.2 was selected to minimize relative shear deformation while ensuring numerical stability using a FLAC code.

Table 4.2 Interface properties from Hatami and Bathurst (2005, 2006)

Interface	*Value*
Soil–block	
δ_{sb} (friction angle) (degrees)	44
ψ_{sb} (dilation angle) (degrees)	11
K_{nsb} (normal stiffness) (MN/m/m)	100
K_{ssb} (shear stiffness) (MN/m/m)	1
Block–block	
δ_{bb} (friction angle) (degrees)	57
c_{bb} (cohesion) (kPa)	46
K_{nbb} (normal stiffness) (MN/m/m)	1000
K_{sbb} (shear stiffness) (MN/m/m)	40
Backfill–reinforcement	
ϕ_b (friction angle) (degrees)	44
s_b (adhesive strength) (kPa)	1000
K_b (shear stiffness) (kN/m/m)	1000

3.3 Interfaces and boundary conditions

The interfaces between other dissimilar materials were modeled as linear spring-slider systems with interface shear strength defined by the Mohr–Coulomb failure criterion. The values of interface stiffness (K_{sbb}) between modular blocks and failure parameters were selected to match physical test results from laboratory direct shear tests (e.g. Bathurst et al. 2008). The value K_{toe} = 4 MN/m/m was selected by Hatami and Bathurst (2005, 2006) based on measurements of horizontal load and displacement at the toe of the instrumented RMC walls (see Figure 4.1). A fixed boundary condition in the horizontal direction was assumed at the numerical grid points on the backfill far-end boundary, representing the bulkheads that were used to contain the soil at the back of the test facility. A fixed boundary condition in both horizontal and vertical directions was used at the foundation level matching the test facility concrete strong floor. Interface shear parameters are summarized in Table 4.2.

3.4 Numerical results

The accuracy of the FLAC model using the three soil constitutive models described earlier and the hyperbolic model for the reinforcement layers is examined in this section by comparing selected predictions with measurements from the RMC wall in Figure 4.1.

Figure 4.5 compares numerical and measured results for the horizontal and vertical toe load components. The physical data show that the vertical toe load is always greater than the facing self-weight. This is due to soil-wall facing interface shear transfer and vertical load transfer at the connections between the reinforcement and the facing column. The horizontal load can be seen to increase with soil height during construction (Figure 4.5a) and with increasing surcharge pressure (Figure 4.5b). All three constitutive soil models are judged to give reasonably similar results during construction. During surcharging the two models with soil dilatancy (Lade and Mohr–Coulomb) can be seen progressively to over-estimate the toe loads with increasing surcharge load level.

Figure 4.6 shows the distribution of the measured and predicted normalized vertical pressure below the wall footing and at the base of the soil backfill. The normalization has been carried out with respect to the magnitude of pressure computed using the fill height and surcharge pressure or, in the case of the footing, using the column self-weight per unit base area. The range bars for the physical data points capture small variations in pressure that were detected at the end of the construction or during constant surcharge pressure increments. All three soil constitutive soil models gave essentially the same results. The predictions captured the trend in the foundation pressure attenuation that occurred immediately behind the facing owing to the down-drag forces described above. The good agreement is

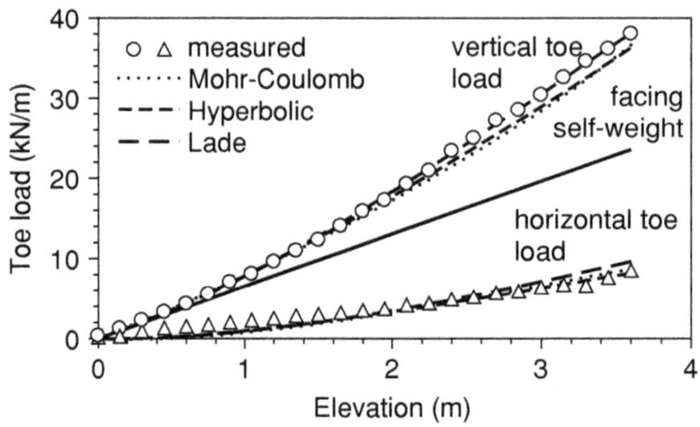

Figure 4.5a Construction

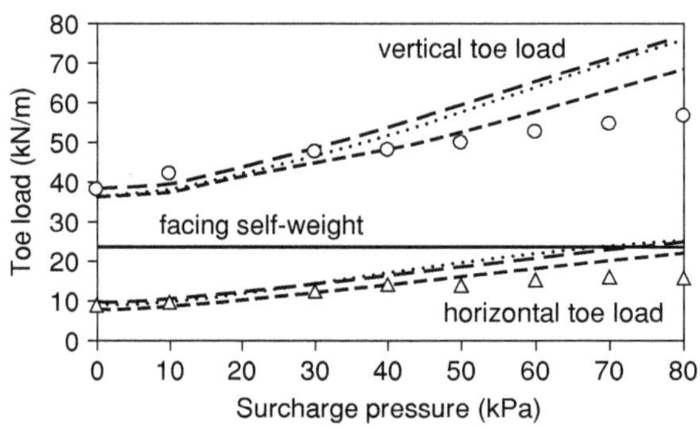

Figure 4.5b Surcharging
Figure 4.5 Toe loads at base of facing column

probably due to the fact that the numerical results are computed largely based on gravity and prescribed surcharge boundary loads, thus the computations are not affected by constitutive soil model type.

The connection loads between the reinforcement layers and the facing column were measured directly during the physical test. Because of the discrete nature of the modular block facing the measured connection, loads vary across the wall face. This can be seen by the large range of bars in Figure 4.7. The accuracy of the connection load predictions varies with model type and load level. However, it can be argued that, for practical purposes, the three methods give values that are in reasonable agreement with the measured values. It should be appreciated that the connection loads at the end of the construction are sensitive to the construction technique for dry-stacked modular block wall systems. Hence differences between measured and predicted results for the two lowermost connections at the end of the construction can also be ascribed to differences in placing and seating

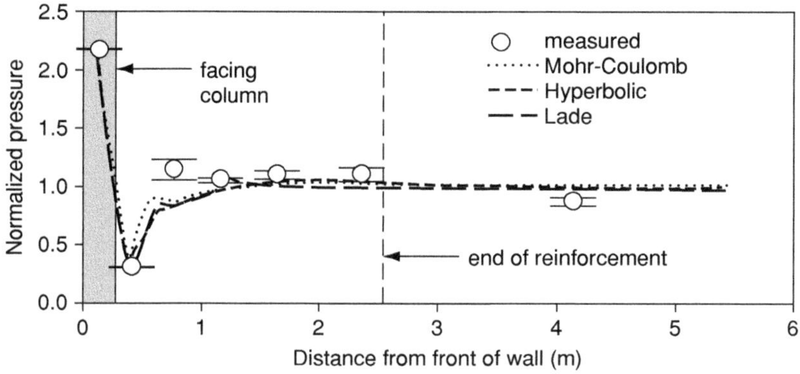

Figure 4.6a End of construction

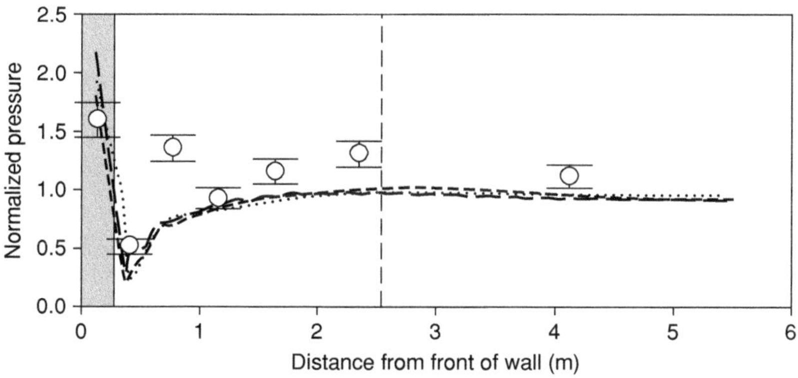

Figure 4.6b Surcharge pressure q = 50 kPa
Figure 4.6 Foundation pressures.
Note: range bars are maximum and minimum pressures

of the blocks during the initial stages of construction of the wall. At the 50 kPa surcharge level, when the effects of the initial variations in construction are masked, almost all three models predict connection loads that fall within the range bars. Slightly lower loads can be visually detected in the plots using the hyperbolic soil model.

Post-construction displacements are plotted in Figure 4.8. All three models capture the trend in post-construction displacements up to the surcharge level of the 80 kPa shown. At higher surcharge levels, strains in the reinforcement layers exceeded 3 percent, which has been recommended as a serviceability limit to prevent the onset of contiguous plasticity in the reinforced soil zone of the geosynthetic reinforced soil walls constructed with granular backfill (Allen et al. 2003, Bathurst et al. 2005).

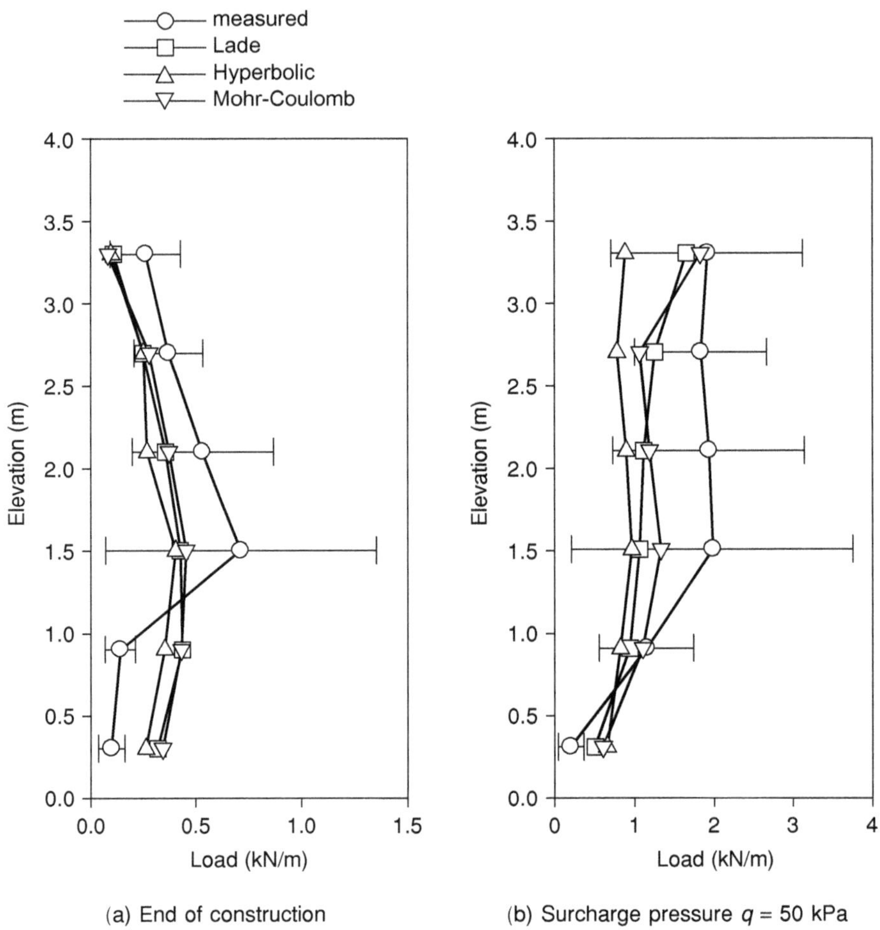

Figure 4.7 Measured and calculated connection loads.
Note: horizontal range bars show the range of connection loads measured across 1 m of wall face at each reinforcement elevation

Finally, Figures 4.9a and 4.9b show the magnitude and distribution of the reinforcement strains at the end of construction and at the 80 kPa surcharge load level respectively. The predicted values are judged to be reasonably close and in practical agreement with measured data at the end of construction, particularly when the scatter in measured values is considered. The peak strain in layer 6 close to the back of the facing during surcharging is clear from the measured strain data, and this is consistent with down-drag forces described earlier. It can be argued that Lade's model gives the best overall fit to the measured data. The hyperbolic model underestimates strain values, particularly close to the wall. There is evidence of

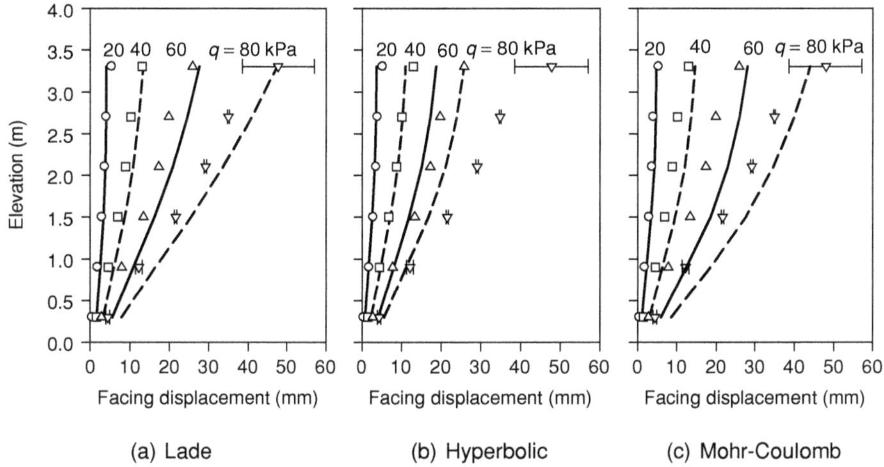

Figure 4.8 Post-construction facing displacements during surcharging
Note: range bars are maximum and minimum recorded values

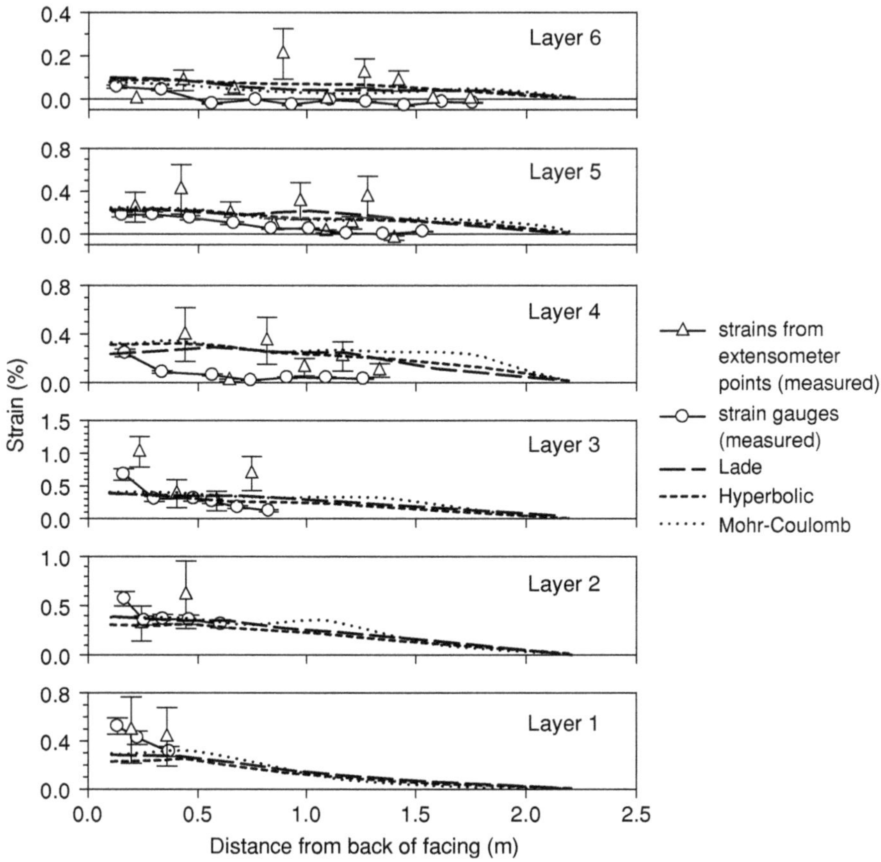

Figure 4.9a End of construction

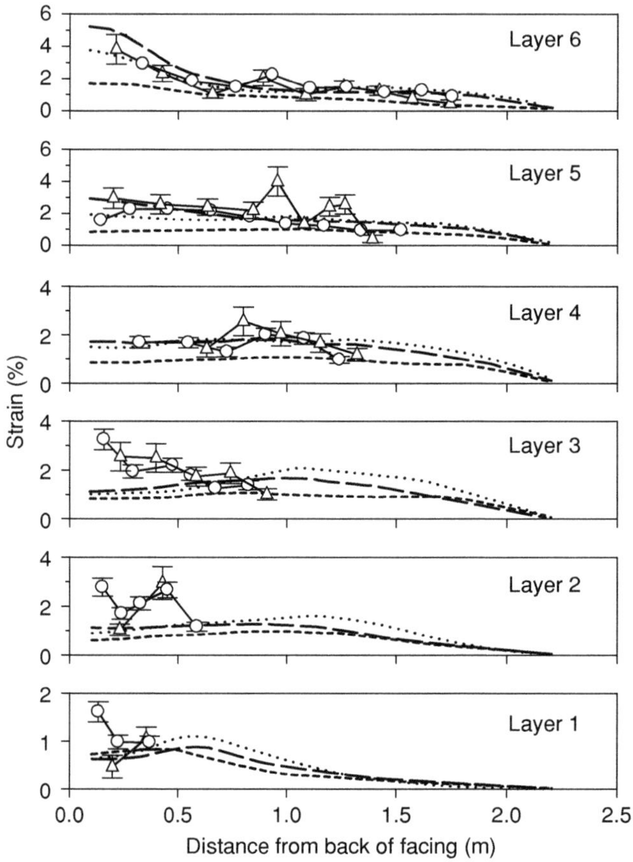

Figure 4.9b Surcharge pressure q = 80 kPa
Figure 4.9 Strain distribution in reinforcement layers

a trend toward the over-estimation of reinforcement strains well within the reinforced soil zone at the end of construction using the Mohr–Coulomb model (see layer 4, Figure 4.9a). This observation has also been noted in simulations for other walls in the RMC test series and usually leads to under-estimation of the surcharge loads levels required to generate a collapse state (Hatami and Bathurst 2005).

Based on the experience gained from the numerical modeling of the RMC walls using the three constitutive models introduced in this chapter, the hyperbolic model is judged to be sufficiently accurate for the prediction of wall response at the end of construction for typical wall heights (i.e. $H \leq 10$ m). This is the in-service condition of a geosynthetic reinforced soil wall in the field and hence can be argued to be of the most interest to design engineers. The hyperbolic model has the advantage that input parameters are reasonably easy to determine from independent laboratory tests and relatively simple to code. Under load conditions leading to greater soil plasticity, Lade's model is a better candidate. However, this model requires more parameters and is more difficult to implement in numerical codes.

4 Example parametric analyses

The utility of a FLAC numerical code using the hyperbolic model for the soil is demonstrated, in this section, to investigate the influence of toe restraint and facing block interface stiffness on wall performance. The same RMC modular blocks with an 8-degree batter are used here. The difference is that the walls are now taken to a height of $H = 6$ m with a soil similar to the RMC sand backfill but with material properties taken from Boscardin et al. (1990). A total of ten reinforcement layers at 0.6 m spacing are used in these simulations. To simplify the interpretation of the results, a typical polyester (PET) geogrid is used with $J_o(t) = 285$ kN/m and $\eta(t) = 0$ in all simulations. An artificially high reinforcement rupture value was used in the code but was only used to complete the input parameters for the model. The focus here is on reinforced soil wall performance under typical operational conditions (end of construction). Therefore, only simulation results with low reinforcement strain values at the end of construction are considered, and non-linear reinforcement load-strain behavior leading to rupture was not a possibility. The height to breadth ratio of the numerical FLAC grid (0.65) and the density of the node points in the simulations to follow were kept the same as in the RMC model that was used to check the accuracy of the general approach.

4.1 Influence of horizontal toe and block interface shear stiffness

In the simulations of the RMC wall, described earlier, a constant horizontal toe stiffness was assumed as $K_{toe} = 4$ MN/m/m. The question arises: What is the consequence of a stiffer or less restrained toe boundary condition on the wall behavior? This effect is investigated here by carrying out simulations with $K_{toe} = 40$, 4, 0.4 and 0.04 MN/m/m. Hence, toe horizontal compliance varies from a factor of 10 to 1/100 of the value used in the RMC experimental walls.

Figure 4.10a shows that, as the toe stiffness decreases, deformation at the base of the wall increases (as expected) and the wall facing profile, taken with respect to the target construction batter of 8 degrees, changes. Figure 4.10b shows that the influence of toe stiffness is sensibly restricted to the bottom third of the wall face. In practice, the toe of hard-faced geosynthetic reinforced soil walls is embedded and (or) seated on a footing, which means that the horizontal toe capacity is available.

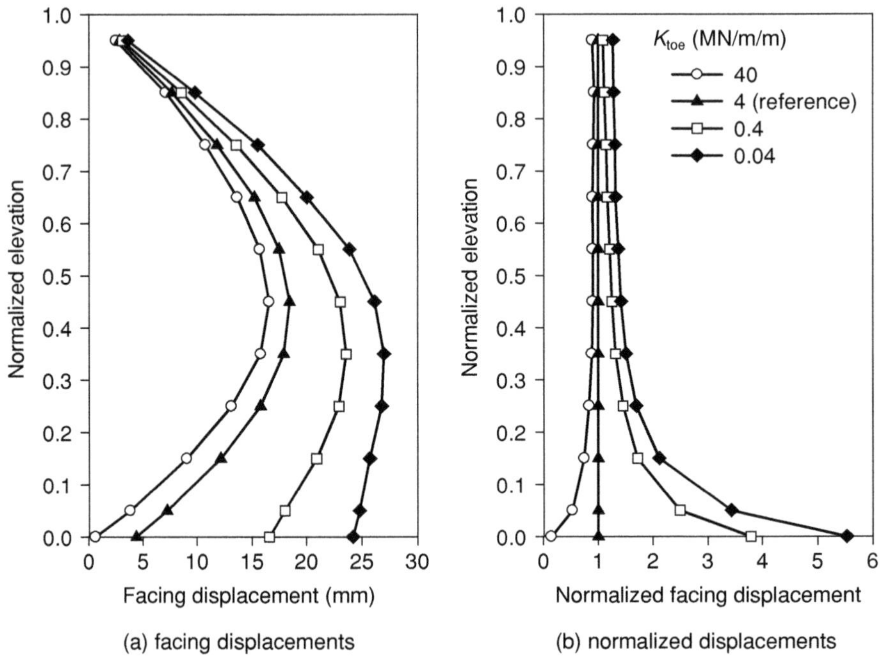

Figure 4.10 Influence of horizontal toe stiffness on wall facing displacements at end of construction (H = 6 m and PET reinforcement)

The effect of increasing wall displacements with decreasing toe stiffness can be detected in the magnitude and distributions of the reinforcement strain shown in Figure 4.11. For brevity, only every other reinforcement layer is plotted. Connection strains (loads) can be seen to increase with decreasing toe stiffness up to about one-third of the wall height. In addition, there is a progressive increase in the reinforcement strain (local peak) within the reinforced soil zone with decreasing toe stiffness. For the lowest toe stiffness case, the trace of local peaks is consistent with zones of higher soil shear strains, which could lead to an internal soil failure in a field wall if the stiffness of the reinforcement was sufficiently low.

The influence of the magnitude of block interface shear stiffness on the distribution of the total horizontal earth force between the connections and facing toe is illustrated in Figure 4.12. For the range of stiffness values investigated, the total load carried by all the connections (reinforcement layers) varies from about 70 percent to 25 percent. This has important implications on design. A stiff facing (defined by high block interface stiffness) will act as a structural component in the geosynthetic reinforced wall system to carry earth loads and thus reduce the reinforcement load demand. In current design methods (AASHTO 2002, BS8006 1995) the load contribution

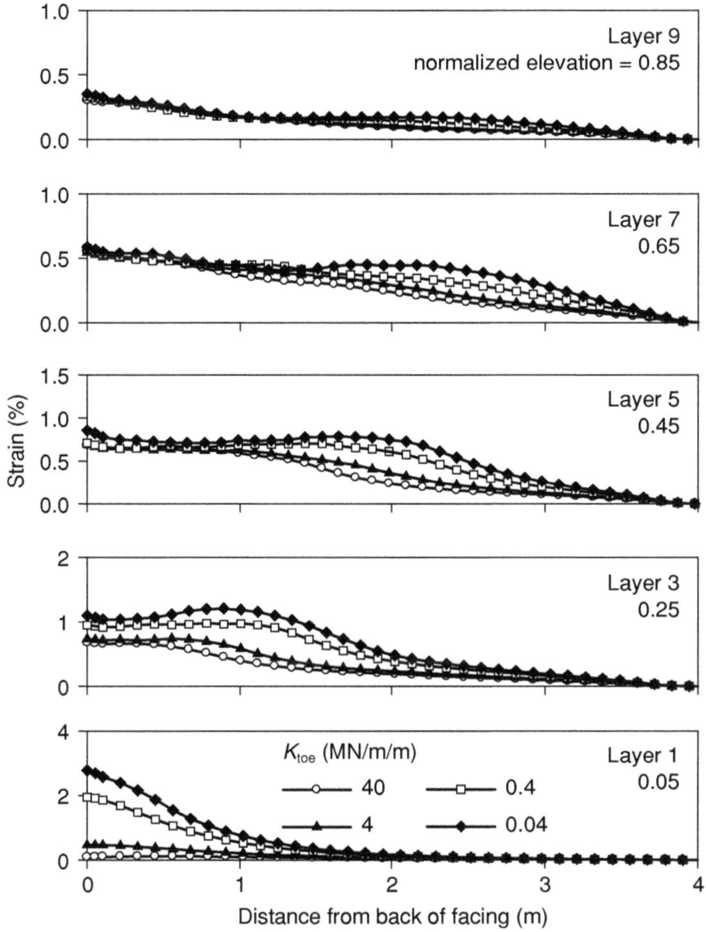

Figure 4.11 Influence of toe stiffness on magnitude and distribution of reinforcement strains (H = 6 m and PET reinforcement)

of the toe restraint and a stiff structural facing is not considered. All internal horizontal earth loads are assumed to be carried by the reinforcement layers alone. This is a source of design conservatism.

5 Conclusions

This chapter uses the example of a carefully instrumented full-scale laboratory test wall to identify numerical modeling issues associated with achieving reasonable predictions of key performance features of geosynthetic reinforced soil walls. The verified numerical approach is then used to investigate the influence of boundary compliance at the toe and facing block

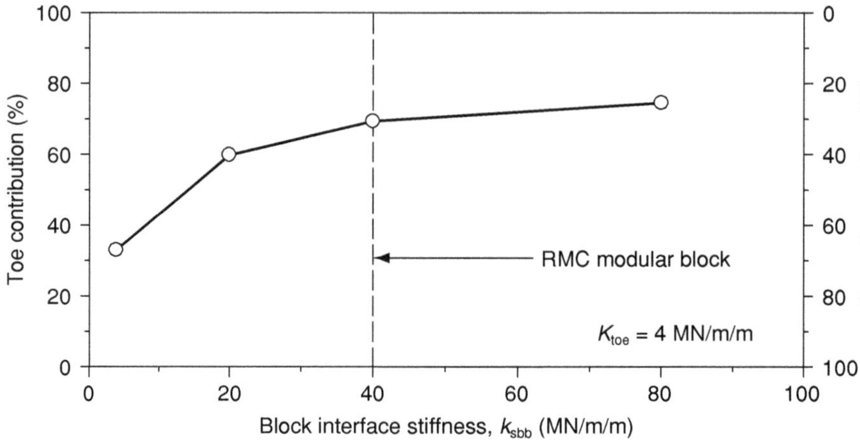

Figure 4.12 Influence of block interface stiffness on the distribution of earth loads to the horizontal toe reaction and reinforcement connections (H = 6 m and PET reinforcement)

interface shear stiffness on a similar structure of greater height. The important issues related to successful numerical modeling of geosynthetic reinforced soil walls, based on the experience of the authors, can be summarized as follows:

(1) Polymeric reinforcement products are visco-elastic-plastic materials and hence have mechanical properties that are strain level and time dependent. Constitutive models for these materials must capture these properties. Simply using a single stiffness value or strain-dependent stiffness values from the results of rapid rate-of-strain tensile tests carried out in the laboratory (e.g. ASTM D 4595, D 6637) will over-estimate the stiffness of the polyolefin geosynthetic reinforcement materials.
(2) The boundaries of the numerical model will have a major influence on the predicted displacements and reinforcement strains (or loads). Any numerical model should be verified against high-quality instrumented walls (e.g. any of the RMC full-scale structures) to ensure that all performance responses are reasonable.
(3) The magnitude of the wall displacements and reinforcement loads in a field wall is very sensitive to quality control during construction (e.g. attention to placement of facing units and reinforcement layers). Capturing the level of care taken on a project-specific structure in the field renders Class A numerical predictions problematic. Nevertheless, parametric analyses of the type demonstrated in this chapter can be used to investigate the sensitivity of wall performance to a range of property values for the structure components.

(4) Geosynthetic walls are constructed in compacted soil layers. It is important that sequential construction and compaction loading be incorporated in the numerical model and soil stiffness parameters adjusted to reflect the effect of compaction energy on soil behavior.

(5) The relatively simple hyperbolic constitutive model for sand described in this paper (or variants) can be used to predict wall performance under operational conditions (i.e. end of construction) provided that the strains in the reinforcement are low enough to preclude the generation of contiguous failure zones (plasticity) through the reinforced soil zone. In practice, for walls with typical height, reinforcement materials, reinforcement spacing and length, this condition is satisfied if maximum strains are less than about 2 percent. For walls designed for higher strains and granular backfills, a more sophisticated soil model that can capture soil dilatancy and possible softening (such as Lade's model) will be required.

References

AASHTO (2002) *Standard Specifications for Highway Bridges*, 17th edn, Washington, DC: American Association of State Highway and Transportation Officials (AASHTO).

AASHTO (2007) *LRFD Bridge Design Specifications*, 4th edn, Washington, DC: American Association of State Highway and Transportation Officials.

Allen, T. M., Bathurst, R. J., Holtz, R. D., Walters, D. L. and Lee, W. F. (2003) A new working stress method for prediction of reinforcement loads in geosynthetic walls, *Canadian Geotechnical Journal*, 40: 976–94.

ASTM D 4595, Standard test method for tensile properties of geotextiles by the wide-width strip method, West Conshohocken, Pa: ASTM International.

ASTM D 6637, Standard test method for determining tensile properties of geogrids by the single or multi-rib tensile method, West Conshohocken, Pa: ASTM International.

Bathurst, R. J., Allen, T. M. and Walters, D. L. (2005) Reinforcement loads in geosynthetic walls and the case for a new working stress design method, *Geotextiles and Geomembranes*, 23 (4): 287–322.

Bathurst, R. J., Althoff, S. and Linnenbaum, P. (2008) Influence of test method on direct shear behaviour of segmental retaining wall units. *ASTM Geotechnical Testing Journal*, 31 (2): 157–65.

Bathurst, R. J. and Hatami, K. (2001) Review of numerical modeling of geosynthetic reinforced soil walls. Invited theme paper, Computer Methods and Advances in Geomechanics, *10th International Conference of the International Association for Computer Methods and Advances in Geomechanics*, 7–12 January, Tucson, Ariz., 2: 1223–32.

Bathurst, R. J. and Kaliakin, V. N. (2005) Review of numerical models for geosynthetics in reinforcement applications. Invited issue paper, Computer Methods and Advances in Geomechanics, *11th International Conference of the International Association for Computer Methods and Advances in Geomechanics*, 19–24 June, Torino, 4: 407–16.

Bathurst, R. J., Miyata, Y., Nernheim, A. and Allen, T. M. (2008) Refinement of K-stiffness method for geosynthetic reinforced soil walls, *Geosynthetics International*, 15 (3).

Bathurst, R. J., Vlachopoulos, N., Walters, D. L., Burgess, P. G. and Allen, T. M. (2006) The influence of facing rigidity on the performance of two geosynthetic reinforced soil retaining walls, *Canadian Geotechnical Journal*, 43 (12): 1225–37.

Bathurst, R. J., Walters, D., Vlachopoulos, N., Burgess, P. and Allen, T. M. (2000) Full scale testing of geosynthetic reinforced walls. Invited keynote paper, ASCE Special Publication No. 103, *Advances in Transportation and Geoenvironmental Systems Using Geosynthetics, Proceedings of Geo-Denver 2000*, 5–8 August, Denver, Colo., pp. 201–17.

Bolton, M. D. (1986) The strength and dilatancy of sands, *Géotechnique*, 36: 65–78.

Boscardin, M. D., Selig, E. T., Lin, R. S. and Yang, G. R. (1990) Hyperbolic parameters for compacted soils, *Journal of Geotechnical Engineering,* 116 (1): 343–75.

BS8006 (1995) *Code of Practice for Strengthened/Reinforced Soil and Other Fills,* Milton Keynes: British Standards Institution.

Duncan, J. M., Byrne, P., Wong, K. S. and Mabry, P. (1980) Strength, stress–strain and bulk modulus parameters for finite element analysis of stresses and movements in soil masses. Department of Civil Engineering, University of California, Berkeley, Calif., Report No. UCB/GT/80-1.

Elias, V., Christopher, B. R. and Berg, R. R. (2001) *Mechanically Stabilized Earth Walls and Reinforced Soil Slopes – Design and Construction Guidelines*, FHWA-NHI-00-043, Washington, DC: Federal Highway Administration.

Guler, E., Hamderi, M. and Demirkan, M. M. (2007) Numerical analysis of reinforced soil retaining wall structures with cohesive and granular backfills, *Geosynthetics International*, 14 (6): 330–45.

Hatami, K. and Bathurst, R. J. (2005) Development and verification of a numerical model for the analysis of geosynthetic reinforced soil segmental walls under working stress conditions, *Canadian Geotechnical Journal*, 42 (4): 1066–85.

Hatami, K. and Bathurst, R. J. (2006) A numerical model for reinforced soil segmental walls under surcharge loading, *Journal of Geotechnical and Geoenvironmental Engineering*, 132 (6): 673–84.

Itasca Consulting Group (2005) FLAC: Fast Lagrangian Analysis of Continua, Version 5.0. Minneapolis, Minn.: Itasca Consulting Group.

Karpurapu, R. G. and Bathurst, R. J. (1995) Behaviour of geosynthetic reinforced soil retaining walls using the finite element method, *Computers and Geotechnics*, 17 (3): 279–99.

Kim, M. K. and Lade, P. V. (1988) Single hardening constitutive model for frictional materials, I. Plastic potential function, *Computers and Geotechnics*, 5: 307–24.

Kongkitkul, W. and Tatsuoka, F. (2007) A theoretical framework to analyse the behaviour of polymer geosynthetic reinforcement in temperature-accelerated creep tests, *Geosynthetics International*, 14 (1): 23–38.

Lade, P. V. and Kim, M. K. (1988a) Single hardening constitutive model for frictional materials, II. Yield criterion and plastic work contours, *Computers and Geotechnics*, 6: 13–29.

Lade, P. V. and Kim, M. K. (1988b) Single hardening constitutive model for frictional materials, III. Comparisons with experimental data, *Computers and Geotechnics*, 6: 30–47.

Lade, P. V. and Lee, K. L. (1976) Engineering properties of soils, Department of Civil Engineering, University of California, Berkeley, Calif., Report No. UCLA-ENG-7652.

Lade, P. V. and Nelson, K. P. (1987) Modelling the elastic behaviour of granular materials, *International Journal for Numerical and Analytical Methods in Geomechanics*, 11: 521–42.

Leshchinsky, D. and Vulova, C. (2001) Numerical investigation of the effects of geosynthetic spacing on failure mechanisms in MSE block walls, *Geosynthetics International*, 8 (4): 343–65.

Ling, H. I. (2003) Finite element applications to reinforced soil retaining walls – simplistic versus sophisticated analyses, *Geomechanics: Testing, Modeling, and Simulation, 1st Japan–US workshop on Testing, Modeling, and Simulation*, ed. J. A. Yamamuro and J. Koseki, ASCE Geotechnical Special Publication No. 143, 27–9 June 2003, Boston, Mass., pp. 77–94.

Ling, H. I. and Leshchinsky, D. (2003) Finite element parametric study of the behaviour of segmental block reinforced-soil retaining walls, *Geosynthetics International*, 10 (3): 77–94.

NCMA (1997) *Design Manual for Segmental Retaining Walls*, 2nd edn, ed. J. Collin, Herndon, Va: National Concrete Masonry Association.

Rowe, R. K. and Ho, S. K. (1997) Continuous panel reinforced soil walls on rigid foundations, *Journal of Geotechnical and Geoenvironmental Engineering*, 123 (10): 912–20.

Walters, D. L., Allen, T. M. and Bathurst, R. J. (2002) Conversion of geosynthetic strain to load using reinforcement stiffness, *Geosynthetics International*, 9 (5–6): 483–523.

Yoo, S. and Song, A. R. (2006) Effect of foundation yielding on performance of two-tier geosynthetic-reinforced segmental retaining walls: a numerical investigation, *Geosynthetics International*, 13 (5): 181–94.

5 Seismic analysis of pile foundations in liquefying soil

D. S. Liyanapathirana and H. G. Poulos

1 Introduction

Pile foundations are widely used to support structures over ground that has inadequate bearing capacity. In seismically active areas with saturated soils, the performance of pile foundations is a complex problem owing to the effects of the progressive buildup of pore water pressure in the soil during earthquake loading. With the increase in pore water pressure, soil stiffness and strength is degraded, and the ability of the soil deposit to support foundations rapidly decreases, leading to soil liquefaction, which is the extreme manifestation of the pore water pressure increase in saturated soils during earthquake loading. The significance of soil-liquefaction-related damage to pile foundations has been clearly demonstrated by major earthquakes, such as those at Niigata in 1964, Loma-Prieta in 1989, Kobe in 1995 and Manzanillo in 1995. Hence the prediction of pile behavior in liquefying soils under earthquake loading is an important issue in the design of pile foundations designed for seismically active regions.

Numerical procedures developed for the analysis of pile behavior in liquefying soil include uncertainties owing to the lack of understanding of the mechanisms involved in pile–soil interaction subject to earthquake loading. However, data recorded during the 1995 Hyogoken-Nambu earthquake, shaking-table tests (Ohtomo 1996, Tamura et al. 2000, Yasuda et al. 2000, Mizuno et al. 2000, Nakamura et al. 2000), and centrifuge tests (Dobry et al. 1995, Abdoun et al. 1997, Horikoshi et al. 1998, Wilson et al. 1999, Wilson et al. 2000) provide valuable information related to pile–soil interaction during pore water pressure generation and subsequent soil liquefaction.

For the analysis of pile foundations in liquefying soil, two-dimensional and three-dimensional finite-element-based numerical models have been developed by many researchers (Hamada et al. 1994, Sakajo et al. 1995, Zheng et al. 1996, Shahrour and Ousta 1998, Finn et al. 2001). Although these models provide better insights into the complex interaction between pile and soil, they are computationally complex and time-consuming. Consequently, use of these models by designers in engineering practice is, for two reasons, restricted: (i) constitutive models used to simulate soil behavior require specific data on soil properties; and (ii) the designers need a more thorough understanding of the numerical procedure and the physical phenomena associated with pile behavior in liquefying soil. Added to this, many commercial software packages used for Geotechnical Engineering applications do not have soil constitutive models that simulate pore pressure generation during cyclic loading.

In recent years, one-dimensional Winkler models based on the finite element and finite difference methods have become popular. In Winkler models, the pile is modeled as a beam, and the interaction between the pile and the soil is modeled using a non-linear spring–dashpot model. These models are computationally efficient. The response of the springs at the pile–soil interface is defined using a *p-y* curve, where *p* is the soil resistance per unit length of the pile and *y* is the pile lateral displacement. These *p-y* curves have, typically, been back-analysed using experimental or model test data. However, for piles founded in liquefying soil, available data are not sufficient to back-analyse *p-y* curves. Therefore, the development of a numerical model, not needing *p-y* curves to obtain the response of the springs used to model the pile–soil interaction, will be helpful for practicing geotechnical engineers.

In liquefying soils, soil stiffness degrades with increasing pore water pressure. Therefore, the stiffness of the springs, which represent the pile–soil interaction, should be degraded based on the amount of pore water pressure generation in the soil. Degraded soil stiffness and strength can be obtained from an effective-stress-based free-field ground response analysis. In the following sections, a free-field ground response analysis method and a one-dimensional Winkler model based on the dynamic finite element method are presented, with the aim of providing a useful numerical tool to analyze this complex problem. The non-linear spring constants for the model are derived by integrating Mindlin's equation. The effect of radiation damping during the earthquake loading is also taken into account.

Finally a pseudo-static approach is outlined, where a static one-dimensional finite element analysis is carried out to obtain the maximum pile displacement, bending moment and shear force in a liquefying soil due to earthquake loading. This method is attractive for design engineers when compared to the more complex dynamic finite element models, which require integration of a system of equations in both the space and the time domain.

2 Free-field ground response analysis

Prior to the seismic analysis of the pile, a free-field ground response analysis should be carried out to determine the ground deformation and the soil stiffness and strength degradation due to pore water pressure generation related to the earthquake loading applied to the soil.

The soil models used for effective stress-based ground response analyses can be divided into four main categories: (i) models based on plasticity theory (Prevost 1985, Pastor et al. 1990, Wang et al. 1990, Ishihara 1993, Muraleetharan et al. 1994, Fukutake and Ohtsuki 1995); (ii) stress path methods (Ishihara and Towhata 1982, Kiku and Tsujino 1996); (iii) correlations between pore pressure response and volume change tendency of dry soils (Finn et al. 1977); and (iv) direct use of experimentally observed undrained pore pressure response (Seed et al. 1976, Sherif et al. 1978, Kagawa and Kraft 1981). The main disadvantage of the first three methods is the large number of parameters required to model a problem. Therefore, a method which belongs to the fourth category is used here, as it requires fewer model parameters. The new method is based on the finite element method and takes into account the progressive buildup of pore water pressure due to earthquake loading, as well as pore water pressure dissipation due to vertical drainage within the soil deposit (Liyanapathirana and Poulos 2002).

During an earthquake, shear stress waves propagate in the vertical direction causing soil deformation in the simple shear manner as illustrated in Figure 5.1. By considering the horizontal equilibrium of the forces acting on the soil element, the equation governing the shear wave propagation in the vertical direction can be derived as follows:

$$\rho \frac{\partial^2 u}{\partial t^2} = \frac{\partial \tau}{\partial z} \tag{1}$$

where u is the horizontal ground motion, t the time, z the distance in the vertical direction and τ the shear stress developed owing to the earthquake loading. The discretization of Equation 1 in space by finite elements produces the following ordinary differential equation:

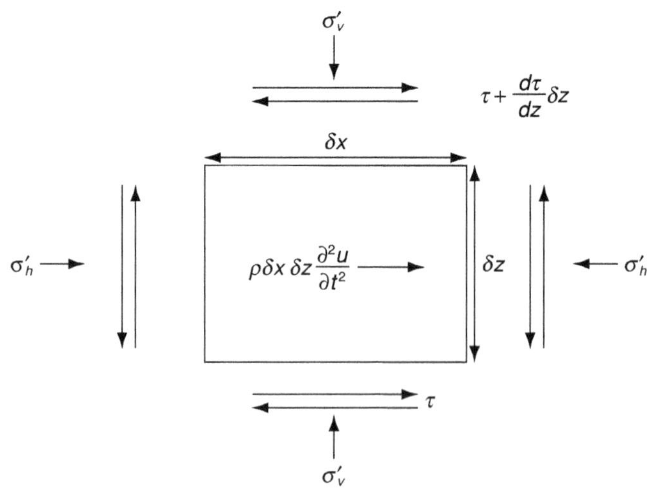

Figure 5.1 Stresses acting on a soil element during an earthquake loading

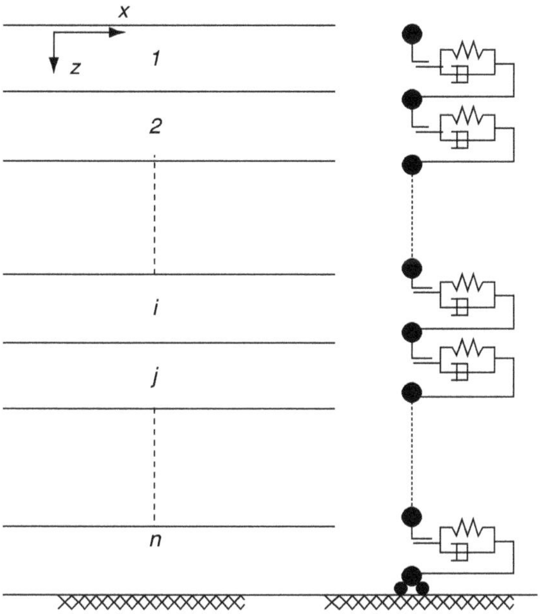

Figure 5.2 Layered soil deposit and the one-dimensional model

$$M\frac{\partial^2 u}{\partial t^2} + C\frac{\partial u}{\partial t} + Ku = F(t) \tag{2}$$

where M, C and K are respectively mass, damping and stiffness matrices. $F(t)$ is the externally applied load. The finite element analysis is carried out by dividing the soil deposit into a number of horizontal layers as shown in Figure 5.2. Uniform soil properties are assigned to each soil layer.

For the free-field ground response analysis, the input motions are applied to the soil deposit at the interface between the soil and the bedrock material through a viscous dashpot with damping $\rho_r V_r$ where ρ_r is the density and V_r the shear wave velocity of the bedrock material shown in Figure 5.1. This viscous dashpot takes into account the energy loss due to dispersion of wave energy (Joyner and Chen 1975). The time integration of the equations of motion is carried out using the constant average acceleration method.

2.1 Soil behavior

The non-linear strain-dependent hysteretic behavior of the soil is modeled using the hyperbolic stress–strain model shown in Figure 5.3. The initial loading phase OA is given by:

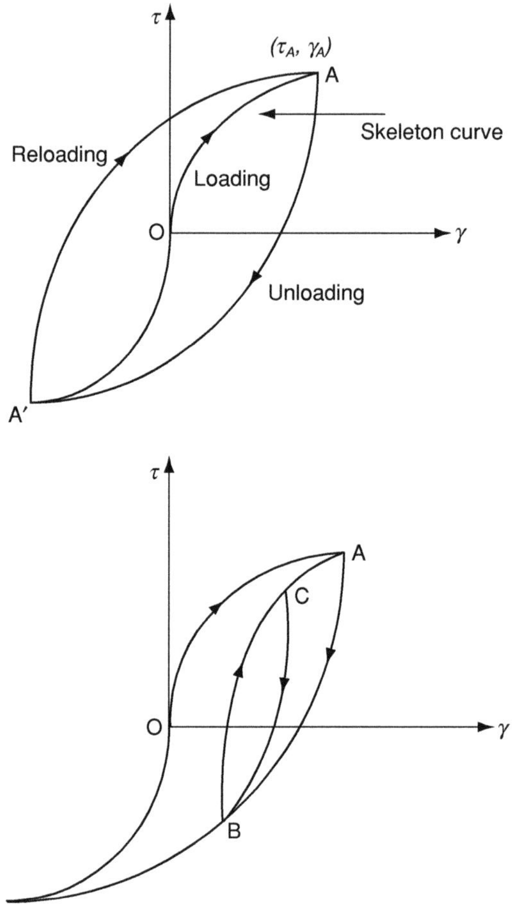

(a)

(b)

Figure 5.3 Hyperbolic stress–strain model used for soil

$$\tau = \frac{Gs_o \gamma}{\left(1 + \left|\frac{\gamma}{\gamma_r}\right|\right)} \tag{3}$$

$$\gamma_r = \frac{\tau_f}{Gs_o} \tag{4}$$

where Gs_o is the initial maximum shear modulus, τ the shear stress when the shear strain amplitude is γ and τ_f the initial maximum shear stress that can be applied to the soil without failure. The stress path AA′ for the unloading or A′A for the reloading shown in Figure 5.3a is given by:

$$\frac{\tau - \tau_f}{2} = \frac{Gs_o\left(\dfrac{\gamma - \gamma_f}{2}\right)}{1 + \left(\dfrac{1}{\gamma_r}\right)\left|\left(\dfrac{\gamma - \gamma_r}{2}\right)\right|} \tag{5}$$

where (γ_r, τ_r) is the stress reversal point (A for the unloading path and A′ for the reloading path). At point A′ the unloading path merges on to the 'skeleton curve', AOA′, and if unloading continues beyond A′ the stress path follows the skeleton curve further down. The computational procedure allows the unloading and reloading paths to follow the skeleton curve, when the magnitude of the previous shear strain exceeds the previous maximum shear strain. If stress reversal takes place before reaching point A′, as at point B shown in Figure 5.3b, the stress path follows BCA up to the next stress reversal at C. If the current loading curve CB intersects the previous loading curve AB, the stress–strain curve follows the previous loading curve ABA′ beyond point B.

The hyperbolic stress–strain relationship used here takes into account the hysteretic damping of the soil, which produces an energy loss per cycle that is frequency-independent but depends on the strain amplitude. In addition to hysteretic damping, an allowance has been made for the viscous damping of the soil, which is a fraction of the critical damping of the soil and is given by:

$$(Damp)_{\text{crit}} = 2.0\sqrt{Gs_o \rho_s} \tag{6}$$

where ρ_s is the density of the soil. Viscous damping takes into account the energy dissipation of the soil due to the visco-elastic properties of the soil. By reviewing previous data concerning equivalent viscous damping, Seed and Idriss (1967) suggested that, under the amplitude of the motions likely to occur during earthquakes, viscous damping should be around 20 percent of the critical damping.

2.2 Pore pressure generation

An effective stress-based pore pressure model is presented to simulate pore pressure generation during the cyclic loading. The new method is based on the equations developed by Seed et al. (1976), but here the equations are used in a different manner. At any time, the cyclic history of the soil can be represented by the pore pressure ratio in the soil (Seed and Idriss 1982), which is given by:

$$r_p = \frac{2}{\pi}\arcsin\left(\frac{N}{N_l}\right)^{\frac{1}{2\theta}} \tag{7}$$

where N is the number of stress cycles applied to the soil, N_l the number of stress cycles required to produce a pore pressure ratio of 100 percent and r_p the ratio of the excess pore water pressure to the initial effective overburden pressure of the soil at any depth in the soil deposit. In the original method developed by Seed et al. (1976), N_l is a constant at each depth of the layered soil deposit. Then the average shear stress ratio, taken as 0.65 τ_{max} at a particular depth, is used to calculate the number of equivalent cycles of the earthquake at each depth of the soil deposit, with τ_{max} being calculated from a total stress analysis without considering any pore pressure effects.

In the new method, different shear stress ratios are used to calculate the pore pressure increment at the end of loading or reloading. The pore pressure increment due to loading or reloading is calculated based on the maximum shear stress (τ_l) reached during the load increase over a quarter cycle. N_l is obtained from the liquefaction strength curve of the particular soil for the shear stress ratio corresponding to $\dfrac{\tau_l}{\sigma'_{v_o}}$ as shown in Figure 5.4. The cyclic history of the soil, at this point, can be obtained from the Equation 7, that is the existing pore pressure ratio of the soil is equivalent to that for N number of cycles, with magnitude τ_l. If N_l number of cycles with magnitude τ_l is applied, then the soil starts to liquefy. If the current cycle rate is $\dfrac{\partial N}{\partial t}$, then the increment can be calculated using the equation for the rate of pore pressure generation given below:

$$\frac{\partial u_g}{\partial t} = \frac{\partial u_g}{\partial N}\frac{\partial N}{\partial t} = \frac{\sigma'_{vo}}{\theta \pi N_l \sin^{2\theta-1}(\pi r_p/2)\cos(\pi r_p/2)}\frac{\partial N}{\partial t} \tag{8}$$

where $\partial u_g/\partial t$ is the rate of pore pressure generation, $\partial N/\partial t$ the rate of application of shear stress cycles to the soil, σ'_{vo} the initial effective overburden pressure, and N_l the number of uniform cycles with magnitude τ_l required

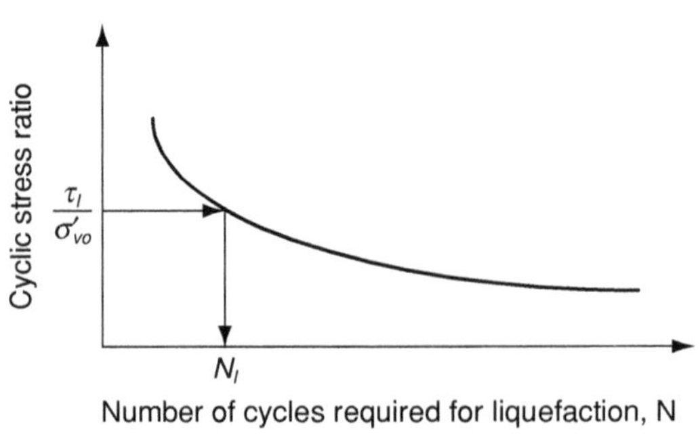

Figure 5.4 Experimental curve of the cyclic stress ratio versus number cycles

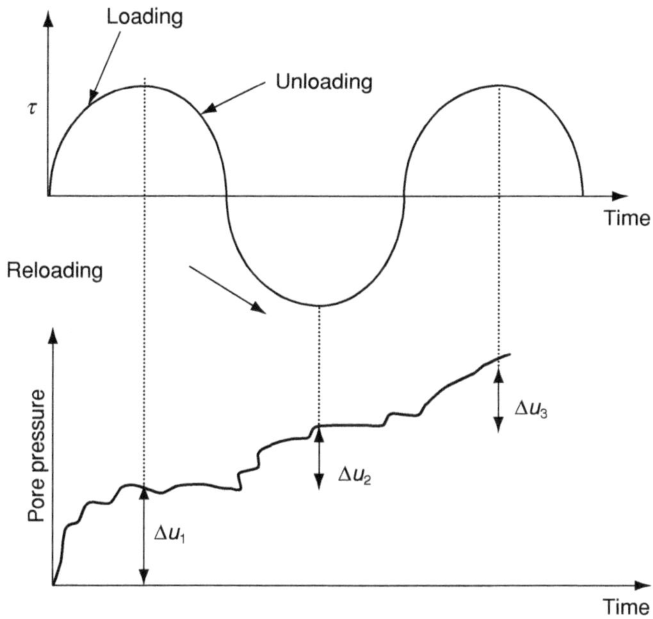

Figure 5.5 Pore pressure rise during cyclic loading

to cause soil liquefaction. In this equation, the cyclic history of the soil is represented by the pore pressure ratio, r_p. Usually the pore pressure buildup is significant only during the loading and reloading phases of each stress cycle (Ishihara and Towhata 1982) as shown in Figure 5.5. Therefore, the pore pressure generation is calculated at the end of each loading or reloading phase.

The method presented in this paper is somewhat similar to the method used when applying the pore pressure model of Seed et al. (1976) for wave loading. In that case the wave load is divided into parcels of different magnitudes, and the equivalent number of cycles is calculated for each wave parcel separately (Booker et al. 1978, Taiebet and Carter 2000).

2.3 Pore pressure dissipation

During each time step of the analysis, pore pressure dissipation due to drainage has also been taken into account using the following equation:

$$\frac{\partial u}{\partial t} = \frac{1}{m_v \gamma_w} \frac{\partial}{\partial z}\left(k \frac{\partial u}{\partial z}\right) + \left(\frac{\partial u_g}{\partial t}\right) \tag{9}$$

where k is the permeability of the soil, m_v the tangent coefficient of volume compressibility, γ_w the unit weight of water and $(\partial u/\partial z)$ the gradient of excess pore pressure in the vertical direction. At moderately high pore pressure levels, m_v is greatly influenced by the relative density of the soil. Therefore, m_v is calculated based on the expression given by Seed et al. (1976) as shown below:

$$\frac{m_v}{m_{vo}} = \frac{e^{Ar_p^B}}{1 + Ar_p^B + 0.5A^2 r_p^{2B}} \tag{10}$$

where $A = 5(1.5\text{-}D_r)$, $B = 3/2^{2Dr}$ and D_r is the relative density of the soil. During pore pressure dissipation, m_v was allowed to remain constant and equal to the maximum value reached during pore pressure buildup.

2.4 *Soil stiffness and strength degradation*

At the end of each loading and reloading phase, the soil stiffness is degraded based on the effective stress level in the soil as shown below:

$$\frac{(Gs_o)_{t=t}}{Gs_o} = \left(\frac{(\sigma'_v)_{t=t}}{\sigma'_{vo}}\right)^n \tag{11}$$

where n is the power exponent used for the effective stress term in the equation for shear modulus of the soil and σ'_v the effective stress of the soil at time t. The shear strength of the soil is also modified progressively at the end of each loading and reloading phase in a similar manner, as shown below:

$$\frac{(\tau_f)_{t=t}}{(\tau_f)_{t=o}} = \left(\frac{(\sigma'_v)_{t=t}}{\sigma'_{vo}}\right)^n \tag{12}$$

where n is the power exponent used for the effective stress term in the equation for the shear strength of the soil.

Although the effective stress of liquefied soil is virtually zero, computationally it is difficult to obtain a stable solution with a near-zero shear modulus and shear strength. By analysing field data recorded at Port Island during the 1995 Kobe earthquake, Davis and Berrill (1998) reported that the shear wave velocity of liquefied soil is about 25 m/s. Ishihara and Towhata (1982) suggested that since the shear stress application during earthquakes is multidirectional, even when shear stresses are reduced to zero in one direction, there will always be some shear stress left in the soil. The rotational simple shear tests performed by Ishihara and Yamasaki (1980) also demonstrated this phenomenon. Therefore, in the following numerical studies, a 2 percent lower limit of the initial effective vertical stress has been set for the effective stress level of the soil. Below this threshold limit,

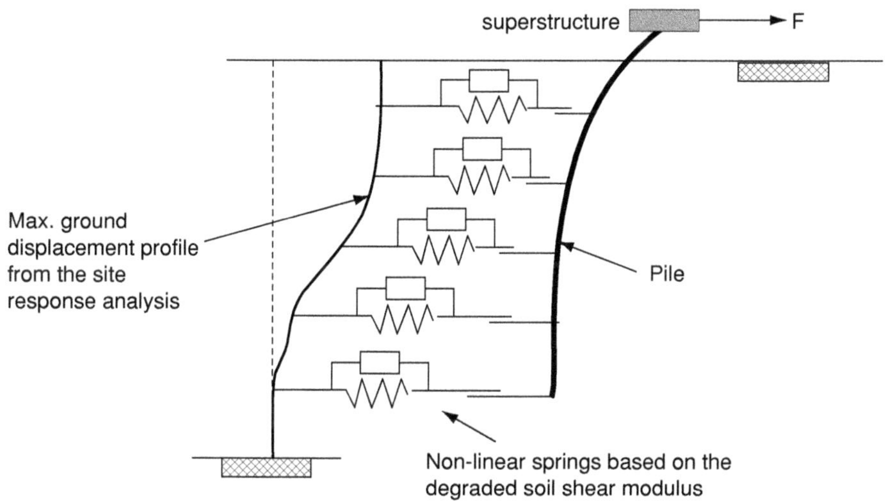

Figure 5.6 Beam on non-linear Winkler foundation model

the effective stress level of the soil is not allowed to decrease and the pore water pressures not allowed to build up.

3 Seismic analysis of piles

The beam on a Winkler foundation model used for the dynamic analysis of piles is shown in Figure 5.6. In this model, the displacement of the soil away from the pile is modeled using the displacements obtained from the ground response analysis at different time steps. In the vicinity of the pile, moving soil interacts with the pile and therefore the soil displacement is different to the displacement of the soil if there were no piles. Therefore, the interaction between the soil away from the pile and the pile is modeled using a non-linear spring–dashpot arrangement. A plastic slider has been incorporated to limit the ultimate lateral pressure at the pile–soil interface. In this model, displacement of the soil at the vicinity of the pile is represented by the movement of the plastic slider, which is different from the displacement of the soil away from the pile represented by the free-field displacements (Liyanapathirana and Poulos 2005a).

The partial differential equation for a beam on a Winkler foundation is given by:

$$E_p I_p \left(\frac{\partial^4 U_p}{\partial z^4} \right) + M_p \left(\frac{\partial^2 U}{\partial t^2} \right) = K_x (U_{ff} - U_p) + C_x \left(\frac{\partial U_{ff}}{\partial t} - \frac{\partial U_p}{\partial t} \right) \tag{13}$$

where E_p is the Young's modulus of the pile material, I_p the inertia of the pile, M_p the mass of the pile, U_p and U_{ff} are respectively pile and free-field displacements, and K_x and C_x are spring and dashpot coefficients of the Winkler model. The above equation can be solved numerically using the finite element method.

The coefficients for K_x are computed by integrating the Mindlin's equation over a rectangular area in the *y-z* plane (Douglas and Davis 1964) at the beginning of each time step. The rectangular area extends $-d/2$ to $+d/2$ in the *y* direction and c_1 to c_2 in the *z* direction, where *d* is the diameter of the pile. The displacement within the rectangular area is uniform and given by:

$$U_{x=0} = \frac{p}{G} f(c_1, c_2, d, y, z, \nu) \tag{14}$$

where *p* is the uniformly distributed load over the rectangular area, $d(c_2 - c_1)$, ν the Poisson's ratio of the soil, and *G* the shear modulus of the soil. The shear modulus varies with time, depending on the amount of pore pressure generation. The variation of *G*, with time, is obtained from the ground response analysis.

If a uniform pressure of P_j is distributed over the *j*th rectangle, then U_{ij} represents the displacement at the centroid of the *i*th rectangle due to P_j. U_{ij} and P_j can be related using an influence coefficient F_{ij} as shown below:

$$U_{ij} = P_j F_{ij} \tag{15}$$

Since the analysis is based on the finite element method, the nodal displacements are of interest. If the pile is divided into *n* elements, by adding the influence from uniform pressures $(p_i)_{i=1 \text{ to } n+1}$ acting on rectangular areas effective for each pile node the displacement at node *i* can be written as:

$$U_i = P_1 F_{i1} + P_2 F_{i2} + \cdots\cdots + P_n F_{in} + P_{n+1} F_{in+1} \tag{16}$$

This can be written in the matrix form as:

$$\{U\} = [\mathrm{F}]\{P\} \tag{17}$$

where *F* is the flexibility matrix of influence coefficients, which is calculated at $y = -d/2$, $y = 0$ and $y = +d/2$ and the average value obtained. Since the analysis is based on the finite element method, the spring coefficients matrix $[K]$ for the Winkler model is obtained by inverting the matrix of influence coefficients $[F]$.

This method takes into account the interaction between the spring coefficients along the pile when calculating displacements. Hence the resulting stiffness matrix $[K]$ which represents the pile–soil interaction is a non-symmetric matrix of size $n \times n$, where n is the number of nodes along the pile. When applying this method for seismic analysis of piles in liquefying soil, the matrix $[K]$ should be formed at the beginning of each time step because the shear modulus of the soil varies with time owing to the progressive buildup of pore water pressures.

The Mindlin equation applied in this method does not automatically take into account radiation damping. This damping is incorporated into the analysis separately, and this term takes into account the radiation damping of the shear waves travelling away from the pile. A radiation damping coefficient of $5\rho_s V_s$ has been used (Tabesh and Poulos 2000), where ρ_s is the density of soil and V_s the shear wave velocity of the soil.

Time integration of the equations of motion is performed using the constant average acceleration method similar to the free-field ground response analysis described in Section 2. The lateral pressure at the pile–soil interface, P, is limited by the equation given by Broms (1964) for sands as shown below:

$$P_y = N_p P_p \tag{18}$$

where N_p is a factor ranging between 3 and 5, and P_p is the Rankine passive pressure given by

$$P_p = \sigma_v' \tan^2\left(45 + \frac{\phi'}{2}\right) \tag{19}$$

where σ_v' is the effective vertical stress and ϕ' the friction angle of the soil. At each time step of the analysis, the lateral pressure is monitored and kept below or equal to the limiting value using an iterative procedure. When the lateral pressure $P < P_y$, the plastic slider attached to the spring–dashpot model is closed, and both the pile and the soil adjacent to the pile move together. When $P = P_y$, the soil adjacent to the pile begins to yield. Then the plastic slider is opened, and the pile and the soil have different motions. The plastic slider closes when $P < P_y$.

4 Pseudostatic approach

Recently, pseudostatic methods have emerged for the seismic analysis of pile foundations (Abghari and Chai 1995, Ishihara and Cubrinowski 1998, Tabesh and Poulos 2001). These methods do not provide the time variation of the bending moment, shear force and displacement of the pile along the depth but they can be used to obtain the maximum pile-bending moment, shear force and displacement during earthquake loading. Compared to more complex dynamic analysis methods, pseudostatic methods are attractive for design engineers.

The method proposed here involves two analysis stages as follows (Liyanapathirana and Poulos 2005b):

(1) A free-field ground response analysis is carried out to obtain, the maximum ground displacement and the minimum effective vertical stress at each depth of the soil deposit and the maximum ground surface acceleration during the earthquake loading.
(2) A static load analysis is carried out for the pile applying maximum free-field ground displacements and a static horizontal load equivalent to the cap-mass multiplied by the maximum ground surface acceleration.

Instead of the maximum ground acceleration, Abghari and Chai (1995) and Tabesh and Poulos (2001) used the spectral acceleration (Dowrick 1977) to calculate the inertial force acting on the pile head. By comparing pseudostatic results with a dynamic pile analysis, Abghari and Chai (1995) concluded that the inertial force should be reduced to 25 percent to obtain the pile displacements and to 50 percent for the bending moments. Tabesh and Poulos (2001) carried out a large number of parametric studies by comparing pseudostatic and dynamic pile analyses, and confirmed that the pseudostatic approach over-estimates the maximum bending moment and the shear force developed in the pile. However, they carried out a linear free-field ground response analysis based on the finite difference method and a pseudostatic pile analysis neglecting the non-linear behavior of the soil.

The inertial force produced by the spectral acceleration is an extremely high value compared to that obtained using the maximum ground surface acceleration, and the numerical analyses carried out during this study have confirmed that the ground surface acceleration is more suitable to calculate the inertial force acting on the pile.

The method proposed by Ishihara and Cubrinovski (1998) is developed to obtain the pile behavior in laterally spreading ground. Lateral spreading can be defined as the lateral displacement of gently sloping ground as a result of soil liquefaction. Usually, soil liquefaction occurs during the earthquake loading and lateral spreading develops subsequent to soil liquefaction. During earthquake loading piles are subjected to kinematic loads from ground displacements caused by earthquake loading and inertial loads from the superstructure. Since inertial loads are small during the lateral spreading phase, which occurs following earthquake loading, they ignored the inertial forces from the superstructure acting on the pile head (Cubrinovski et al. 2006). However, during shake-table tests (Cubrinovski et al. 2006) and centrifuge tests (Brandenberg et al. 2007) it was observed that large ground deformations occur concurrently with the high accelerations. These observations demonstrate the need to consider the combination of inertial loads and kinematic loads when predicting pile damage using simplified numerical tools.

The method presented here is a reduced version of the previous method described in Section 3 for the seismic analysis of piles, where the pile–soil interaction is modeled using the method of a beam on a non-linear

Winkler foundation. The partial differential equation for this analysis is given by:

$$E_pI_p\left(\frac{\partial^4 U_p}{\partial z^4}\right) = K_x(U_{ff} - U_p) + M(a_{\max}) \tag{20}$$

where E_p is the Young's modulus of the pile material, I_p the inertia of the pile, U_p the pile displacement, U_{ff} the free-field lateral soil displacement, M the cap-mass, K_x the spring coefficients of the Winkler model, and $a_{\max}$ the maximum ground surface acceleration. Equation 20 is solved numerically using the finite element method. The spring coefficients K_x is obtained by integrating the Mindlin's equation, as discussed in Section 3, based on the minimum vertical effective stress level obtained from the free-field ground response analysis. The 'cap-mass', M, represents the weight of the superstructure supported by the pile foundation. Although the superstructure is a multi-degree of freedom system, in the design of pile foundations it is reduced to a single degree of freedom system to simplify the analysis.

5 Analysis of centrifuge data

In this section, centrifuge tests carried out for single piles at the Rensselaer Polytechnic Institute (RPI) have been modeled using the numerical method described in the previous sections. These tests were carried out at a centrifuge acceleration of 50 g. The three-layer soil profile used for the test is shown in Figure 5.7, where all dimensions are in prototype scale.

The slightly cemented sand layers used at the top and bottom of the model are pervious but non-liquefiable, while the sand layer in the middle is liquefiable with a relative density of 40 percent. The earthquake event simulated for this centrifuge test is a sine wave with an amplitude of 0.25 g and a frequency of 2 Hz over a period of 20 seconds. Since the cemented sand layers at the top and the bottom of the model are non-liquefiable, in the numerical analysis it was assumed that there were not any pore water pressures generated in these two layers.

Variation of shear modulus of the soil along the depth of the soil deposit is assumed to be given by (Popescu and Prevost 1993):

$$G_s = G_o\left(\frac{\frac{1+2K_o}{3}\sigma'_v}{p_o}\right)^{0.7} MPa \tag{21}$$

where K_o is the coefficient of earth pressure at rest, p_o the reference normal stress and G_o the low strain shear modulus of the soil. The middle Nevada sand layer has a relative density of 40 percent, mass density (solid)

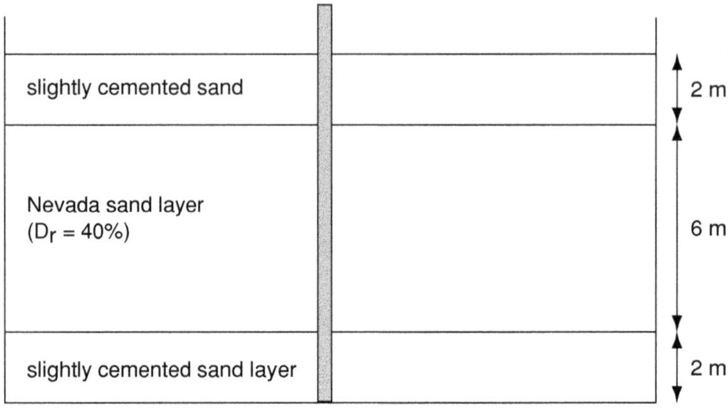

Figure 5.7 Layout of the centrifuge model

of 2670 kg/m^3, porosity of 0.424, low strain shear modulus of 25 MPa, reference mean effective normal stress of 100 kPa, friction angle of 33° and permeability of 6.6×10^{-5} m/s. The cemented sand used for the test has cohesion of 0.65 Mpa, but other properties are not available. Therefore it is assumed that the shear modulus of the cemented sand has a density of 2500 kg/m^3 and shear modulus ten times higher than that of the middle Nevada sand layer.

Figure 5.8 shows the pore pressures predicted during the test and calculated using the new model. The middle Nevada sand layer completely liquefied a few seconds after the application of the input motion. Within the period of the earthquake loading, the pore water pressures obtained from the new model agreed well with the centrifuge data, but after the shaking event the pore pressures recorded during the centrifuge test are very low. When t = 60 s, the recorded pore pressure ratio was 0.2 but the prediction from the numerical model was 0.8.

After 20 s, the pore pressure changes should occur only due to the consolidation of the soil, but the rate of pore pressure decrease recorded during the centrifuge test is too high to be attributed only to pore pressure dissipation. During the VELACS (verification of liquefaction analysis by centrifuge studies) project, a similar trend was also observed. One reason for the lower pore pressures recorded during pore pressure dissipation may be leakage of water through the walls of the strong box or instrumentation wires during the test. Popescu and Prevost (1993) suggested that the permeability of the soil should be increased by a factor of four to carry out the consolidation analysis subsequent to earthquake loading. However, more centrifuge and field test data needs to be analysed to come to a conclusion about this phenomenon.

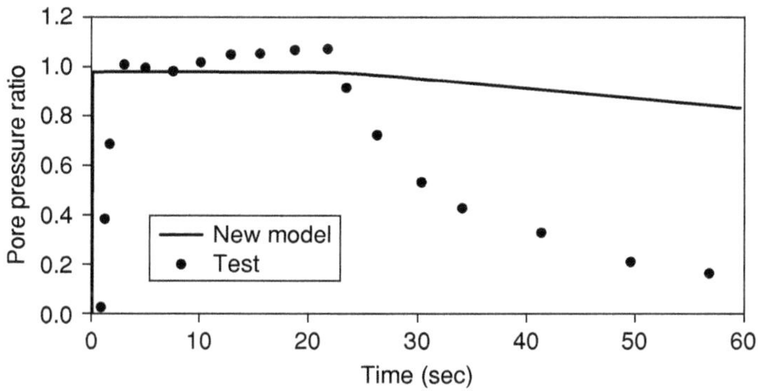

*Figure 5.8*a Pore pressure distribution at 7.8 m below the surface

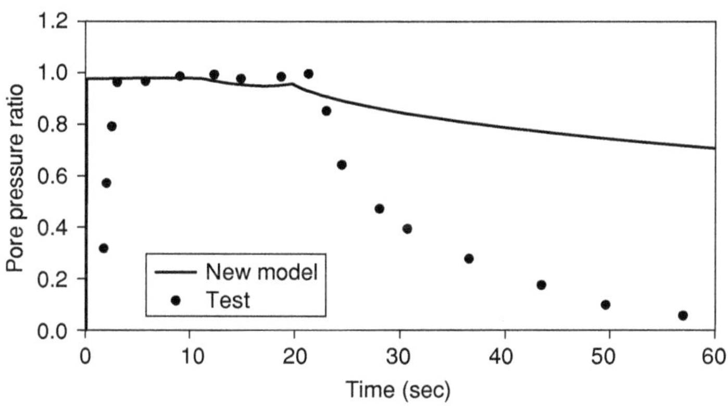

*Figure 5.8*b Pore pressure distribution at 6 m below the surface

Figure 5.9 shows the lateral displacement along the depth of the soil deposit at different times during and after the earthquake event. The soil has been displaced by 0.25 m during the first 4 seconds of earthquake loading and reached 0.8 m at t = 24 s. The displacement during the earthquake shaking period agrees well with the centrifuge data, but at t = 24 s (which is 4 s after the shaking event) the centrifuge test shows a stiffer response than that obtained from the numerical model. If the pore pressure ratios are compared at t = 24 s, the numerical model gives a pore pressure ratio of 0.9, but the pore pressure ratio recorded during the centrifuge test is 0.6.

The seismic analysis of the pile is performed using the numerical procedure described in Section 3. The pile used for the centrifuge test is 10 m long in prototype scale with a bending stiffness of 8000 kN/m^2 and a diameter of 0.47 m. For the pile analysis, ground deformations and effective stresses of the soil along the depth of the soil deposit at each time step of the analysis are recorded during the free-field ground response analysis.

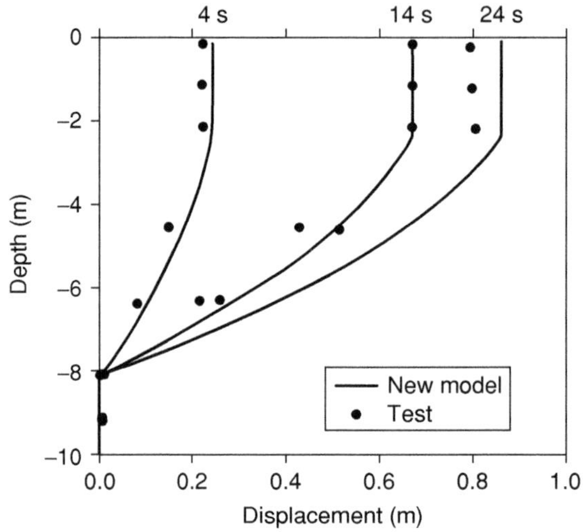

Figure 5.9 Lateral displacement along depth at different times

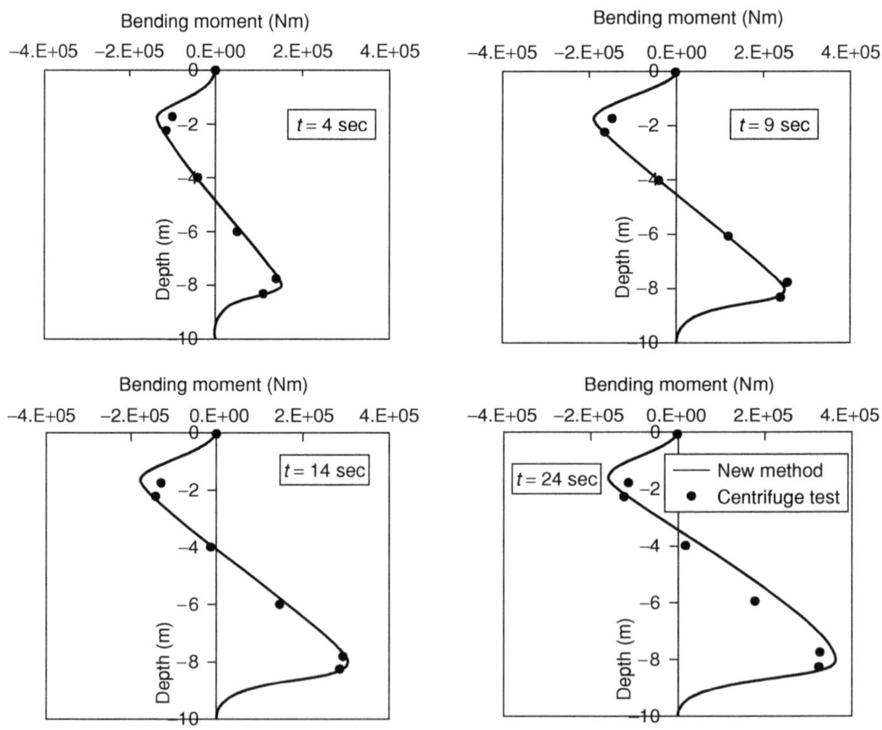

Figure 5.10 Bending moment profiles along pile depth

Figure 5.10 shows the bending moment along the pile obtained from the numerical and centrifuge tests at $t = 4$ s, 9 s, 14 s and 24 s. The bending moment becomes a maximum at the interface between the liquefying and the non-liquefying soils. Figure 5.11 shows the variation of pile-bending moment with time at different points along the pile during the shaking. These figures clearly demonstrate the ability of the numerical model to simulate pile behavior in a liquefying soil.

In this analysis, the effect of sloping ground has not been incorporated, because the slope of the centrifuge model is 2°. Analysis of piles founded in sloping ground subjected to soil liquefaction is discussed by Liyanapathirana and Poulos (2006) using the numerical modeling and centrifuge tests carried out by Brandenberg et al. (2005). In that case, the soil is subjected to an initial shear stress due to the self-weight of the soil before the application of the earthquake load.

Finally, the proposed pseudostatic method has been used to predict the maximum bending moment developed in the pile. Since this pile does not carry a cap-mass, there will not be any inertial force applied at the pile head. The pile analysis has been carried out by applying the maximum ground deformations along the depth of the pile obtained from the free-field ground response analysis. In this case, the pile–soil interaction is represented only by springs and plastic sliders. The spring coefficients for the stiffness matrix and capacities of plastic sliders are calculated using the minimum effective stress level during the seismic event recorded at each depth of the soil deposit. Figure 5.12 shows the maximum bending moment profile along the pile obtained from the pseudostatic approach and clearly demonstrates the ability of the pseudostatic approach in predicting the maximum bending moment during the seismic event.

6 Conclusions

A dynamic effective-stress-based free-field ground response analysis method and a numerical procedure for the seismic analysis of pile foundations in liquefying soil have been outlined. The history of soil stiffness and strength degradation due to the net effect of pore water pressure generation and dissipation has been incorporated into the pile analysis, and these values are obtained from the free-field ground response analysis. The ground deformations recorded during the free-field ground response analysis have been applied to the pile dynamically to obtain the variation of pile deformation, bending moment and shear force with time. A pseudostatic analysis method has also been presented to obtain the maximum pile deformation, bending moment and shear force due to the earthquake event, where a non-linear static load analysis is carried out for the pile. Centrifuge models have been simulated using the proposed numerical model, and the pile behavior computed using the dynamic Winkler model, and the pseudostatic method generally agrees well with the centrifuge data.

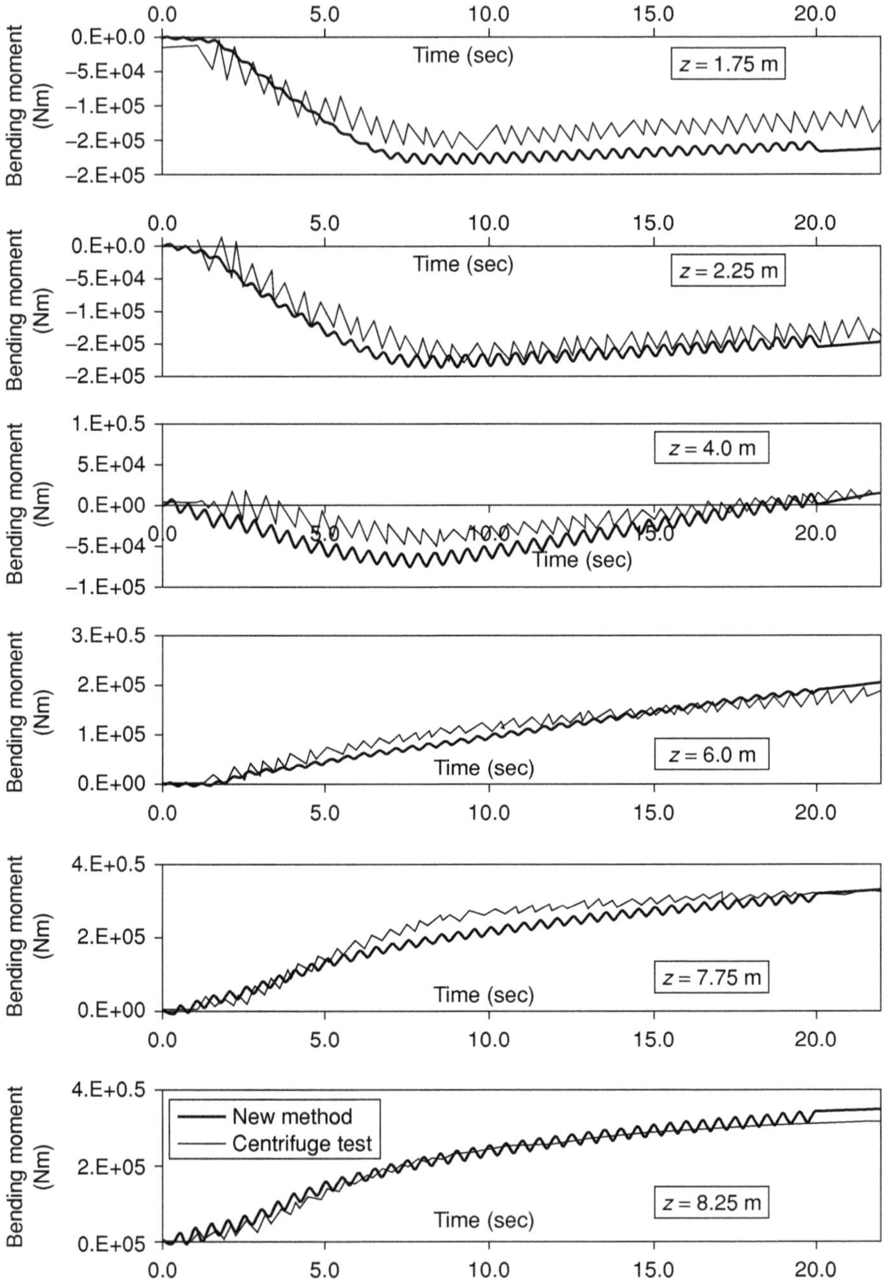

Figure 5.11 Time histories of bending moments along pile

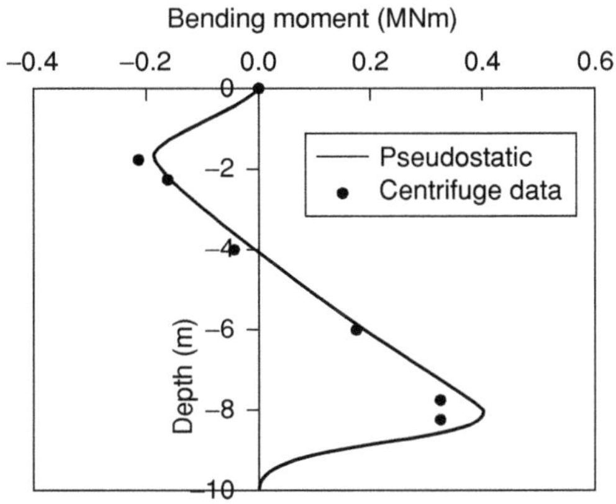

Figure 5.12 Maximum bending moment along the pile

References

Abdoun, T., Dobry, R. and O'Rouke, T. D. (1997) Centrifuge and numerical modelling of soil–pile interaction during earthquake induced soil liquefaction and lateral spreading, *Observation and Modelling in Numerical Analysis and Model Tests in Dynamic Soil-Structure Interaction Problems – Proceedings of Sessions Held in Conjunction with Geo-Logan '97*, Logan, Utah, pp. 76–90.

Abghari, A. and Chai, J. (1995) Modelling of soil–pile–superstructure interaction for bridge foundations, *Proceedings of Performance of Deep Foundations under Seismic Loading*, ed. J. P. Turner, New York: ASCE, pp. 45–9.

Booker, J. R., Rahman, M. S. and Seed, H. B. (1978) GADFLEA: a computer program for the analysis of pore pressure generation and dissipation during cyclic or earthquake loading, Report No. EERC 76-24, College of Engineering, University of California, Berkeley, Calif.

Brandenberg, S. J., Boulanger, R. W., Kutter, B. L. and Chang, D. (2005) Behaviour of pile foundations in laterally spreading ground during centrifuge tests, *Journal of Geotechnical and Geoenvironmental Engineering*, ASCE, 131 (11): 1378–91.

Brandenberg, S. J., Boulanger, R. W., Kutter, B. L. and Chang, D. (2007) Liquefaction-induced softening of load transfer between pile groups and laterally spreading crust, *Journal of Geotechnical and Geoenvironmental Engineering*, ASCE, 133 (1): 91–103.

Broms, B. B. (1964) Lateral resistance of piles in cohesionless soils, *Journal of Soil Mechanics and Foundations Division*, ASCE, 90 (SM3): 123–56.

Cubrinovski, M., Ishihara, K. and Poulos, H. G. (2006) Pseudo-static analysis of piles subjected to lateral spreading. Geotechnical earthquake engineering workshop, University of Canterbury, New Zealand.

Davis, R. O. and Berrill, J. B. (1998) Energy dissipation and liquefaction at Port Island, Kobe, *Bulletin of New Zealand National Society for Earthquake Engineering*, Waikanae, 31: 31–50.

Dobry, R., Taboada, V. and Liu, L. (1995) Centrifuge modelling of liquefaction effects during earthquakes, *Proceedings of the First Conference on Earthquake Geotechnical Engineering*, Tokyo, pp. 1291–1324.

Douglas, D. J. and Davis, E. H. (1964) The movement of buried footings due to moment and horizontal load and the movement of anchor plates, *Géotechnique*, 14: 115–32.

Dowrick, D. J. (1977) *Earthquake Resistant Design: A Manual for Engineers and Architects*, New York: John Wiley.

Finn, W. D. L., Lee, K. W. and Martin, G. R. (1977) An effective stress model for liquefaction, *Journal of the Geotechnical Engineering Division*, ASCE, 103 (GT6): 517–33.

Finn, W. D. L., Thavaraj, T. and Fujita, N. (2001) Piles in liquefiable soils: seismic analysis and design issues, *Proceedings of the 10th International Conference on Soil Dynamics and Earthquake Engineering*, Philadelphia, Pa, p. 48.

Fukutake, K. and Ohtsuki, A. (1995) Three-dimensional liquefaction analysis of partially improved ground, *Proceedings, First International Conference on Earthquake Geotechnical Engineering*, Japan, pp. 851–6.

Hamada, M., Sato, H. and Nakamura, T. (1994) An experimental and numerical study on liquefaction-induced ground displacement, *Proceedings of the 5th US National Conference on Earthquake Engineering*, 4: 169–78.

Horikoshi, K., Tateishi, A. and Fujiwara, T. (1998) Centrifuge modelling of a single pile subjected to liquefaction-induced lateral spreading, *Soils and Foundations, Special Issue No. 2*, Tokyo, pp. 193–208.

Ishihara, K. (1993) Liquefaction and flow failure during earthquakes, *Geotechnique*, 43 (3): 351–415.

Ishihara, K. and Cubrinowski, M. (1998) Performance of large-diameter piles subjected to lateral spreading of liquefied deposits, *Thirteenth Southeast Asian Geotechnical Conference*, Taipei.

Ishihara, K. and Towhata, I. (1982) Dynamic response analysis of level ground based on the effective stress method, *Soil Mechanics – Transient and Cyclic Loads*, ed. G. N. Pande and O. C. Zienkiewicz, New York: John Wiley, pp. 133–72.

Ishihara, K. and Yamasaki, F. (1980) Cyclic simple shear tests on saturated sand in multi-directional loading, *Soils and Foundations*, 20 (1): 45–59.

Joyner, W. B. and Chen, A. T. F. (1975) Calculation of nonlinear ground response in earthquakes, *Bulletin of Seismological Society of America*, 65 (5): 1315–36.

Kagawa, T. and Kraft, L. M. (1981) Modelling the Liquefaction Process, *Journal of the Geotechnical Engineering Division*, ASCE, 107 (GT12): 1593–1607.

Kiku, H. and Tsujino, S. (1996) Post liquefaction characteristic of sand, *Proceedings, 11th World Conference on Earthquake Engineering*, Paper No. 1088.

Liyanapathirana, D. S. and Poulos, H. G. (2002) A numerical model for dynamic soil liquefaction analysis, *Soil Dynamics and Earthquake Engineering*, 22: 1007–15.

Liyanapathirana, D. S. and Poulos, H. G. (2005a) Seismic lateral response of piles in liquefying soil, *Journal of Geotechnical and Geoenvironmental Engineering*, ASCE, 131 (12): 1466–79.

Liyanapathirana, D. S. and Poulos, H. G. (2005b) Pseudostatic approach for seismic analysis of piles in liquefying soil, *Journal of Geotechnical and Geoenvironmental Engineering*, ASCE, 131 (12): 1480–7.

Liyanapathirana, D. S. and Poulos, H. G. (2006) Behaviour of pile groups in liquefying soil, *Proceedings of the GeoCongress 2006: Geotechnical Engineering in the Information Technology Age*, Atlanta, Ga, pp. 1–6.

Mizuno, H., Sugimoto, M., Mori, T., Iiba, M. and Hirade, T. (2000) Dynamic behaviour of pile foundation in liquefaction process – shaking table tests utilising big shear box, *Proceedings of the 12th World Conference on Earthquake Engineering,* Auckland, Paper No. 1883.

Muraleetharan, K. K., Mish, K. D. and Arulanandan, K. (1994) A fully-coupled nonlinear dynamic analysis procedure and its verification using centrifuge test results, *International Journal of Numerical and Analytical Methods in Geomechanics*, 18: 305–24.

Nakamura, T., Sugano, T., Oikawa, K. and Mito, M. (2000) An experimental study on the pier damaged by 1995 Hyogoken-Nanbu earthquake, *Proceedings of the 12th World Conference on Earthquake Engineering*, Auckland, Paper No. 1038.

Ohtomo, K. (1996) Effects of liquefaction induced lateral flow on a conduit with supporting piles, *Proceedings of the 11th World Conference on Earthquake Engineering*, Paper No. 386.

Pastor, M., Zienkiewicz, O. C. and Chan, A. H. C. (1990) Generalised plasticity and the modelling of soil behaviour, *International Journal of Numerical and Analytical Methods in Geomechanics*, 14: 151–90.

Popescu, R. and Prevost, J. H. (1993) Centrifuge validation of a numerical model for dynamic soil liquefaction, *Soil Dynamics and Earthquake Engineering*, 12: 73–90.

Prevost, J. H. (1985) A simple plasticity theory for frictional cohesionless soils, *Soil Dynamics and Earthquake Engineering*, 4: 9–17.

Sakajo, S., Chai, J. C., Nakajima, K. and Maeda, M. (1995) Effect of group pile on liquefaction resistance of sandy ground, *Proceedings of the First Conference on Earthquake Geotechnical Engineering, Tokyo*, pp. 755–60.

Seed, H. B. and Idriss, I. M. (1967) Analysis of soil liquefaction: Niigata earthquake, *Journal of Soil Mechanics and Foundation Division*, ASCE, 93 (SM3): 83–108.

Seed, H. B. and Idriss, I. M. (1982) On the importance of dissipation effects in evaluating pore pressure changes due to cyclic loading, *Soil Mechanics – Transient and Cyclic Loads*, ed. G. N. Pande and O. C. Zienkiewicz, New York: John Wiley.

Seed, H. B., Martin, P. P. and Lysmer, J. (1976) Pore water pressure changes during soil liquefaction, *Journal of the Geotech. Eng. Division*, ASCE, 102 (GT4): 323–46.

Shahrour, I. and Ousta, R. (1998) Numerical analysis of the behaviour of piles in saturated soils under seismic loading, *Proceedings of the 11th European Conference on Earthquake Engineering.*

Sherif, M. A., Ishibashi, I. and Tsuchiya, C. (1978) Pore pressure prediction during earthquake loadings, *Soils and Foundations*, JSSMFE, 18 (4): 19–30.

Tabesh, A. and Poulos, H. G. (2000) A simple method for the seismic analysis of piles and its comparison with the results of centrifuge tests, *Proceedings of the 12th World Conference on Earthquake Engineering, Auckland*, Paper No. 1203.

Tabesh, A. and Poulos, H. G. (2001) Pseudostatic approach for seismic analysis of single piles, *Journal of Geotechnical and Geoenvironmental Engineering*, ASCE, 127 (9): 757–65.

Taiebet, H. and Carter, J. P. (2000) A semi-empirical method for the liquefaction analysis of offshore foundations, *Int. J. Num. Anal. Methods in Geomechanics*, 24 (13): 991–1010.

Tamura, S., Suzuki, Y., Tsuchiya, T., Fujii, S. and Kagawa, T. (2000) Dynamic response and failure mechanisms of a pile foundation during soil liquefaction by shaking table test with a large scale laminar shear box, *Proceedings of the 12th World Conference on Earthquake Engineering, Auckland*, Paper No. 0903.

Wang, Z. L., Dafalias, Y. F. and Shen, E. K. (1990) Bounding surface hypoplasticity model for sand, *Journal of Engineering Mechanics*, ASCE, 116 (5): 983–1001.

Wilson, D. W., Boulanger, R. W. and Kutter, B. L. (1999) Lateral resistance of piles in liquefying sand, Geotechnical Special Publication No. 88, *Analysis, Design, Construction and Testing of Deep Foundations*, pp. 165–79.

Wilson, D. W., Boulanger, R. W. and Kutter, B. L. (2000) Observed seismic lateral resistance of liquefying sand, *Journal of Geotechnical and Geoenvironmental Engineering*, ASCE, 126 (10): 898–906.

Yasuda, S., Ishihara, K., Morimoto, I., Orense, R., Ikeda, M. and Tamura, S. (2000) Large-scale shaking table tests on pile foundations in liquefied ground, *Proceedings of the 12th World Conference on Earthquake Engineering, Auckland*, Paper No. 1474.

Zheng, J., Susuki, K. and Ohbo, N. (1996) Evaluation of sheet pile-ring countermeasure against liquefaction for oil tank site, *Soil Dynamics and Earthquake Engineering*, 15 (6): 369–79.

6 The effect of negative skin friction on piles and pile groups

C. J. Lee, C. W. W. Ng and S. S. Jeong

1 Introduction

Piles are often designed to resist axial loads acting on the pile head through the development of positive shaft resistance (PSR) and end-bearing resistance. In addition to the pile head loading condition, shear stresses are mobilized at the pile–soil interface when the soil settles more than the pile in consolidating ground owing to the development of relative movement between the soil and the pile. A proportion of the weight of the surrounding soil is transferred to the pile causing additional compressive load (dragload) on the pile shaft and pile settlement (downdrag). In this situation, the mobilized shear stresses act downward and are called negative skin friction (NSF). Although the basic mechanism and the solution to the NSF problem are well established, there are, among engineers, debates regarding various crucial aspects of the NSF phenomena, which can lead to confusion and misunderstanding (Lee 2001). Hence the basic terms related to the NSF problem are adopted according to the recent definitions proposed by Fellenius (1999):

(1) *Negative skin friction (NSF)*: soil resistance acting downward along the pile shaft as a result of a downdrag and inducing compression in the pile
(2) *Downdrag*: downward movement on a deep foundation unit due to NSF and expressed in terms of settlements
(3) *Dragload*: load transferred to a deep foundation unit from NSF
(4) *Neutral plane (NP)*: location where equilibrium exists between the sum of the downward-acting permanent loads applied to the pile and dragload due to negative skin friction and the sum of upward-acting positive shaft resistance and mobilized toe resistance. The neutral plane is also where the relative movement between the pile and the soil is zero.

There have been a number of investigations into the behavior of isolated single piles and piles in groups subject to NSF, including experimental observations and theoretical studies (Shibata et al. 1982, Jeong et al. 1997,

Maugeri et al. 1997, Lee et al. 2002, Jeong et al. 2004, Lee and Ng 2004, Comodromos and Bareka 2005, Hanna and Sharif 2006, Lee et al. 2006). Various causes have been reported, mainly related to the increase in effective vertical stress in the soil (Phamvan 1989, Little 1994, Lee et al. 1998, Fellenius 2006). The development of dragload and downdrag could cause difficulty in the construction and the maintenance of the structure supported.

However, the majority of current design approaches are based on simplified methods and are thus not satisfactory. Dragload predictions for a single test pile presented by various researchers at the Wroth Memorial Symposium varied within a range of 98 percent to 515 percent of the measured value (Little and Ibrahim 1993). It is also known that less dragload develops on piles within a group, owing to pile–soil interaction, but exaggerated dragloads in groups are predicted using current design methods, especially for central piles (Terzaghi and Peck 1948, Combarieu 1985, Briaud et al. 1991, Jeong 1992). However, lesser group effects have been reported from previous experimental measurements (Koerner and Mukhopadhyay 1972, Denman et al. 1977, Shibata et al. 1982, Little 1994, Thomas 1998), although reliable measurements are rather limited.

Contrary to common design approaches, where elastic analysis is thought to present approximate estimations of pile behavior, soil slip is very likely to develop at the pile–soil interface, owing to large soil movements for piles in consolidating ground. Nishi and Esashi (1982) and Phamvan (1989) pointed out the significance of the consideration of soil slip at the interface for single piles. Kuwabara and Poulos (1989), Chow et al. (1990), Teh and Wong (1995) and Chow et al. (1996) also reported that the effect of soil slip at the pile–soil interface is a key factor affecting pile group behavior. Lee et al. (2002) and Lee and Ng (2004) indicated that elastic solutions generally overpredict dragloads and downdrag for single piles and the shielding effect in pile groups.

Although the study of dragload on piles has been investigated and well documented, the investigation of downdrag has received less attention. Failure of the foundations in terms of serviceability criteria due to downdrag is not uncommon in practice (Davisson 1993). There are limited settlement measurements from field monitoring (Bjerrum et al. 1969, Endo et al. 1969, Lambe et al. 1974, Okabe 1977, Fellenius 2006), even though several unserviceable pile foundations caused by downdrag were reported (Chellis 1961, Lambe et al. 1974, Inoue et al. 1977, Davisson 1993, Acar et al. 1994, Chan 1996). Recently it was proposed by Poulos (1997) and Fellenius (1997) that downdrag should be included in the design of piles in consolidating soil.

Several experimental measurements have shown that the application of axial loads on the pile head reduced dragload since the relative displacement between the pile and the soil changes (Fellenius 1972, Okabe 1977, Bozuzuk 1981, Leung et al. 2004, Chan et al. 2006). Furthermore, Wong (1991) and Indraratna et al. (1992) reported a reduction in dragload with axial load from numerical analyses due to the change of the positions of the neutral plane. However, Lee (2001) reported that the effects of axial loading on dragload changes remain ambiguous among engineers (Alonso et al. 1984). There has not been any relevant work concerning the effects of axial loading on dragload changes for piles in groups.

To date, the behavior of piles in groups connected to a cap in consolidating soil has attracted less attention among engineers than the study of isolated piles in groups. There are only a limited number of relevant case histories on the behavior of piles connected to a cap (Chellis 1961, Garlanger and Lambe 1973). These authors have reported the development of tensile forces in the outer piles, leading, in some cases, to pulling the outer piles from the pile cap. Kuwabara and Poulos (1989) and Chow et al. (1996) reported distributions of axial forces along piles obtained from elastic solutions using Mindlin's elastic theory. They showed the development of tensile forces in outer piles near the pile head.

In this chapter, the results of two-dimensional (2D) and 3D Finite Element Analysis (FEA), incorporating the effects of soil slip, are presented. The numerical parametric analyses include no-slip elastic analyses and elasto-plastic slip analyses for single piles and pile groups. The results are compared to elastic solutions. In the no-slip elastic analysis, the nodal compatibility relationship is always satisfied. Parametric studies are presented examining the major factors affecting soil slip behavior and soil–pile interaction. Finally, the results from the FEA are compared to field observations and example case histories. In particular, several issues mentioned above, such as dragload and downdrag, piles connected to a cap in a group and the effect of axial loading on pile behavior have been studied.

2 Finite element modelling

2.1 Finite element mesh and boundary conditions

The behavior of single piles and pile groups was investigated by carrying out a 2D axisymmetric and a 3D numerical analysis respectively. The finite element package ABAQUS (1998) was used for calculation, and the program PATRAN used for mesh generation. Although a 3D problem can often be simplified to a 2D model, it is not possible for this research as the effects of the relative location of piles within a group cannot be modeled properly. Figure 6.1 shows a representative 2D axisymmetric mesh and a representative 3D finite element mesh used in the numerical analyses respectively. In addition, Figure 6.2 shows a 3D mesh used in the analysis of piles with a cap. The 0.5 m thick pile cap was located 0.5 m above the top of the soil surface. The piles were assumed to rest at the boundary between the clay and the bearing layer (i.e. pile length is equal to the thickness of the clay). Various sensitivity studies were carried out to design the most appropriate finite element mesh for a 3D analysis. A

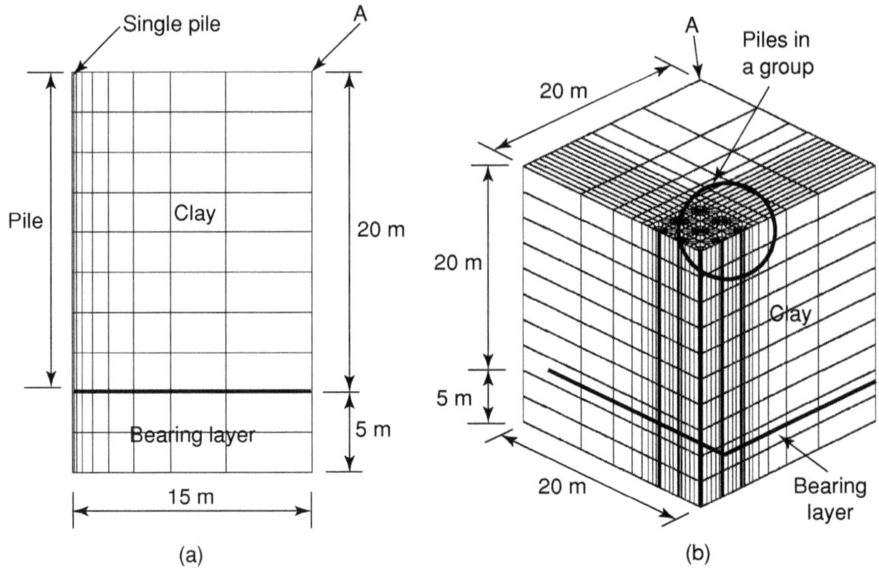

Figure 6.1 Representative FE meshes used in (*a*) 2D (axisymmetric) and (*b*) 3D (5 × 5 piles at 2.5D spacing) analyses

single pile and 5 × 5 piles in a group with pile length L = 20 m, pile diameter D = 0.5 m and center-to-center pile spacing 2.5D are illustrated in Figures 6.1 and 6.2. Owing to the plane of symmetry, only a quarter of the whole mesh was used in the 3D analysis. A fine mesh was used near the pile–soil interface as large shear strain variations were expected there. The mesh became coarser further away from the pile. Eight noded second-order quadrilateral elements and 27 noded second-order brick elements were used for the 2D and 3D analyses respectively. The bottom of the mesh was pinned, and lateral boundaries were supported by rollers. The positions of the piles in the groups are shown in Figure 6.3.

2.2 *Types of analyses*

No-slip elastic analyses (called no-slip analysis) and elasto-plastic slip analyses (called slip analysis) were carried out. In the no-slip analyses, perfect bonding was assumed between the soil elements next to the pile and the pile elements, neglecting soil slip at the interface. Different pile lengths and pile group configurations were considered in the parametric numerical analyses. A certain amount of surface loading, Δp, was applied at the top of the clay layer. Piles in groups were analysed to study several issues such as interaction among piles, bearing layer stiffness, surface loading and pile group configuration. All analyses were carried out under uncoupled

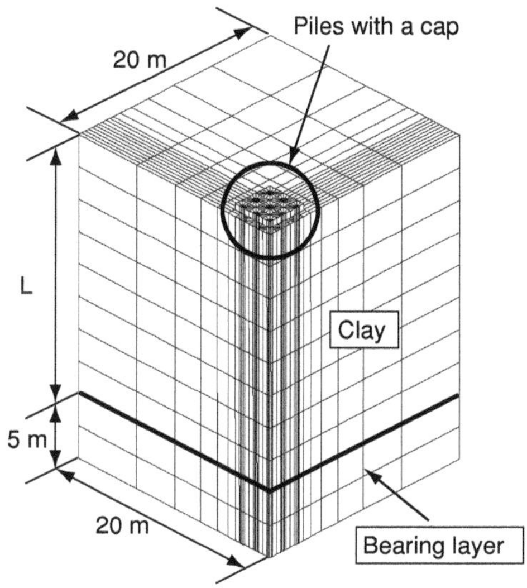

Figure 6.2 Representative FE mesh used in the analysis for piles with a cap

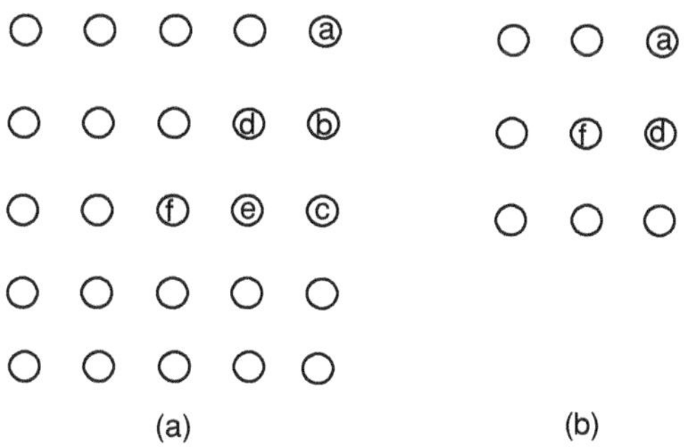

Figure 6.3 Locations of piles in groups: (*a*) 5 × 5 group, (*b*) 3 × 3 group

and fully drained conditions with the groundwater table located at the clay surface assuming a hydrostatic pore pressure distribution.

2.3 Constitutive models and material parameters

In this study two different types of numerical analyses were carried out: no-slip elastic analysis and elasto-plastic slip analyses. In each no-slip elastic analysis, the soil elements were assumed to connect to the pile perfectly, i.e. no slippage between the pile and the soil. An isotropic elastic model

Table 6.1 Representative material parameters used in the numerical analyses

Material	*Model*	*E(MPa)*	*c(kPa)*	ν	$\phi_c(°)$	$\psi(°)$	K_o	$\gamma_t(kN/m^3)$
pile	Isotropic elastic	12,500	.	0.25	.	.	0.01	5
clay	Mohr–Coulomb	5	3	0.3	20	0.1	0.65	18
Bearing layer		50	0.1	0.3	35	10	0.5	20

was used for the pile, clay and bearing layer. In order to compare to previous elastic solutions, the material parameters used in the no-slip elastic analyses were selected according to the parameters adopted in the studies of Poulos and Davis (1980), Chow et al. (1990) and Lee (1993). Table 6.1 summarizes the material parameters used in the numerical analyses.

In each elasto-plastic slip analysis, an isotropic elastic model was used for the pile and a non-associated elasto-plastic model with Mohr–Coulomb failure criterion for the clay and bearing layer. For clay the internal angle of friction ϕ is set at 20°, typical of a critical state angle with a very small dilation angle ψ as large shear deformation develops at the interface. For sand the peak internal angle of friction was set at 45° with a dilation angle of 10° consistent with the critical state angle being about 35°. The pile–soil interaction in each elasto-plastic slip analysis was governed by varying a limiting shear displacement and an interface friction coefficient μ, where $\mu = \tan(\delta)$ and δ is the interface friction angle. A limiting shear displacement of 5 mm was assumed to achieve full mobilization of the interface friction.

2.4 *Numerical modeling procedures*

After initial equilibrium, the NSF was initiated by applying a uniform surface loading on the top of the clay surface (Figure 6.4). Obviously, an increase in the vertical effective stress due to Δp leads to soil settlement and NSF along the upper length of the pile in equilibrium with PSR and toe resistance. In addition, the effects of axial loading on the pile head on the dragload changes were modeled by applying uniform stress on the pile head after the full development of the NSF. All analyses were carried out under uncoupled drained conditions with the groundwater table located at the surface. In each elasto-plastic slip analysis, duplicated nodes were used to form an interface of zero thickness to allow soil slip at the soil–pile interface (ABAQUS 1998). The nodes of the soil elements in contact with the pile slip along the pile when soil yielding occurs if the relative shear displacement is greater than or equal to 5 mm, i.e. full interface friction is developed. In the analysis, pile installation effects on different soil types

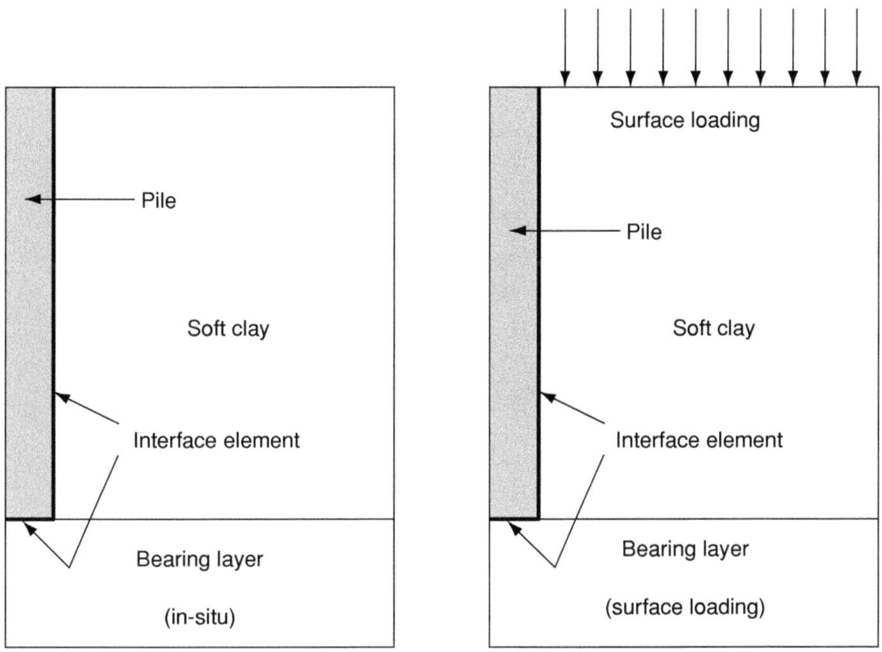

Figure 6.4 Numerical analysis sequences

were not included and hence so-called wished-in-place pile was modeled. This is because the pile installation effect is extremely difficult to quantify scientifically, if not impossible in numerical analysis (Baguelin and Frank 1982).

Two sets of interface surface (i.e. pile side and pile toe) were specified at the pile–soil interface in the slip analysis, while continuum elements were used at the pile–soil interface in the no-slip analysis (Figure 6.4). The interface elements were composed of 2D quadratic 18-node elements, each element of two 9-node surfaces compatible with the adjacent solid elements (the two surfaces coincide initially). This model was selected from the element library of ABAQUS (1998). ABAQUS uses the Coulomb frictional law where frictional behavior is specified by an interface friction coefficient μ and a limiting displacement γ_{crit} (see Figure 6.5). The compressive normal effective stress p' between two contact surfaces was multiplied by an interface friction coefficient μ to give a limiting frictional shear stress $\mu \times p'$. As shown in Figure 6.6, the interface elements of zero thickness can only transfer shear forces across their surfaces when p' acts on them. When contact occurs, the relationship between the shear force and the normal pressure is governed by a modified Coulomb's friction theory. Thus, these elements are completely defined by their geometry, a friction coefficient μ,

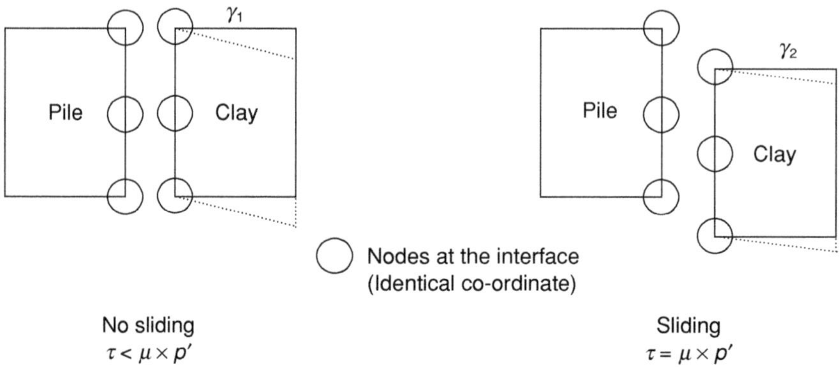

Figure 6.5 Interface modeling

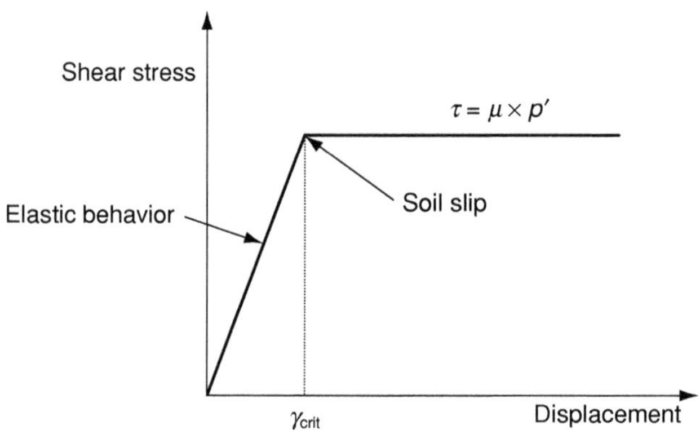

Figure 6.6 Behavior of interface element

where $\mu = \tan(\delta)$, an elastic stiffness and a limiting displacement γ_{crit} used to provide convergence. If the shear stress applied along the surfaces was less than $\mu \times p'$ the surfaces would stick. The nodes of the soil elements in contact with a pile could slide along it when soil slip occurs. The limiting shear displacement and interface friction coefficient values were typical values obtained from the literature (Broms 1979, Lee 2001). For detailed information on the numerical analyses and the adopted input parameters, see Lee (2001).

2.5 Post analysis

On completion of the analysis, the dragload was calculated from the summation of the vertical stress in the pile element:

$$\text{Dragload} = \pi r^2 \sigma_v \quad (1)$$

where r is the pile radius and σ_v is the vertical stress in the pile element. In the case of piles in the 3D analysis, the vertical stress was averaged at the same elevation. The pile–soil interaction effect among the piles (called the shielding effect) was expressed in terms of the reduction of the dragload (the compressive force on the piles), P_r, and the reduction of downdrag (the pile settlement), W_r, following Teh and Wong (1995) and Lee and Ng (2004).

$$P_r = (P_{s,\max} - P_{g,\max})/P_{s,\max} \text{ and } W_r = (W_s - W_g)/W_s$$

in which $P_{\max}$ is the maximum dragload and W is the downdrag at the pile head in the central pile. The suffixes s and g represent a single pile and piles in a group respectively. In addition, the maximum tensile force at the outer piles, $T_{\max}$, was normalized to be $T_{\max}/P_{s,\max}$ to quantify the development of tensile forces for the outer piles.

3 Comparison with elastic solution

3.1 Normalized dragload

Figure 6.7 shows the distribution of the normalized dragload (P_a/E_cS_oL) (where P_a is maximum dragload), with respect to the Young's modulus of clay E_c, the surface settlement S_o and pile length L, based on the formal elastic solutions for an end-bearing pile by Poulos and Mattes (1969). Three relative clay stiffnesses K were considered ($K = E_p/E_c$, Young's modulus of the pile E_p = 2 GPa). It also shows the computed results based both on a no-slip elastic model with an interface friction coefficient $\mu = \infty$ and an elasto-plastic slip model based on initially isotropic stress conditions (i.e. $K_o = 1$). The no-slip elastic simulations agree closely with the elastic solutions, although both methods overestimate the dragload. When soil slip is taken into account, a smaller dragload is predicted, such that the interface shear stress exceeds neither the clay nor the interface shear strength. Since the slip analysis ($\mu = 0.3$) requires 5 mm of relative displacement prior to sliding, the slip analysis also shows a reduced dragload in stiff soil ($K = 50$, $K = E_p/E_c$) due to partial shear mobilization. The correction of these two effects, one numerical and one physical in origin, can be seen in Figure 6.7 to produce a very significant effect on the dragload P_a. Figure 6.7 also shows the dragload predicted from the β-method (Burland 1973), assuming a β-value of 0.3, which produces a single curve regardless of K. This method would result in the shear stress being somewhat overestimated, since the maximum shear stress is assumed to act along the entire length of the pile and hence partial mobilization of skin friction near the pile toe could not be included. In addition the actual vertical effective stress at the interface would be smaller than expected by the β method, owing to the

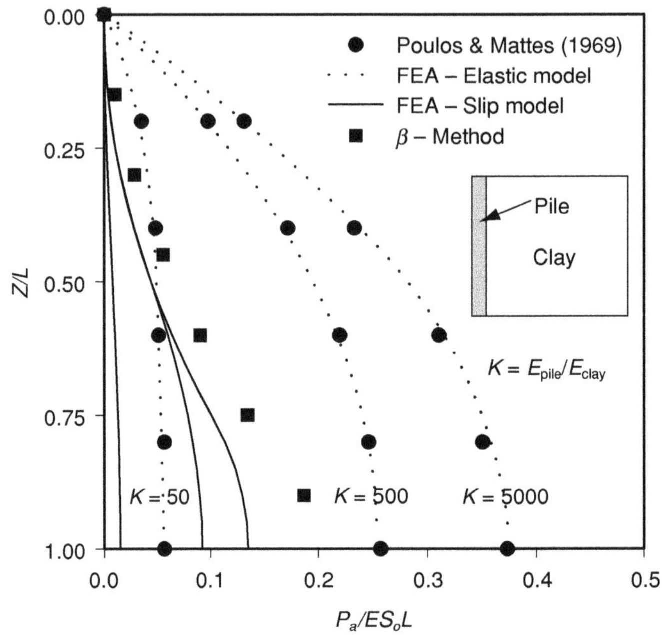

Figure 6.7 Distribution of dragload as determined by various methods ($K_o = 1.0$, $\mu = 0.3$, $\beta = 0.3$)

transfer of some of the soil weight to the pile (Zeevaert 1983, Jeong 1992, Bustamante 1999). In summary, both the β method and the elastic solutions predict excessively large dragloads, as would be expected.

3.2 Effects of relative stiffness between the pile and the soil (E_p/E_c)

Figure 6.8 illustrates the variations in the normalized downdrag, $WE_p/2\Delta p r_o$ (Poulos and Davis 1980), with the relative pile stiffness, E_p/E_c, for an end-bearing pile, where W is the pile head settlement for a single pile and r_o is the pile radius. The results from existing elastic solutions (Poulos and Davis 1980, Lee 1993) are included in the figure for comparisons. The Young's modulus of the pile (12.5 GPa), E_p, was kept constant in the parametric analyses, whereas the Young's moduli of the clay, E_c, was varied from 1.25 MPa to 125 MPa. Hence, the relative pile stiffness ratio, E_p/E_c, ranged from 100 to 10,000. It should be noted that E_p/E_c ranging from 100 to 1000 corresponded to the stiffness of hard to stiff clays (E_c = 12.5 MPa to 125 MPa), whereas E_p/E_c ranging from 1000 to 10,000 corresponded to the stiffness of stiff to soft clay (E_c = 1.25 MPa to 12.5 MPa). Since a single end-bearing pile was considered in the study, the development of downdrag was directly related to the elastic compression of the pile since pile tip settlement was not permitted.

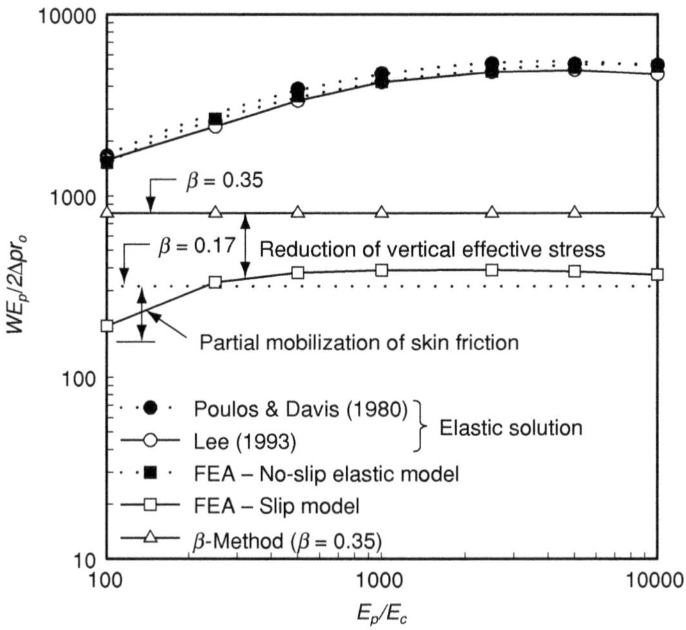

Figure 6.8 Variations of normalized downdrag with E_p/E_c (K_o = 1.0, E_c = 1.25–125 MPa, E_p = 12.5 GPa)

It can be seen from Figure 6.8 that the normalized downdrag from the no-slip elastic analysis shows a linear increase in magnitude with the pile stiffness ratio, E_p/E_c, until the ratio reaches about 500. However, when E_p/E_c is greater than 2000, the normalized downdrag approaches a limiting value. This suggests that the development of downdrag is governed by the degree of shear strain mobilization at the pile–soil interface. For stiffer soils (where E_p/E_c is small), the downdrag of the pile increases linearly with the shear strain developed at the interface. For softer soils (where E_p/E_c is large), the downdrag increases with the E_p/E_c ratio but at a reduced rate, indicating that a limiting shear strain condition is reached at the pile–soil interface, particularly near the pile toe. The computed results from the no-slip elastic analysis agree closely with the elastic solutions.

Although the general trend is similar, the computed downdrag from slip analysis is significantly smaller than that from the no-slip elastic analysis and analytical elastic solutions. For a given stiffness ratio, the slip analysis predicts only about 10 percent (between eight and fourteen times smaller) of the computed downdrag from no-slip elastic analysis as only limited shear stress is permitted to transfer from the consolidating soil to the pile to drag the pile down in the former analysis. The larger the stiffness ratio, the bigger the difference in the computed downdrag between the two types of analyses. This is because the no-slip elastic analysis does not allow soil yielding to take place at the interface, and thus substantially large shear stress can be developed. The large downward shear stress mobilized at the interface leads to large dragload and downdrag, especially in relatively soft soil (when E_p/E_c is large). On the other hand, the slip analysis predicts nearly full soil slip along the entire pile shaft, and hence a limited shear stress transfer from the soil to the pile is controlled by the limiting shear displacement of 5 mm. The substantially large computed downdrag from the no-slip analyses is consistent with the current understanding that dragload is generally overpredicted by elastic solutions (Lee et al. 2002).

The normalized downdrag calculated by the β method (Burland 1973) using a β value of 0.35 is 802, regardless of E_p/E_c, as shown in Figure 6.8. The calculated normalized downdrag is substantially smaller than the results from elastic analyses, but it is 2.2 to 4.2 times larger than the computed values from slip analyses. It should be noted that the higher the clay stiffness (E_p/E_c is small), the larger the discrepancy between the β method and slip analysis due to partial mobilization of the interface skin friction in slip analysis. This may be attributed to the use of maximum shear stress in the β method along the entire pile length and a reduction in the vertical effective stress near the pile. Thus, there is a decrease in the horizontal effective stress in slip analysis as a result of shear stress transfer from the consolidating soil to the pile. A β value of 0.17 is back-calculated (as the dotted line shown in Figure 6.8) to match the results from slip analysis when full slip at the pile–soil interface develops for E_p/E_c larger than 500. However, when E_p/E_c is smaller than 250, a smaller β value is required as a result of the partial mobilization of skin friction.

3.3 Comparison with elastic analysis for piles in a group

Figure 6.9 shows the development of dragload for end-bearing piles in a 5×5 pile group with a pile spacing of 5D as determined from numerical analysis using a no-slip elastic model. Figure 6.9 also shows an elastic solution presented by Kuwabara and Poulos (1989). The positions of piles referred to in Figure 6.9 are shown in Figure 6.3a. The piles are modelled as free-headed piles in a group, which can be regarded as similar to piles in a flexible pile cap. No constraint is provided at the pile heads, and no axial load is applied on the pile heads. It is assumed that piles are not connected to each other, so that each pile can respond separately in response to NSF and in accordance with its magnitude. The trends of dragload distributions are very similar from both analyses, although the solutions from Kuwabara and Poulos (1989) predict smaller dragloads, particularly at the pile toes, a discrepancy also reported by Chow et al. (1996). These small dragloads, which amount to only 10–50 percent of that predicted for a single pile, are not supported by previous observations with similar pile configurations (Denman et al. 1977). Figure 6.9 also presents the results from an elasto-plastic analysis, which included soil slip ($\mu = 0.3$). Relatively

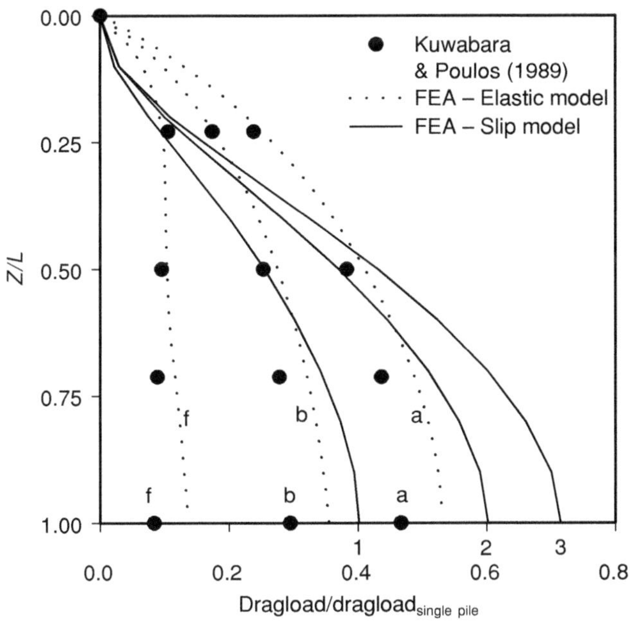

Figure 6.9 A comparison of dragload with and without slip ($S = 5.0D$, $E_c = 20$ MPa, $\mu = 0.35$)

small reductions in the dragload (40–70 percent compared with a single pile) are observed.

3.4 Comparison with elastic analysis for piles in a group with a cap

Kuwabara and Poulos (1989) reported the development of axial forces in end-bearing piles with a depth of *Z/L* on end bearing 5 × 5 piles at a 5D spacing that were connected to a rigid cap from the elastic solution, where *Z* was the pile depth and *L* the pile length. No-slip and slip analyses were conducted. Figures 6.10a and 6.10b show the distributions of the normalized axial forces $P_g/P_{s,max}$ for these piles in each group for the elastic solution and the numerical analyses. The suffixes *s* and *g* represent a single pile and piles in a group respectively.

Some tensile forces were computed at the corner pile (pile a) near the pile head, whereas the compressive forces were computed at the center pile (pile f) in Figure 6.10a. The computed results from the no-slip analysis matched those from the elastic solution at the upper part of the piles (*Z/L* = 0.0 to 0.4). However, some discrepancies existed at the lower part of the piles (*Z/L* = 0.4 to 1.0). For the center pile, a gradual decrease in the normalized dragload was obtained from the elastic solution on the lower

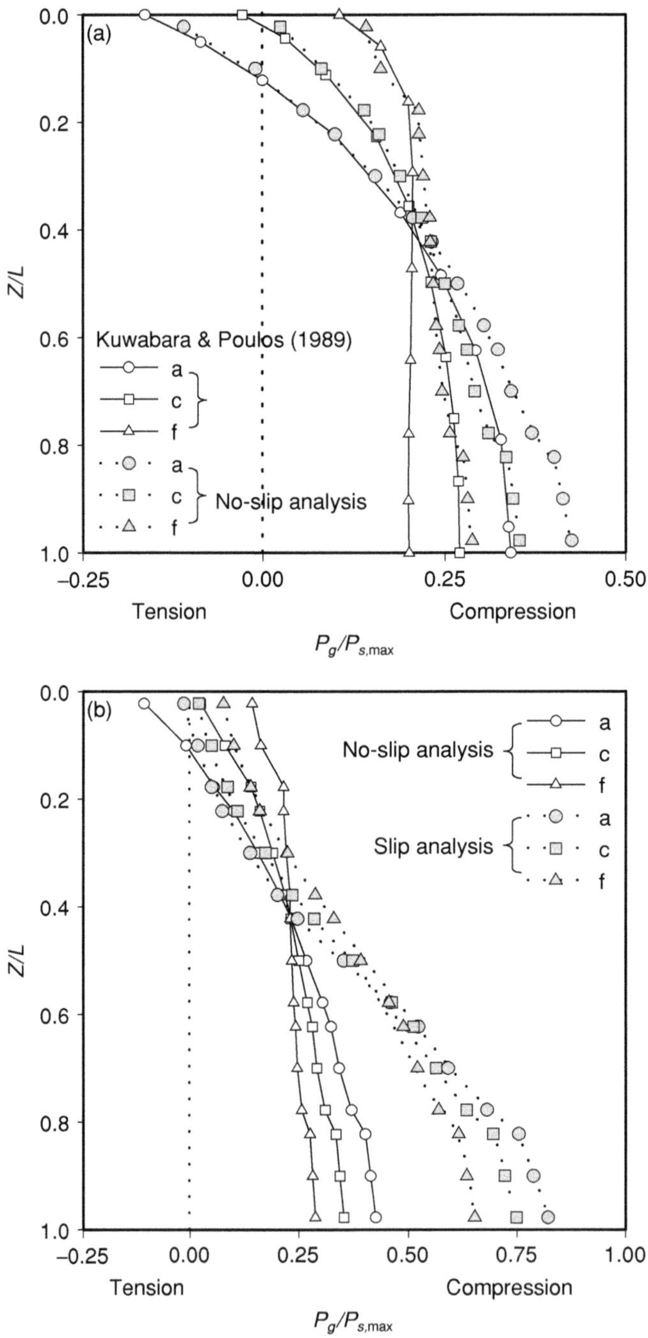

Figure 6.10 Distributions of the normalized axial forces on piles in a 5 × 5 pile group at 5.0D spacing

part of the pile, implying mobilization of PSR. Because pile movement was prevented except for elastic pile compression, PSR may not have been mobilized at the mid-depth of the pile. Contrary to the elastic solution, no-slip analysis showed a gradual increase in the axial force with depth, consistent with the results reported by Chow et al. (1996). This implies the continuous development of NSF along the pile shaft. At the lower part of the piles, the outer piles showed a larger increase in the normalized dragload with depth compared to the inner piles, as expected, owing to the shielding effect.

It was noted that smaller normalized tensile forces were computed near the pile head from the slip analysis for the corner pile, compared to the no-slip analysis, as shown in Figure 6.10b [$T_{max}/P_{s,max}$ = 11% (no-slip analysis) and 2% (slip analysis)]. In addition, the P_r for the piles computed from slip analysis was much larger (no-slip analysis: 57–71%, slip analysis: 18–35%). This was expected as the slip analysis predicted a smaller shielding effect among the piles (Lee and Ng 2004). This is probably because the elastic solutions overemphasized the shielding effects, excluding soil slip, which reduced the protection offered to the piles inside the group by the piles outside the group.

4 Behavior of single piles

4.1 Dragload and downdrag

Figures 6.11a and 6.11b show different distributions of typical shear stress (Figure 6.11a) and dragload (Figure 6.11b) for a friction pile and an end-bearing pile. In this analysis different stiffnesses ($E_{bearing\ layer}$) are used to

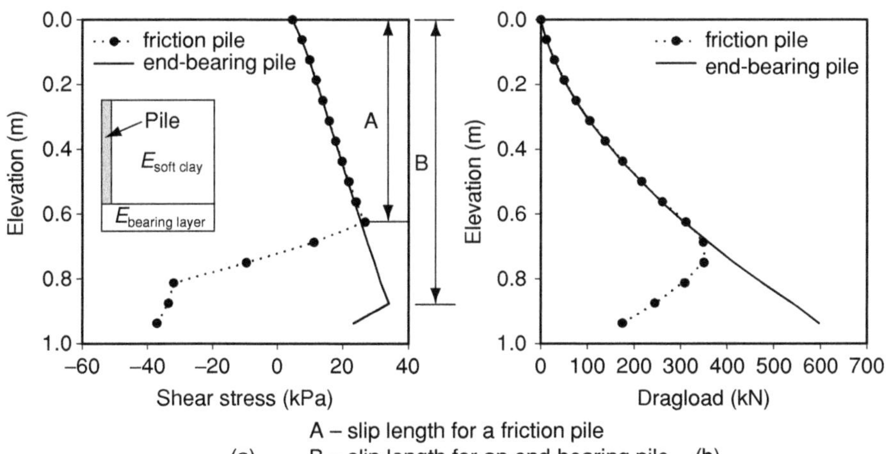

Figure 6.11 Shear stress and dragload distributions ($E_{soft\ clay}$ = 5 MPa, μ = 0.35) $E_{bearing\ layer}$ = 5 MPa (friction pile), 5 GPa (end-bearing pile)

represent the bearing layer (noted as a bearing layer in Figure 6.4) to model a friction pile or end-bearing pile. The different stiffness ratios ($E_{\text{bearing layer}}/E_{\text{soft clay layer}}$) of 1 and 1000 are used for a friction pile and an end-bearing pile respectively. For a friction pile, NSF is developed along a section extending from the top of the pile to the neutral plane, which is located at around 70 percent of the pile length measured from the surface. For an end-bearing pile, NSF is developed along the entire length of the pile. This can be explained based on the relative settlement between the soil and the pile. When a friction pile settles more than the soil, PSR develops along the lower part of the pile shaft. However, only very small pile movement is possible in the case of end-bearing piles. NSF is therefore developed along the entire length of the pile. Partial mobilization of negative skin friction and positive shaft resistance is observed near the neutral plane, where the shear stress is less than the limiting shear stress $\mu \times \sigma'$, owing to a small relative movement between the soil and the pile. Slip lengths of about 63 percent and 93 percent of the total pile length are observed for friction and end-bearing piles respectively (see Figure 6.11a).

More dragload (596 kN) and less downdrag (1.5 mm) are developed for an end-bearing pile, and less dragload (350 kN) and more downdrag (79.0 mm) are developed for a friction pile. It has been shown that, in most cases, dragload is not significant on piles shorter than 30 m (UniNews 1998). However, there are still many cases where failure of structures is reported due to excessive pile settlement (downdrag) (Acar et al. 1994). Excessive pile settlement is very likely developed for friction piles, therefore such piles should be installed deep into a stiff layer where possible. The structural integrity and drivability of the pile and the piling system must be thoroughly investigated. Where it is not possible to found the pile in a competent layer, an alternative engineering solution should be considered.

4.2 Effects of the interface friction coefficient

Figure 6.12a shows distributions of the computed shear stresses with the relative depth *Z/L*, where *Z* is the depth below the ground surface, based on the different analysis conditions. Interface friction coefficients of 0.2 and 0.4 were used. Results from no-slip analysis and the β method were included for comparison. The development of the shear stresses was heavily dependent on the interface friction coefficient. In slip analysis, it was observed that, the smaller the interface friction coefficient (i.e. $\mu = 0.2$), the longer the slip length and vice versa. Slip length is defined as the distance from the pile head to the point at which partial mobilization of the shear strength begins. The distributions of shear stresses from slip analysis were nearly linear as the β method predicted. However, partial mobilization of the shear stress near the neutral plane was observed owing to smaller relative displacement between the pile soil than the limiting shear displacement (i.e. 5 mm).

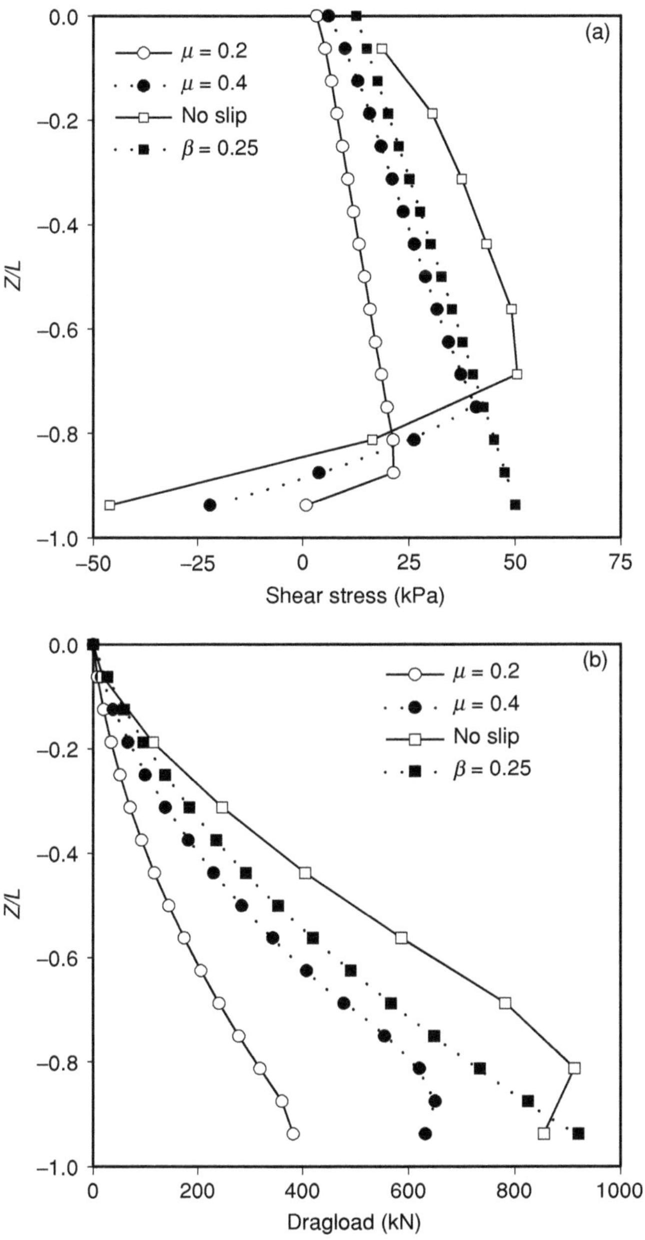

Figure 6.12 Effects of interface friction coefficient on (*a*) shear stress, (*b*) dragload

From the no-slip analysis, where the compatibility relation was valid even with the very large shear strains, large shear stresses occurred. These were even larger than the shear stresses predicted by the β method (i.e. $\beta = 0.25$). The shear stress predicted by the β method was also overestimated since maximum shear stress was assumed for the entire length of the pile. Partial mobilization of skin friction near the neutral plane and the reduction in the vertical stress at the pile–soil interface could not be included.

Dragload distributions from the same analysis conditions are illustrated in Figure 6.12b. The increase in the interface friction coefficient from 0.2 to 0.4 resulted in a 65 percent increase in the maximum dragload. Compared to the slip analysis in which the interface friction coefficient was 0.2, no-slip analysis showed a significant increase in dragload of about 140 percent. The estimated maximum dragload in both the no-slip analysis and the β method showed similar dragloads even though the shear stresses from the no-slip analysis were larger than those from β method except near the pile tip. This was due to the development of the positive shaft resistance (PSR) near the pile tip in the no-slip analysis. Because PSR could not be developed near the pile tip, this suggested that more dragload would be computed for an end-bearing pile from the no-slip analysis than that from the β method.

4.3 Effects of surface loading

Distributions of the computed shear stress at different surface loadings are given in Figure 6.13a to show the effects of surface loading. The mobilized shear stresses τ_m were normalized with the maximum shear strength at the interface with depth, defined by $\mu \times p'$. When the soil slip developed, the interface shear strength could be fully mobilized, whereas the shear strength could be only partially mobilized when the relative displacement was not sufficient to develop soil slip. Near the pile head, the shear strength was fully mobilized (i.e. $\tau_m/\mu \times p' = 100\%$) for all surface loadings considered up to *Z/L* of up to –0.4. However, when the interface shear strength was partially mobilized (i.e. $\tau_m/\mu \times p' < 100\%$), much reduced shear stresses were computed. The arrows in Figure 6.13a indicate the development of the partial mobilization of the skin friction. When surface loading was small (10 kPa), soil slip could only develop at the upper half of the pile length since relative displacement was smaller than the limiting shear displacement. However, with an increase in the surface loading, slip length gradually increased. Soil slip developed nearly on the entire surface of the pile except when surface loading was smaller than 25 kPa. This means that, for ordinary situations, full soil slip nearly develops at the pile–soil interface. Under the application of large surface loading of 150 kPa, soil slip was developed for almost the entire pile length. Below the neutral plane, the shear strength could only be partially mobilized, leading to less dragload. Near the pile tip, owing to the development of PSR,

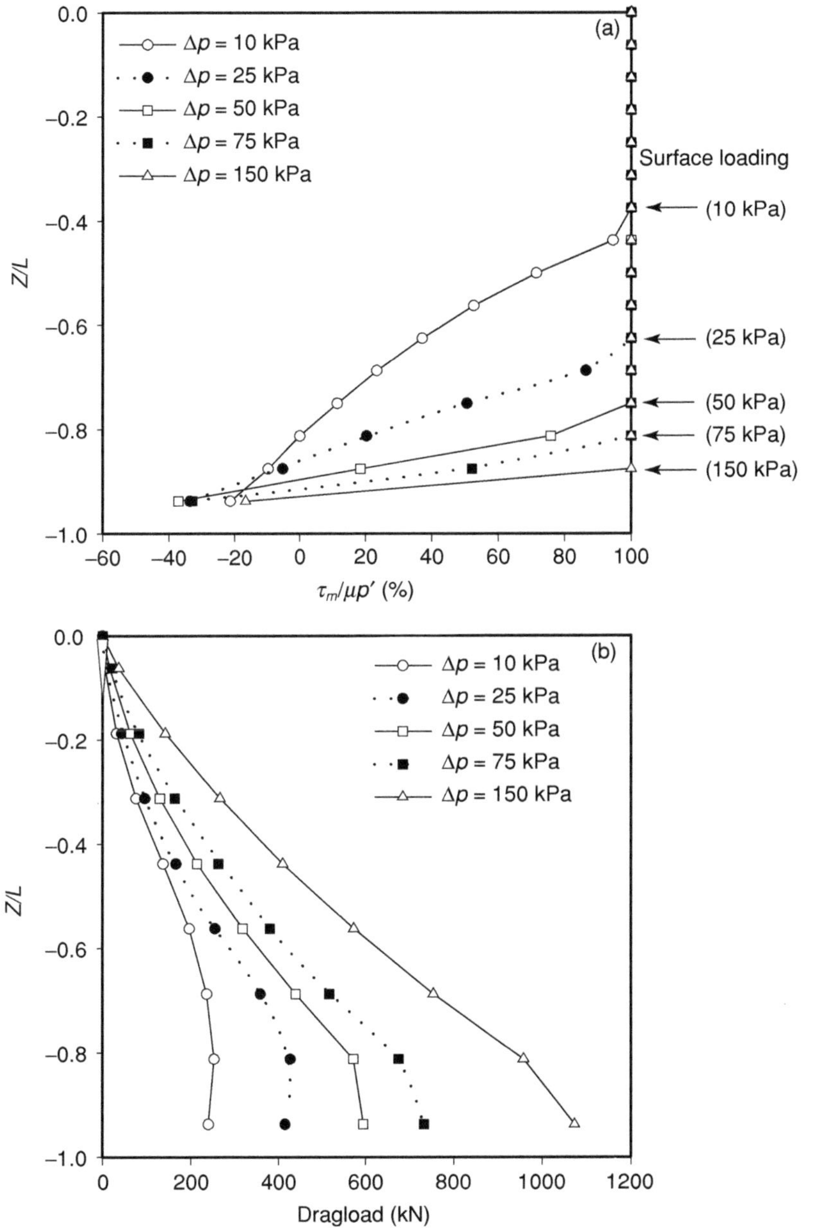

Figure 6.13 Effects of surface loading (*a*) shear stress mobilization, (*b*) dragload

the direction of the shear stress changed and was still partially mobilized. However, the maximum PSR was about 20–40 percent of the interface shear strength.

Dragload distributions in Figure 6.13b show that the increase in dragload was more affected by increased vertical stress than by the development of the soil slip. An increase in the surface loading from 10 kPa to 25 kPa results in about a 70 percent increase in dragload. A similar amount of increase in dragload is computed when the surface loading was increased from 50 kPa to 150 kPa. Therefore, it could be said that, once soil slip develops for nearly the entire length of the pile, dragload development is governed mainly by the increase in the vertical soil stress.

4.4 Effects of axial loading on changes of dragload

In many cases, a pile subjected to dragload also has to carry simultaneous vertical dead load. Research has been carried out to examine the reduction in dragload using experimental tests (Fellenius 1972, Okabe 1977, Bozuzuk 1981, Shen et al. 2002, Leung et al. 2004, Chan et al. 2006) and numerical analyses (Wong 1991, Indraratna, 1993, Jeong et al. 2004). To clarify the effect of external axial loading on dragload, axial loading was applied in steps to the pile top after the full development of dragload and compared with the PSR case of an axially loaded pile.

Figure 6.14a shows the two load-transfer characteristics of a pile. This curve demonstrates the reductions in dragload with a gradual increase in axial load. Assuming the maximum pile capacity is 3000 kN using $\beta = 0.25$, beyond the working load level (about 1000 kN) only positive skin friction is mobilized along the pile. This is because the dragload and the axial load combined increase the pile settlement relative to the soil and consequently change the location of the neutral plane. It is to be noted in Figure 6.14(b) that the position of the neutral plane changes toward the pile head with increasing axial loads, as shown in Figure 6.14(a). Therefore, to determine the location of the neutral plane, an analysis of the load distribution in the pile must first be performed. From this analysis, illustrated in Figures 6.14(a) and 6.14(b), a reduction of the dragload on the pile results in a lowering of the location of the neutral plane but has a proportionally smaller effect on the magnitude of the maximum load in the pile.

Since changes of the relative displacement due to the axial loading heavily depend on pile settlement, different bearing layer stiffnesses E_b were considered, with a range of 1 E_c to 100 E_c (E_c = 5 MPa). The former and the latter could be considered as a floating pile and an end-bearing pile respectively. Figure 6.15 shows reductions in dragload under the axial loading at different relative bearing stiffnesses (E_b/E_c). The reductions in dragload were expressed as the reduction in maximum dragload under axial loading (i.e. $P_{D\text{ no axial loading}} - P_{D\text{ axial loading}}$) compared with maximum dragload without axial loading (i.e. $P_{D\text{ no axial loading}}$), where P_D is the maximum

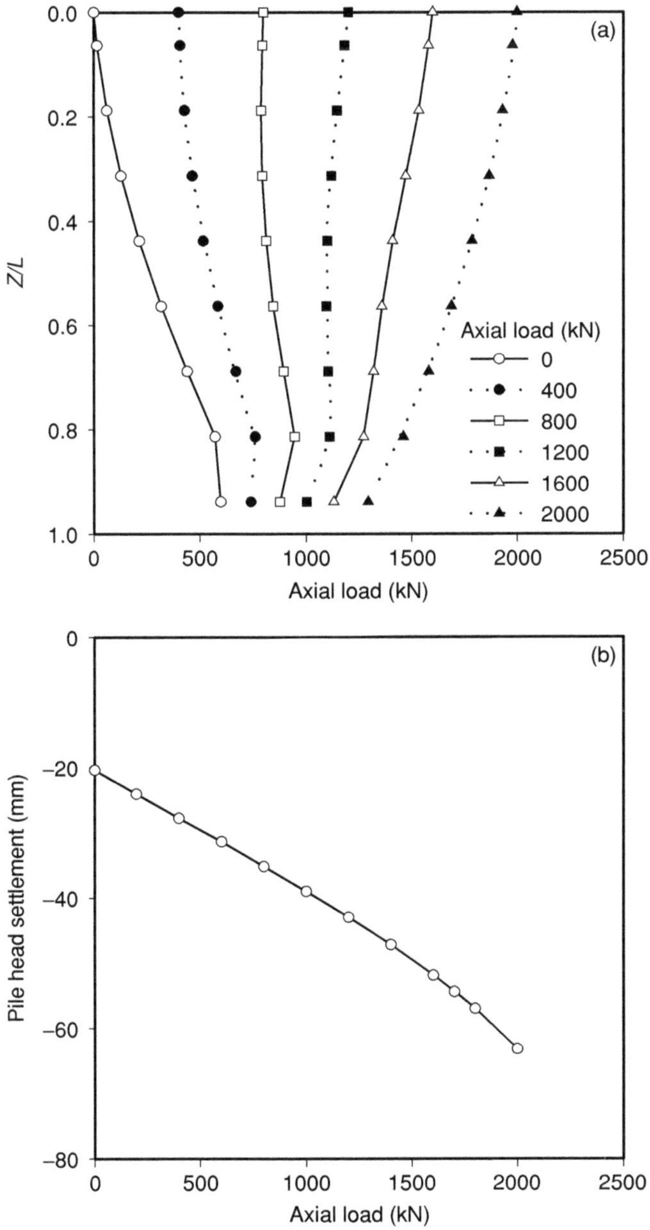

Figure 6.14 Effects of axial load on (*a*) dragloads, (*b*) pile head settlemet

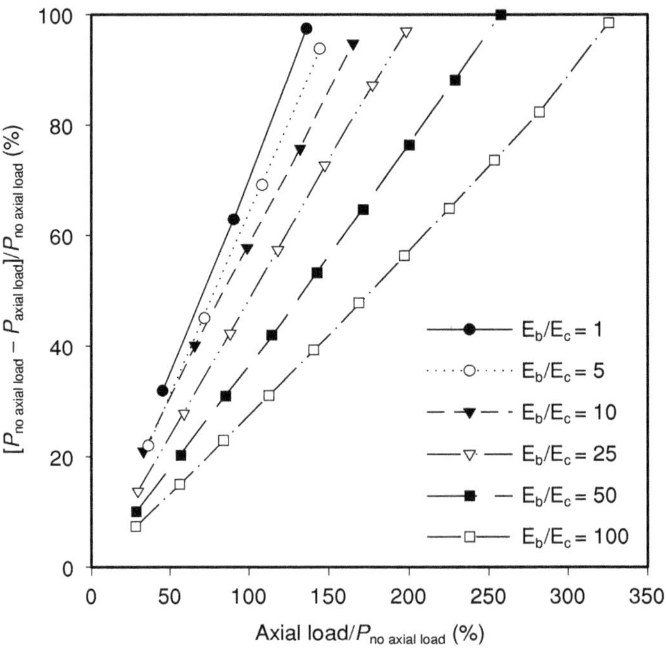

Figure 6.15 Distributions of reductions in dragload with axial loads (E_c = 5 MPa)

dragload. The axial loading was normalized with the maximum dragload $P_{D \text{ no axial loading}}$ at a given relative bearing stiffness. Results showed that, the stiffer the bearing layer, the less the reductions in dragload at the given axial loading. This was because settlement changes in the stiffer layer under the axial loading were very small compared to that of the piles in the softer ground. In addition, the stiffer the bearing layer, the higher the axial loading to eliminate dragload. Generally, in order to eliminate dragload, axial loads of about 125–325 percent of the maximum dragload were required. This finding was different from that of Alonso et al. (1984), who discussed reductions in dragload for an end-bearing pile. However, it should be noted that, although axial load reduces dragload, it also induces an increase in pile settlement. Thus, reductions in dragload and increases in pile settlement must be thoroughly checked in practical pile design.

5 Downdrag on piles in groups

5.1 Interaction effects in the pile group

Behavior of an individual pile and piles in a 5 × 5 group with pile spacing of 2.5D was analysed to investigate the interaction effects in a pile group. Figure 6.16a shows distributions of the normalized soil settlement with the relative depth, Z/L, at the different pile locations, where Z is the depth below the ground surface. The exact locations of piles a, d, and f in the pile group are shown in Figure 6.3a (a: corner pile; d and f: inner piles). In addition, pile head settlements are also indicated in Figure 6.16a. The soil settlement shown is an averaged soil settlement at the pile–soil interface. The soil and pile settlements are normalized with the maximum soil settlement, S_o (i.e. 151 mm), at the far end of the mesh (nodal point A in Figure 6.1a or 6.1b), representing the green field settlement. It can be seen from Figure 6.16a that the computed soil surface settlements at the interfaces, a, d and f, inside the pile group are about 51–77 percent of the maximum soil settlement, while that of the single pile is about 96 percent of the maximum soil settlement. The computed results are consistent with a small-scale laboratory test of 5 × 5 pile groups at 2D and 3D spacing (Ergun and Sonmez 1995). The reduced soil surface settlements inside the pile group are attributed to the pile–soil interaction effect, which is called the shielding effect on the inner piles by the outer piles, leading to a decrease in the downdrag on piles inside the group. Owing to the consolidation of the clay, the computed soil settlement is much larger than the downdrag on each pile, except near the pile tip. The downdrag on each pile head is

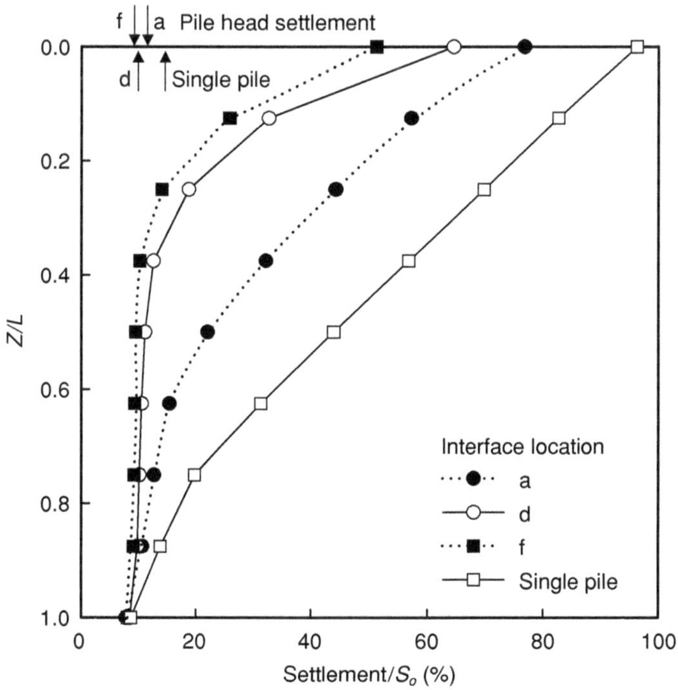

Figure 6.16(a) Distributions of normalized soil settlement for a single pile and a 5 × 5 pile group

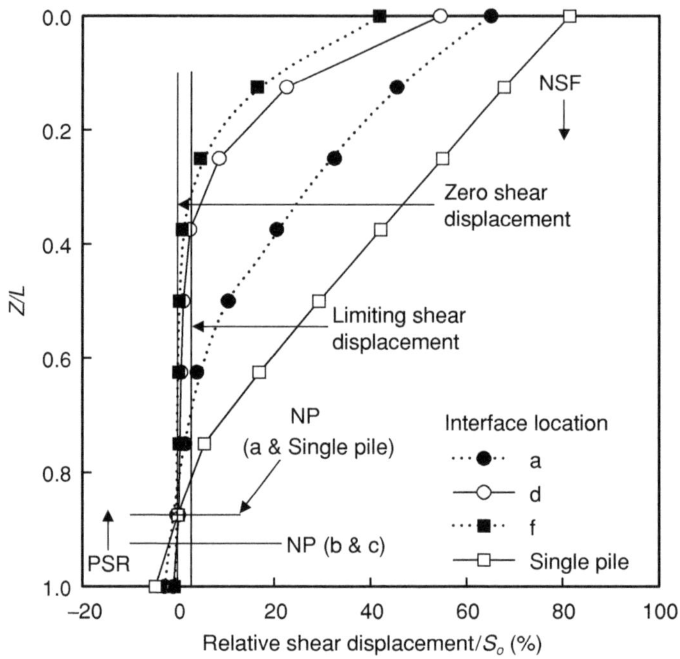

Figure 6.16(b) Distributions of normalized relative shear displacement for a single pile and a 5 × 5 pile group

about 9–12 percent of the maximum soil settlement. For piles located at a, d, and f, the maximum computed downdrag is 79 percent, 68 percent and 63% of the maximum downdrag of the single pile respectively. The computed settlement of the bearing stratum (i.e. $Z/L = 1.0$) shown in Figure 6.16 is about 10 percent of the maximum soil settlement.

Figure 6.16b shows the normalized relative shear displacements at the pile–soil interface of piles at different locations within the pile group. The relative shear displacement (i.e. soil settlement minus pile settlement at the interface) is normalized with S_o. The positive normalized relative shear displacements demonstrate a larger soil settlement than the pile settlement whereas a negative value signifies a larger pile settlement than soil settlement. Obviously, different relative shear displacements induce different degrees of mobilization of the skin friction at the interface. The figure shows that there is a change in the direction of the side resistance along the pile: i.e. from NSF to PSR via a neutral plane (NP), at which the relative shear displacement is zero. For a given depth, the relative shear displacements at the inner piles (d and f) are smaller than those at the corner (a) and at the single pile. The smaller mobilization of relative shear displacement at the inner piles (d and f) is attributed to the shielding effects (or pile interaction effects) within the pile group. On the other hand, the normalized relative shear displacements at the corner pile (a) and at the single pile are larger than the limiting value of 5 mm for more than 50 percent of the pile length near the surface. Based on the defined condition of the interface slip (the shear displacement reaches the maximum limiting shear displacement of 5 mm for the full mobilization of skin friction in the slip analysis), a slip length can be found along which the skin friction is fully mobilized. It can be seen from Figure 6.16b that the smallest soil slip develops at the center pile (f) whereas the largest soil slip is computed for the single pile. The computed slip lengths at the different locations of a, d and f are 63 percent, 31 percent and 25 percent of the pile length of 20 m respectively, whereas the computed slip length of the single pile is 75 percent of the pile length. The reduction of the slip length at the inner piles (d and f) is attributed to the shielding effects within the pile group.

To study the shielding effects of piles on the degree of shear strength mobilization along a pile, Figure 6.17 shows the relationship between the normalized shear stress ratio at the interface, τ_m/τ_p, with depth, where τ_m is the mobilized shear stress and τ_p the interface shear strength. It is clear

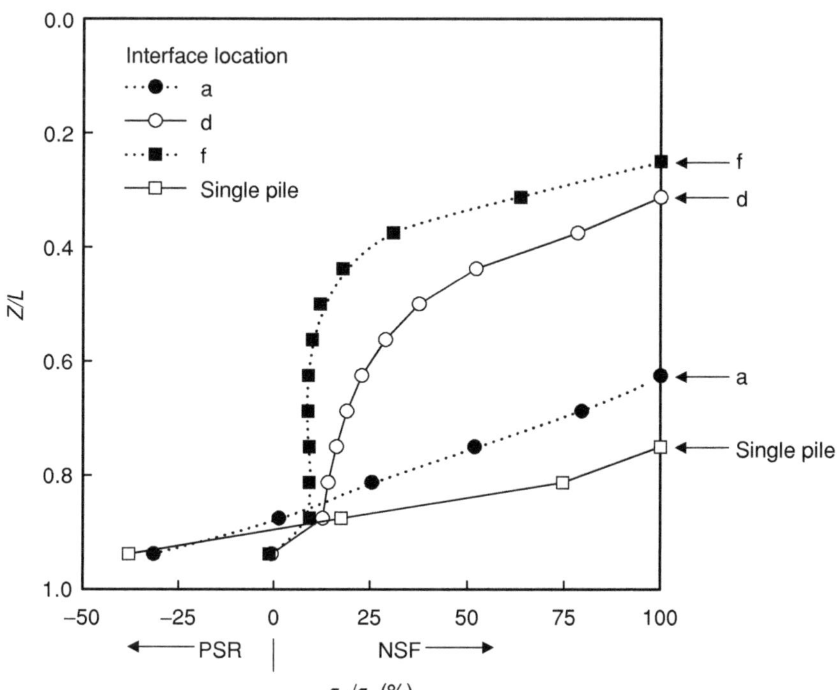

Figure 6.17 Distributions of degree of shear strength mobilization for a single pile and a 5 × 5 pile group

that the magnitude of τ_m/τ_p is governed by the relative shear displacement at the interface. It can be seen from Figure 6.17 that the depth of full mobilization (or slip length) of the interface shear strength (i.e. $\tau_m/\tau_p = 100\%$) depends on the pile location (for single pile: from $Z/L = 0$ to $Z/L = 0.75$; at a: from $Z/L = 0$ to $Z/L = 0.63$; at d: from $Z/L = 0$ to $Z/L = 0.31$; at f: from $Z/L = 0$ to $Z/L = 0.25$). The arrows in Figure 6.17 point to the beginning of partial mobilization (i.e. $\tau_m/\tau_p < 100\%$) of the shear strength toward the pile tip. Clearly, the inner piles are well shielded by other outer piles along most of the pile. Near the tip of the inner piles, negligible PSR is computed in contrast to the large mobilized values for the outer and single piles to resist their large mobilized NSF at the upper part of the piles.

When soil slip is not sufficiently developed, owing to a smaller mobilized relative displacement than the maximum allowable shear displacement at the interface, a smaller shear stress than the interface shear strength will be induced. This is consistent with the previous discussion of Figure 6.8. Different degrees of shear strength mobilization in the piles lead to different slip lengths. The degree of shear strength mobilization at the inner piles, d and f, is smaller than that from the corner and the single pile since the inner piles are shielded (or protected) by the outer piles. This suggests that sacrificial piles can be designed and built to protect pile groups in consolidating soils.

Following Poulos and Mattes's approach (1969), the distributions of normalized dragload (P_a), P_a/E_cS_oL with depth are shown in Figure 6.18 for the four piles. When the shear strength is fully mobilized near the pile head (i.e. $Z/L = 0$ to 0.25) for all the piles, as expected, there is no clear difference in the dragload distribution. However, when the shear strength is partially mobilized to different degrees in the lower part of the piles, it is clear that the single and corner (a) piles attract significantly higher dragload than do the inner and centre (d and f) piles, resulting in a large PSR mobilized to maintain equilibrium. The dragloads mobilized for the group piles at a, d and f are 72 percent, 43 percent and 35 percent of dragload for the single pile respectively. It can be seen from Figure 6.18 that, for piles d and f, NP is located at about $Z/L = 0.88$, whereas those for pile a and the single pile are located at about $Z/L = 0.93$. On the other hand, owing to very small PSR at the inner piles (d and f), the maximum dragload is induced at the pile tip (NP is near the pile tip). For a given bearing layer, the computed normalized dragload near the pile tip at the single pile and pile a indicate a larger increase in toe resistance below the pile tip as compared to those at piles d and f.

5.2 Effects of the relative bearing layer stiffness (E_b/E_c) on the shielding effects

Figure 6.19 shows the variations of the computed reduction in the downdrag, W_r, with relative bearing stiffness, E_b/E_c, for piles at different pile loca-

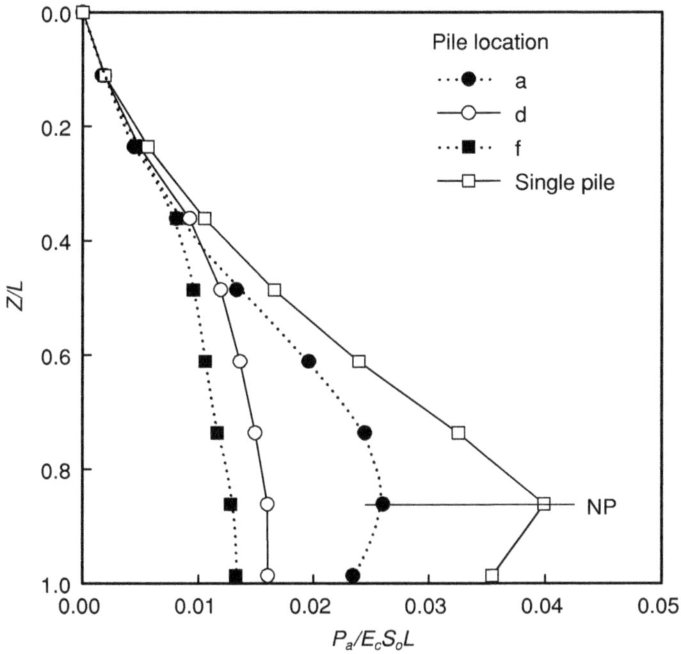

Figure 6.18 Distributions of normalized dragload for a single pile and a 5 × 5 pile group

tions in the group (refer to Figure 6.3a), where E_b is the Young's modulus of the bearing layer. An identical finite element mesh for the 5 × 5 pile group was used and elasto-plastic slip analyses were carried out. In these analyses, E_c was kept constant (5 MPa) whereas E_b was varied from 1 E_c to 100 E_c (5 MPa to 500 MPa). A low E_b/E_c value of 1 and a high base stiffness ratio of 100 represent a floating and an end-bearing pile respectively. It can be seen from the figure that W_r in all piles non-linearly increases with an increase in E_b/E_c from 3 percent to 55 percent. The change in W_r depends on the pile location in the group. For a given stiffness ratio, the maximum W_r is induced at the centre pile (f) owing to the shielding effects whereas the minimum W_r is developed at the corner pile (a). The difference in W_r between the center and corner piles can be up to 25 percent for a given E_b/E_c. An increase in E_b/E_c results in an increase in the difference of W_r for each pile. This shows that, the stiffer the bearing stratum, the greater the reduction in W_r between the centre and corner piles and, hence, the shielding effects on downdrag for an end-bearing pile group are larger than for a floating pile group.

The variations of P_r with E_b/E_c for piles in a group at different pile locations are shown in Figure 6.19. Compared with the downdrag reductions in the group, the dragload reductions are relatively insensitive to changes

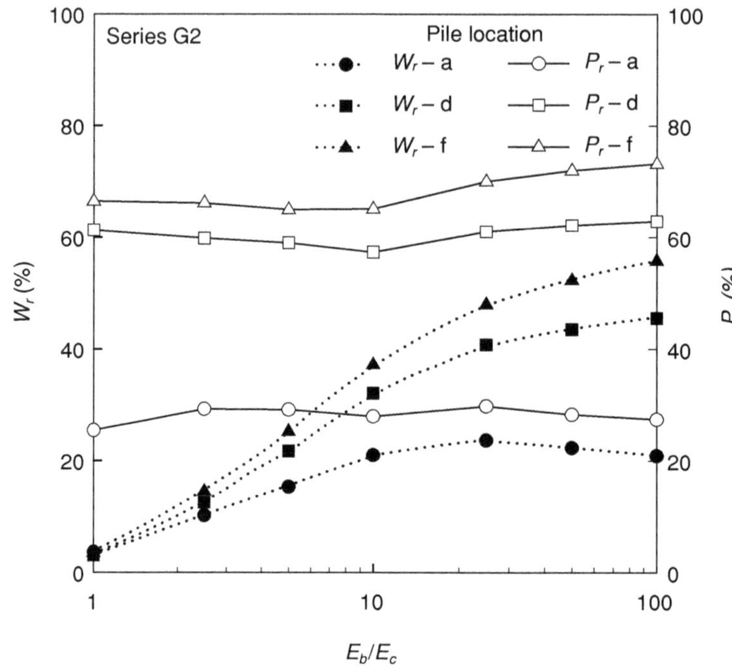

Figure 6.19 Variations of W_r and P_r with E_b/E_c for a 5 × 5 pile group (E_c = 5 MPa)

in E_b/E_c. Relatively, the centre pile has the largest where the corner pile has the smallest change in P_r. Irrespective of the value of E_b/E_c, the difference between the largest and the smallest change can be up to 35 percent, illustrating the importance of the shielding effects on dragload. The computed results are consistent with a previous study by Chow et al. (1990).

The gradual increase in W_r with E_b/E_c is related to the development of downdrag and dragload at the pile tip. The normalized dragload distributions with depth for piles with E_b/E_c equal to 1 and 100 are shown in Figure 6.20. It can be seen that dragloads for end-bearing piles (i.e. E_b/E_c = 100) gradually increase with depth for the entire pile length, while those for the floating piles (i.e. E_b/E_c = 1) increase with depth initially but reduce from NP (the location of NP for a: Z/L = 0.67, for d and f: Z/L = 0.81) toward the pile tip. Since the dragloads near the pile tip are practically the same for the floating piles, similar downdrag at the pile tip is expected. On the other hand, the dragload near the pile tip for each end-bearing pile is very different. The corner pile has the smallest P_r and the center pile has the largest P_r owing to the difference in W_r for each pile as shown in Figure 6.20. Irrespective to the locations of the piles, a larger toe resistance is mobilized at the pile tip for end-bearing piles (E_b/E_c = 100) than for the floating piles (E_b/E_c = 1). Moreover, for end-bearing piles, the magnitude

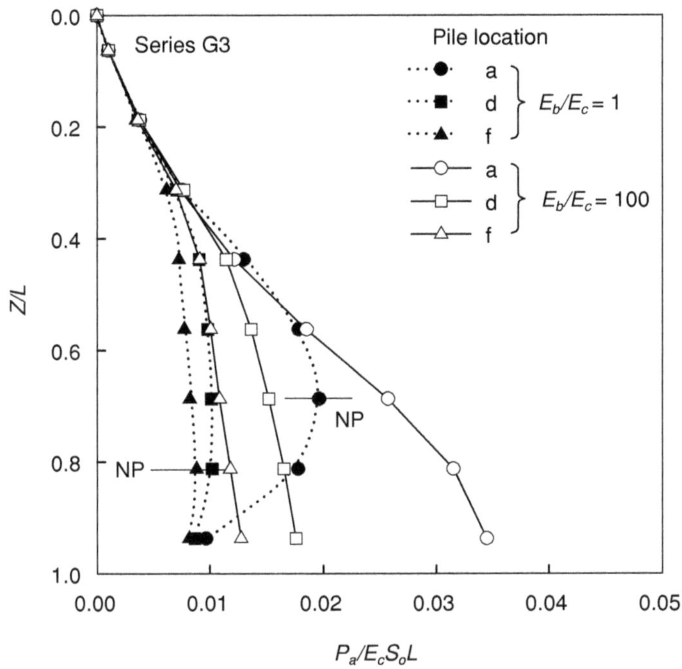

Figure 6.20 Distributions of normalized dragload for floating and end-bearing piles in a 5 × 5 pile group (E_c = 5 MPa)

of mobilized toe resistance strongly depends on the location of the pile in the group. On the contrary, for floating piles, the magnitude of the mobilized toe resistance appears to be independent of the pile locations in the group.

5.3 Effects of surface loading on the shielding effects in the pile group

Figure 6.21 illustrates the variations in W_r and P_r with the normalized surface loading, $\Delta pL/\gamma_{crit}E_c$, at different pile locations from the elasto-plastic slip analyses. The normalized surface loading represents the ratio of the maximum soil settlement, $\Delta pL/E_c$, to the limiting shear displacement, γ_{crit}, which is equal to 5 mm. As illustrated in the figure, both W_r and P_r decrease non-linearly with an increase in normalized surface loading. The changes in W_r and P_r are attributed to the extent of soil slip developed at the interface. The larger the surface loading, the greater the extent of the soil slip and hence the higher the shielding effects developed in the pile group, resulting in smaller W_r and P_r. As discussed previously, the center and corner piles develop the largest and smallest group effects respectively. For the range of surcharge loading considered from 12.5 kPa to 100 kPa in the analyses,

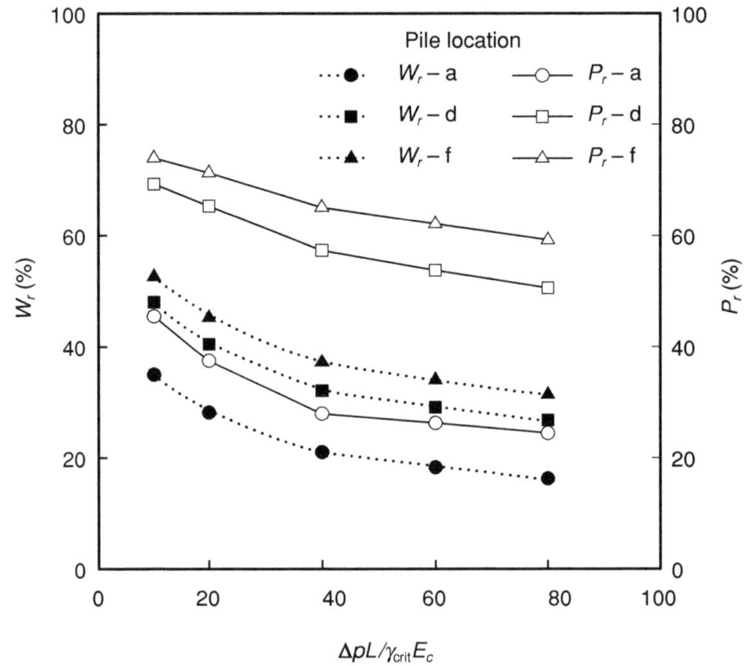

Figure 6.21 Variations of W_r and P_r with normalized surface loading for a 5 × 5 pile group

the computed group effects lead to a reduction of W_r and P_r of the order of 20 percent, irrespective of the location of the pile. Compared with P_r, W_r at the corner pile (pile a) is less affected by the group interaction for a given surcharge loading, indicating that the use of sacrificing piles outside the pile group will be more effective on P_r than on W_r. Since P_r is larger than W_r at the center pile, this also implies that shielding effects are more effective on P_r than on W_r.

5.4 Effects of the pile group configuration on the shielding effects in pile groups

In order to quantify the effects of the pile group configuration (i.e. the total number of piles and pile spacing) on downdrag development, three series of elasto-plastic slip analyses were carried out. Figure 6.22 shows the computed W_r at three selected pile locations, a, d and f, for two 5 × 5 pile groups at 2.5D and 5.0D spacing and a 3 × 3 pile group at 2.5D spacing (see Figures 6.3a and 6.3b). At the given pile spacing of 2.5D, a larger reduction in W_r is computed for the 5 × 5 pile group than for the 3 × 3 pile

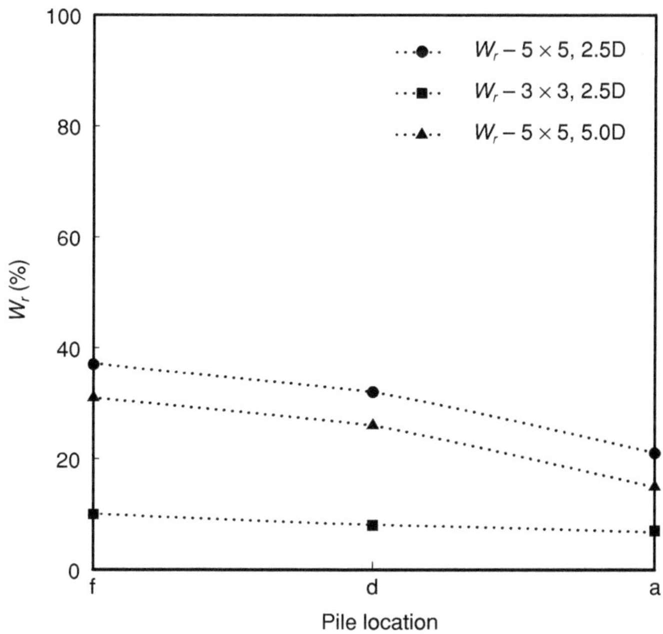

Figure 6.22 Changes in W_r with different pile group configurations

group. The computed W_r for the 5 × 5 pile group ranges from 21 percent to 37 percent, being the smallest at the corner and the largest at the center pile respectively. On the other hand, for the 3 × 3 pile group, there is very small variation in W_r from 7 percent to 10 percent at the three different pile locations, suggesting practically a uniform pile group settlement. Based on the three cases studied, the larger the number of piles in a group, the greater the shielding effects on W_r. The computed W_r for the 3 × 3 piles at 2.5D are much smaller than those obtained from the elastic solutions (W_r of 50 percent) reported by Lee (1993) and Teh and Wong (1995) for the same pile group configuration. This is probably due to the elastic solutions over-emphasizing the shielding effects by failing to account for soil slip, which reduces the protection offered to a pile inside a group by its neighbours. By considering the different shielding effects in two pile spacings on W_r within the 5 × 5 pile groups, it can be seen from Figure 6.22 that the computed results from the analysis on the 5 × 5 pile group at 5.0D spacing show only a small reduction in W_r of 6 percent to 7 percent compared with those from the analysis of the 5 × 5 piles at 2.5D spacing. These computed results suggest that W_r is more sensitive to the total number of piles than to the pile spacing within the pile group.

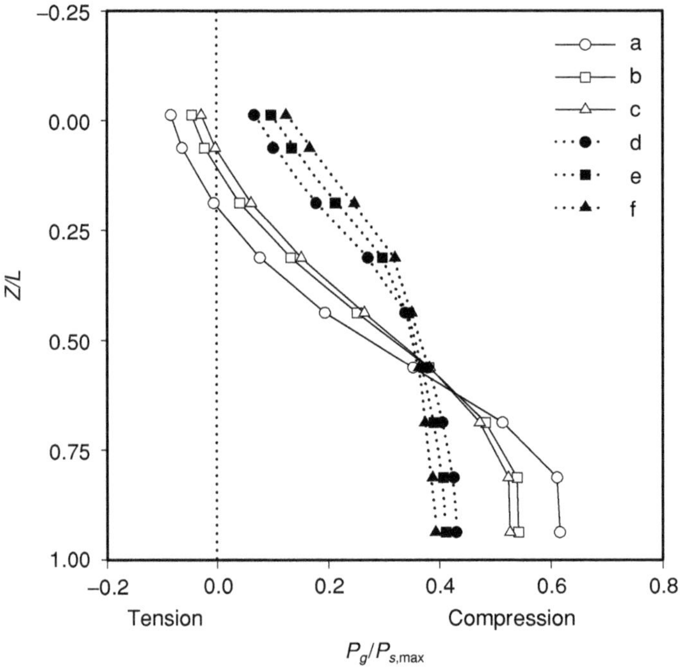

Figure 6.23 Distributions of the normalized axial forces on piles in a 5 × 5 pile group at 2.5D spacing with slip allowed

5.5 Interaction between the piles and the cap

Figure 6.23 illustrates the distributions of the normalized axial forces $P_g/P_{s,max}$ on the piles with depth in a 5 × 5 group at a 2.5D spacing, from the slip analysis. It was observed that tensile forces developed near the top of the outer piles (a, b, c), while compressive force developed near the top of the inner piles (d, e, f). The normalized maximum tensile forces $T_{max}/P_{s,max}$ at the outer piles were 3 percent to 8 percent. The tensile forces may have resulted in the pulling of the piles from the pile cap, implying the necessity of proper reinforcement at the pile head and the cap connection (Figure 6.24). The maximum tensile force developed at the corner pile (a), while the maximum compressive force developed at the center pile (f). It is interesting to note that the net increase in the axial forces from the pile head to the pile tip for the outer piles was larger than that at the inner piles. The axial forces at the outer piles gradually increased with pile depth, while the increase in the axial force at the inner piles from the mid-depth (Z/L = 0.4) to the pile tip was insignificant. This is probably related to the shielding effects based on the pile position. The computed W_r at the

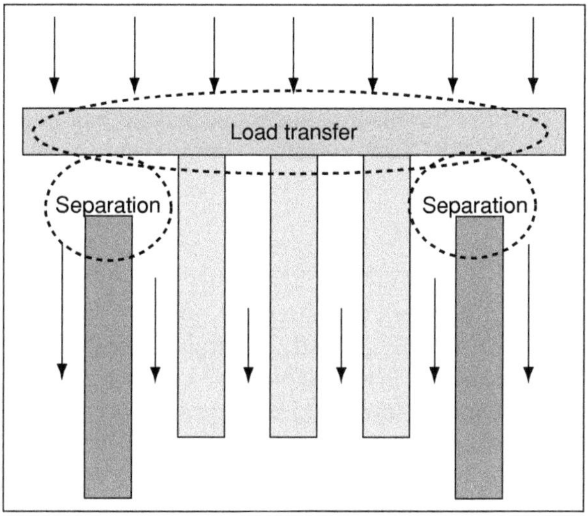

Figure 6.24 Pull-out of outer piles due to tensile force

central pile was 29 percent, which is smaller than its P_r of 38 percent at the same pile, reported by Lee and Ng (2004).

The distributions of the normalized axial forces among the piles heavily depend on the pile location, thereby leading to a different pile head settlement. The computed results show that the corner pile settled slightly more than the center pile ($W_a/W_f = 1.014$) since the dragload on the outer piles was larger than on the inner piles. Thus, the outer piles tended to drag the cap down, whereas the inner piles resisted the cap movement. This is because the pile cap rigidity induced nearly uniform pile settlements (Kuwabara and Poulos 1989). The interaction between the outer piles, the inner piles and the cap resulted in the development of tensile forces at the outer piles and compressive forces at the inner piles through the redistribution of axial forces among the piles (Teh and Wong 1995).

6 Comparisons between measured and computed dragloads and group effects

Two examples reported by Combarieu (1985) and Jeong (1992) were analysed based on the no-slip and slip analysis. In addition, comparisons were performed between computed and measured dragloads for single piles and group effects for piles in a group from experimental tests (Okabe 1977, Phamvan 1989, Little 1994, Lee et al. 1998). Since detailed information regarding material parameters and stress histories is, in some case studies, not provided, numerical simulations were made based on the best assumption of appropriate soil parameters, stress histories and boundary conditions. However, when modeling the case studies reported by Phamvan (1989) and Lee et al. (1998), the Modified Cam Clay soil model was used since all the required information was provided by the respective researchers from the experimental measurements and the Mohr–Coulomb soil model has been used in other case studies by Okabe (1977), Combarieu (1985), Jeong (1992) and Little (1994).

6.1 Example analysis by Jeong

In order to clarify the difference between slip analysis and no-slip continuum analysis for piles in consolidating ground, a simple example case reported by Jeong (1992) is discussed here. Geometry, soil parameters and boundary conditions of a single isolated end-bearing pile are shown in Figure 6.25. The rectangular-shaped pile measured 0.6 m × 0.6 m and was 30 m in length. A surface loading Δp of 250 kPa was applied on the surface of the soft clay layer. A rigid boundary condition was assumed for the bearing layer to simulate an end-bearing pile. In the original FEA, an effective friction angle of 25° and an effective cohesion of 3 kPa were used based on the Drucker–Prager soil model.

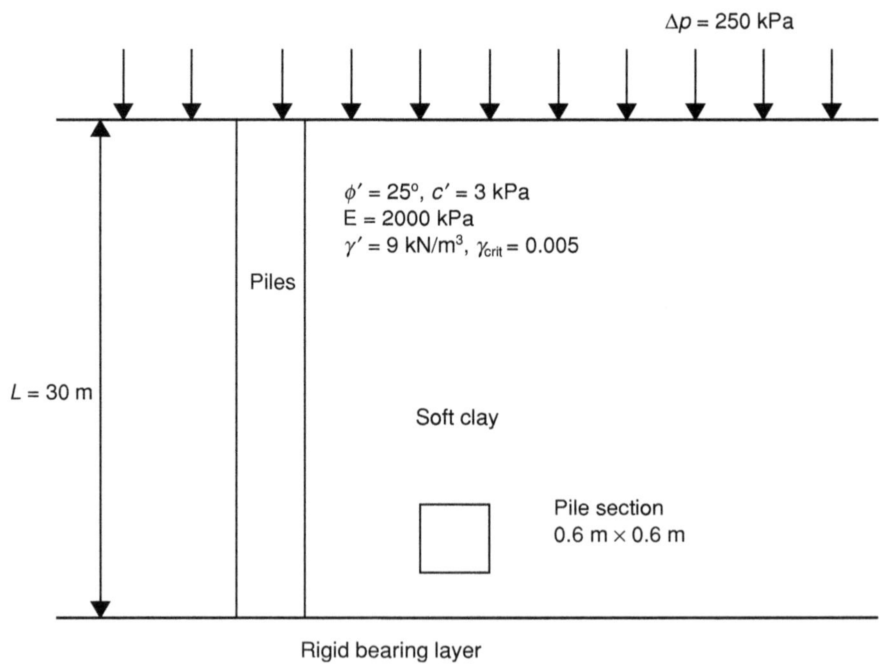

Figure 6.25 Configuration of a single pile reported by Jeong (1992)

However, in this analysis, a circular pile having the same sectional area was adopted in a 2D axisymmetric condition. A non-associated Mohr–Coulomb model was used for the clay. Since Jeong (1992) did not consider the interface friction coefficient μ, it was estimated as a value less than the soil friction angle ($\mu = \tan(25°) = 0.466$) because the interface friction angle was generally smaller than the internal friction angle of the soil. Therefore, in these analyses, two interface friction coefficients of 0.3 and 0.4 were used, which was similar to the measured interface friction coefficient from field observation by Phamvan (1989).

Figure 6.26 demonstrates computed dragload distributions obtained from the slip model ($\mu = 0.3, 0.4$) and the β method ($\beta = 0.25$), together with the no-slip continuum analysis and previous predictions by Jeong (1992). From the slip analysis, dragloads of 4857 kN and 6412 kN were obtained with $\mu = 0.3$ and 0.4 respectively. Figure 6.26 also shows the prediction from the no-slip continuum analysis, taking into account the central integration points of the soil elements next to the pile. A very close correlation was observed (11,748 kN) since Jeong (1992) also considered only the central integration points (12,821 kN). From the calculation based on the β method, a dragload of 6930 kN was obtained, slightly larger than that of slip analysis. Compared to the calculation based on the β method, the continuum analysis over-predicted NSF by about 200 percent. Since the surface loading

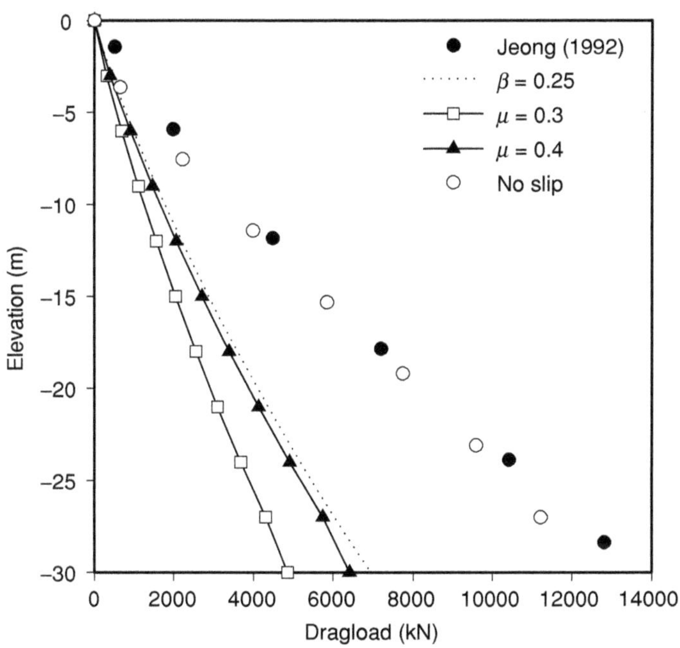

Figure 6.26 Dragload distributions computed by various methods

was very large (250 kPa) and the soil was very soft (E = 2 MPa), soil slip was likely to develop over the entire pile length. Therefore, it was not surprising that the result from the continuum analysis was not appropriate for such a large strain problem. In the absence of soil slip, excessive dragload was computed to have occurred. It was discovered that the increase in horizontal stress due to the surface loading was excessively large, thereby leading to very large shear stress. Furthermore, the results from the continuum analysis depended on the position of the integration point considered and the thickness of the soil element next to the pile as reported by Lee et al. (2002).

6.2 Example case study by Combarieu (1985), later extended by Jeong (1992)

In this example a rectangular pile group of 3 × 4 presented by Combarieu (1985) is considered. Information on the configuration of the pile group, as well as on the soil parameters, is shown in Figure 6.27. It is assumed that a relative displacement of 5 mm would fully mobilize skin friction. Predictions of the dragload on single piles and piles in a group determined from various design methods have been presented based on Combarieu (1985) and Jeong (1992). Additional results from the numerical analysis described in this paper as well as work by Shibata et al. (1982), Chow et al. (1990), Teh and Wong (1995) and Chow et al. (1996) are discussed. Since detailed

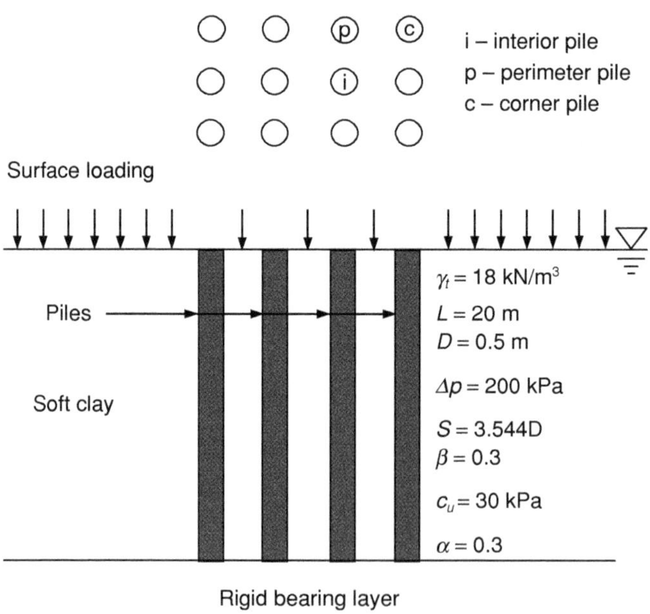

Figure 6.27 Configuration of a pile group by Combarieu (1985)

soil parameters are not presented, typical soil parameters are assumed. Assuming a K_o value of 0.6 to 0.7 for normally consolidated clay and a β value of 0.3 given by Combarieu (1985), an estimate of the interface friction coefficient μ of 0.4 and 0.5 is obtained. The stiffness modulus E of the clay is taken as 5 MPa and 10 MPa respectively. Both 2D and 3D numerical analyses were carried out for a single pile and piles in a group respectively.

As shown in Table 6.2, less dragload is computed for a single pile from the analyses described here than from the conventional β method. Since the pile tip is located on a rigid bearing layer, the neutral plane is very close to the pile toe. Therefore, reduced skin friction due to partial mobilization near the neutral plane does not have a significant effect. Also, since the surface loading is very large (200 kPa), the maximum shear stress could have been developed along the entire pile length. However, when considering the effective vertical stress near the pile, smaller than expected stresses are observed owing to the transfer of soil weight to the pile shaft, and hence less shear stress is developed along the pile shaft. Therefore, the dragload estimate based on the β method should give an upper-bound estimate as discussed previously.

For piles in a group, very large group effects are predicted from conventional approaches [1–5] (refer to Table 6.2), 3D numerical analysis based on continuum analyses [6], a simplified method [7] and a graphic method [8]. Extremely large group effects P_r (54–84 percent) are predicted for the central pile. However, such large group effects would only be possible for small pile spacing, large surface loading and when pile numbers are large, as shown in the previous comparison. A smaller reduction in dragload (2–14 percent) is obtained from the slip model. Similar observations are made by various researchers [9, 10, 11], although slightly different pile configurations are considered, i.e. 3 × 3 pile group and a pile spacing of 3D. Despite the number of piles being larger (12), pile spacing is wider (3.5D) in this example. Therefore, it could be assumed that the difference would be insignificant. An overall small group reduction (2–25 percent) is obtained from these analyses [9–11], which is similar to the prediction based on the slip model. Furthermore, it had been shown from the experimental observations that the group effect was, for most cases, relatively small. It was found that conventional design methods normally over-predict dragload and group effects.

6.3 *Study of Okabe (1977)*

Okabe (1977) reported the results from full-scale field measurements of dragload in a pile group, resulting from a combination of dewatering and surcharge loading. The pile group consisted of 38 piles spaced at 2.1D (Figure 6.28). There were 14 external piles for protection, which were to take most of the negative skin friction and were free to move, and 24 internal

Table 6.2 Estimated dragloads and group effects

	Group(total) kN	*Dragload (kN) and group effect (%)*			*Single pile (kN)*
		Corner	*Perimeter*	*Interior*	
β-method					2640
1 Terzaghi and Peck (1948)	β-Method – 33,130				
	α-Method – 11,832				
2 Zeevaert (1957)	26,620	2640	2480	590	
		0%	6%	78%	
3 Broms (1966)	Case 1 – 31,680	(4775)[1]	1730	590	
	Case 2 – 33,130	2640			
	Case 3 – 22,120	0%	34%	78%	
4 Broms (1976)	27,580	2640	2640	590	
		0%	0%	78%	
5 Combarieu (1985)	10,448	1265	758	420	
		52%	71%	84%	
6 Jeong (1992)	18,746	2054	1541	642	2568
		20%	40%	75%	
7 Briaud et al. (1991)	17,096	1980	1320	628	
		25%	50%	76%	
8 Shibata et al. (1982)	17,636	1663	1426	1214	
		37%	46%	54%	
9 Teh and Wong (1995)		2%		10%	
10 Chow et al. (1990)	Pile configuration 3 × 3, 3.0D	15%	18%	22%	
11 Chow et al. (1996)		10–25%			
12 Present study	21,768	1878	1802	1722	1916
	$\mu = 0.5$, $E = 5$ MPa	2%	6%	10%	
	21,198	1809	1769	1674	1945
	$\mu = 0.5$, $E = 10$ MPa	7%	9%	14%	
	23,478	1491	1474	1445	1534
	$\mu = 0.4$, $E = 5$ MPa	2%	4%	6%	
	17,618	1486	1465	1442	1558
	$\mu = 0.4$, $E = 10$ MPa	5%	6%	7%	

Notes:
1 Computed dragloads for the corner pile (4775 kN) is larger than the single pile (2640 kN).
2 Computed dragloads for the case single pile and [1]–[6] refer to Combarieu (1985) and Jeong (1992).

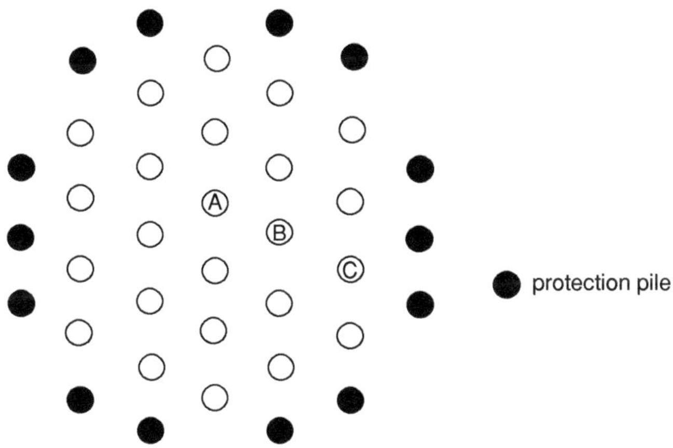

Figure 6.28 Configuration of group of 38 piles Okabe (1977)

end-bearing piles, connected to a rigid pile cap. This observation was found in the literature reporting the largest group effects for piles in a pile group. A group effect of roughly 85–8 percent was reported on the inside piles. Almost no skin friction developed along the length of the piles, with little compressive force on the pile head. A 51 percent group effect P_r was also measured for the external protection piles. A tensile force of 605 kN was approximately measured at the pile head of the protection piles. However, since the protection piles were free to move, little or no tensile force should have been developed. The reliability of this measurement is therefore questionable. This has also been discussed by Teh and Wong (1995).

Since information regarding soil parameters, loading sequence, exact pile configuration, structural load on the footing and interaction between the topsoil and the footing are not available, some simplifications have to be made for the purpose of the FE analysis. In this analysis the pile cap is not in contact with the soil surface for the purpose of simplicity in the FE simulation and for comparison with a previous theoretical study by Kuwabara and Poulos (1989). Therefore, the pile cap is above the soil surface at the beginning of the analysis. No external load is applied on the pile cap or pile heads, but dead weight from the cap has been included. In this analysis, the pile group is modeled as a 6 × 6 group with a pile spacing of 2.1D (Figure 6.29). Therefore only the most important features of the real situation could be approximately simulated. The heads of the inner pile (i) and the central pile (c) were connected to a rigid footing, while the outer protection piles were vertically separated from the footing, as reported by Okabe (1977). For predictions, the soil properties and surface loading were estimated by fitting the measured dragload distribution, pile head settlement (70 mm) and elastic compression (10 mm) of the pile to

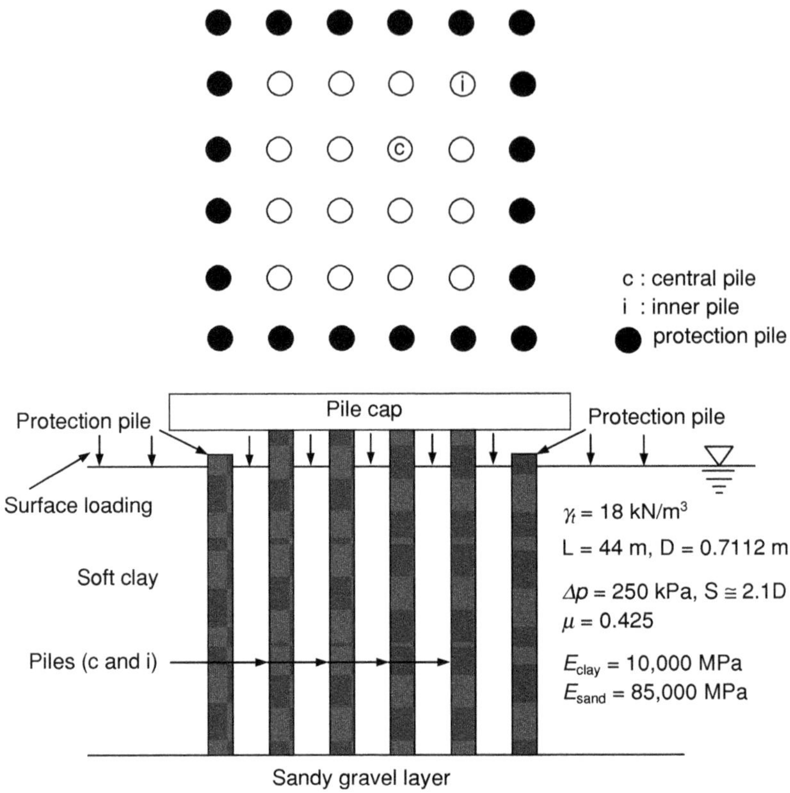

Figure 6.29 Configuration of pile group simulated in the FEA to model the full-scale experiment of Okabe (1977)

that computed by the FEA. Therefore, a soil stiffness modulus E of 10 MPa (clay), E_{sand} 85 MPa, E_{pile} 30 GPa (steel pipe pile), an interface friction coefficient μ of 0.425 and a soil surface loading of 250 kPa, resulting from combined effects of an increase in effective vertical stress in the soil due to dewatering and embankment loading, are assumed. 3D analyses were carried out to model the rigid pile caps properly.

The dragload of piles in a group from the field measurement and the numerical analysis is shown in Figure 6.30. Some tension develops at the pile head for the inner piles (i) and compressive force is observed along the central piles (c) since the outer piles try to move further compared to the inner piles as shown in Figure 6.30 (see Figures 6.28 andand 6.29 for the position of piles). Group effects P_r varying from 44 percent to 72 percent are computed. The maximum dragload is computed for the protection piles, whereas the central pile has the least dragload. The computed group effect for the protection piles (44 percent) is very similar to the values

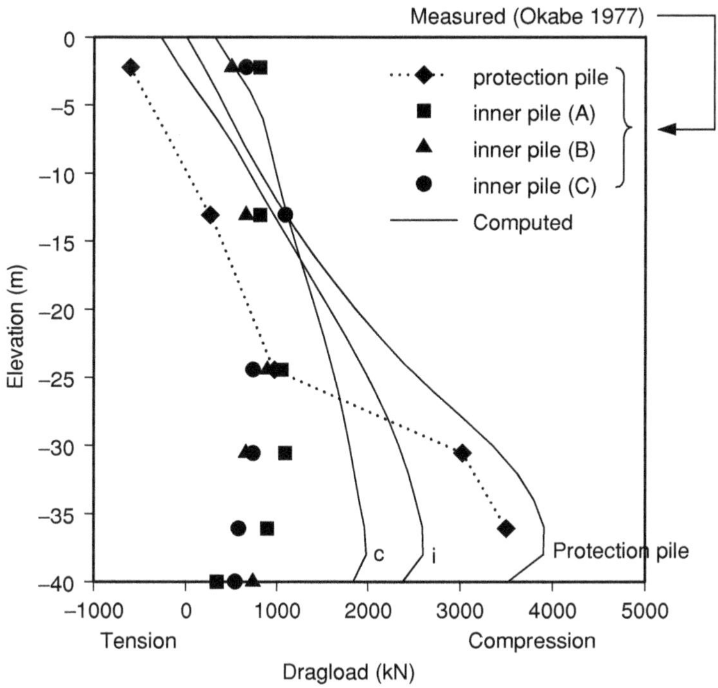

Figure 6.30 The distribution of dragload

reported by Okabe (1977) (51 percent) whereas smaller group effects (more dragload) are computed for inner (60 percent) and central piles (72 percent) than the range 85–8 percent measured by Okabe (1977).

6.4 Study of Phamvan (1989)

Phamvam (1989) reported the development of dragload on a single pile due to embankment loading. After construction of a 2 m high embankment, a pile was driven through the weathered crust, soft and medium stiff clay until the pile toe located on a bearing layer of stiff clay. Detailed material parameters and stress histories were measured which enabled a FE simulation to be conducted (Phamvan 1989). Figure 6.31 shows the distribution of dragload as determined from the field measurement as well as from the numerical analyses. The no-slip continuum analysis and the conventional β method, taking an average β value of 0.2 as measured in the field, over-estimated the dragloads. The β method may have produced a better prediction if the partial mobilization of NSF and PSR near the neutral plane had been considered. A good agreement with the field measurement is obtained when soil slip is taken into account in the numerical

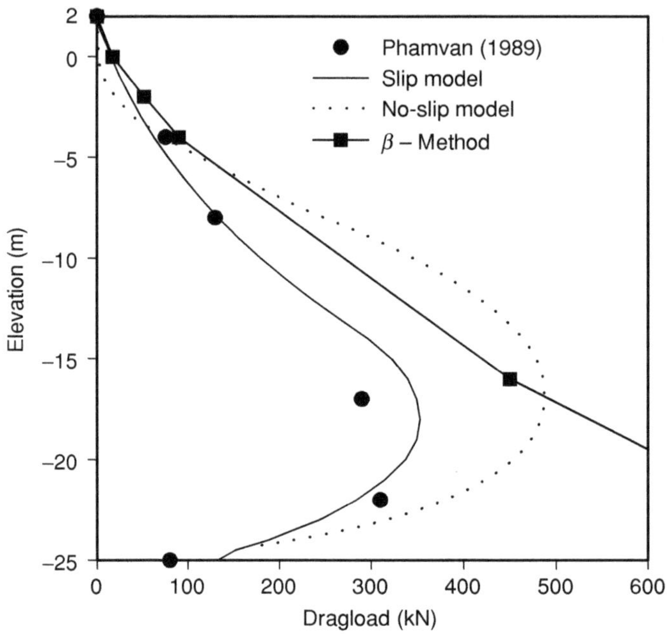

Figure 6.31 Comparisons of the development of dragload

simulation (slip analysis), although the position of the predicted neutral plane is slightly above its measured location.

6.5 Case study by Little (1994)

Little (1994) presented measurements of dragloads for two pile groups (friction piles and end-bearing piles) of nine piles (3 × 3) with spacing of 4.0D (D = 0.406 m). One group (friction piles) was driven to 20.4 m, approximately 1 m above the gravel layer. The other group of piles (end-bearing piles), driven to set on to the gravel layer, were slightly longer than the friction piles (20.8 m). An embankment loading of 40 kPa was applied on the top of the clay after driving the piles. A soil modulus E of 3.5 MPa was back-calculated for the clay layer from the measured ground settlement (180 mm). An interface friction coefficient μ of 0.35 was used, typical for a driven concrete pile (Teh and Wong 1995). Since soil parameters and detailed boundary conditions were not presented, only a 2D FEA was carried out to estimate dragloads for a friction pile and an end-bearing pile, respectively. The maximum dragloads for the piles inside the groups were interpolated based on the 3D numerical results reported by Lee et al. (2002).

Table 6.3 shows both measured and computed dragloads for the center pile and the normalized group effects ((dragload$_{\text{corner pile}}$ – dragload$_{\text{centre pile}}$)/

Table 6.3 Comparison of the dragload and group effects

	Measured	*Predictions by various methods*			
		Jeong (1992)	*Briaud et al. (1991)*	*Shibata et al. (1982)*	*Present study*
Dragload for central pile (kN)	friction (187) bearing (202)	friction (72–90) bearing (90–108)	106	friction (103–29) bearing (111–134)	friction (223) bearing (347)
Normalized group effect (%)	friction (14) bearing (18)	(51)	friction (29–43) bearing (53–60)	friction (24) bearing (26)	(15)

dragload$_{\text{corner pile}}$) between a center pile and a corner pile as the dragload for a single pile was not reported. The measured dragloads on the corner piles were 220 kN and 250 kN for friction and end-bearing piles respectively. In order to predict dragloads for a central pile, based on methods presented by Shibata et al. (1982) and Jeong (1992), the measured dragload for a single pile was assumed as 200–50 kN and 250–300 kN for a friction pile and a bearing pile respectively, based on the measured dragload distributions on piles in a group. The normalized group effects based on no-slip analysis by Jeong (1992) are interpolated for a pile spacing of 4.0D. Larger normalized group effects (51 percent) and hence smaller dragloads (72–90 kN for friction piles and 90–108 kN for end-bearing piles) were estimated for the central piles using the continuum approach. The predictions from Briaud et al. (1991) and Shibata et al. (1982) also underestimated dragload for the central pile. However, smaller normalized group effects (15 percent), which match reasonably with the field observations (14–18 percent), are computed from the slip approach compared to those from other approaches.

Figure 6.32 shows the measured and computed distributions of dragload on a center pile. The distributions of dragload from the numerical analyses were based on the results from the single pile analysis. Dragload distributions for the center piles were estimated based on the interpolated group effects. Since the measured neutral plane was located at 55 percent and 60 percent of the pile length for the friction pile and the end-bearing pile respectively, fewer dragloads developed. However, positions of the neutral plane determined from FEA varied around 70 percent and 90 percent of the pile length for the friction and the end-bearing piles respectively. The field observations implied that the pile tip was not located into a stiff bearing layer, leading to large downdrag. Therefore, larger dragloads (223 kN

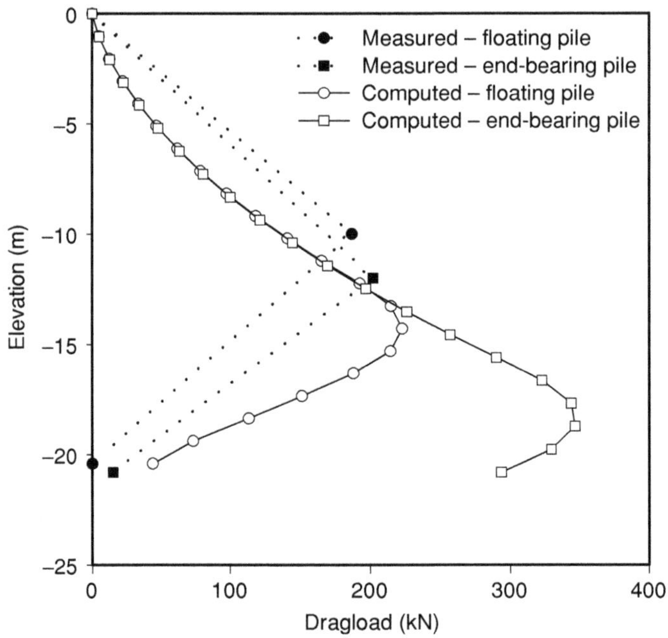

Figure 6.32 Distributions of measured and computed dragload

and 347 kN for friction and end-bearing piles respectively) were computed for the central piles from FEA rather than from the measured dragloads (187 kN for friction piles and 202 kN for end-bearing piles respectively) since a larger dragload was computed for a single pile.

6.6 *Study of Lee et al. (1998)*

Lee et al. (1998) reported measurements of dragload in a model pile from a single centrifuge test. The model pile had a diameter of 0.03 m and a length of 0.45 m. Complete data regarding the soil parameters and stress histories of the soil and the boundary conditions during the test were reported. A large interface friction angle of 25.8° was measured from the experiment. However, since the pile was installed at 1 g (*g* is the centrifugal gravity), the pile behavior would have been similar to that of a bored pile (Fioravante et al. 1994). Therefore, in this analysis, friction coefficients of 0.3 and 0.4 were used to investigate a possible reduction in the interface friction coefficient. Two factors contribute to the development of NSF mechanism. First, a small amount of dragload develops owing to the increase of self-weight of the soil during the acceleration of the centrifuge test package from 1 g to 50 g. Then, after consolidation had taken placed,

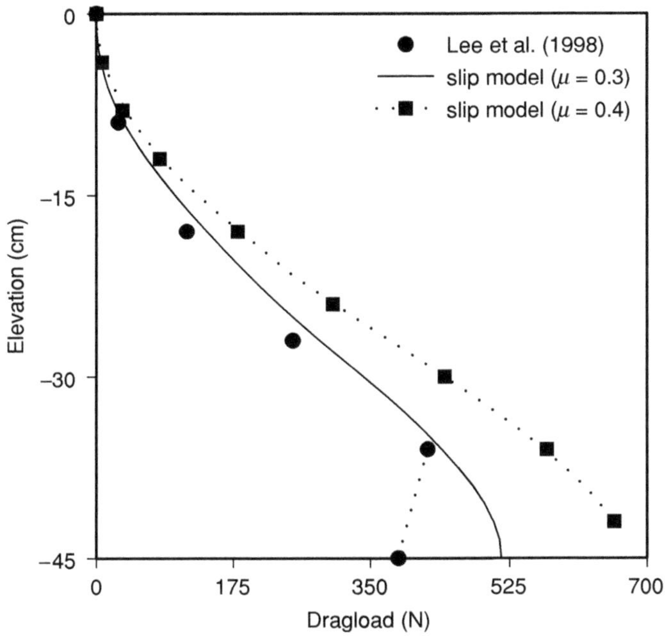

Figure 6.33 Fitting the slip model to centrifuge data

due to dewatering, the effective vertical stress in the soil increased by approximately 62 kPa.

Figure 6.33 demonstrates the distributions of dragloads as obtained from the centrifuge test and numerical analyses. A closer prediction is observed when a friction coefficient of 0.3 is adopted. Overall, a reasonable prediction of dragload is obtained from the slip model. Although the model pile is intended to have a fixed base, a pile movement of 1.5 mm has been measured during the centrifuge test. Therefore, some positive shaft resistance is measured near the pile tip where soil settlement is smaller than that of the pile. Hence, the maximum dragload is developed at 80–5 percent of the pile length measured from the top of the pile. In the numerical analysis the model pile is assumed to be end-bearing resting on a rigid base. Therefore, by allowing some pile movement (by modeling a fictitious soil layer below the pile tip), a better prediction of the distribution of dragload would be expected (shown in Figure 6.32 by the dotted line).

7 Conclusions

The development of negative skin piles and the shielding effects (group effect) on piles in pile groups in consolidating soil was investigated by carrying out 2D axisymmetric and 3D FE parametric analyses, with and without considering soil slip at the interface. The computed results from the no-slip elastic and elasto-plastic slip analyses were compared with the reported elastic solutions, example analyses and several experimental measurements. The effects of axial load on dragload changes were also studied.

The results of the numerical analyses described here are well compared to elastic solutions and recent theoretical studies. It has been found that the estimation of dragload and group effects from current design methods is neither satisfactory nor realistic. Dragload is normally over-estimated from empirical methods, elastic and continuum analyses. Current methods also over-emphasize group effects by failing to account for soil slip (plastic soil yielding), which reduces the protection offered to a pile inside a group by its neighbours. A larger dragload was computed using the no-slip continuum analyses, where slip was not included.

Numerical analysis showed that, the smaller the interface friction coefficient and the higher the surface loading, the more the soil slip at the interface and vice versa. Therefore, it could be said that the development of the negative shear stress heavily depended on the slip, which was mainly governed by the interface friction angle and the surface loading. Extremely large reductions in dragload were computed for piles inside the group, but the slip analysis showed smaller group effects. The computed group effects from the continuum analysis failed to account for soil slip that reduced the protection offered to a pile inside a group by their neighbors. These group effects are significantly smaller than previous research works, which should therefore be reconsidered. The reduced soil surface settlements and downdrag inside the pile group are attributed to the pile–soil interaction effect or the so-called shielding effects on the inner piles by the outer piles of the pile group. The depth of full mobilization (slip length) of the interface shear strength along a pile within pile groups depends on the pile location. The smallest soil slip develops at the center pile whereas the largest soil slip is computed for the single pile.

The gradual increase in the axial load after the full development of dragload resulted in reductions in dragload. The numerical analysis demonstrated that, the stiffer the bearing layer, the smaller the dragload reductions. An overall axial loading of about 125–325 percent of maximum dragload was required to eliminate dragload, depending on the relative bearing layer stiffness.

For piles connected to a cap, tensile forces may develop near the pile head for the outer piles. The elastic model over-estimates tensile forces at the outer pile and the shielding effect. The tensile forces are associated with the redistribution of axial forces among piles through the shielding effect. The tensile force development on piles is more sensitive to the number of piles than the pile spacing. The greater the shielding effect, the higher the tensile force in the outer piles. It was shown that the influence of NSF on the pile behavior is more uniformly distributed among piles with a cap than among free-headed piles. The issue on the development of tensile forces on the pile head at the outer piles needs to be carefully considered in pile design to prevent pull-out of the piles from the cap.

The example case histories demonstrated that the simple β method might predict a better estimate of dragload compared to no-slip continuum analysis. Numerical back-analysis of previous experimental observations compared well with slip analysis. Experimental observations confirmed that more realistic interface and plastic yielding behavior of the soil at the interface must be introduced if analyses were to be accurate.

Should the potential exist for the development of NSF on piles in soft ground, dragload (compressive force) is normally not a major problem in terms of the design strength of the pile material. However, downdrag (pile settlement) could present some difficulties from a serviceability viewpoint. Piles should therefore be installed to a stiff layer in order to reduce downdrag, depending on drivability and dragload (Lee 2001). Friction piles should be used with great care, since a strong connection to a stiff pile cap or raft may be required to prevent differential settlement. The general lesson to be drawn from this work is that the pile–soil interactions within a pile group, together with corresponding stiffness and soil slip, must be considered if the serviceability of the foundation system is to be properly assured.

References

ABAQUS (1998) ABAQUS User's and Theory Manuals, Version 5.8, Hibbitt, Karlsson & Sorensen Inc., Rhode Island.

Acar, Y. B., Avent, R. and Taha, M. R. (1994) Downdrag on friction piles: a case history, *Proceedings of Settlement 94: Vertical and Horizontal Deformations of Foundations and Embankment*, Geotechnical Special Publication, No. 40, ASCE, New York, pp. 986–99.

Alonso, E. E., Josa, A. and Ledesma, A. (1984) Negative skin friction on piles: a simplified analysis and prediction procedure, *Géotechnique*, 34 (3): 341–57.

Baguelin, F. and Frank, R. (1982) Theoretical studies of piles using the finite element method, *Foundation Engineering*, ed. Georges Pilot, Presses Ponts et Chaussées.

Bjerrum, L., Johannessen, I. J. and Eide, O. (1969) Reduction of negative skin friction on steel piles to rock, *7th ICSMFE, Mexico City*, Rotterdam: Balkema, pp. 27–34.

Bozuzuk, M. (1981) Bearing capacity of pile preloaded by downdrag. *10th ICSMFE, Stockholm*, Rotterdam: Balkema, pp. 631–6.

Briaud, J. L., Jeong, S. and Bush, R. (1991) Group effect in the case of downdrag, Geotechnical Engineering Congress, Geotechnical Special Publication, No. 27, ASCE, New York, pp. 505–18.

Broms, B. B. (1979) Negative skin friction, *Proc, 6th Asian Regional Conference on SMFE, Singapore*, 2: 41–75.

Burland, J. B. (1973) Shaft friction of piles in clay – a simple fundamental approach, *Ground Engineering*, 6 (3): 30–42.

Bustamante, M. (1999) Current French design practice for axially loaded piles, *Ground Engineering*, 38–44.
Chan, S. F. (1996) Foundation failures, *Short Course on Deformation of Soils and Rocks and Displacement of Structures – Soil and Rock Structure Interaction*, Asian Institute of Technology, Thailand, pp. 5–6.
Chan, S. H., Lam, S. Y., Ng, C. W. W. and Leung, C. F. (2006) Axial loading effect on piles with locked-in negative skin friction in consolidating soil, *Physical Modelling in Geotechnics, 6th ICPMG 06*, ed. C. W. W. Ng, Zhang and Wang, pp. 807–13.
Chellis, R. D. (1961) *Pile Foundations*, New York: McGraw-Hill.
Chow, Y. K., Chin, J. T. and Lee, S. L. (1990) Negative skin friction on pile groups, *International Journal for Numerical and Analytical Method in Geomechanics*, 14: 75–91.
Chow, Y. K., Lim, C. H. and Karunaratne, G. P. (1996) Numerical modeling of negative skin friction on pile groups, *Computers and Geotechnics*, 18 (3): 201–24.
Combarieu, O. (1985) *Frottement négatif sur les pieux*, Rapport de recherche LPC No 136, Laboratoire Centrale des ponts et chaussées.
Comodromos, E. M. and Bareka, S. V. (2005) Evaluation of negative skin friction effects in pile foundations using 3D nonlinear analysis, *Computers and Geotechnics*, 32: 210–21.
Davisson, M. T. (1993) Negative skin friction in piles and design decision, *3rd International Conference on Case Histories in Geotechnical Engineering, Rolla, Missouri*, 3: 1793–1801.
Denman, K. J., Nicholls, R. A. and Symons, M. V. (1977) Model studies of negative friction on pile groups, *Proc. Large Ground Movements and Structures*, ed. J. D. Geddes, pp. 252–71.
Endo, M., Minou, A., Kawasaki, T. and Shibata, T. (1969) Negative skin friction acting on steel pipe pile in clay, *7th ICSMFE, 2, Mexico City*, pp. 85–92.
Ergun, M. U. and Sonmez, D. (1995) Negative skin friction from surface settlement measurements in model group tests, *Canadian Geotechnical Journal*, 32: 1075–9.
Fellenius, B. H. (1972) Down-drag on piles in clay due to negative skin friction, *Canadian Geotechnical Journal*, 9: 323–37.
Fellenius, B. H. (1997) Piles subjected to negative friction: a procedure for design, Discussion, *Geotechnical Engineering, Journal of the Southeast Asian Geotechnical Society*, 28 (1): 277–81.
Fellenius, B. H. (1999) Piling terminology. http://www.geoforum.com/info/pileinfo/terminology.asp.
Fellenius, B. H. (2006) Results from long-term measurement in piles of drag load and downdrag, *Canadian Geotechnical Journal*, 43: 409–30.
Fioravante, V., Jamiolkowski, M. and Pedroni, S. (1994) Modelling the behaviour of piles in sand subjected to axial load, *Proc. Centrifuge 1994*, pp. 455–60.
Garlanger, J. E. and Lambe, T. W. (1973) Proceedings of a symposium on downdrag on piles, Research Report R-73-56, Soil-331, Department of Civil Engineering, MIT, Cambridge, Mass.
Hanna, A. M. and Sharif, A. (2006) Drag force on single piles in clay subjected to surcharge loading, *International Journal of Geomechanics*, ASCE, 130 (6): 89–96.
Indraratna, B. (1993) Development of negative skin friction on driven piles in soft Bangkok clay, Reply, *Canadian Geotechnical Journal*, 30: 887–8, 890–1.

Indraratna, B., Balasubramaniam, A. S., Phamvan, P. and Wong, Y. K. (1992) Development of negative skin friction on driven piles in soft Bangkok clay, *Can. Geotech. J.*, 29: 393–404.
Inoue, Y., Tamaoki, K. and Ogai, T. (1977) Settlement of building due to pile downdrag, *9th ICSMFE, 1, Tokyo*, pp. 561–4.
Jeong, S. (1992) Nonlinear three dimensional analysis of downdrag on pile groups, PhD thesis, Texas A & M University.
Jeong, S., Kim, S. and Briaud, J. (1997) Analysis of downdrag on pile groups by the finite element method, *Computers and Geotechnics*, 21 (2): 143–61.
Jeong, S., Lee, J. H. and Lee, C. J. (2004) Slip effect at the pile–soil interface on dragload, *Computer and Geotechnics*, 31 (2): 115–26.
Koerner, M. K. and Mukhopadhyay, C. (1972) Behaviour of negative skin friction on model piles in medium-plastic silt, *Highway Research Record*, 405: 34–44.
Kuwabara, F. and Poulos, H. G. (1989) Downdrag forces in group of piles, ASCE, *Journal of Geotechnical Engineering*, 115 (6): 806–18.
Lambe, T. W., Garlanger, J. E. and Leifer, S. A. (1974) Prediction and field evaluation of downdrag forces on a single pile, Research report R74-27, Soils publication 339, Department of Civil Engineering, MIT, Cambridge, Mass.
Lee, C. J. (2001) The influence of negative skin friction on piles and in pile groups, PhD thesis, Cambridge University.
Lee, C. J., Bolton, M. D. and Al-Tabbaa, A. (2002) Numerical modelling of group effect on the distribution of dragloads in pile foundations, *Géotechnique*, 52 (5): 325–35.
Lee, C. J., Chen, H. T. and Wang, W. H. (1998) Negative skin friction on a pile due to excessive ground water withdrawal, *Proc. Centrifuge 1998*, pp. 513–18.
Lee, C. J., Lee, J. H. and Jeong, S. (2006) The influence of soil slip on negative skin friction in pile groups connected to a cap, *Géotechnique*, 56 (1): 53–6.
Lee, C. J. and Ng, C. W. W. (2004) Development of downdrag on piles and in pile groups in consolidating soil, *Journal of Geotechnical and Geoenvironmental Engineering*, ASCE, 130 (9): 905–14.
Lee, C. Y. (1993) Pile groups under negative skin friction, *Journal of Geotechnical Engineering*, ASCE, 119 (10): 1587–1600.
Leung, C. F., Liao, B. K., Chow, Y. K., Shen, R. F. and Kog, Y. C. (2004) Behaviour of pile subject to negative skin friction and axial load, *Soils and Foundations*, 44 (6): 17–26.
Little, J. A. (1994) Downdrag on piles: review and recent experimentation. *Proc. of Settlement 94: Vertical and Horizontal Deformations of Foundations and Embankment*, Geotechnical Special Publication, No. 40, ASCE, pp. 1805–26.
Little, J. A. and Ibrahim, K. I. (1993) Predictions associated with the pile downdrag study at the SERC soft clay site at Bothkennar, Scotland, *Predictive Soil Mechanics (Wroth Memorial Symposium)*, London: Thomas Telford, 796–818.
Maugeri, M., Castelli, F. and Motta, E. (1997) Settlement of a piled foundation due to negative skin friction: a case history, *14th ICSMFE, Hamburg*, Balkema, Rotterdam: Balkema, pp. 1111–14.
Nishi, K. and Esashi, Y. (1982) Field measurement and prediction of negative skin friction in piles, *Proc. International Symposium on Numerical Models in Geomechanics, Zurich*, pp. 776–84.
Okabe, T. (1977) Large negative friction and friction-free pile methods, *9th ICSMFE, 1, Tokyo*, pp. 679–82.

Phamvan, P. (1989) Negative skin friction on driven piles in Bangkok subsoils, PhD thesis, AIT, Bangkok.

Poulos, H. G. (1997) Piles subjected to negative friction: a procedure for design, *Geotechnical Engineering, Journal of the Southeast Asian Geotechnical Society*, 28 (1): 23–44.

Poulos, H. G. and Davis, E. H. (1980) *Pile Foundation Analysis and Design*, New York: John Wiley.

Poulos, H. G. and Mattes, N. S. (1969) The analysis of downdrag in end-bearing piles, *7th ICSMFE, Mexico City*, pp. 203–8.

Shen, W. Y. and Teh, C. I. (2002) A variational solution for downdrag force analysis of pile groups, *International Journal of Geomechanics*, 2 (1): 75–91.

Shibata, T., Sekiguchi, H. and Yukitomo, H. (1982) Model test and analysis of negative friction acting on piles, *Soils and Foundations*, 22 (2): 29–39.

Teh, C. I. and Wong, K. S. (1995) Analysis of downdrag on pile groups, *Géotechnique*, 45 (2): 191–207.

Terzaghi, K. and Peck, R. B. (1948) *Soil Mechanics in Engineering Practice*, New York: John Wiley.

Thomas, J. (1998) Performance of piles and pile groups in clay, PhD thesis, University of Western Australia.

UniNews (1998) Downdrag and dragload – aren't they synonyms?

http://www.unisoftltd.com/news98.htm.

Wong, Y. K. (1991) Numerical simulation of settlements associated with driven piles in Bangkok clay, MEng Thesis, AIT, Bangkok.

Zeevaert, L. (1983) *Foundation Engineering for Difficult Subsoil Conditions*, 2nd edn, New York: Van Nostrand Reinhold.

7 Semi-analytical approach for analyzing ground vibrations caused by trains moving over elevated bridges with pile foundations

Y. B. Yang and Yean-Seng Wu

Abstract

A *semi-analytical approach* is presented for the analysis of three-dimensional ground vibrations induced by trains traveling over a multi-span elevated bridge with pile foundations. The train is modeled as two sets of moving loads, with one accounting for the front wheelsets and the other for the rear wheelsets, the bridge as a series of elastically supported beams, and the ground as a viscoelastic halfspace. By the present semi-analytical technique, the entire vibration problem is divided into three *subproblems*, each dealing with the dynamic behavior of the superstructure of the bridge, the interaction between the bridge foundations and surrounding soils, and the wave propagation through the halfspace by point sources. Extensive parametric studies were conducted to evaluate the influence of some key parameters on the ground vibrations caused by moving trains. The numerical results indicated that *train–bridge resonance* can result in drastically amplified ground responses, which decay in an oscillatory manner as the site-to-bridge distance increases. Moreover, there exists a *saturation phenomenon* in the ground acceleration response spectra when the train speed exceeds a certain limit. In addition, a comparatively lower level of ground vibrations exists for certain combinations of bridge girder span length and train speed, which suggests the existence of some *optimal designs* for the bridge concerning mitigation of train-induced ground vibrations. Finally, the effect of *elastic bearings* of the bridge on the ground responses to the moving trains is also studied.

1 Introduction

In order to reduce land usage, while providing an exclusive right of way, increasing numbers of bridges and elevated bridges are nowadays constructed as the supporting structure for mass rapid-transit and high-speed railway systems. However, the ground vibrations caused by the trains moving over the bridges may annoy the occupants of nearby residential buildings, while aggravating the function of equipment or production lines of adjacent high-precision factories. As the vibration or noise standards for residential buildings and production lines have become stricter, the problem of vehicle-induced vibrations has received increasing attention from researchers and engineers, especially for high-speed railways, since a train with higher speeds and heavier axle loads tends to induce larger ground vibrations.

Although there exists a large amount of research on train-induced ground vibrations (Hanazato et al. 1991; Krylov and Ferguson 1994; Krylov 1995; Madshus et al. 1996; Takemiya 1997), most of these works were aimed at the vibrations caused by trains moving on the ground or under the ground surface. For the problem of ground vibrations caused by trains moving over bridges or elevated bridges, however, relatively few attempts have ever been made. Hung et al. (1999) performed a parametric study on the ground vibration due to trains traveling over a bridge in a two-dimensional manner, where the effects of piers, piles, layered soils, and rock beds were considered. Based on the finite element method, Yoshioka (2000) studied the basic characteristics of ground vibration caused by the SKS trains and discussed the effectiveness of various measures in reducing the ground vibrations. By employing the absorbing boundary conditions, Ju (2002) conducted a three-dimensional analysis for the ground vibrations caused by high-speed trains moving over an elevated bridge of seven spans, along with discussion on the effectiveness of some countermeasures in mitigating the ground vibrations. Recently, the train-induced wave propagation in layered soils was investigated by Yang et al. (2003) using the 2.5D finite/infinite element approach, in which the load-moving effect in the third dimension was taken into account. Owing to the large computational efforts required in modeling the various components of the vehicle–bridge–soil system, only a limited part of the bridge structure and ground medium has been considered in most previous studies, which may not yield a complete three-dimensional picture for the train-induced ground vibrations. On the other hand, a fully three-dimensional modeling of the vehicle–bridge–soil system is difficult, or at least expensive, for the purpose of conducting a comprehensive parametric study for all the parameters involved.

In view of the limitations of the full numerical modeling approaches, Wu et al. (2002) proposed a *semi-analytical approach* that can efficiently simulate the train-induced ground vibrations in a three-dimensional sense. This approach hinges on the combined use of the analytic solution to the problem of an elastically supported beam being traveled over by a series of moving loads and Green's function for an elastic halfspace subjected to a harmonic point load. By such an approach, both the dynamic response of the bridge to the moving loads and the wave propagation properties of the ground can be easily taken into account, when compared with the approaches that rely on full numerical modeling. Later, such an approach was enhanced through the inclusion of pile–soil interactions, refinement of the train model and Green's function used to represent the halfspace (Wu and Yang 2004). In this chapter, basically the same semi-analytical approach will be employed, but with further improvements made for modeling of the bridges and the trains. Parametric studies will be conducted for evaluating the effects of the key parameters involved in the train–bridge–foundation–soil system on the ground response. Besides, the paper will look into the mechanism of ground response in relation to the train–bridge resonance, as well as the influence of the elastic bearings installed on the bridges for the purpose of reducing the earthquake forces transmitted from the ground.

2 Train, bridge, foundation and ground models

Figure 7.1a shows a train moving over a bridge that consists of multiple girder units supported by elastic bearings, which in turn are supported by piers resting on pile foundations. Each pile foundation consists of a number of floating piles covered by a rectangular pile cap below the ground surface. The train comprises several identical cars connected in a series and moves over the bridge with constant speeds. To simplify the formulation and, most importantly, to help identify the key parameters involved in the problem, some assumptions or simplifications will be made in the theoretical formulation.

The train is modeled as two sets of moving loads of equal intervals, with the first set referring to the loading action of the front wheelsets and the second set to that of the rear wheelsets. The second set of moving loads falls behind the first set by a distance equal to that between the front and rear wheelsets of the train car, as shown in Figure 7.1b. The girders and elastic bearings of the bridge units are modeled by uniform beams and by linear springs respectively. Therefore, the superstructure of the bridge, including the girders and elastic bearings, can be simulated as a series of *elastically supported beams*, of which only the first two modes are considered important. As for the substructure of the bridge, each of the piers and the supporting pile cap are both idealized as rigid bodies, while the piles within each group are collectively modeled as a frequency-dependent *linear spring–dashpot system* (see Figure 7.1b). Finally, the ground is assumed to be a *viscoelastic halfspace* composed of a homogeneous and an isotropic soil medium.

In this chapter, the effect of the inertia of the train and the track system, and the interaction between the train, the track and the bridge are all neglected, since they have little influence on the forces generated on the pier tops, which are the source for ground vibrations. An *equivalent lumped-parameter model* is adopted to represent, jointly, the pier and the foundation–soil interaction system, as shown in Figure 7.1b, with the

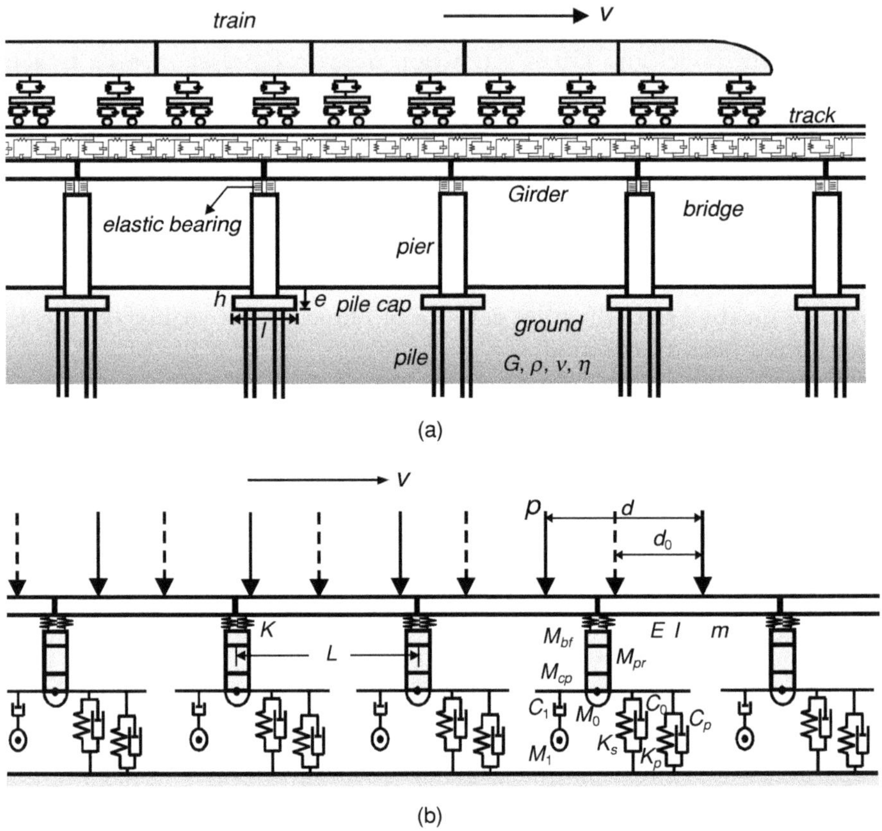

Figure 7.1 Multi-span bridge subjected to a train traveling with constant speed v: (*a*) schematic; (*b*) simplified model used in analysis

embedment effect of the pile cap being considered. Also indicated in Figure 7.1b are the physical parameters involved in the modeling, which will be defined in the sections when they first appear. Only the vertical vibrations of the train, the bridge and the ground are of concern to this study.

3 Method of analysis

For a train moving over an elevated bridge, the transmission of the bridge vibrations to the ground involves *three successive mechanisms*: (*a*) the dynamics of the superstructure of the bridge, (*b*) the interaction between the bridge foundations and the surrounding soil, and (*c*) the wave propagation in the halfspace generated by the point sources. The logistics of the formulation of the problem considered can be developed on the basis of the three subproblems, as is outlined below. First, the force acting on the top of each pier of the bridge, due to the moving train, is determined with due account being taken of the dynamic properties of the girders and the elastic bearings. Next, the interactive force between the foundation of each pier and the surrounding soil generated by the pier-top force is obtained by an equivalent lumped-parameter model for simulating the related components. Then the vibration response of the ground surface to the foundation–soil interactive force of each pier is computed by using existing Green's functions for a homogeneous and viscoelastic halfspace. Finally, by superimposing all the ground vibrations generated by the foundation–soil interactive forces of all the piers of the bridge, the total ground response can be obtained.

3.1 Forces acting on the top of each pier due to the moving train

When a train moves over the bridge, forces will be generated on the top of each pier, which will be referred to as the *pier-top forces*, through the vibration of adjacent girders. As was stated in Section 2, each girder of the bridge is modeled as an *elastically supported beam* and the train moving over the bridge as two sets of moving loads of equal intervals. By so doing, the problem considered can be reduced to one with an elastically supported beam subjected to two sequences of moving loads, which are different by a *time lag*. Because the solution to the problem under consideration can be regarded as the summation of the individual contributions by the two sets of moving loads, the problem of an elastically supported beam subject to the action of a single set of moving loads will first be formulated.

Consider a beam of length L, supported by two vertical linear springs of stiffness K at the two ends as shown in Figure 7.2. A sequence of loads of magnitude p spaced at distance d, i.e. car length, moves over the beam with a constant speed v. The *equation of motion* for the beam traveled by the moving loads can be written as (Yang et al. 1997):

$$m\ddot{u} + c\dot{u} + EIu'''' = p\sum_{k=1}^{N}\delta[\varsigma - v(t - t_k)] \times [H(t - t_k) - H(t - t_k - L/v)] \quad (1)$$

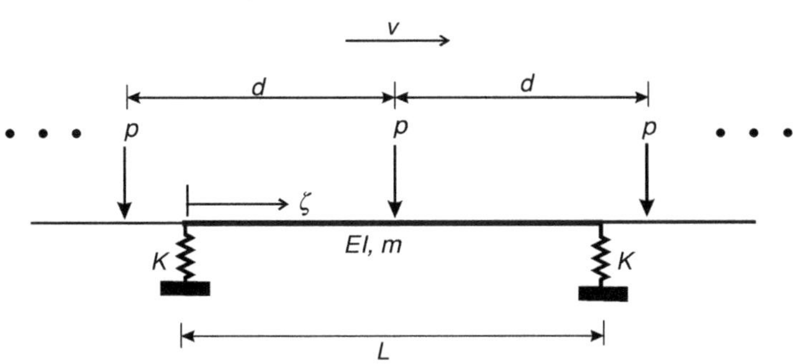

Figure 7.2 Elastically supported beam subjected to uniform moving loads

where an overdot denotes a derivative with respect to time t and a prime denotes a derivative with respect to the coordinate ζ of the beam ($0 \le \zeta \le L$), u is the vertical deflection, m the per-unit mass, c the damping coefficient, E the elastic modulus, and I the moment of inertia of the beam. Also, δ is the delta function, H the unit step function, N the total number of moving loads, and $t_k = (k - 1)d/v$ represents the arrival time for the kth load on the beam. Correspondingly, the boundary conditions for the elastically supported beam are:

$$\begin{aligned} &EIu''(0, t) = 0, && EIu''(L, t) = 0 \\ &EIu'''(0, t) = -Ku(0, t), && EIu'''(L, t) = Ku(L, t) \end{aligned} \tag{2}$$

and the initial conditions are:

$$u(\varsigma, 0) = \dot{u}(\varsigma, 0) = 0 \tag{3}$$

By modal superposition, the deflection of the beam, $u(\zeta,t)$, can be expressed as:

$$u(\varsigma, t) = \sum_{i=1}^{n} q_i(t)\varphi_i(\varsigma) \tag{4}$$

where $q_i(t)$ and $\varphi_i(\zeta)$ respectively denote the *generalized coordinate* and *shape function* of the ith mode of vibration, and n the number of vibration modes considered. In this study, only the first two modes of the beam will be considered, of which the vibration shapes are:

$$\varphi_1(\varsigma) = \frac{\sin(\pi\varsigma/L) + \kappa}{1 + \kappa}$$

$$\varphi_2(\varsigma) = \frac{\sin(2\pi\varsigma/L) - 2\kappa\varsigma/L + \kappa}{1 + \kappa} \tag{5a,b}$$

where κ is the *stiffness ratio* of the beam to the linear spring, defined as (Yau et al. 2001):

$$\kappa = \frac{EI\pi^3}{KL^3} \tag{6}$$

As can be seen from Equation (5), the first mode is symmetric and the second mode anti-symmetric. They satisfy the following orthogonality condition:

$$\int_0^L \varphi_1(\varsigma)\varphi_2(\varsigma)d\varsigma = 0 \tag{7}$$

By substituting Equations (4) and (5) into Equation (1), multiplying both sides by $\varphi_1(\zeta)$, integrating over the length L, and utilizing the orthogonality condition, the equation of motion for the first mode of the beam (Yau et al. 2001) can be found, from which the generalized coordinate $q_1(t)$ can be solved as:

$$q_1(t) = \frac{2pL^3(1+\kappa)}{\pi^4 EI(1+4\kappa/\pi)}\left\{\frac{1}{1-S_1^2}\sum_{k=1}^{N}[g_{11}(v, t-t_k)H(t-t_k) + g_{11}(v, t-t_k-L/v)H(t-t_k-L/v)] + \kappa\sum_{k=1}^{N}[g_{12}(t-t_k)H(t-t_k) - g_{12}(t-t_k-L/v)H(t-t_k-L/v)]\right\} \quad (8)$$

where S_1 is the *speed parameter*, $S_1 = \pi v/\omega_1 L$, with ω_1 indicating the first natural frequency of the beam:

$$\omega_1 = \left[\frac{\omega_0^2(1+4\kappa/\pi)}{1+8\kappa/\pi+2\kappa^2}\right]^{\frac{1}{2}}, \quad \omega_0 = \frac{\pi^2}{L^2}\left(\frac{EI}{m}\right)^{\frac{1}{2}} \quad (9)$$

and $g_{11}(v, t)$ and $g_{12}(t)$ are given as follows:

$$g_{11}(v, t) = \sin(\pi vt/L) - S_1 e^{-\xi_1\omega_1 t}\sin(\omega_1 t)$$
$$g_{12}(t) = 1 - e^{-\xi_1\omega_1 t}\cos(\omega_1 t) \quad (10a,b)$$

Here, ξ_1 indicates the damping ratio of the first mode, and ω_0 is the fundamental frequency of the corresponding beam with hinged supports. Note that, in deriving Equation (8), the damping ratio ξ_1 is assumed to be very small, i.e. $\xi_1 \ll 1$, such that the damped frequency can be approximated by the first frequency, i.e. $\omega_d = \omega_1(1-\xi_1^2)^{1/2} \approx \omega_1$.

By a similar procedure, the generalized coordinate $q_2(t)$ for the second mode of the beam can be solved as follows:

$$q_2(t) = \frac{pL^3(1+\kappa)}{8\pi^4 EI(1+2\kappa/\pi)}\left\{\frac{1}{1-4S_2^2}\sum_{k=1}^{N}[g_{21}(v, t-t_k)H(t-t_k) - g_{21}(v, t-t_k-L/v)H(t-t_k-L/v)] + \frac{2\kappa S_2}{\pi}\sum_{k=1}^{N}[-g_{22}(t-t_k)H(t-t_k) + g_{22}(t-t_k-L/v)H(t-t_k-L/v)] + \kappa\sum_{k=1}^{N}[g_{23}(t-t_k)H(t-t_k) + g_{23}(t-t_k-L/v)H(t-t_k-L/v)]\right\} \quad (11)$$

where $S_2 = \pi v/\omega_2 L$, with the second frequency ω_2 of the elastically supported beam given as:

$$\omega_2 = \left[\frac{16\omega_0^2(1 + 2\kappa/\pi)}{1 + 4\kappa/\pi + 2\kappa^2/3}\right]^{\frac{1}{2}} \tag{12}$$

and $g_{21}(v, t)$, $g_{22}(v, t)$, and $g_{23}(t)$ are:

$$g_{21}(v, t) = \sin(2\pi vt/L) - 2S_2 e^{-\xi_2\omega_2 t}\sin(\omega_2 t)$$

$$g_{22}(t) = \omega_2 t - e^{-\xi_2\omega_2 t}\sin(\omega_2 t)$$

$$g_{23}(t) = 1 - e^{-\xi_2\omega_2 t}\cos(\omega_2 t) \tag{13a–c}$$

Here, ξ_2 denotes the damping ratio of the second mode of the beam. The deflection u of the beam under the action of a series of moving loads considering the first two modes of vibration can be determined as follows:

$$u(\varsigma, t) = q_1(t)\varphi_1(\varsigma) + q_2(t)\varphi_2(\varsigma) \tag{14}$$

where the functions $q_1(t)$, $\varphi_1(\zeta)$, $q_2(t)$, $\varphi_2(\zeta)$ have been given in Equations (8), (5a), (11) and (5b) respectively.

By the fact that each pier of the bridge is shared by two adjacent girders, the force $F_s^f(t)$ acting on the top of each pier due to the action of the first set of moving loads, i.e. the front wheelsets of the train, via two adjacent girders can be expressed as:

$$\begin{aligned} F_s^f(t) &= \underline{Ku(L, t + L/v)H(t + L/v)} + \underline{Ku(0, t)H(t)} \\ &= K\{q_1(t + L/v)\varphi_1(L) + q_2(t + L/v)\varphi_2(L)\}H(t + L/v) \\ &\quad + K\{q_1(t)\varphi_1(0) + q_2(t)\varphi_2(0)\}H(t) \\ &= K\left\{q_1(t + L/v)\frac{\kappa}{1+\kappa} + q_2(t + L/v)\frac{-\kappa}{1+\kappa}\right\}H(t + L/v) \\ &\quad + K\left\{q_1(t)\frac{\kappa}{1+\kappa} + q_2(t)\frac{\kappa}{1+\kappa}\right\}H(t) \\ &= \frac{\kappa EI\pi^3}{\kappa(1+\kappa)L^3}\{[q_1(t + L/v) - q_2(t + L/v)]H(t + L/v) + [q_1(t) + q_2(t)]H(t)\} \end{aligned} \tag{15}$$

The first and second terms on the first line of Equation (15) represent the forces from the left and right girders of the pier respectively. The force $F_s^r(t)$ acting on the top of the same pier due to the second set of moving loads, i.e., the rear wheelsets of the train, has a form identical to Equation (14), but with a time lag of d_0/v, where d_0 denotes the distance between the front and rear wheelsets of the train car:

$$F_s^r(t) = F_s^f(t - d_0/v) \tag{16}$$

Summing Equations (14) and (15) yields the *total force* $F_s(t)$ acting on the top of *each pier* of the bridge due to the moving train as follows:

$$\begin{aligned} F_s(t) &= F_s^f(t) + F_s^r(t) = F_s^f(t) + F_s^f(t - d_0/v) \\ &= \frac{\kappa EI\pi^3}{\kappa(1+\kappa)L^3}\{[q_1(t + L/v) - q_2(t + L/v)]H(t + L/v) + [q_1(t) + q_2(t)]H(t) \\ &\quad + [q_1(t + L/v - d_0/v) - q_2(t + L/v - d_0/v)]H(t + L/v - d_0/v) \\ &\quad + [q_1(t - d_0/v) + q_2(t - d_0/v)]H(t - d_0/v)\} \end{aligned} \tag{17}$$

in which the condition $t = 0$ corresponds to the instant at which the leading wheelset of the train arrives right at the pier of interest.

3.2 *Interactive force between foundation and surrounding soil*

Because the interactive force between the foundation of each pier and the surrounding soil, referred hereafter to as the *foundation excitation force*, is generated by the forces acting on the top of the corresponding pier, it can be obtained from the dynamic analysis of the pier and foundation–soil system, given the pier-top forces. As depicted in Figure 7.3, the *equivalent lumped-parameter model* representing the pier and the foundation–soil system has two nodes, i.e. Node-0 and Node-1, to which the corresponding degrees of freedom (DOFs) are u_0 and u_1 respectively. There is a total of eight parameters involved in such a model: (1) M_r represents the inertia of the pier, pile cap and backfill soil on the pile cap; (2) M_0 and M_1 account for

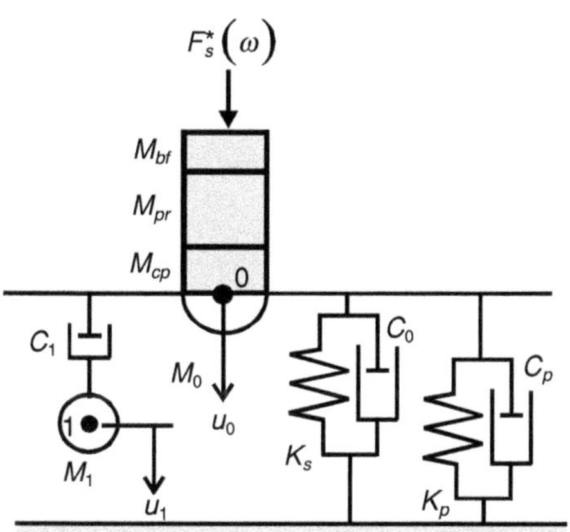

Figure 7.3 Lumped-parameter model adopted for the foundation and soil

the inertia effects of the added mass to the foundation and soil respectively; (3) K_s represents the vertical stiffness of the lumped-parameter model, which is equal to the static-stiffness coefficient of the embedded pile cap on the soil halfspace; (4) C_0 and C_1 account for the radiation damping of the model; and (5) K_p and C_p are the equivalent dynamic-stiffness and damping coefficients for the group pile–soil system, respectively. All the eight parameters, mentioned above, are defined as follows:

$$M_r = M_{pr} + M_{cp} + M_{bf} \tag{18}$$

with M_{pr} denoting the mass of the pier, M_{cp} the mass of the pile cap, and M_{bf} the mass of the backfill soil, $M_{bf} = 4bl\rho(e - h)$, for $e > h$, in which h = the height of the pile cap, e = the embedment depth of the pile cap, $2l$ and $2b$ = the length and width of the pile cap respectively and ρ = the density of the soil medium:

$$M_0 = \frac{b^2}{c_s^2}\mu_0 K_s, \quad M_1 = \frac{b^2}{c_s^2}\mu_1 K_s, \quad C_0 = \frac{b}{c_s}\gamma_0 K_s, \quad C_1 = \frac{b}{c_s}\gamma_1 K_s \tag{19a}$$

with

$$K_s = \frac{Gb}{1-\nu}\left[3.1\left(\frac{l}{b}\right)^{0.75} + 1.6\right]\left[1 + 0.25\left(1 + \frac{b}{l}\right)\left(\frac{e}{b}\right)^{0.8}\right] \tag{19b}$$

where G and ν = the shear modulus and Poisson's ratio of the soil medium respectively. It is known that $G = E/2(1 + \nu)$ with E denoting the elastic modulus of the soil. The shear-wave velocity of the soil is $c_s = (\mu/\rho)^{1/2}$, with μ denoting the *Lamé constant*, and μ_0, μ_1, γ_0 and γ_s = the dimensionless coefficients (Wolf 1994):

$$\mu_0 = \begin{cases} 0 & \text{if } \nu \le \frac{1}{3} \\ 0.9\left(\nu - \frac{1}{3}\right) & \text{if } \nu > \frac{1}{3} \end{cases},$$

$$\mu_1 = 0.38, \quad \gamma_0 = 0.80 + 0.35\frac{e}{r_{eq}}, \quad \gamma_1 = 0.32 - 0.01\left(\frac{e}{r_{eq}}\right)^4 \tag{20}$$

Here, r_{eq} is the radius of the equivalent disk of the pile cap:

$$r_{eq} = \sqrt{\frac{4bl}{\pi}} \tag{21}$$

The above parameters and associated dimensionless coefficients were determined by fitting the dynamic-stiffness coefficients of the lumped-parameter model into the corresponding exact values of the equivalent disk on the soil halfspace with account being taken of the embedding effect of the pile cap. Note that M_0 represents the effect of added mass to the pile cap as the foundation vibrates, which is only present for the cases of nearly incompressible soils, i.e. with $\upsilon > 1/3$.

The frequency-dependent dynamic *stiffness* and *damping coefficients* for the group pile–soil system, K_p and C_p, can be determined as:

$$K_p(\omega) = \frac{\sum_{i=1}^{n} k_i(\omega)}{\sum_{i=1}^{n} \alpha_i}, \quad C_p(\omega) = \frac{\sum_{i=1}^{n} c_i(\omega)}{\sum_{i=1}^{n} \alpha_i} \tag{22}$$

where $k_i(\omega)$ and $c_i(\omega)$ are the dynamic stiffness and damping coefficients for the ith pile respectively, α_i the interaction factor for the ith pile relative to a specific (reference) pile in the same group, and n the total number of piles in the group. The coefficients $k(\omega)$ and $c(\omega)$ can be computed by using Novak's (1977) procedure, also available in Prakash and Sharma (1990). The factor α can be calculated from the figure given by Poulos (1968) or Prakash and Sharma (1990), given the length, radius and spacing of the pile and Poisson's ratio of the soil.

The *equations of motion* for the lumped-parameter model shown in Figure 7.1(b) subject to the total pier-top force can be expressed in frequency domain as:

$$[-\omega^2 M_1 + i\omega C_1]u_1^*(\omega) = i\omega C_1 u_0^*(\omega) \tag{23a}$$

$$\left\{-\omega^2 M + i\omega[C_0 + C_1 + C_p(\omega)] + [K + K_p(\omega)] + \frac{\omega^2 C_1^2}{-\omega^2 M_1 + i\omega C}\right\} u_0^*(\omega) = F_s^*(\omega) \tag{23b}$$

where $M = M_r + M_0$, $u_0^*(\omega)$ and $u_1^*(\omega)$ = the displacement amplitudes for the DOFs of the two nodes, ω = the circular frequency (rad/s), and $F_s^*(\omega)$ = the Fourier transform of the total pier-top force $F_s(t)$ given in Equation (17). From Equations (23a) and (23b), one can solve for the displacement amplitude $u_0^*(\omega)$ of the pile cap as follows:

$$u_0^*(\omega) = \left\{K + K_p(\omega) - \frac{\omega^2 C_1^2 M_1}{\omega^2 M_1^2 + C_1^2} - \omega^2 M + i\omega\left[C_0 + C_p(\omega) + \frac{\omega^2 C_1 M_1^2}{\omega^2 M_1^2 + C_1^2}\right]\right\}^{-1} F_s^*(\omega) \tag{24}$$

The vibration of the pile cap as given above is obtained by considering the contribution of the 'structural part' of the system, which consists of the pier, pile cap and grouped piles, and the 'soil part' of the system, i.e. the soil surrounding the pile cap and group piles, via calculation of the coefficients K_p and C_p in Equation (22).

Since the vibrations of the ground surface due to train actions are generated mostly by the pile cap near the ground surface, the contributions of the pile bodies that are embedded deeply in the soil will be neglected to simplify the formulation. By utilizing the interactive force-displacement relationship between the pile cap and the surrounding soil, the *foundation excitation force*, or more specifically the *interactive force between the pile cap and soil*, $P_{in}^*(\omega)$, in the frequency domain can be determined for *each pier* as follows (Wolf 1994):

$$P_{in}^*(\omega) = (\tilde{K} + i\omega\tilde{C})u_0^*(\omega) \tag{25}$$

where

$$\tilde{K} = \frac{4Gr_{eq}}{1-v}\left(1 + 0.54\frac{e}{r_{eq}}\right), \quad \tilde{C} = \begin{cases} \pi\rho r_{eq}^2\left(c_p + c_s\dfrac{2e}{r_{eq}}\right) & for\ v \le \dfrac{1}{3} \\ 2\pi c_s\rho r_{eq}^2\left(1 + \dfrac{e}{r_{eq}}\right) & for\ v > \dfrac{1}{3} \end{cases} \tag{26a,b}$$

Notice that Equation (26a) is valid only for $e/r_{eq} \le 2$, beyond which significant errors may arise. The foundation excitation force or the interactive force induced between the pile cap and soil by the train action, as shown in Equation (25), is the *primary source* of vibration for the ground surface.

3.3 Ground vibration response to foundation excitation force of a single pier

From the viewpoint of ground vibration, the interactive force between the pile cap of each pier of the bridge and the surrounding and underlying soil can be regarded as an external excitation to the ground acting at the depth of embedment of the pile cap. Although the excitation is not a point load, but acting over an area equal to that of the pile cap, it can be treated as a *point load*, if the distance from the pier to the point of interest on the ground is larger than twice the equivalent radius of the pile cap, i.e. $r/r_{eq} > 2$ (Wolf 1994). Such a treatment will be adopted in this study, since it can largely simplify the complexity of the problem, while maintaining the accuracy of solutions.

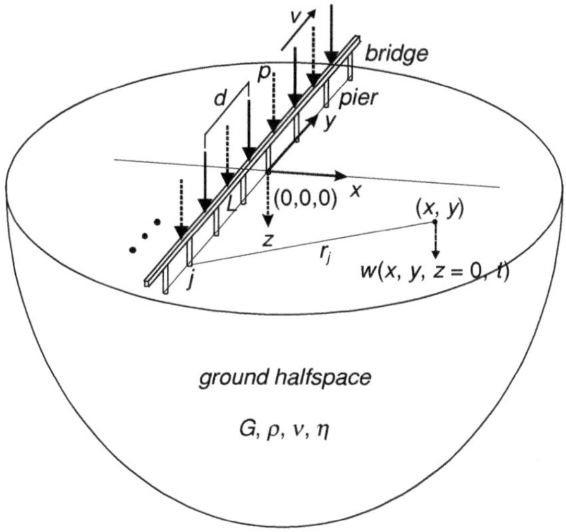

Figure 7.4 Three-dimensional illustration of the train loads moving over a multi-span bridge built on a homogenous and isotropic half space

With reference to Figure 7.4 for the definition of the coordinate system, the frequency spectrum of *vibration of the ground surface* (i.e. with $z = 0$) due to the excitation of a pile cap of the bridge can be computed as:

$$\begin{aligned} w_s^*(x, y, \omega) &= \tilde{w}_s^*(r, \omega) = P_{in}^*(\omega) G_{zz}(r, \omega), \\ v_s^*(x, y, \omega) &= i\omega w_s^*(x, y, \omega), \quad \text{for} \quad r = \sqrt{(x - x_0)^2 + (y - y_0)^2} > 0 \\ a_s^*(x, y, \omega) &= -\omega^2 w_s^*(x, y, \omega), \end{aligned} \tag{27}$$

where (x_0, y_0) and (x, y) denote the positions of the excitation (i.e. the centre of the pile cap) and a generic point on the ground surface respectively; $w_s^*(x, y, \omega)$, $v_s^*(x, y, \omega)$, $a_s^*(x, y, \omega)$ are the displacement, velocity and acceleration responses of the point (x, y) in frequency (ω) domain; $G_{zz}(r, \omega)$ is *Green's function* for a homogeneous, isotropic and viscoelastic halfspace subject to a harmonic point load on the plane surface (Miller and Pursey 1954; Wolf 1994):

$$G_{zz}(r, \omega) = \frac{1 - \nu}{2\pi G(r - 2r_{eq}/\pi)} e^{-i\left[\frac{\omega(r - 2r_{eq}/\pi)(1 - i\eta)}{\gamma c_R}\right]},$$

$$\text{if} \quad r - 2r_{eq}/\pi \leq \beta\lambda_R \tag{28a}$$

$$G_{zz}(r,\omega) = \frac{1-\nu}{2\pi G\sqrt{\beta\lambda_R(r-2r_{eq}/\pi)}}\, e^{-i\left[\frac{\omega(r-2r_{eq}/\pi)(1-i\eta)}{\gamma c_R}\right]},$$

$$\text{if}\quad \beta\lambda_R < r - 2r_{eq}/\pi \leq 1.5\lambda_R \tag{28b}$$

$$G_{zz}(r,\omega) = \frac{1-\nu}{2\pi G\sqrt{\beta\lambda_R(r-2r_{eq}/\pi)}}\, e^{-i\left[\frac{\pi}{4}+\frac{\omega(r-2r_{eq}/\pi)(1-i\eta)}{c_R}\right]},$$

$$\text{if}\quad 1.5\lambda_R < r - 2r_{eq}/\pi \tag{28c}$$

where c_R = the *Rayleigh-wave* velocity of the soil, $\lambda_R = 2\pi c_R/\omega$ = the wavelength of the Rayleigh waves, η = the material damping ratio, $\gamma = 12/13$ = a modification factor for the Rayleigh-wave velocity in the near field, and $\beta = (2/3)\ \beta^*$ = a parameter used to define the boundary between the near and far fields, with β^* defined as follows (Rücker 1982):

$$\beta^* = \left(\frac{1-\nu}{2\pi}\right)^2 \frac{\{8q_R - [48 - 32q_p^2]q_R^3 + 48[1 - q_p^2]q_p^5\}^2}{[2q_R - 1]^4[q_R^2 - q_p^3]} \tag{29}$$

where $q_R = c_s/c_R$ and $q_p = c_s/c_p$.

It should be noted that: (1) the contribution of the *shear* and *dilatational waves* to the response of the ground surface was assumed negligible, since the Rayleigh waves carry most of the vibration energy of the ground surface; (2) the term $(r - 2r_{eq}/\pi)$ in Equation (28) accounts for the discrepancy in the spatial patterns of the ground response resulting from the point-source of the real-type loading; (3) only the *material damping* associated with the velocity of the Rayleigh waves is considered in the present study. The effect of velocity dispersion is neglected for its smallness in magnitude for a homogenous halfspace, as is the case considered herein.

3.4 Ground vibration response to foundation excitation forces of multiple piers

As mentioned previously, the total response of the ground surface is computed as the superposition of the response to the excitation of each of the piers of the bridge traveled over by the train. It is assumed that the train travels over the bridge from $y = -\infty$ to $y = \infty$ (with reference to Figure 7.4) with a constant speed v. Based on the assumption of point loads, the *total foundation excitation force* $F(x, y, t)$ acting on the ground contributed by *all* of the piers along the bridge is simply:

$$F(x,y,t) = \sum_{j=-\infty}^{\infty} P_{in}(t - jL/v)\delta(x)\delta(y - jL) \tag{30}$$

where $P_{in}(t - jL/v)$ is the foundation excitation force of the jth pier. Performing the Fourier transform to Equation (30) yields the frequency spectrum of the total foundation excitation force, $F^*(x,y,\omega)$:

$$F^*(x, y, \omega) = \int_{-\infty}^{\infty} F(x, y, t)e^{-i\omega t}dt = \sum_{j=-\infty}^{\infty} P_{in}^*(\omega)e^{-i\omega(jL/v)}\delta(x)\delta(y - jL) \quad (31)$$

where $P_{in}^*(\omega)$ is exactly the same as given in Equation (25). Using Green's function for the halfspace, i.e. Equation (28), and integrating over the x-y domain, the frequency spectra of the ground vibrations under the action of the total foundation excitation force of the bridge is obtained as:

$$\{w^*, v^*, a^*\}(x, y, \omega) = \{1, i\omega, -\omega^2\}P_{in}^*(\omega) \sum_{j=-\infty}^{\infty} e^{-i\omega(jL/v)}G_{zz}(r_j, \omega) \quad (32a)$$

where

$$r_j = \sqrt{x^2 + (y - jL)^2} > 0 \quad (32b)$$

and appropriate Green's function $G_{zz}(r, \omega)$ should be selected from Equations (28a–c) for different values of r_j. It should be noted that the number of bridge piers involved in computation of the ground response using Equation (32a) can not be infinite as the computation must be carried out by numerical procedures. The number of the piers used is determined by considering the accuracy and convergence characteristics of the solution. The time-history response of the ground surface can be determined as the inverse Fourier transform of Equation (32a), that is:

$$\{w, v, a\}(x, y, t) = \frac{1}{2\pi}\int_{-\infty}^{\infty} \{w^*, v^*, a^*\}(x, y, \omega)e^{i\omega t}d\omega \quad (33)$$

4 Some remarks on implementation of the analysis procedure

Because the force acting on the top of each pier of the bridge, i.e. $F_s(t)$, as given in Equation (17), has a complicated form in most cases, the discrete Fourier transform is employed to compute the frequency spectrum of the excitation $F_s^*(\omega)$, namely:

$$F_s^*(\omega_k) = \Delta t \sum_{n=0}^{N_s-1} F_s(t_n)e^{-i\omega_k t_n}, \quad -\frac{N_s}{2} \leq k \in z \leq \frac{N_s}{2} - 1 \quad (34)$$

where Δt = the time interval, $N_s = T/\Delta t$ = the number of sampling points, T = the time duration, $\omega_k = k\Delta\omega = k(2\pi/T)$, the discrete circular frequency (rad/s), and $t_n = n\Delta t$ = the nth discrete time instant. The duration T used in the computation should be no less than $[(N-1)d+d_0+L]/v$, since the pier-top force $F_s(t)$ equals zero for $t \le -L/v$, and reduces gradually to zero for $t \ge [(N-1)d+d_0+L]/v$. Also, the time interval Δt used should be made small, such that the number of sampling points is large enough to ensure the accuracy of analysis. For convenience, N_s is always set to be even.

It should be noted that the discrete frequency spectrum $F_s^*(\omega_k)$ obtained after the discrete Fourier transform is complex, of which the real and imaginary parts, i.e. $\mathrm{Re}(F_s^*(\omega_k))$ and $\mathrm{Im}(F_s^*(\omega_k))$, are symmetric and anti-symmetric respectively with respect to ω. In obtaining the frequency response of the ground surface by Equation (32), only the responses for the positive frequencies, i.e. for $\omega_n = n\Delta\omega$, with $n = 0, 1, 2, \ldots, N_s/2-1$, need be computed, while those for the negative frequencies are obtained by the concept of symmetry and anti-symmetry, that is:

$$\begin{aligned}&\mathrm{Re}[w_s^*(x,y,\omega_{-n})] = \mathrm{Re}[w_s^*(x,y,\omega_n)] \quad \text{for} \quad n = 1, 2, \ldots, N_s/2-1\\ &\mathrm{Im}[w_s^*(x,y,\omega_{-n})] = -\mathrm{Im}[w_s^*(x,y,\omega_n)]\end{aligned} \tag{35a}$$

and

$$\mathrm{Re}[w_s^*(x,y,\omega_{-N_s/2})] = 0, \quad \mathrm{Im}[w_s^*(x,y,\omega_{-N_s/2})] = 0 \tag{35b}$$

The time-history response of the ground surface to the multiple foundation excitation forces is obtained by superimposing the response to individual foundation excitation force over all the piers of the bridge considered, through the use of Equations (33) and (32a),

$$\begin{aligned}\{w, v, a\}(x,y,t_n) &= \sum_{j=N_0}^{N_f} \{w_{sj}, v_{sj}, a_{sj}\}(x_j, y_j, t_n)[H(t_n - t_{j,0}) - H(t_n - t_{j,f})]\\ &= \sum_{j=N_0}^{N_f}\left[\sum_{k=-N_s/2}^{N_s/2-1} \frac{1}{N_s\Delta t}\{1, i\omega_k, -\omega_k^2\}P_{in}^*(\omega_k)G_{zz}(r_j,\omega_k)e^{i\omega_k t_n}\right][H(t_n - t_{j,0})\\ &\quad - H(t_n - t_{j,f})]\end{aligned} \tag{36a}$$

with

$$\begin{aligned}&x_j = x, \quad y_j = y - jL, \quad r_j = \sqrt{x_j^2 + y_j^2}, \quad t_n = n\Delta t,\\ &t_{j,0} = (j-1)L/v, \quad t_{j,f} = (j-1)L/v + T\end{aligned} \tag{36b}$$

where $\{w, v, a\}(x, y, t_n)$ denote the complete time-history responses of a point (x, y) on the ground surface at discrete time t_n, $\{w_{sj}, v_{sj}, a_{sj}\}(x_j, y_j, t_n)$ the ground response to the foundation excitation force of the jth pier, (x_j, y_j) the position of the point (x, y) relative to that of the jth pier. The Heaviside function H has been included to indicate the time-lag effect of the moving loads on each pier, and N_0 and N_f denote the starting and ending pier numbers of the bridge respectively, considered for computation of the ground response. As a result, the number of piers considered in the computation, denoted by N_{pier}, is equal to $N_f - N_0 + 1$, i.e. $N_{pier} = N_f - N_0 + 1$. The value of $P_{in}^*(\omega_k)$ for each ω_k is computed by using Equation (25), together with Equations (24) and (26).

The vibration response of the ground surface obtained in this study includes the displacement, velocity and acceleration responses, of which the latter two are of particular interest. The acceleration response is presented in the form of a time history or maximum value in gal, while the velocity response will be studied primarily in the form of the *1/3 octave band spectrum* in decibels (dB) (Gordon 1991). The procedure for calculation of the 1/3 octave band spectrum of the ground surface velocity is outlined as follows (Ju 2002):

(1) Calculate the *power spectrum density function* (PSDF) from the frequency response of the ground velocity:

$$S(f) = \frac{2|v^*(f)|^2}{T} \tag{37}$$

where $S(f)$ = the power spectrum density function, $v^*(f)$ = the frequency response of the ground velocity, T = the time duration and f = the frequency in Hz.

(2) Determine the *cumulative power spectrum density* at the central frequency by integration of the PSDF $S(f)$ over each 1/3 octave band frequency interval:

$$E(f_c) = \int_{f_l}^{f_u} S(f)\mathrm{d}f \tag{38}$$

where $E(f)$ = the cumulative power spectrum density function, f_c, f_u and f_l = the central, upper and lower frequencies of the 1/3 octave band frequency intervals, as defined in Table 7.1.

(3) Compute the root mean square value of the cumulative power spectrum density values $E(f_c)$:

$$\sigma(f_c) = \sqrt{E(f_c)} \tag{39}$$

where $\sigma(f_c)$ = the root mean square value of $E(f_c)$. (4) Calculate the 1/3 octave band value in decibels (dB) from $\sigma(f_c)$:

$$L(f_c) = 20\log_{10}\frac{\sigma(f_c)}{\sigma_0} \tag{40}$$

where $L(f_c)$ = the 1/3 octave band value in decibels and σ_0 = the reference velocity of 2.54×10^{-8} m/s.

Table 7.1 Central, lower and upper frequencies of the 1/3 octave band

Lower frequency (Hz)	*Central frequency (Hz)*	*Upper frequency (Hz)*
0.71	0.8	0.89
0.89	1	1.12
1.12	1.25	1.41
1.41	1.6	1.78
1.78	2	2.24
2.24	2.5	2.82
2.82	3.15	3.55
3.55	4	4.47
4.47	5	5.62
5.62	6.3	7.08
7.08	8	8.91
8.91	10	11.22
11.22	12.5	14.1
14.1	16	17.8
17.8	20	22.4
22.4	25	28.2
28.2	31.5	35.5
35.5	40	44.7
44.7	50	56.2
56.2	63	70.8
70.8	80	89.1
89.1	100	112
112	125	141
141	160	178
178	200	224
224	250	282
282	315	354
354	400	447
447	500	562
562	630	707

5 Verification of the analysis method

The method of analysis developed in the preceding sections will first be verified by comparing the results of the analysis for a seven-unit bridge traversed by the train with those reported in the literature using the finite element method (Ju 2002). The following data are adopted in this example: $N = 16$, $p = 218$ kN, $d = 25$ m, $d_0 = 17.5$ m, and $v = 80$ m/s for the train loads; $E = 2 \times 10^7$ kN/m^2, $I = 8.72$ m^4, $m = 18.5$ t/m, $L = 30$ m, $\xi_1 = \xi_2 = 0.025$, $M = 1390$ t, $b = l = 10.8$ m, $h = 3.6$ m, $e = 4.6$ m, $E_p = 2 \times 10^7$ kN/m^2, $r_0 = 0.9$ m, $n = 5$ for the bridge; and $\alpha = 0.6$ for the side piles, $\alpha = 1.0$ for the central pile. A total of eight piers, i.e. $j = -3 \sim 4$ or $N_{pier} = 8$, is considered. A very small value is used for the stiffness ratio κ, i.e. $\kappa = 0.001$, to simulate the simple support condition for the girders. The data adopted for the ground are as follows: $G = 5.0 \times 10^4$ kN/m^2 (for computing the Green's function) and 14.7×10^4 kN/m^2 (for computing the pile dynamic stiffness), $\rho = 2.0$ t/m^3, $v = 0.48$, $c_p = 274$ m/s, $c_s = 158$ m/s, $c_R = 145$ m/s and $\eta = 0.01$. The duration of computation is taken as $T = 20$ s and the time increment as $\Delta t = 0.005$ s. The origin of the coordinate system (x,y,z) is located at the ground surface ($z = 0$) right under the midpoint of the middle unit (between $j = 0$ and $j = 1$) of the bridge.

As can be seen from Figure 7.5, the results obtained by the present method agree generally with those obtained by Ju (2002), concerning both the

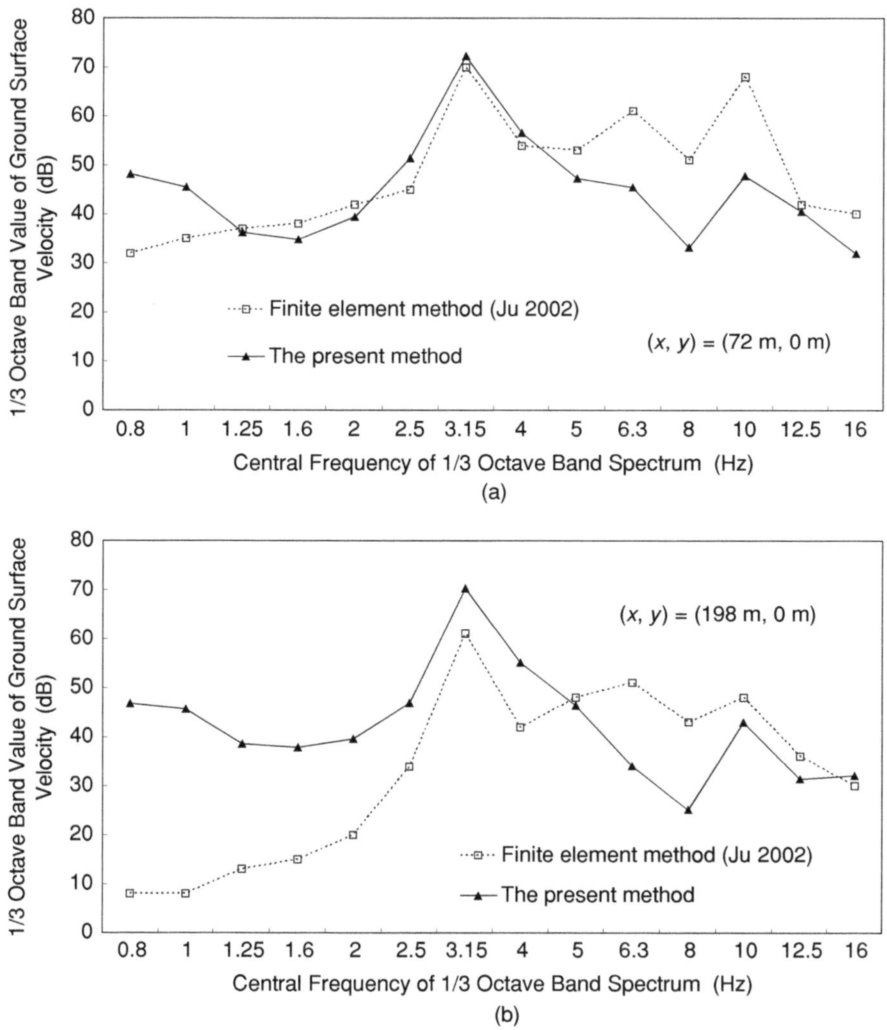

Figure 7.5 Comparison of the present method with the finite element method (Ju 2002): (*a*) ground surface response at $(x, y) = (72$ m, 0 m$)$; (*b*) ground surface response at $(x, y) = (198$ m, 0 m$)$

dominant frequencies and the corresponding dB values. The discrepancies between the two results can be attributed to the difference implied in the modeling of the system and in the shear modulus and material damping adopted for the halfspace in the two studies. It should be noted that the effort and computation time required in obtaining the results, as those shown in Figure 7.5, by the present method are drastically less than that by the finite element method, since the time-consuming mesh generation procedure has been avoided, while a relatively small amount of computation is required for the operations involving the system matrices. Owing to the above-mentioned advantages over the finite element or other numerical methods, the present method should prove suitable for comprehensive investigations on train-induced ground vibrations.

6 Numerical study on three-dimensional train-induced ground vibrations

In this section, the ground vibrations induced by trains traveling over elevated bridges will be parametrically studied. The general layout of the elevated

Table 7.2 Data used for the train, bridge and ground

Train loads

1.1.1 Type	*1.1.2 N*	p *(kN)*	d *(m)*	d_0 *(m)*	*Total length (m)*
SKS	16	276	25.0	17.5	452.5

*Bridge**

L *(m)*	E *(kN/m²)*	I *(m⁴)*	m *(t/m)*	$\xi_1 = \xi_2$	M *(t)*	b = l *(m)*	H *(m)*	e *(m)*	r_0 *(m)*	l_p *(m)*
20	2.82×10^7	3.81	34.6	0.25	500.5	7.75	2.1	1.0	0.9	50
25	2.82×10^7	5.98	35.8	0.25						
30	2.82×10^7	8.72	36.7	0.25	n	α_1	α_2	α_3	α_4	
35	2.82×10^7	12.97	37.5	0.25	4	1.0	0.55	0.45	0.55	
40	2.82×10^7	17.90	38.7	0.25						

Ground

G *(kN/m²)*	ρ *(t/m³)*	ν	c_p *(m/s)*	c_s *(m/s)*	c_R *(m/s)*	η
2.1×10^4	2.0	0.33	177	102	94	0.02
9.8×10^3	2.0	0.33	121	70	65	0.02
8.0×10^4	2.0	0.33	346	200	184	0.02

* Identical dimensions and properties are assumed for the substructure of the bridge with different span lengths.

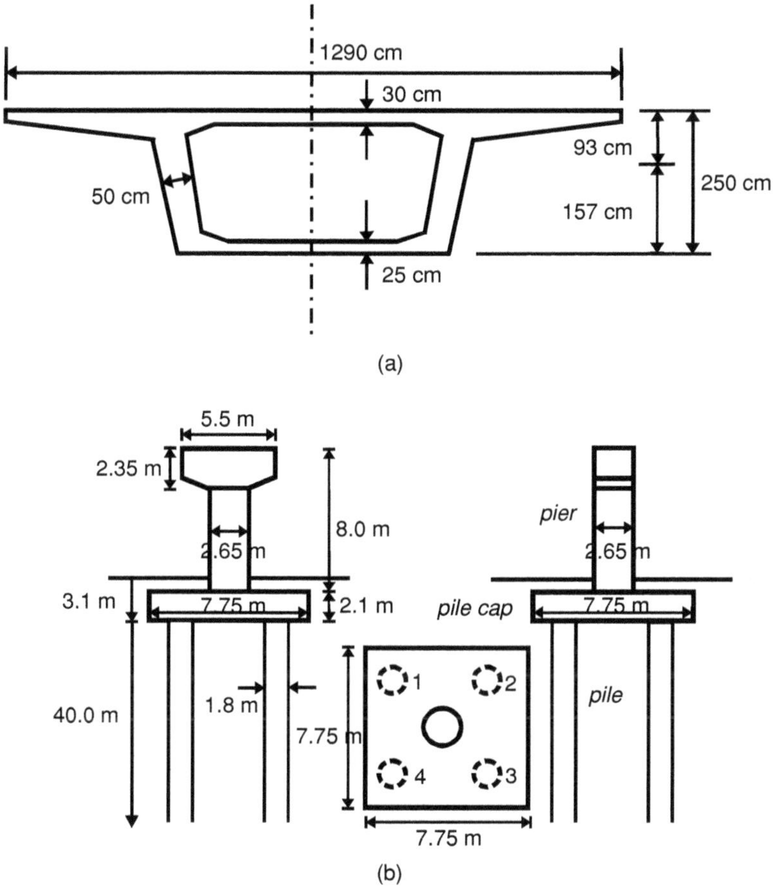

Figure 7.6 Geometry and dimensions of the bridge: (*a*) typical box-girder cross-section; (*b*) pier and pile foundation

bridge considered is shown in Figure 7.6, and the data for the train, the bridge and the ground are listed in Table 7.2. In the following analyses, the duration T considered for the passage of the train over each pier is 20 s, and the time increment Δt used is 0.005 s.

6.1 Determination of minimum number of bridge piers used in analysis

Before the analysis is conducted, the *minimum number of bridge piers*, or equivalently the minimum number of bridge girders, required for achieving an accurate solution is first determined. To this end, the following scenarios are considered: (1) the girder span length $L = 30$ m, the soil shear modulus for the ground $G = 2.1 \times 10^4$ kN/m², and the train speed $v = 200$, 300 and 400 km/h, respectively; (2) the train speed $v = 300$ km/h, the soil shear modulus for the ground $G = 2.1 \times 10^4$ kN/m², and the girder span length $L = 20$, 30 and 40 m, respectively; (3) the girder span length $L = 30$ m, the train speed $v = 300$ km/h, and the soil shear modulus for the ground $G = 9.8 \times 10^3$, 2.1×10^4 and 8.0×10^4 kN/m² respectively. Simple

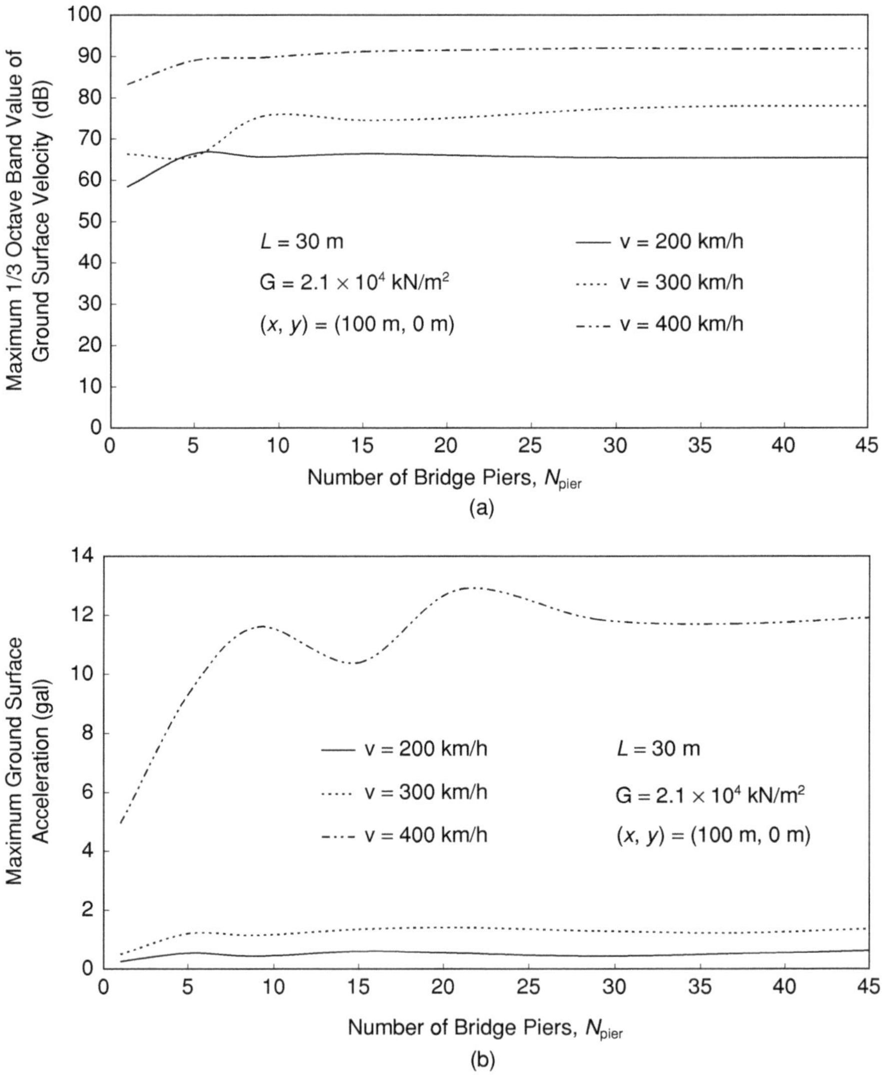

Figure 7.7 Maximum response of the ground surface vs the number of piers under different train speeds: (*a*) 1/3 octave band value (in dB); (*b*) acceleration (in gal)

support condition, i.e. $\kappa = 0.001$, is assumed for the bridge girders in all the scenarios.

The responses of the ground surface at the point of $(x, y) = (100\ \text{m}, 0\ \text{m})$ with respect to the number of bridge piers for the three scenarios considered are shown in Figures 7.7–9 respectively, with part (a) denoting the maximum 1/3 octave band value (dB) and part (b) the maximum acceleration (gal). Apparently, the maximum 1/3 octave band response approaches an

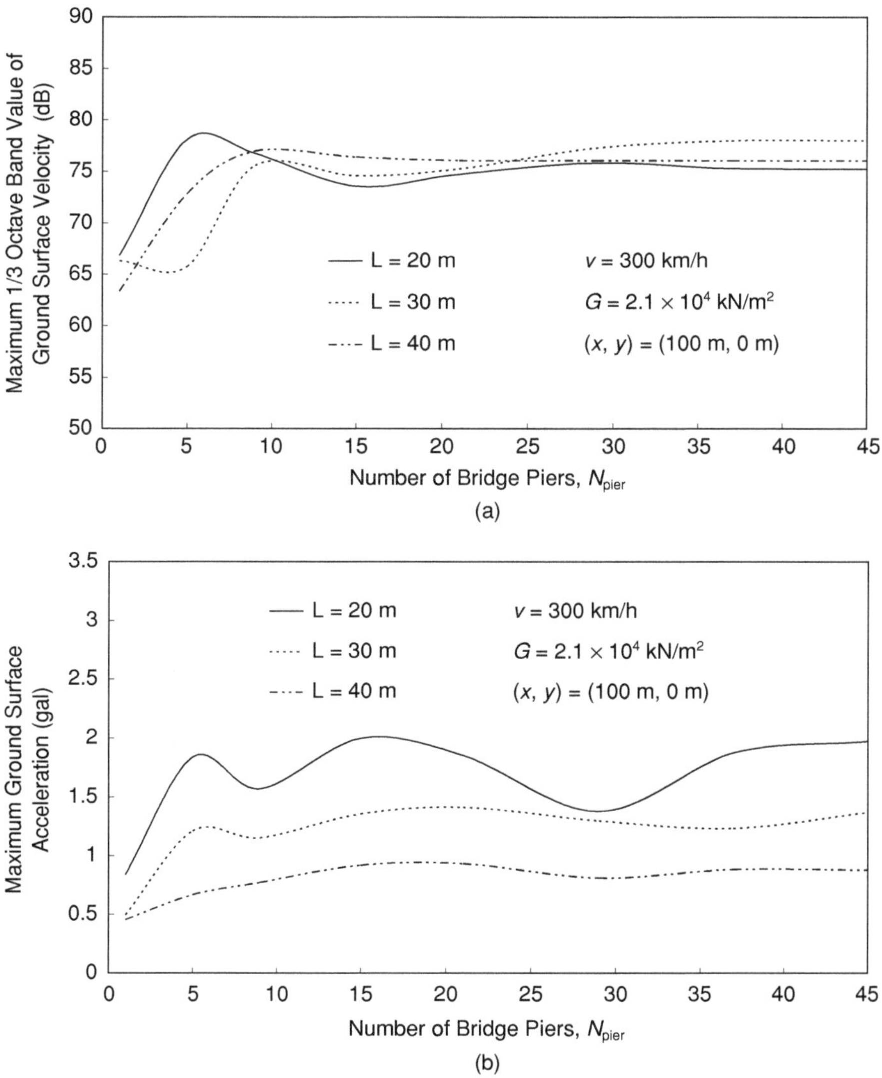

Figure 7.8 Maximum response of the ground surface vs the number of piers under different girder span lengths: (*a*) 1/3 octave band value (in dB); (*b*) acceleration (in gal)

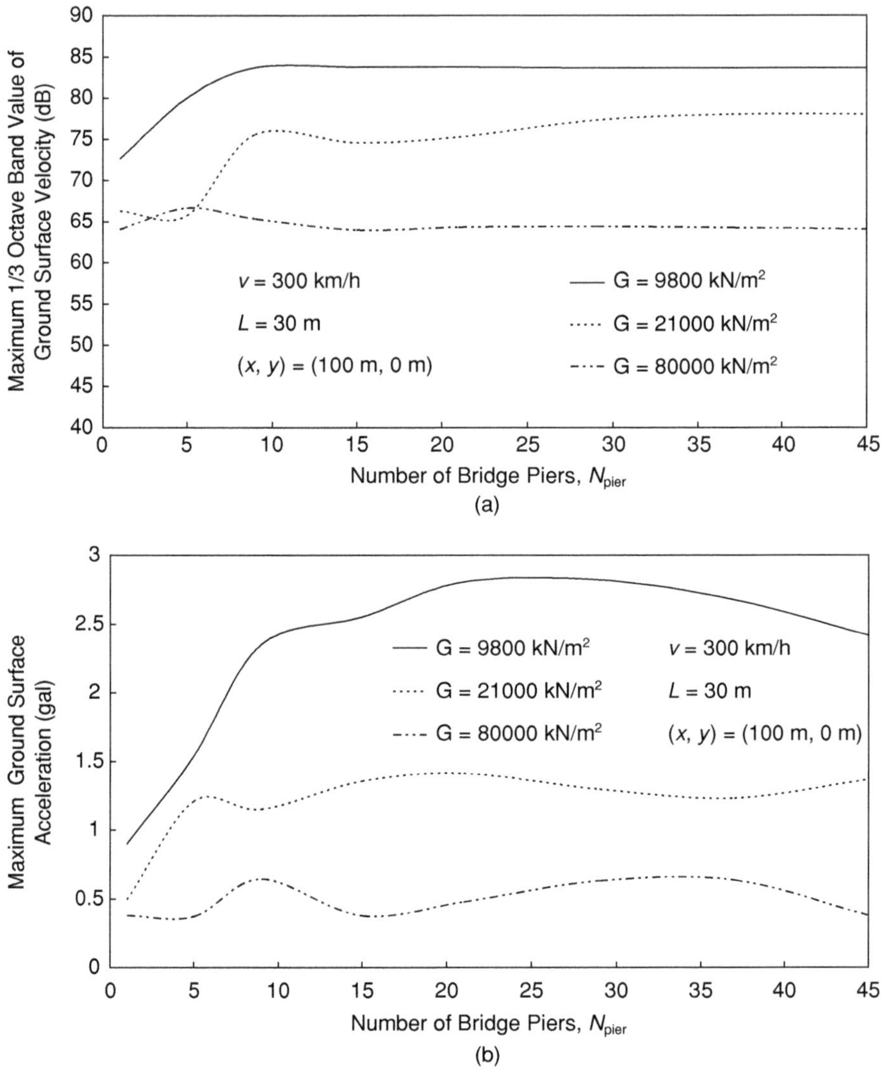

Figure 7.9 Maximum response of the ground surface vs the number of piers under different soil properties: (*a*) 1/3 octave band value (in dB); (*b*) acceleration (in gal)

asymptote as the number of bridge piers reaches 30 for all of the scenarios. Also, the maximum acceleration response exhibits a similar trend, in general, as the number of bridge piers reaches 45, although the responses due to different numbers of bridge piers display some oscillations, which makes the asymptotic trend not so obvious. Based on the above observations, the minimum number of bridge piers to be used in analysis for reasonably capturing the actual responses is 45, which implies $j = -22$ to 22 or $N_{pier} = 45$, with $j = 0$ denoting the central pier or the origin of the coordinates.

6.2 Pier-top force, foundation excitation force and ground vibration response

Figure 7.10 shows the typical bridge and ground vibration responses generated by the train as it travels over the multi-span bridge with $v = 200$, 300 and 400 km/h, in which (a) represents the pier-top force, (b) the foundation excitation force, (c) and (d) respectively the frequency response and 1/3 octave band spectrum of velocity of the ground surface at the point of $(x, y) = (100\text{ m}, 0\text{ m})$, and (e) and (f) respectively the time-history response and response spectrum acceleration of the ground surface for 5 percent damping ratio at the same point. In all cases, the span length of each girder of the bridge is $L = 30$ m.

As can be seen from Figure 7.10a, the pier-top forces under the three train speeds vary cyclically during the train passage over the pier, resulting from the repetitive action of the train axle loads. Of interest is that the peak values of the pier-top force for $v = 200$ and 300 km/h are comparable, while that for $v = 400$ km/h is much higher (by about 160 percent). Furthermore, the extent of the cyclical varying of the pier force for $v = 400$ km/h increases as there are more train axle loads passing the pier of concern, a phenomenon not existing for the cases of $v = 200$ and 300 km/h. By these facts, it is indicated that the so-called *train–bridge resonance* occurs with the girders of the bridge at the train speed $v = 400$ km/h. According to Yang et al. (1997), the train–bridge resonance occurs if the train speed v (m/s), the train car length d (m) and the fundamental frequency of the bridge girder ω_1 (rad/s) satisfy the following conditions:

$$\frac{\omega_1 d}{2v} = n\pi, \quad n = 1, 2, 3, \ldots \tag{41}$$

By substituting the data for the train and the bridge, i.e. $d = 25$ m and $\omega_1 = 28.4$ rad/s, into Equation (41), the primary ($n = 1$) resonant speed as 407 km/h (= 112.9 m/s) is obtained, which is very close to the train speed of $v = 400$ km/h under discussion. Consequently, it is confirmed that train–bridge resonance has occurred with the bridge girders.

The foundation excitation force shown in Figure 7.10b, resulting from the action of the pier-top force through the pier, exhibits nearly the same phenomenon as that of the pier-top force, but with a much smaller intensity, which can be attributed primarily to the dissipation and suppression effects of the damping and mass inertia of the pier, foundation and surrounding soil. The reduction of the foundation excitation force from the pier-top force is about 65 percent ~70 percent in both peak and average values, where

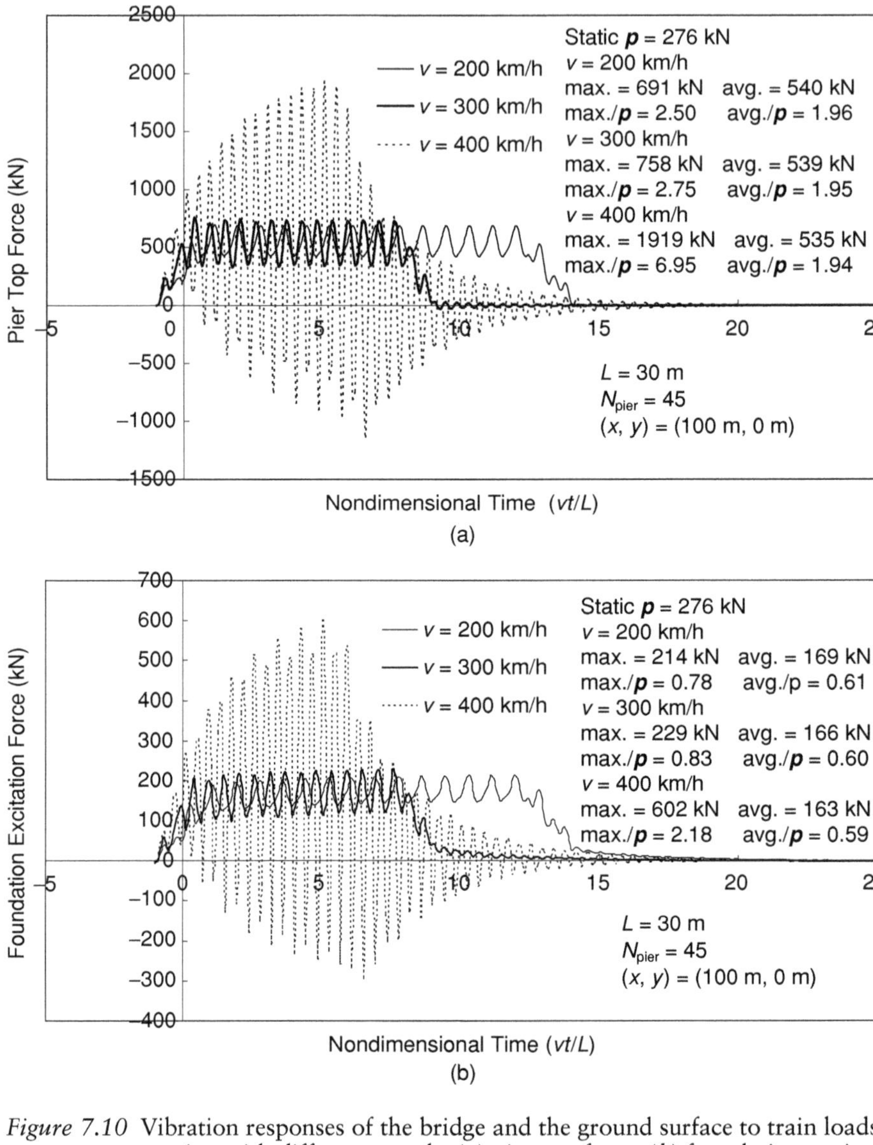

Figure 7.10 Vibration responses of the bridge and the ground surface to train loads moving with different speeds: (*a*) pier top force; (*b*) foundation excitation force; (*c*) frequency spectrum of ground velocity; (*d*) 1/3 octave band spectrum of ground velocity; (*e*) ground acceleration time history; (*f*) response spectrum of ground acceleration with 5% damping ratio

1 percent, 1 percent and 2 percent damping ratios are assumed for the pier, foundation and soil medium respectively.

As for the ground response, it can be seen from Figure 7.10c that the frequency contents of velocity responses of the ground surface caused by

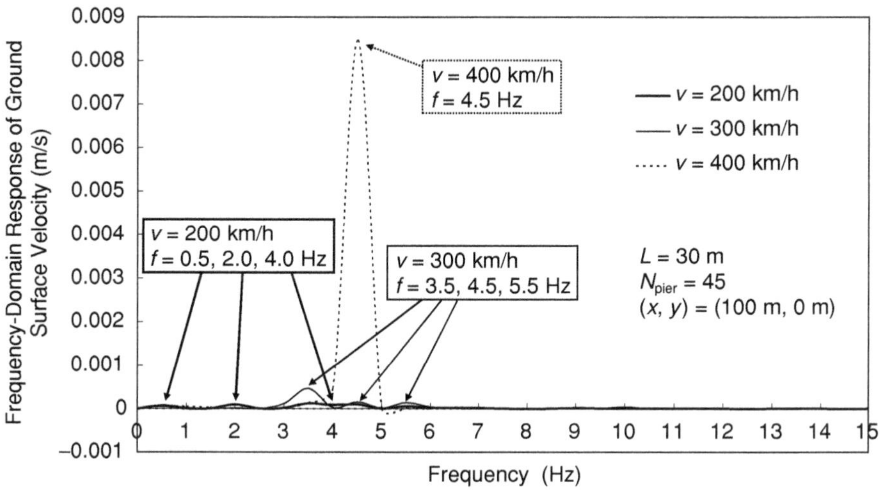

(c)

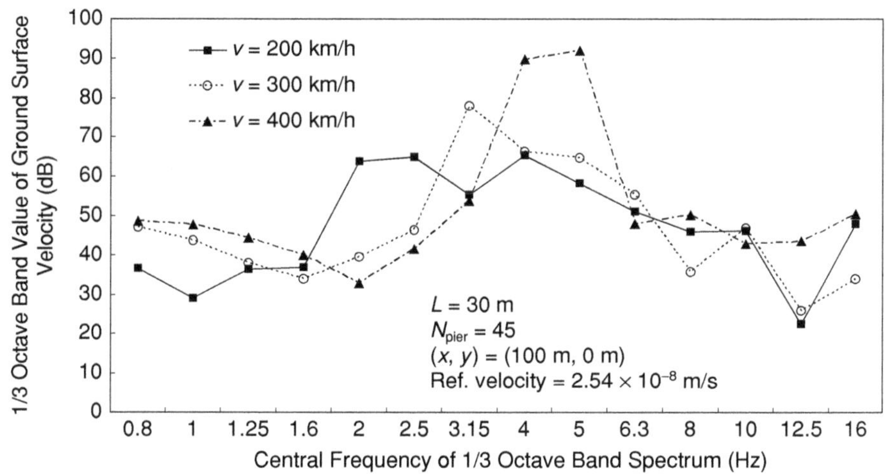

(d)

Figure 7.10 (cont'd)

the three train speeds are rather different. Specifically, higher frequency responses appear at $f = 0.5$, 2.0 and 4.0 Hz for $v = 200$ km/h and at $f = 3.5$, 4.5 and 5.5 Hz for $v = 300$ km/h. In contrast, for $v = 400$ km/h most of the frequency response occurs in the narrow region around $f = 4.5$ Hz, indicating that the vibration for this case propagates in the ground mainly via single-harmonic waves. In general, larger frequency responses of velocity of the ground surface to the three train speeds occur mainly at $f < 10$ Hz, particularly within the range from $f = 2.0$ to 6.0 Hz. On the other

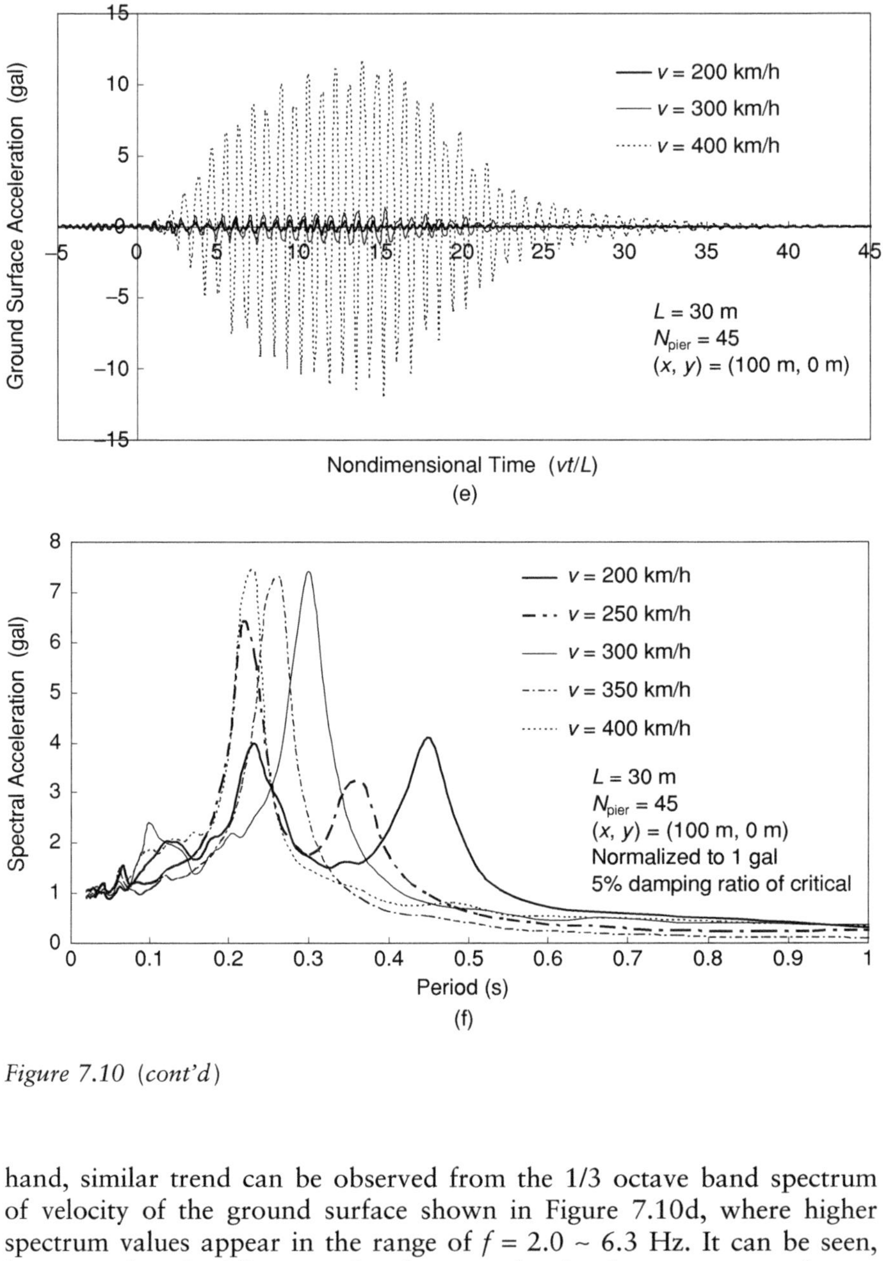

Figure 7.10 (cont'd)

hand, similar trend can be observed from the 1/3 octave band spectrum of velocity of the ground surface shown in Figure 7.10d, where higher spectrum values appear in the range of $f = 2.0 \sim 6.3$ Hz. It can be seen, however, that the 1/3 octave band spectra for the three cases are of some difference. The predominant spectrum responses for $v = 200$ km/h occur within a central frequency band spanning from 2.0 ~ 5.0 Hz, and those for $v = 300$ km/h within a central frequency band of 3.15 ~ 6.3 Hz, while those for $v = 400$ km/h occur at a central frequency band of 4.0 and 5.0

Hz. Additionally, the maximum spectrum values for the three cases are 65.5, 78.0 and 92.0 dB respectively. It can be found that, the lower the train speed, the more uniform the 1/3 octave band spectrum becomes.

From the ground acceleration shown in Figure 7.10e, it can be seen that the time-history response of acceleration of the ground surface under v = 400 km/h exhibits substantial difference in terms of temporal distribution and magnitude of the response from those under the other two speeds, mainly owing to the occurrence of train–bridge resonance at that speed. The peak ground accelerations (PGAs) due to the three train speeds are 0.6, 1.4 and 11.9 gals respectively.

The response spectra of the ground surface accelerations shown in Figure 7.10e were plotted in Figure 7.10f, where each acceleration response has been normalized to have a maximum value of 1.0 gal. It can seen that the response spectrum for v = 200 km/h has two pronounced peak values at natural periods of 0.23 s and 0.45 s, meaning that the modal responses of a structure corresponding to the two natural periods will be significantly greater than those of the other periods under the ground motion. The response can even be higher if the predominant natural period of the structure is equal to or close to one of the two periods. In other words, the ground surface acceleration for v = 200 km/h is more harmful to structures with a predominant natural period of 0.23 s or 0.45 s. In contrast, the response spectrum for v = 300 km/h exhibits only one peak value at the period of 0.3 s, indicating that the ground motion will cause considerable vibration on structures with a predominant natural period of 0.3 s. The response spectrum for v = 400 km/h is similar to that for v = 300 km/h, but with the peak value appearing at the natural period of 0.23 s.

It should be added that the peak value of the response spectra moves toward the short-period (or high-frequency) side as the train speed increases. Moreover, higher peak spectrum values exist for v = 300 and 400 km/h than for v = 200 km/h, which reveals that the former two ground motions are much more harmful than the latter to structures with respective predominant natural periods. Finally, by the fact that the peak spectrum values for v = 300 and 400 km/h are nearly the same, it is inferred that there exists a *saturation phenomenon* with the peak spectrum value when the train speed exceeds a specific limit. To confirm this inference, two more cases, i.e. v = 250 and 350 km/h, were added in Figure 7.10f. Evidently, the saturation phenomenon does exist, and the specific limit can be identified as around v = 300 km/h for the present study.

6.3 Ground vibration response with respect to site-to-bridge distance

In Figure 7.11(a) the maximum 1/3 octave band (dB) response of the ground surface vibrations due to three train speeds of v = 200, 300 and 400 km/h were plotted with respect to the distance from the point of interest to the bridge (along the x axis), referred to as the *site-to-bridge distance* d_{sb}. The

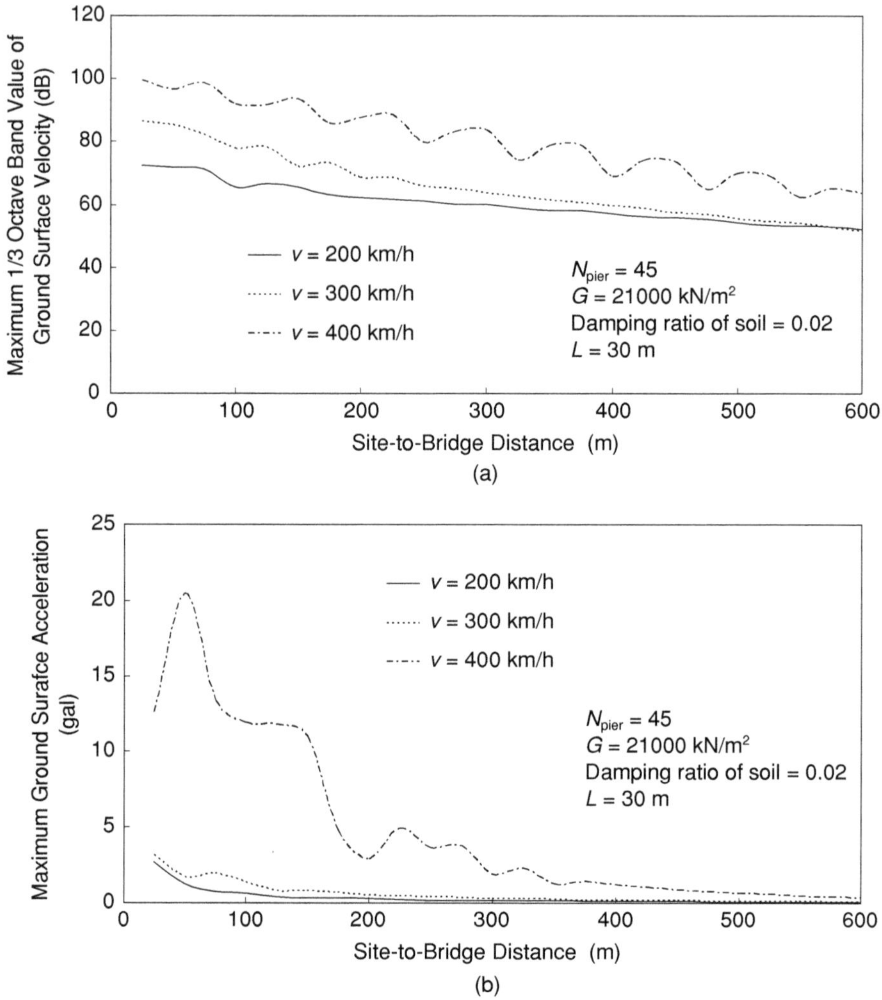

Figure 7.11 Maximum response of the ground surface vs the site-to-bridge distance under different train speeds: (*a*) 1/3 octave band value (in dB); (*b*) acceleration (in gal)

condition of $L = 30$ m and $G = 2.1 \times 10^4$ kN/m^2 was used in generating the results shown in the figure. As can be seen, the maximum dB responses for $v = 200$ and 300 km/h decay linearly with increasing site-to-bridge distance, while that for $v = 400$ km/h (the resonant speed) attenuates in a regularly oscillatory manner as the site-to-bridge distance increases. As for the ground acceleration, the maximum acceleration responses with respect to the site-to-bridge distance under the three train speeds were displayed in Figure 7.11b. Again, the maximum acceleration response for $v = 400$ km/h

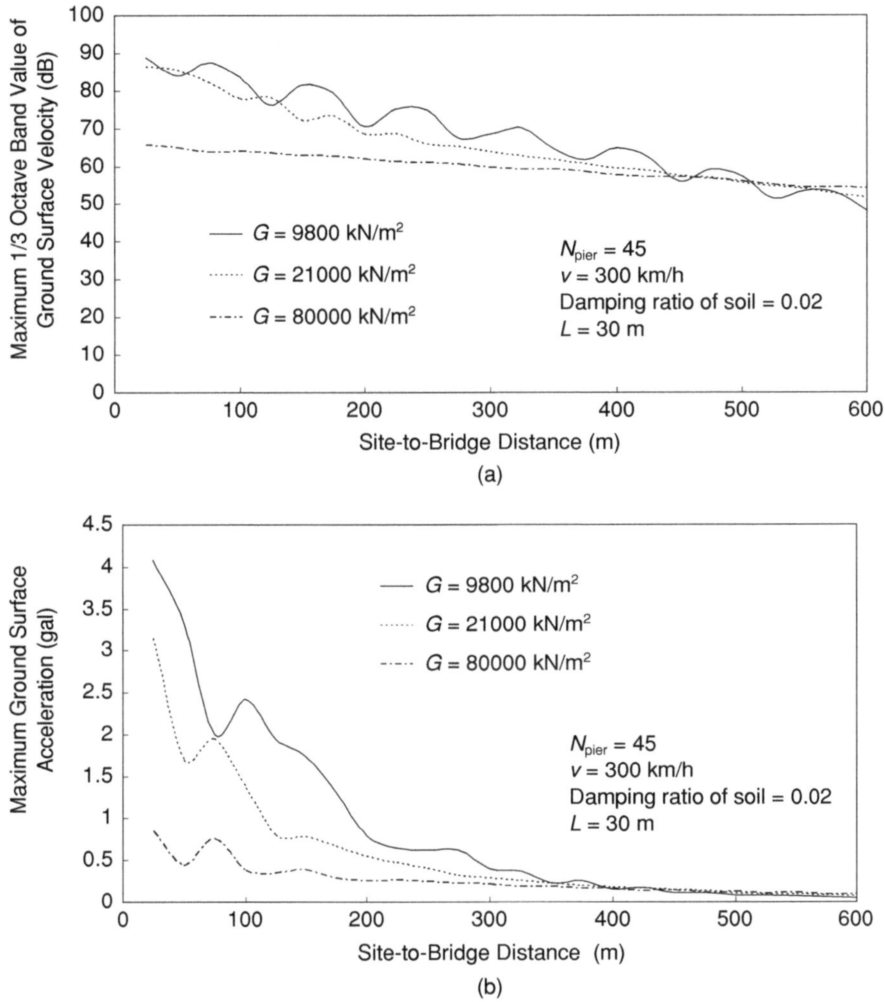

Figure 7.12 Maximum response of the ground surface vs the site-to-bridge distance under different soil properties: (*a*) 1/3 octave band value (in dB); (*b*) acceleration (in gal)

attenuates more irregularly and rapidly than those under the other two speeds. Besides, the maximum acceleration response for v = 400 km/h increases as the site-to-bridge distance increases from 25 m to 50 m, a phenomenon not observed for the other two cases. In general, the acceleration response decays more rapidly than the 1/3 octave band response does.

Figure 7.12a shows the maximum 1/3 octave band (dB) response of the ground vibrations due to three different soil shear moduli, i.e. $G = 9.8 \times 10^3$, 2.1×10^4 and 8.0×10^4 kN/m^2, for the ground with respect to the site-to-bridge distance. The condition of $L = 30$ m and $v = 300$ km/h was assumed in obtaining the results shown in the figure. It can be observed that the maximum dB response for soils with smaller shear moduli decays more rapidly. Furthermore, the maximum dB response for soils with smaller shear moduli will be higher if the site-to-bridge distance is less than 500 m and will be lower otherwise. Of interest is that the maximum dB response for $G = 9.8 \times 10^3$ kN/m^2 attenuates in an oscillatory manner with increasing site-to-bridge distance, which is similar to the case described in Section 6.2 for $v = 400$ km/h and $L = 30$ m under the train–bridge resonance. This phenomenon reveals that some kind of resonance has occurred with the bridge or the ground, or both. Since there is no evidence for the occurrence of the train–bridge resonance under the present condition of $L = 30$ m and $v = 300$ km/h by checking the histogram of the associated pier-top force (not shown) or by using Equation (41), the resonance should occur with the ground. It is therefore inferred that the resonance indicated above results from the coincidence of the primary frequency of the propagating waves and the fundamental frequency of the ground. As far as the ground acceleration response is concerned, it can be seen from Figure 7.12b that the attenuation of the maximum acceleration response (in gal) is generally similar to that of the maximum 1/3 octave band response stated above. It can be seen that the maximum acceleration responses for the three soil properties considered all display a local fluctuation in the range of short site-to-bridge distances, e.g. for $d_{sb} = 50$, 75 and 100 m respectively, which makes the attenuation not completely monotonic. Moreover, for softer soil media, the fluctuation can occur at a longer site-to-bridge distance.

As shown in Figures 7.13, as expected, the ground response decays more rapidly for larger damping ratios of the soil. Furthermore, the maximum 1/3 octave band response of the ground surface decays more rapidly in regions near the source. Such a phenomenon is more obvious for higher soil damping ratios.

From Figures 7.11, 7.12 and 7.13, it can be found that the ground response attenuates with the site-to-bridge distance in an approximately exponential manner, that is, the attenuation curve can be expressed by a function of the type:

$$Y = a(d_{sb})^{-m} \tag{42}$$

where Y is the ground response, a the coefficient, and m the exponent. Figure 7.14 shows the exponent of the attenuation function for the ground surface response with respect to the damping ratio of the soil. Obviously, the exponential value of the attenuation function for the acceleration response is higher than that for the 1/3 octave band response. Also, the exponential value is larger for soils with higher damping ratios. It can be seen that

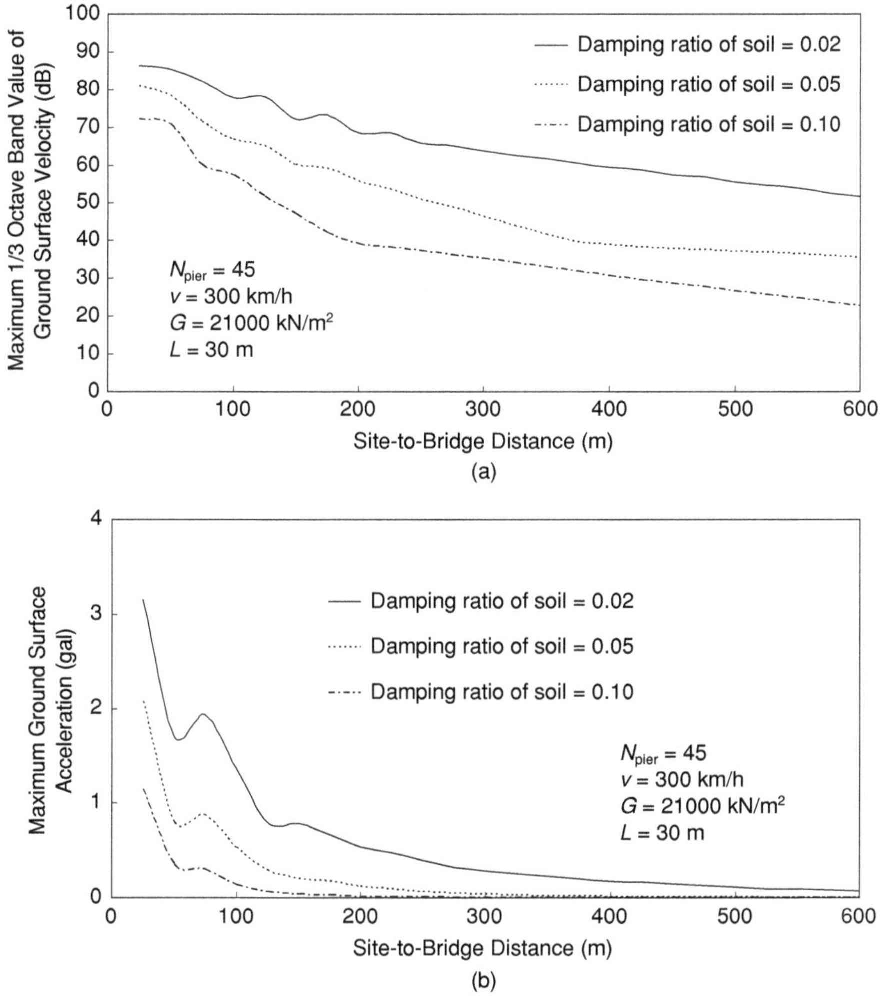

Figure 7.13 Maximum response of the ground surface vs the site-to-bridge distance under different damping ratios of the soil: (*a*) 1/3 octave band value (in dB); (*b*) acceleration (in gal)

the exponential value does not increase linearly with the soil damping ratio, but at a gradually decreasing rate.

6.4 Maximum ground responses under different train speeds and girder span lengths

To investigate further the characteristics of the ground vibrations induced by trains traveling over elevated bridges, the maximum bridge and ground

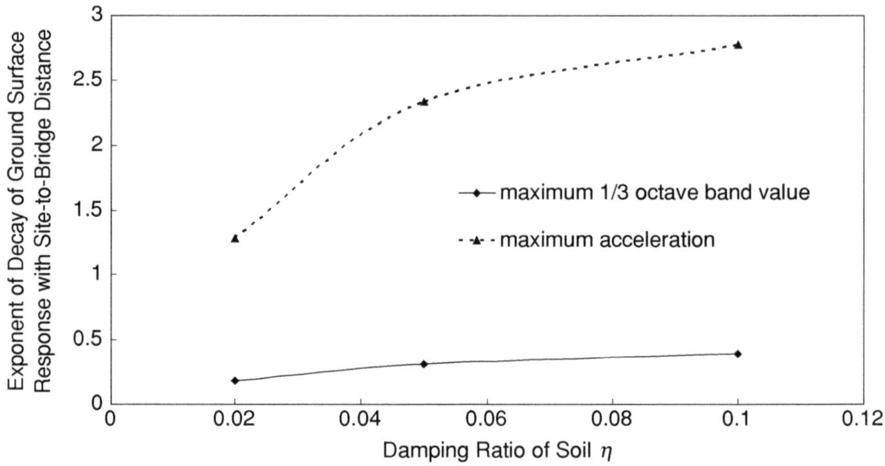

Figure 7.14 Exponent value of the attenuation function for the ground surface response vs the damping ratio of soil

responses under various combinations of train speeds and girder span lengths were computed and plotted in Figures 7.15, 7.16 and 7.17. Figure 7.15 shows the maximum foundation excitation forces, Figure 7.16 the maximum 1/3 octave band values (dB), and Figure 7.17 the peak ground accelerations (gal), all presented in both three-dimensional and contour forms. Five different girder span lengths, i.e. $L = 20$, 25, 30, 35 and 40 m, were considered for the bridge, and for each girder span length the train speed v ranges from 50 to 500 km/h. As can be seen from Figure 7.15, relatively higher maximum foundation excitation forces are generated by the train for $v > 350$ km/h and $L < 35$ m, with extremely large values appearing at $v = 500$ km/h and $L = 25$ m. The presence of the extreme force at $v = 500$ km/h and $L = 25$ m is primarily attributed to the occurrence of train–bridge resonance, which can be verified by Equation (41) using the related bridge and train loads data. Similar resonance phenomenon can also be found at $v = 400$ km/h and $L = 30$ m, as mentioned in Section 6.2, but the corresponding foundation excitation force is much smaller than that under $v = 500$ km/h and $L = 25$ m, since the train–bridge resonances are more intensive for shorter bridge girder spans (Yang et al. 1997). Moreover, it is observed that the maximum foundation excitation force exhibits comparatively lower values for all the train speeds considered when the girder span length is around 35 m. From the results, it is generally concluded that the higher the train speed or the shorter the girder span length, the larger the maximum foundation excitation force.

As far as the ground response is concerned, it can be seen from Figure 7.16 that the maximum 1/3 octave band response due to the train action has a

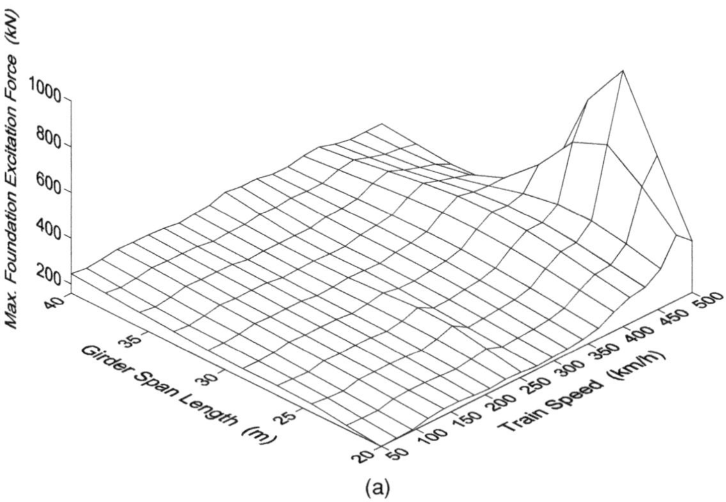

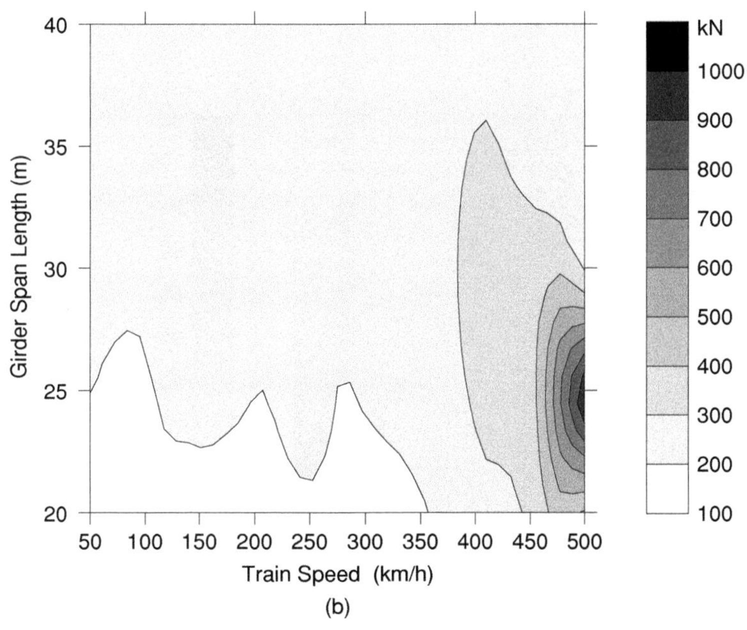

Figure 7.15 Maximum foundation excitation forces (kN) of the bridge due to the train loads under various train speeds and girder span lengths

main region of peak values for $v > 300$ km/h and $L < 35$ m, as well as two secondary regions of medium values centered at $v = 150$ and 250 km/h for $L = 20 \sim 25$ m. Of interest is that relatively smaller spectrum values occur for all the train speeds considered when the girder span length is

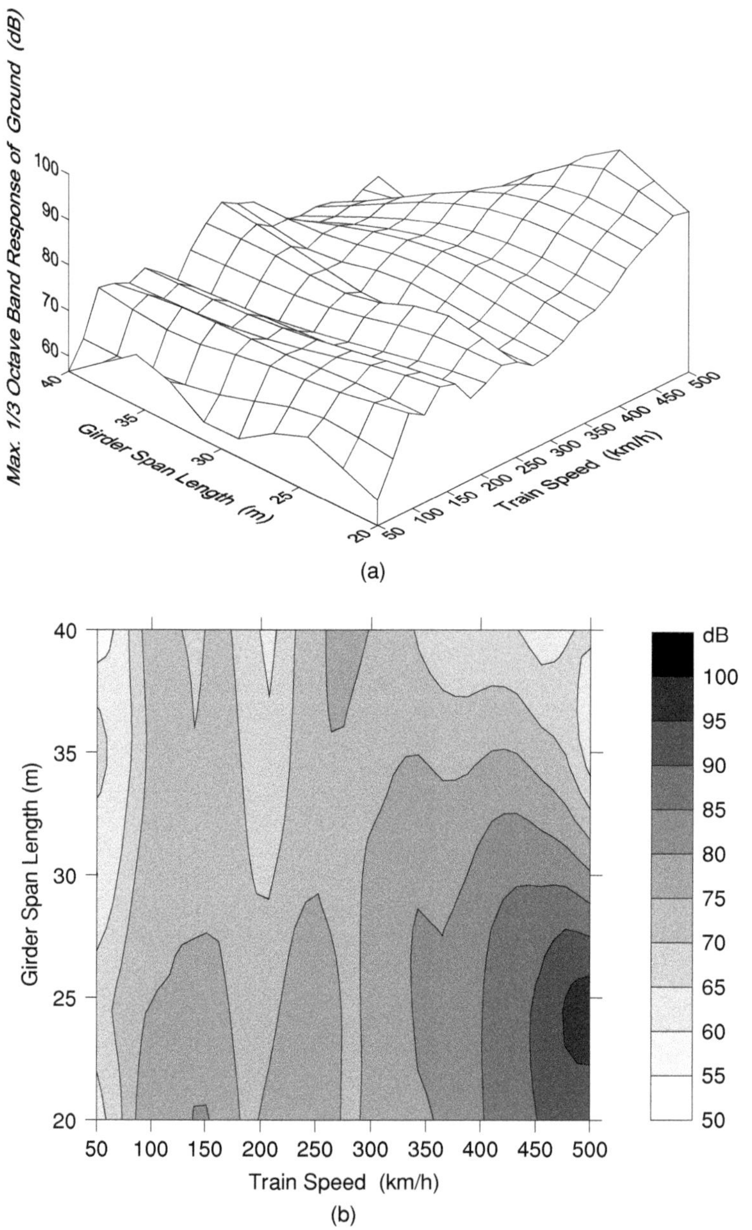

Figure 7.16 Maximum 1/3 octave band responses (dB) of the ground surface due to the train loads under various train speeds and girder span lengths

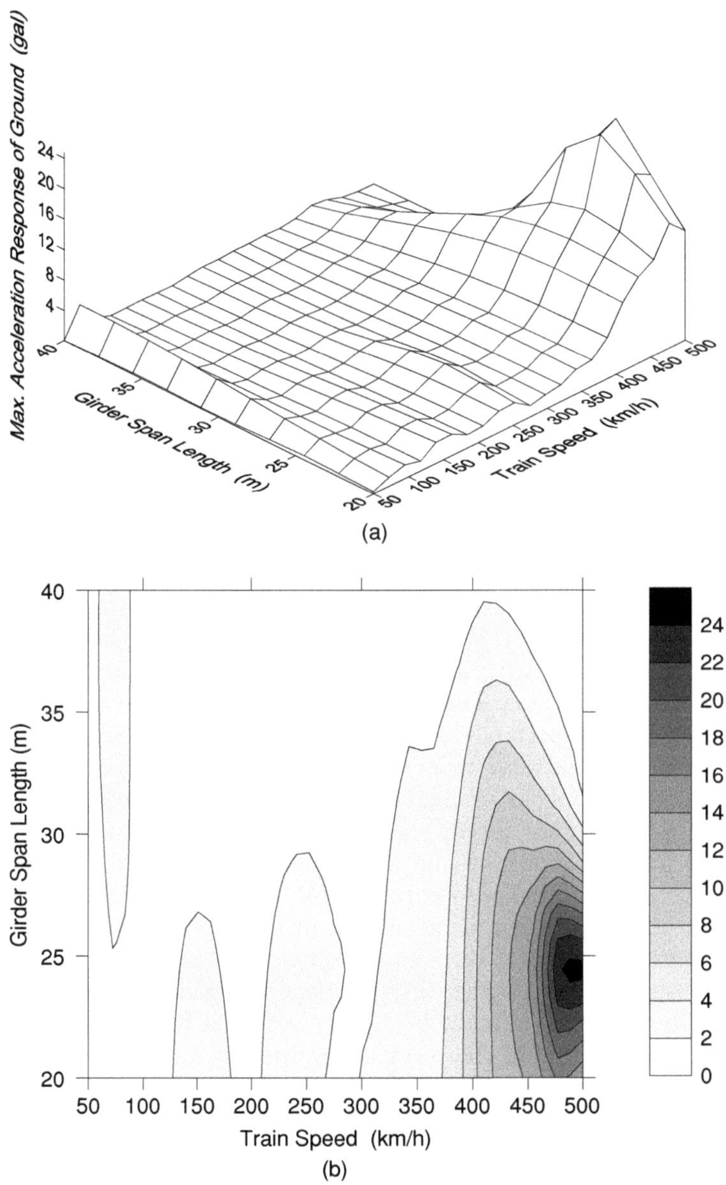

Figure 7.17 Maximum acceleration responses (gal) of the ground surface due to train loads under various train speeds and girder span lengths

L = 35 m, which is consistent with the phenomenon observed for the 1/3 octave band response discussed above. These findings indicate that it is possible to mitigate the train-induced ground vibrations through a proper selection of the span length for the girders of the bridge. Additionally, by comparison of Figures 7.16 with 7.15, it can be found that a large ground vibration response is not necessarily caused by high foundation excitation force and vice versa, as evidenced by the region of L = 20 ~ 25 m and v = 100 ~ 250 km/h of the two figures.

As can be seen from Figure 7.17, the peak ground acceleration (PGA) responses to the train action exhibit larger values at higher train speeds (v > 350 km/h) and shorter girder span lengths (L < 30 m). Besides, the extreme PGA can reach a value of as high as 25 gal (at L = 25 m and v = 500 km/h), which is comparable to the intensity of the ground motions induced by a perceivable earthquake.

6.5 Displacement profile of ground surface due to moving trains

Figures 7.18, 7.19 and 7.20 show the displacement profiles of the ground surface under the train speeds of v = 200, 300 and 400 km/h respectively. Note that the positive z direction in these figures is opposite to that shown in Figure 7.4, in which the positive displacement represents the downward deflection of the ground surface. The following observations can be made from the figures: (1) As expected, the displacement response is larger for higher train speeds. (2) The wave-front of the wave propagations extending backwards at the two sides of the bridge is closer to the bridge axis as the train moves at a higher speed. (3) The tail trajectories of the wave propagations are more pronounced and gradually prevail toward the two sides of the bridge as the train speed is raised. (4) There exists a region of negative displacements, i.e. upward deflection, in front of the wave-front of the wave propagations. Also, the region of negative displacements is wider for higher train speeds. (5) Higher displacement responses concentrate more noticeably in the region near the bridge for trains with higher speeds, especially on the region around the piers under the instantaneous action area of the train on the bridge. (6) At higher train speeds the concentration region of larger displacement responses falls more behind the front ridge of the train, as indicated in the figures. (7) The displacement distribution for v = 400 km/h is obviously fluctuant and scattered owing to the train–bridge resonance.

6.6 Effect of elastic bearings of bridges on ground vibrations

Since elastic bearings are frequently installed on bridges as a means of reducing upward transmitting forces during an earthquake, the effect of the elastic bearings on the train-induced ground vibrations needs to be explored

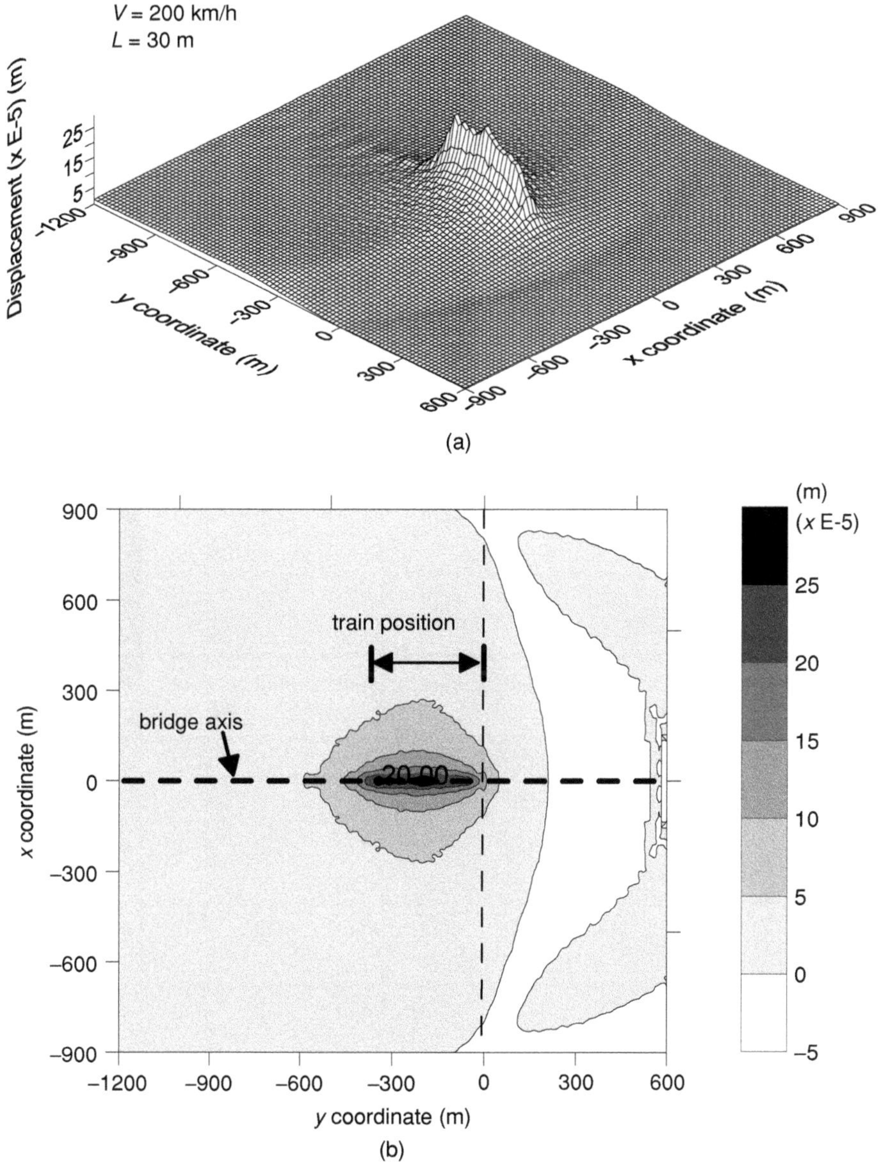

Figure 7.18 Displacement profile of the ground surface generated by the train moving at v = 200 km/h: (*a*) three-dimensional plot; (*b*) contour plot

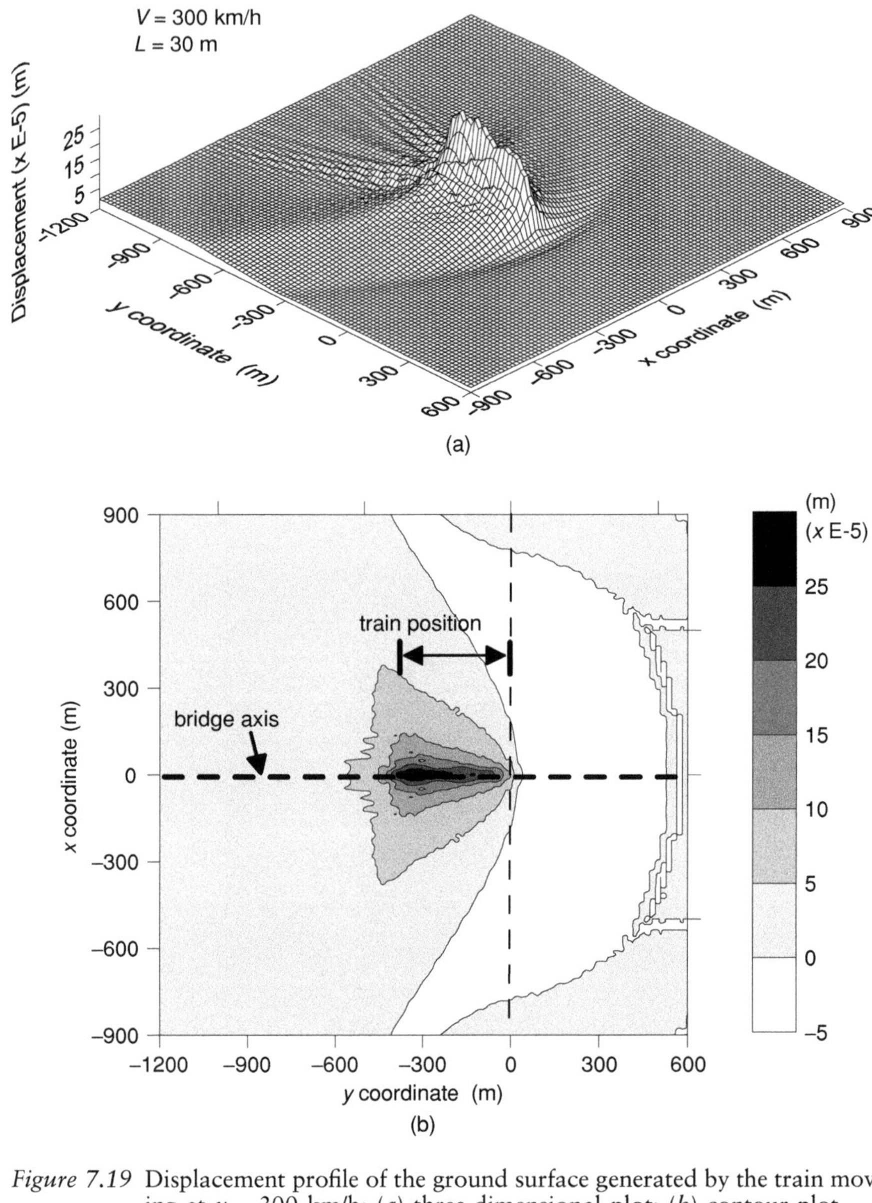

Figure 7.19 Displacement profile of the ground surface generated by the train moving at v = 300 km/h: (*a*) three-dimensional plot; (*b*) contour plot

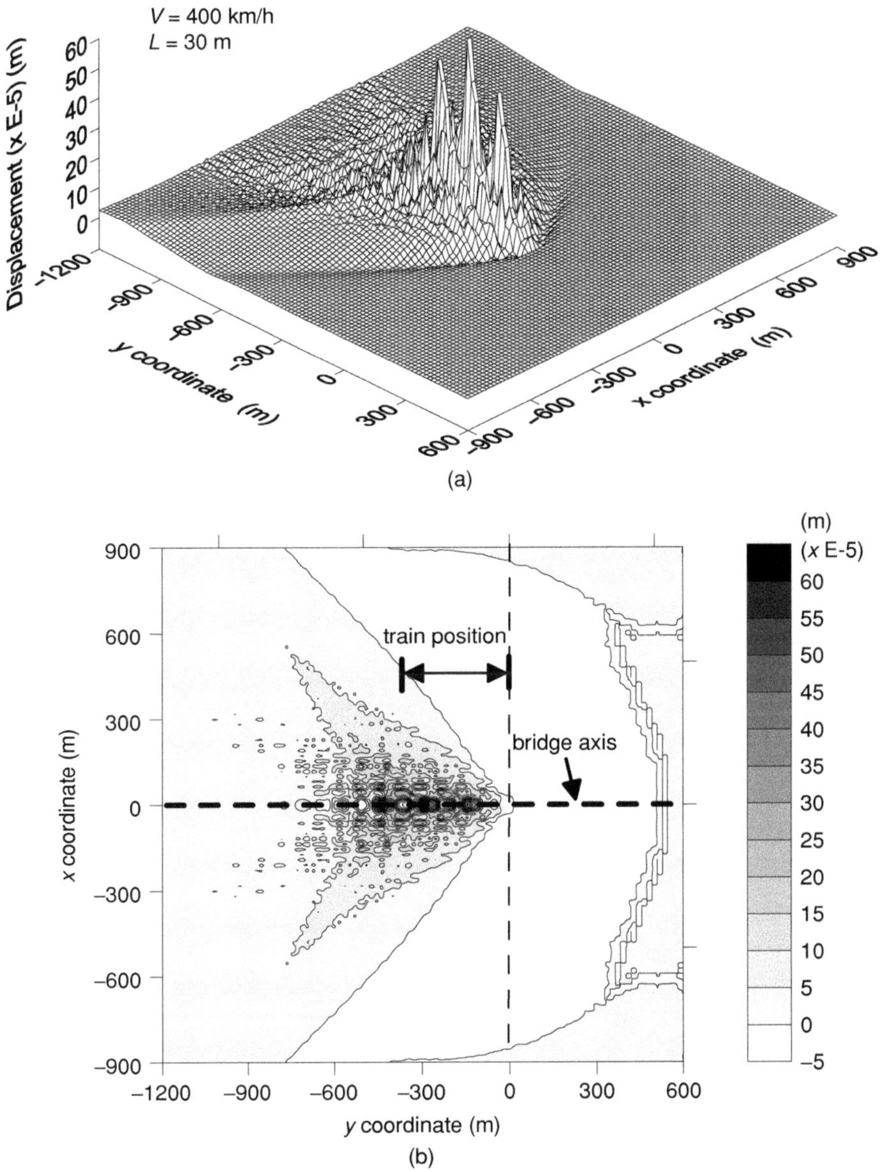

Figure 7.20 Displacement profile of the ground surface generated by the train moving at v = 400 km/h: (*a*) three-dimensional plot; (*b*) contour plot

further. Figure 7.21 shows the responses of the ground surface to the action of the train as it travels over a bridge with various girder span lengths equipped with elastic bearings, in which three girder span lengths $L = 20$, 30 and 40 m are considered. For each girder span length, six stiffness ratios of the elastic bearings, i.e. $\kappa = 0.001$ (simple support), 0.10, 0.15, 0.20, 0.30 and 0.45, are considered, with larger κ values representing softer elastic bearings. Note that the κ values considered sufficiently cover those encountered in engineering practice. Three train speeds, i.e. $v = 200$, 300 and 400 km/h, are considered for each combination of girder span length and stiffness ratio.

As can be seen from Figure 7.21, the maximum ground responses are, in general, of less difference under different κ values for all the girder span lengths and train speeds considered, except for the case of $L = 30$ m and $v = 400$ km/h, in which the maximum ground response decreases significantly as the κ value increases. For the case of $L = 30$ m, $v = 400$ km/h, train–bridge resonance occurs under the condition of simple supports, i.e. with $\kappa = 0.001$. As the κ value increases, the train–bridge resonance becomes less obvious, thereby leading to reduced ground vibration intensities. The above discussion indicates that the installation of elastic bearings on the bridge can effectively mitigate the train-induced ground vibrations under the condition of train–bridge resonance.

It can also be seen from the figure that the maximum ground response for $L = 40$ m and $v = 300$ km/h increases with increasing κ values in the

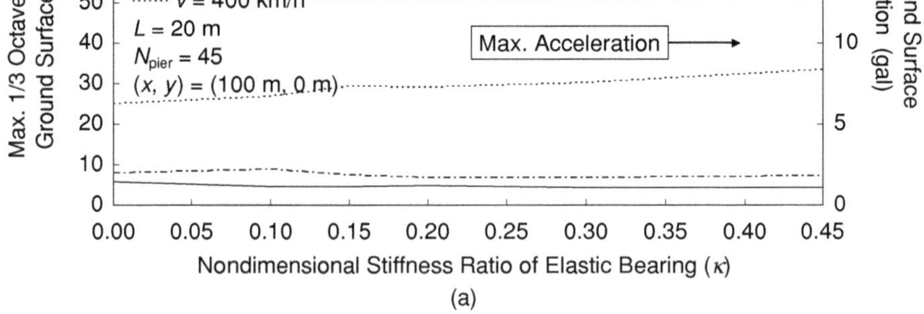

Figure 7.21 Maximum responses of the ground surface under different stiffness ratios of the elastic bearings and train speeds for: (*a*) $L = 20$ m; (*b*) $L = 30$ m; (*c*) $L = 40$ m

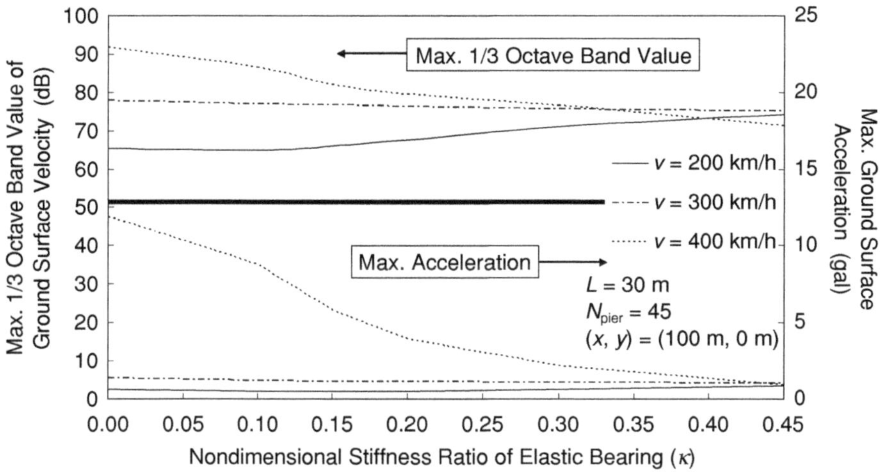

(b)

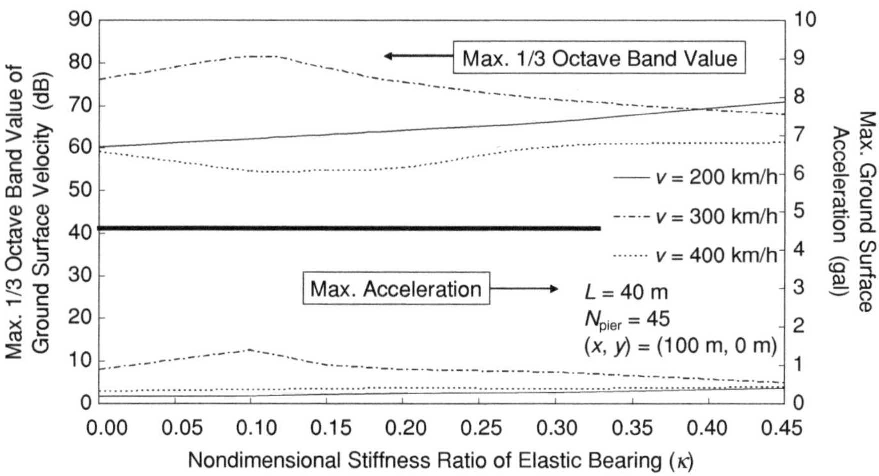

(c)

Figure 7.21 (*cont'd*)

range of $\kappa \leq 0.1$, and decreases as the κ value increases in the range of $\kappa > 0.1$.

On the other hand, the maximum ground responses under different κ values and girder lengths of $L = 20$, 30 and 40 m were presented in Figures 7.22(a)–(c) for train speeds of $v = 200$, 300 and 400 km/h respectively. For the case with the train speed equal to v = 200 km/h, it is indicated that the presence of elastic bearings can result in a reduction of the maximum ground responses for L = 20 m, as compared with the maximum ground

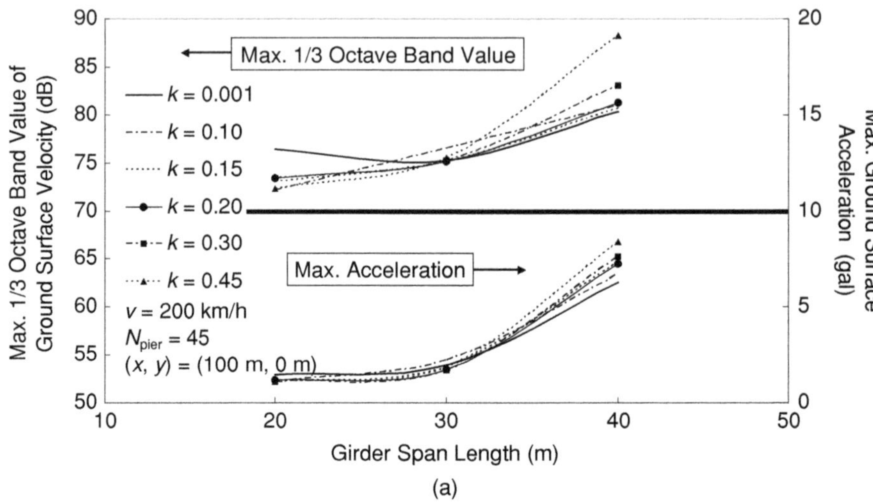

Figure 7.22 Maximum responses of the ground surface under different girder span lengths and stiffness ratios of the elastic bearings for: (*a*) v = 200 km/h; (*b*) v = 300 km/h; (*c*) v = 400 km/h

responses under simple support condition, i.e. $\kappa = 0.001$, and that it will amplify the maximum ground responses for $L = 40$ m. Moreover, the extent of the reduction or amplification in the ground response due to the elastic bearings increases as κ increases. The maximum ground responses for $L = 30$ m remain almost unaffected by the presence of elastic bearings under the train speed of $v = 200$ km/h. It can also be seen that the maximum 1/3 octave band response of the ground surface is generally more sensitive to the presence of elastic bearings than the maximum acceleration response.

For the case with the train speed raised to $v = 300$ km/h, the variation of the maximum ground responses due to the effect of elastic bearings is quite different from that for $v = 200$ km/h, as shown in Figure 7.22b. At this train speed, the ground vibration intensity will be slightly amplified for $L = 20$ m and reduced for $L = 40$ m owing to the presence of elastic bearings, which is exactly opposite to the phenomenon observed for $v =$ 200 km/h discussed above. Also, the extent of the amplification or reduction is greater for larger κ values.

For the train speed of $v = 400$ km/h, the variation of the maximum ground responses to elastic bearings is more complicated, as shown in Figure 7.22c. The presence of elastic bearings can have amplification and suppression effects on the maximum ground responses for $L = 20$ and 40 m respectively. On the other hand, when $L = 30$ m the maximum ground response will be amplified as $0.001 \le \kappa \le 0.20$ and will be reduced as $\kappa > 0.20$. The

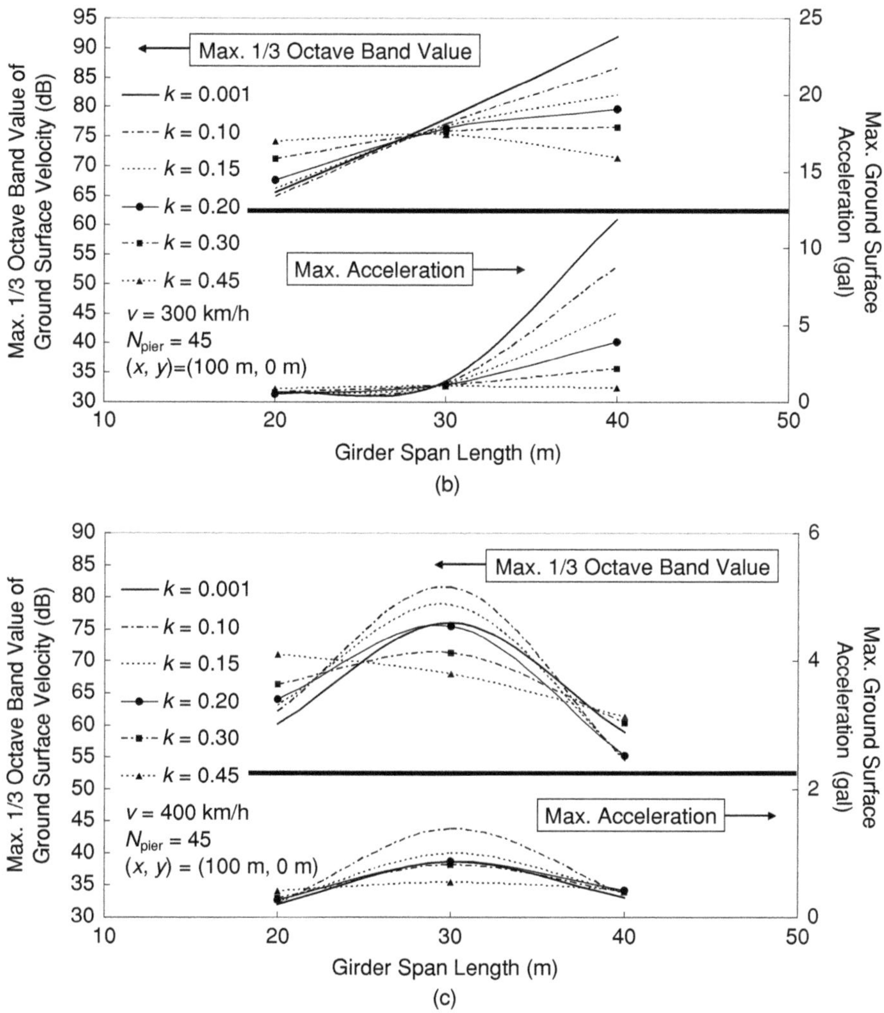

Figure 7.22 (cont'd)

reason is that for lower κ values, i.e. $\kappa \leq 0.20$, the train speeds at which the train–bridge resonance occurs are closer to v = 400 km/h than that for $\kappa = 0.001$, by which the train–bridge resonance is enhanced and thus results in the increase of the maximum ground responses. As for higher κ values, i.e. $\kappa > 0.20$, the resonant train speeds deviate much more from $v = 400$ km/h than that for $\kappa = 0.001$, by which the train–bridge resonance is weakened and leads to the reduction of the maximum ground responses.

Figure 7.23 shows the time-history response of the ground surface under the train–bridge resonance (L = 30 m, v = 400 km/h) concerning the effect of the elastic bearings. As can be seen, the lower the stiffness (the larger κ value) of the elastic bearings, the higher the extent of reduction in the ground

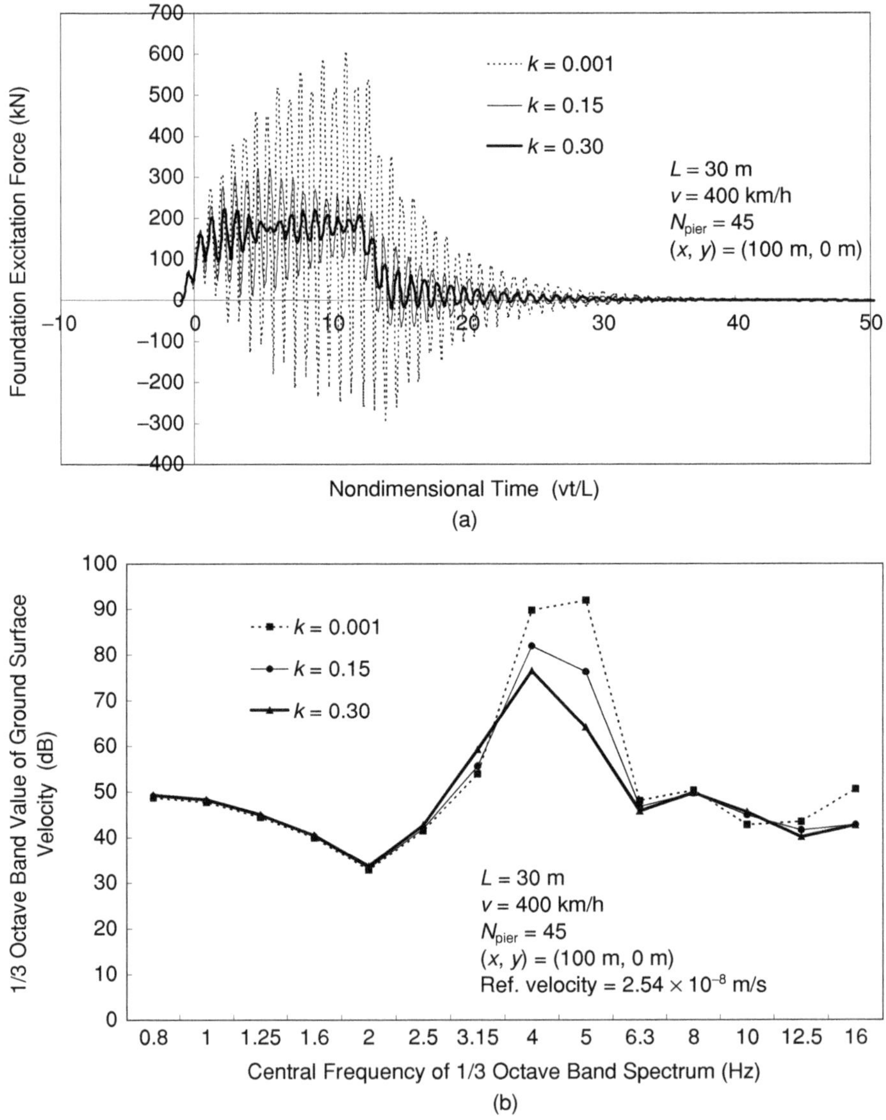

Figure 7.23 Responses of the ground surface under the train–bridge resonance due to the effect of elastic bearings: (*a*) foundation excitation force; (*b*) 1/3 octave band spectrum; (*c*) acceleration

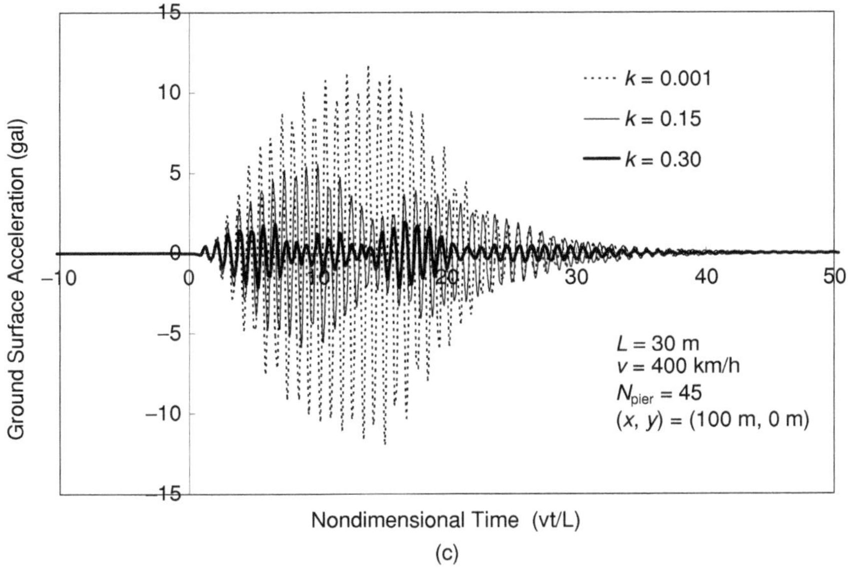

Figure 7.23 (cont'd)

response. This is especially true for the responses around the dominant frequency, as indicated in the 1/3 octave band spectrum. Also, the presence of elastic bearings makes the peak 1/3 octave band response move toward the low-frequency side. Notice that in the range of very low frequencies, i.e. for $f < 2.5$ Hz, the ground vibration level is almost unaffected by the existence of elastic bearings. This means that there is no help in suppressing the very low frequency ground vibrations induced by trains moving over the bridge by the installation of elastic bearings.

7 Concluding remarks

This chapter presented an investigation into the ground vibrations induced by trains traveling over a multi-span elevated bridge with pile foundations. To simulate more realistically the ground vibrations induced, the structural dynamics of the bridge, the foundation–soil interaction, and the wave propagation of the ground were considered. Based on the models used for the train, the bridge and the ground, the analysis was carried out in a semi-analytical manner. Owing to the partial analytical nature of the approach, much less time and effort are required in the analysis compared with those that rely fully on numerical modeling, using, say, the boundary or finite element methods. The validity of the present approach was verified through the study of an example reported in the literature.

From the results of the parametric study presented in the chapter, some important conclusions can be drawn as follows: (1) For a girder span length of $L = 30$ m and train speeds of $v = 200$ to 400 km/h, as commonly encountered in modern high-speed railways, the dominant frequency of the train-induced ground vibrations is between 2–6 Hz. (2) There exists a saturation phenomenon in the acceleration response spectra of the ground accelerations as the train speed exceeds a specific limit, i.e. 300 km/hr for the cases studied. (3) Dramatically large ground vibrations will be induced as resonance occurs between the train and bridge. (4) Ground vibration under the train–bridge resonance attenuates in an oscillatory manner with the increase in the site-to-bridge distance, which is substantially different from that under non-resonant conditions. Also, the ground vibration under train–bridge resonance decays more rapidly than those under non-resonant conditions. (5) Comparatively lower levels of ground vibration exist for some combinations of girder span length and train speed. This implies the existence of an optimal design for the bridge concerning the mitigation of train-induced ground vibrations. (6) The installation of elastic bearings on the bridge can effectively alleviate the train-induced ground vibrations under the circumstance of train–bridge resonance. However, under the non-resonant conditions, the presence of the elastic bearings may reduce or amplify the ground vibration responses, depending on the girder span length and the train speed.

Although reasonable models and practical data have been adopted in the study to simulate the train-induced ground vibrations, there are inevitably simplifications and assumptions made in the modeling of the various subsystems, which is especially true for the ground model. For example, to simplify the problem, the ground has been idealized as a homogenous and isotropic halfspace, which may deviate from the real ground conditions. Besides, some properties of the train, the bridge and the ground, such as the material damping of the bridge foundation and the soil medium, are wide-ranging, and it is not easy to determine suitable values for use in the analysis. For this reason, the emphasis of the present study has been placed on the qualitative aspects of the train-induced ground vibrations, rather than on quantitative results. In addition, it should be noted that the conclusions drawn in this study remain strictly only for the data, theory and assumptions adopted.

The method presented in the chapter can be enhanced through the use of more delicate models for the train, the bridge and the foundation–soil interaction system. It can also be generalized to deal with the problem of train-induced vibrations on stratified soils, if Green's functions specific to layered media reported elsewhere, Kausel and Peek (1982), Luco and Apsel (1983) are used. By using the enhanced or generalized models, the effects of further influential factors, not covered in the study, such as the horizontal train actions or stratification of the soil, on the train-induced ground vibrations can be investigated.

References

Gordon, C. G. (1991) Generic criteria for vibration sensitive equipment, *SPIE*, 1619: 71–5.

Hanazato, T., Ugai, K., Mori, M. and Sakaguchi, R. (1991) Three-dimensional analysis of traffic-induced ground vibrations, *Journal of Geotechnical Engineering Division*, ASCE, 117: 1133–51.

Hung, H. H., Yau, J. D. and Yang, Y. B. (1999) Ground vibrations induced by high speed trains, *Structural Engineering*, 14: 3–20. In Chinese.

Ju, S. H. (2002) Finite element analyses of wave propagations due to a high-speed train across bridges, *International Journal for Numerical Methods in Engineering*, 54: 1391–1408.

Kausel, E. and Peek, R. (1982) Dynamic loads in the interior of a layered stratum: an explicit solution, *Bulletin of the Seismological Society of America*, 72 (5): 1459–81.

Krylov, V. V. (1995) Generation of ground vibration by superfast trains, *Applied Acoustics*, 44: 149–64.

Krylov, V. V. and Ferguson, C. (1994) Generation of low frequency ground vibrations from railway trains, *Applied Acoustics*, 42: 199–213.

Luco, J. E. and Apsel, R. J. (1983) On the Green's functions for a layered half-space. Part I, *Bulletin of the Seismological Society of America*, 73 (4): 909–29.

Madshus, C., Bessason, B. and Hårvik, L. (1996) Prediction model for low frequency vibration from high speed railways on soft ground, *Journal of Sound and Vibration*, 193: 195–203.

Miller, G. F. and Pursey, H. (1954) The field and radiation impedance of mechanical radiators on the free surface of a semi-infinite isotropic solid, *Proceedings Royal Society A*, 223: 521–41.

Novak, M. (1977) Vertical vibration of floating piles, *Journal of Engineering Mechanics Division*, ASCE, 103 (EM1): 153–68.

Poulos, H. G. (1968) Analysis of the settlement of the pile groups, *Géotechnique*, 18: 449–71.

Prakash, S. and Sharma, H. D. (1990) *Pile Foundations in Engineering Practice*, New York: John Wiley.

Rücker, R. (1982) Dynamic behaviour of rigid foundations of arbitrary shape on a halfspace, *Earthquake Engineering and Structural Dynamics*, 10: 675–90.

Takemiya, H. (1997) Prediction of ground vibration induced by high-speed train operation, *Proceedings of the 18th Sino-Japan Technology Seminar*, Chinese Institute of Engineers, pp. 1–10.

Wolf, J. P. (1994) *Foundation Vibration Analysis Using Simple Physical Models*, Englewood Cliffs, NJ: Prentice Hall.

Wu, Y. S., Hsu, L. C. and Yang, Y. B. (2002) Ground vibrations induced by trains moving over a series of elevated bridges, *Proceedings of the Tenth Sound and Vibration Conference, Taipei*, pp. 1–7. In Chinese.

Wu, Y. S. and Yang, Y. B. (2004) A semi-analytical approach for analyzing ground vibrations caused by trains moving over elevated bridges, *Soil Dynamics and Earthquake Engineering*, 24 (12): 949–62.

Yang, Y. B., Hung, H. H. and Chang, D. W. (2003) Train-induced wave propagation in layered soils using finite/infinite element simulation, *Soil Dynamics and Earthquake Engineering*, 23 (4): 263–78.

Yang, Y. B., Yau, J. D. and Hsu, L. C. (1997) Vibration of simple beams due to trains moving at high speeds, *Engineering Structures*, 19: 936–44.

Yau, J. D., Wu, Y. S. and Yang, Y. B. (2001) Impact response of bridges with elastic bearings to moving loads, *Journal of Sound and Vibration*, 248: 9–30.

Yoshioka, O. (2000) Basic characteristics of Shinkanshen-induced ground vibration and its reduction measures, *Proceedings of the International Workshop WAVE2000, Bochum*, pp. 219–37.

8 Efficient analysis of buildings with grouped piles for seismic stiffness and strength design

Izuru Takewaki and Akiko Kishida

Abstract

An efficient method is presented for the analysis of buildings with grouped piles for seismic stiffness and strength design. A continuum model composed of a dynamic Winkler-type soil element and a pile is used to describe the dynamic behavior of the pile–soil system within a reasonable accuracy. The pile-group effect is considered by introducing the influence coefficients among piles which are defined for interstory drifts and pile-head bending moments. It is shown that, while the pile-group effect generally decreases the interstory drift of buildings, it may increase the bending moment at the head of piles in some cases. This fact implies that the procedure without the pile-group effect leads to the conservative design for superstructures and requires a modification of member design for piles. It is concluded that a detailed and efficient examination of the pile-group effect is absolutely necessary in the practical seismic design of buildings from the simultaneous viewpoints of stiffness and strength.

Keywords: structure–pile–soil system, dynamic Winkler-spring, pile-group effect, bedrock earthquake input, hysteretic and radiation damping, stiffness and strength design, performance-based design.

1 Introduction

Pile foundations are often used for the design of buildings at the site where a soft-surface ground rests on rather stiff bedrock. Uneven settlement of foundation mats and lack of bearing capacity of the resisting soils may be the principal reasons for the use of pile foundations. It should be pointed out that, in most of the previous studies on the dynamic resistance of piles or in their practical design, the pile-group effect has never been considered explicitly except through a general approximate procedure including a coefficient to be multiplied on the sum of the stiffness quantities for single piles [1–3]. This approximate procedure may lead to an acceptable and practical design when the so-called inertial interaction effect only is considered in the design of building-pile systems. However, it is not made clear whether this procedure is also acceptable in the case of the design of building-pile systems for various parameters of pile spacing, pile–soil stiffness ratios, etc., under the so-called kinematic interaction effect. Because it has been pointed out recently that both the inertial and kinematic interaction effects have to be taken into account simultaneously in soft grounds, an efficient and fairly accurate procedure for the evaluation of these effects together with the pile-group effects is strongly desired [2, 4].

The purpose of this chapter is to explain an efficient and fairly accurate analysis procedure for the stiffness and strength design of building–pile–soil systems including the pile-group effects under the simultaneous consideration of inertial and kinematic interaction effects. While both stiffness and strength are the principal design objectives in the design of superstructures, strength is the only principal design objective in foundations. It seems that a practically acceptable design procedure does not have to be rigorous or use large amounts of computation but does require procedures to be both efficient in use and robust with parameter uncertainties.

A continuum model composed of a dynamic Winkler-type soil element and a pile is used to express, fairly accurately, the dynamic behavior of the pile–soil system [4–8]. The pile-group effect is considered by using the influence coefficients among piles which are defined separately under the inertial interaction effect and under the kinematic interaction effect. The influence coefficients are introduced here both for interstory drifts and for pile-head bending moments. While the influence coefficients were defined for the foundation mat displacement and the pile-head forces, those have never been introduced for the interstory drifts of buildings and the pile-head bending moments [9]. It is shown here that, while the pile-group effect decreases the interstory drift of buildings in general, it may increase the bending moment at the head of piles. This fact implies that the procedure without the pile-group effect results in a conservative design for superstructures and requires a modification of the member design for piles.

2 Building–pile–soil system

A building–pile–soil interaction system is considered as shown in Figure 8.1. It is assumed that the pile foundation is composed of multiple piles and that the pile head is fixed to the foundation beam; i.e. the nodal rotation of the pile head is vanishing. The stiffness of the Winkler-type soil element is taken from the value defined for the fixed pile head [5, 6]. The damping of the Winkler-type soil element is presumed to be a combination of radiation damping into the horizontal direction and linear hysteretic damping [5–8]. This treatment means the introduction of a frequency-dependent damping coefficient.

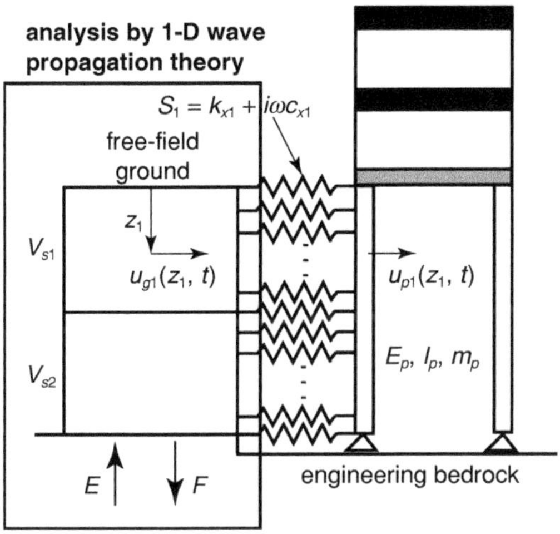

Figure 8.1 Building-pile system supported by the free-field ground through the Winkler-type soil element

The building is modeled using a shear-type building model. The mass in the i-th floor of the building and that in the foundation mat are denoted by m_i and m_0, respectively. Let k_i and c_i denote the story stiffness and the corresponding damping coefficient respectively of the i-th story. The so-called dynamic stiffness in the i-th story is expressed by $K_i = k_i + i\omega c_i$. The symbol i indicates the imaginary unit and ω the excitation frequency. It is also assumed that the surface ground of a horizontal layer rests on the semi-infinite uniform bedrock and that the building-pile system is connected with the free-field ground through the Winkler-type soil element [4–8]. The aspect ratio of the building is assumed to be so small that the rocking motion is negligible.

The numbering of the soil layer begins from the ground surface. This means that the surface soil layer is the first soil layer, and the coordinate z_1 in the first soil layer is directed downward from the ground surface. L_1 denotes the thickness of the first soil layer, and V_{s1} and ρ_1 denote the shear wave velocity and the mass density respectively in the first soil layer. The shear wave velocity and mass density in the bedrock are denoted by V_{s2}, ρ_2 respectively. Let V_{s1}^* and V_{s2}^* denote the complex shear wave velocity including the linear hysteretic damping ratio in the surface soil and in the bedrock respectively. E_p, I_p, m_p denote the Young's modulus, the second moment of area and the mass per unit length of the pile. The impedance of the Winkler-type soil element in the surface soil layer is described by $S_1 = k_{x1} + i\omega c_{x1}$ in terms of the stiffness and damping coefficients, k_{x1} and c_{x1}, of the Winkler-type soil element in the surface soil layer.

3 Impedance of single pile

In this section, it is assumed that a harmonic horizontal force $Pe^{i\omega t}$ acts at the pile head of a single pile. It is also assumed that the damping of the pile is negligible and the semi-infinite property of the bedrock can be modeled by a viscous boundary at the pile tip. The damping coefficient of the viscous boundary per unit area is expressed as $C = \rho_2 V_{s2}$.

The horizontal displacement of the pile in the surface soil layer is denoted by $u_{p1}(z_1, t)$. The equation of motion of the pile in the surface soil layer under a harmonic horizontal force $Pe^{i\omega t}$ can be expressed by

$$E_p I_p \frac{\partial^4 u_{p1}}{\partial z_1^4} + m_p \frac{\partial^2 u_{p1}}{\partial t^2} + S_1 u_{p1} = 0 \tag{1}$$

Because the excitation is harmonic, let us express u_{p1} as

$$u_{p1}(z_1, t) = \hat{U}_{p1}(z_1) e^{i\omega t} \tag{2}$$

Substitution of Equation (2) into Equation (1) and comparison of the coefficients yield the following differential equation for amplitude:

$$(d^4 \hat{U}_{p1}/dz_1^4) - \lambda_1^4 \hat{U}_{p1} = 0 \tag{3}$$

In Equation (3), the parameter λ_1 is given by

$$\lambda_1 = ((m_p \omega^2 - S_1)/E_p I_p)^{0.25} \tag{4}$$

The general solution of Equation (3) can be constructed through the linear combination of four linearly independent basis functions and may be expressed as

$$\hat{U}_{p1}(z_1) = D_1 e^{-\lambda_1 z_1} + D_2 e^{\lambda_1 z_1} + D_3 e^{-i\lambda_1 z_1} + D_4 e^{i\lambda_1 z_1} \tag{5}$$

The amplitude of the horizontal displacement at the pile head is then described by

$$\hat{U}_{p1}(0) = D_1 + D_2 + D_3 + D_4 \tag{6}$$

Let us introduce the following parameters $D_i^* \equiv D_i/P$ for later manipulation. Then

$$\hat{U}_{p1}(0) = (D_1^* + D_2^* + D_3^* + D_4^*)P \tag{7}$$

The dynamic stiffness (impedance) $k_x^{(1)}$ of a single pile is then derived as

$$k_x^{(1)} = P/\hat{U}_{p1}(0) = 1/(D_1^* + D_2^* + D_3^* + D_4^*) \tag{8}$$

This impedance is obtained from the following four boundary conditions:

(1) equilibrium of the shear force at the pile head with the applied force $Pe^{i\omega t}$
(2) vanishing of the angle of rotation at the pile head
(3) equilibrium of the shear force at the pile tip (bottom of the pile) with the damping force at the viscous boundary, and
(4) vanishing of the bending moment at the pile tip.

These conditions may be described mathematically by

$$[\{d_1\} \quad \{d_2\} \quad \{d_3\} \quad \{d_4\}]\{D_1 \quad D_2 \quad D_3 \quad D_4\}^T = \{0 \quad -P \quad 0 \quad 0\}^T \tag{9a}$$

where $(\ )^T$ indicates transpose of a vector and $\{d_1\}$, $\{d_2\}$, $\{d_3\}$, $\{d_4\}$ indicate

$$\begin{aligned}
\{d_1\} &= \{-\lambda_1 \quad E_P I_P \lambda_1^3 \quad (E_P I_P \lambda_1^3 + Ci\omega)e^{-\lambda_1 L_1} \quad -E_P I_P \lambda_1^2 e^{-\lambda_1 L_1}\}^T \\
\{d_2\} &= \{\lambda_1 \quad -E_P I_P \lambda_1^3 \quad (-E_P I_P \lambda_1^3 + Ci\omega)e^{\lambda_1 L_1} \quad -E_P I_P \lambda_1^2 e^{\lambda_1 L_1}\}^T \\
\{d_3\} &= \{-i\lambda_1 \quad -E_P I_P i\lambda_1^3 \quad (-E_P I_P \lambda_1^3 + Ci\omega)e^{-i\lambda_1 L_1} \quad E_P I_P \lambda_1^2 e^{-i\lambda_1 L_1}\}^T \\
\{d_4\} &= \{i\lambda_1 \quad E_P I_P i\lambda_1^3 \quad (E_P I_P \lambda_1^3 + Ci\omega)e^{i\lambda_1 L_1} \quad E_P I_P \lambda_1^2 e^{i\lambda_1 L_1}\}^T
\end{aligned} \tag{9b–e}$$

It is interesting to note that the parameters D_i^* are obtained from D_i by substituting $P = 1$ in Equation (9a).

4 Comparison with other models and recorded data during an earthquake

4.1 Comparison with other models

For verification of the accuracy of the present Winkler-type soil element model and its interaction with the pile, the thin-layer method has been used for the analysis of single piles [9, 10]. The model parameters are shown in Figure 8.2(a). Figure 8.2(b) shows the comparison of the real and imaginary parts of the impedance. The solid line indicates the result by the present method, and the dotted line presents the result by the thin-layer method. It can be concluded that the present method has an acceptable accuracy.

In addition, an investigation of the accuracy of the bedrock input has been made. Figure 8.3 shows the comparison of the transfer functions of the bending moments and the shear forces in a pile by the present multi-input continuum model with those by the single-input FEM model (see Figure 8.6 shown later). It can be concluded that the present model is acceptable also for the bedrock input. Figure 8.4 presents the peak bending moments and shear forces in a pile subjected to a design response spectrum. It can be observed that the peak bending moments and shear forces have a tendency similar to the corresponding transfer functions.

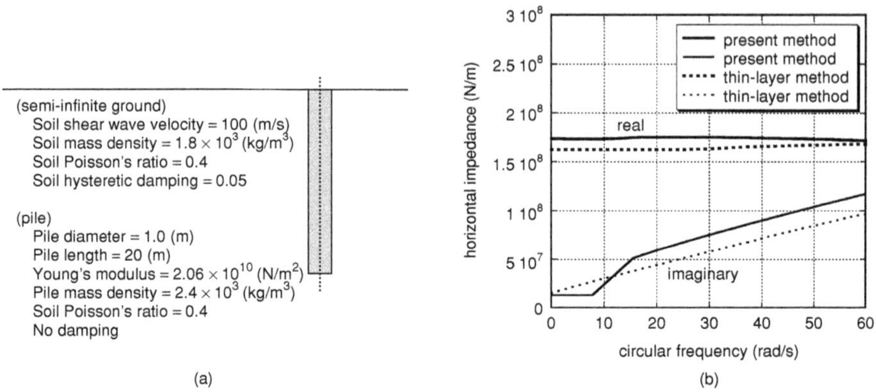

Figure 8.2 Comparison of the impedance by the present Winkler-type spring method with that by the thin-layer method

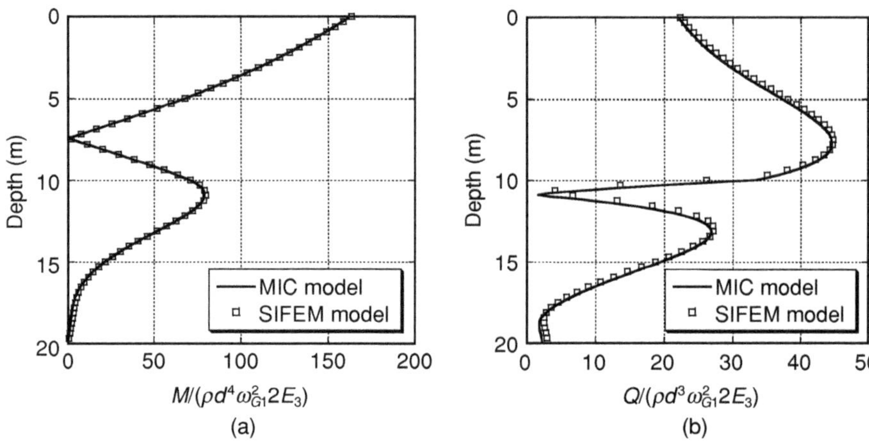

Figure 8.3 Comparison of the transfer function amplitude by the multi-input continuum model with that by the single-input finite-element model

4.2 Comparison with recorded data during an earthquake

For verification, in a real way, of the present Winkler soil element model, a model has been constructed and a numerical simulation has been implemented for an actual building with piles [4, 8]. The building, illustrated in Figure 8.5, is located in Yokohama in Japan. This building is composed of a twelve-story steel frame with twenty cast-in-place reinforced-concrete piles, 35 m long and 1.7 m in diameter. For simulation of the peak response of the bending strains in the piles, a finite-element model, as shown in Figure 8.6, has been utilized [3, 8]. This finite-element model contains

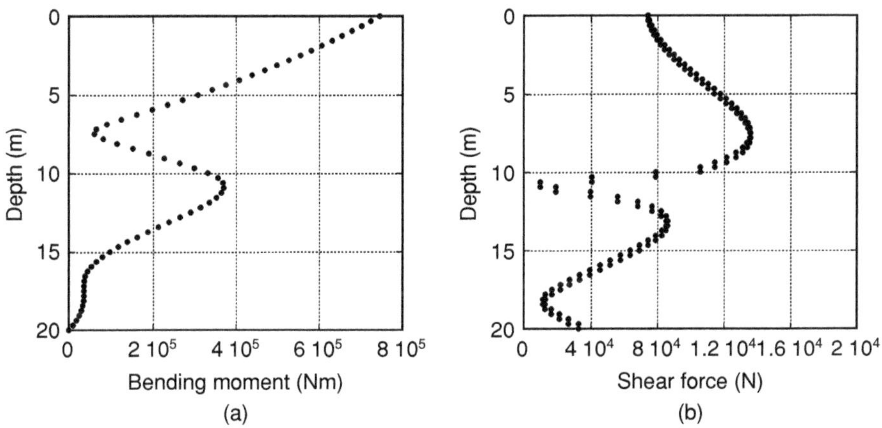

Figure 8.4 Peak bending moment by the single-input finite-element model

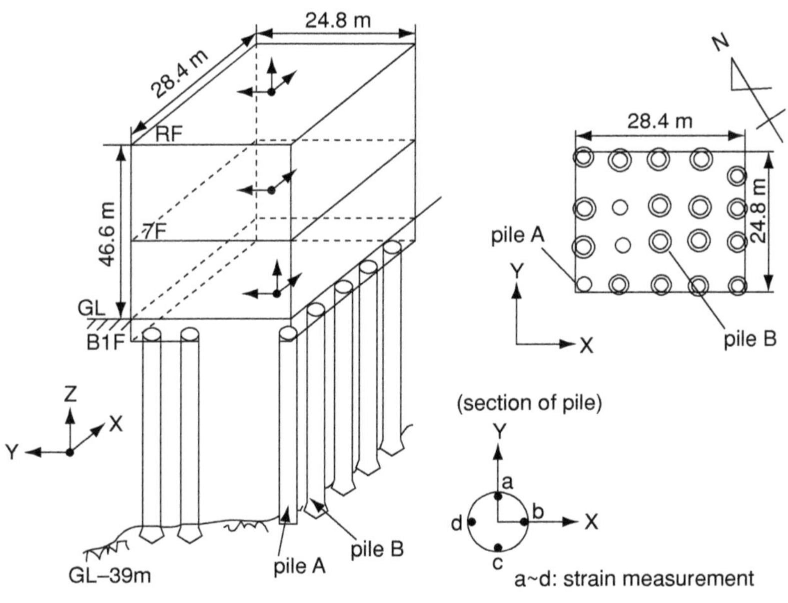

Figure 8.5 12-story steel building with 20 piles at Yokohama in Japan

the present Winkler soil element. It is noted that the shape function for the free-field ground and that for the piles are different, i.e. a linear function for the free-field ground and a cubic function for the piles. Figure 8.7(a) illustrates the profile of the shear wave velocity of the ground. Figure 8.7(b) shows the comparison of the peak pile bending strain computed by the analytical model containing the present Winkler soil element with that

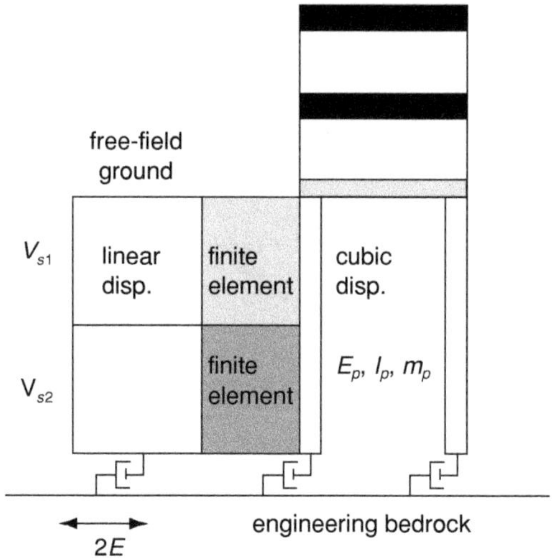

Figure 8.6 Finite-element model for the building–pile–soil system

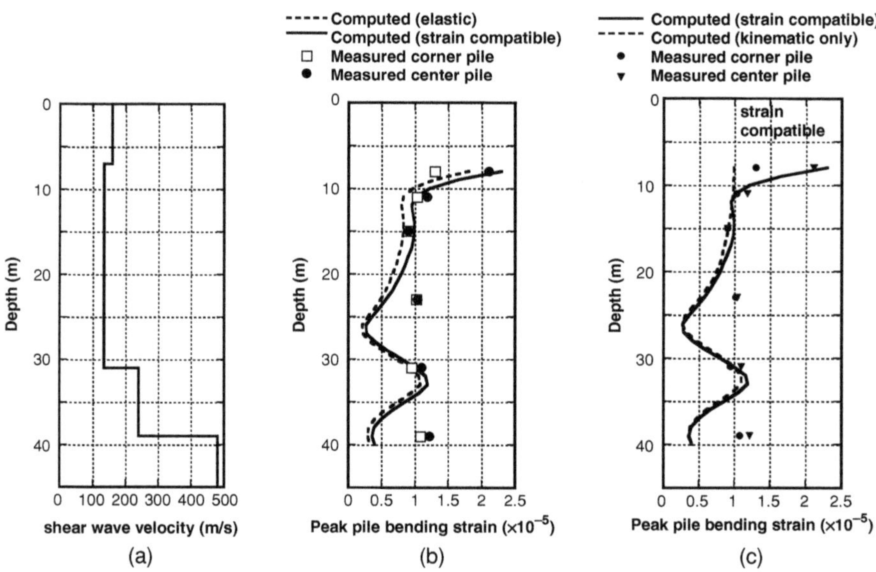

Figure 8.7 (*a*) Shear wave velocity profile of the ground, (*b*) comparison of the peak pile bending strain computed by the analytical model including the present Winkler-type soil element with that recorded during an earthquake in 1992, (*c*) effect of kinematic interaction on the peak pile bending strain

recorded during an earthquake in 1992. A reasonably good agreement can be observed near the pile head. This demonstrates the validity of the present continuum model including the Winkler soil element. The bending strain contains both the inertial effect and the kinematic effect. It has been confirmed from the analytical simulation that both the inertial effect and the kinematic effect are contained, almost in the same magnitude, in this case. This is shown in Figure 8.7(c). It may also be stated that the pile-group effect is, in the present case, rather small.

5 Influence coefficients for efficient analysis of pile-group effect

5.1 Inertial effect

For simple presentation of the essential feature of the procedure, the case is considered where the building is supported by four piles, i.e. pile 1, 2, 3 and 4. Following References 5–8 and 11, the stiffness and damping coefficients of the Winkler soil element in the surface soil layer are evaluated by

$$k_x = 1.2E_s \tag{10a}$$

$$c_x = 6a_0^{-0.25}\rho_s V_s d + 2\beta_s \frac{k_x}{\omega} \tag{10b}$$

In Equations (10a, b), E_s, $a_0 = \omega d/V_s$, d and β_s are the Young's modulus of the soil, non-dimensional frequency, pile diameter and hysteretic damping ratio of the soil respectively. The subscript number 1 for designating the surface soil layer is removed for simplicity of expression. The damping ratio of the Winkler soil element is assumed to be the combination of the hysteretic damping ratio and the radiation damping ratio, as shown in Figure 8.8.

For a simple explanation of the basis, the influence coefficient of pile 1 on pile 2 is explained here. The other cases can be treated almost in the same manner. A polar coordinate is employed here. The ratio a_{21} of the additional

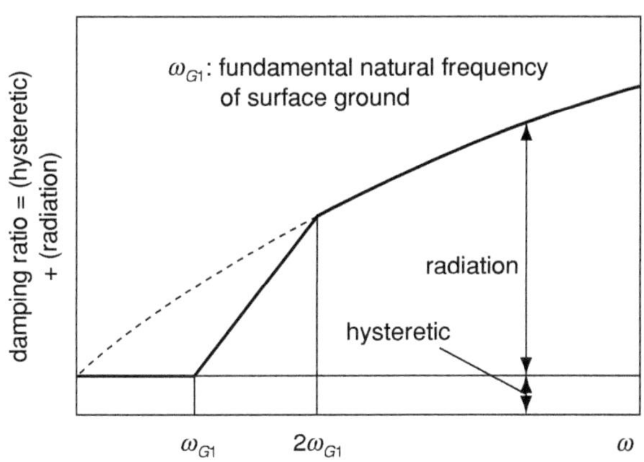

Figure 8.8 Damping ratio of the Winkler-type soil element: combination of hysteretic damping and radiation damping

pile-head horizontal displacement of pile 2 resulting from the inertial loading at the pile head of pile 1 to the pile-head horizontal displacement of pile 1 due to the inertial loading has been obtained in Reference 11.

$$\alpha_{21} = \frac{3}{4}\psi(r,\theta)\left(\frac{k_x + i\omega c_x}{k_x + i\omega c_x - m_p\omega^2}\right) \tag{11}$$

In Equation (11), $\psi(r, \theta)$ is the ratio of the ground-surface horizontal displacement of the point at (r, θ) to the pile-head horizontal displacement at the origin. This is described as

$$\psi(r,\theta) \approx \psi(r,0)\cos^2\theta + \psi(r,\frac{\pi}{2})\sin^2\theta \tag{12a}$$

$$\psi(r,0) = \frac{\sqrt{r_0}}{\sqrt{r}}\exp\left(\frac{-\beta_s\omega(r-r_0)}{V_{La}}\right)\exp\left(\frac{-i\omega(r-r_0)}{V_{La}}\right) \tag{12b}$$

$$\psi(r,\frac{\pi}{2}) = \frac{\sqrt{r_0}}{\sqrt{r}}\exp\left(\frac{-\beta_s\omega(r-r_0)}{V_s}\right)\exp\left(\frac{-i\omega(r-r_0)}{V_s}\right) \tag{12c}$$

V_{La}, r_0 indicates the so-called Lysmer's analogue's wave velocity and the radius of the pile respectively.

The validity and accuracy of the method in Reference 11 has been clarified by the present authors through comparison with the results of a rigorous model [12]. This will be shown later in Section 6.1.

5.2 Kinematic effect

The influence coefficient of pile 1 on pile 2 is considered and explained here as in Section 5.1, and other cases can be dealt with almost in the same way. The ratio $\bar{\alpha}_{21}$ of the additional pile-head horizontal displacement of pile 2 resulting from the kinematic loading for pile 1 to the pile-head horizontal displacement of pile 1 due to the kinematic loading has been explained in Reference 11.

$$\bar{\alpha}_{21} \approx \psi(r,\theta)(\Gamma - 1) \tag{13}$$

In Equation (13), the parameter Γ indicates the following quantity

$$\Gamma = \frac{k_x + i\omega c_x}{E_pI_p\delta^4 + k_x + i\omega c_x - m_p\omega^2} \tag{14a}$$

where

$$\delta = \omega / V_s^* \tag{14b}$$

and

$$V_s^* = V_s\sqrt{1 + 2i\beta_s} \tag{14c}$$

5.3 *Total response of interstory drift*

Consider the total response of the building–pile–soil system, as shown in Figure 8.1, under the dynamic forced displacement $u_g = U_g e^{i\omega t}$ at the bedrock surface. Assume that the building does not exist when considering the kinematic interaction. $\bar{U}_{11}$, $\bar{U}_{22}$, $\bar{U}_{33}$, $\bar{U}_{44}$ denote the horizontal pile-head displacements of piles 1, 2, 3 and 4 respectively, without interaction among the piles under the dynamic forced displacement $u_g = U_g e^{i\omega t}$ at the bedrock surface. In the case where the pile size is the same in all the piles, $\bar{U}_{11} = \bar{U}_{22} = \bar{U}_{33} = \bar{U}_{44}$ holds. Introduced here are the different influence coefficients for the inertial and kinematic loadings. In compliance with Section 5.1, let α_{ij} denote the ratio of the additional pile-head horizontal displacement of pile i resulting from the inertial loading at the pile head of pile j to the pile-head horizontal displacement of pile j without pile-group effects due to the inertial loading. On the other hand, following Section 5.2, let $\bar{\alpha}_{ij}$ denote the ratio of the additional pile-head horizontal displacement of pile i resulting from the kinematic loading for pile j to the pile-head horizontal displacement of pile j without pile-group effects due to the kinematic loading.

Four-pile models are considered. Using the superposition principle, the total pile-head horizontal displacement of a pile under the pile-group effect may be described as the sum of the following effects:

(1) pile-head displacement as a single pile under the dynamic forced displacement at the bedrock surface (without the building)
(2) additional pile-head displacement resulting from the other three piles under the dynamic forced displacement at the bedrock surface (without the building)
(3) pile-head displacement as a single pile due to the pile-head loading with the amplitudes P_1, P_2, P_3, P_4
(4) additional pile-head displacement resulting from the other three piles due to the pile-head loading with the amplitudes P_1, P_2, P_3, P_4

The total pile-head displacement of pile 1 may then be obtained as

$$U_1 = \bar{U}_{11} + \bar{\alpha}_{12}\bar{U}_{22} + \bar{\alpha}_{13}\bar{U}_{33} + \bar{\alpha}_{14}\bar{U}_{44} + \frac{P_1}{k_x^{(1)}} + \alpha_{12}\frac{P_2}{k_x^{(1)}} + \alpha_{13}\frac{P_3}{k_x^{(1)}} + \alpha_{14}\frac{P_4}{k_x^{(1)}} \tag{15}$$

Let $U^{(2\times2)}$ denote the horizontal displacement of the foundation mat. The displacement compatibility condition is then described by $U_1 = U^{(2\times2)}$ and is expressed by

$$U_1 = U^{(2\times2)} = \bar{U}_{11} + \bar{\alpha}_{12}\bar{U}_{22} + \bar{\alpha}_{13}\bar{U}_{33} + \bar{\alpha}_{14}\bar{U}_{44} + \frac{P_1}{k_x^{(1)}} + \alpha_{12}\frac{P_2}{k_x^{(1)}} + \alpha_{13}\frac{P_3}{k_x^{(1)}} + \alpha_{14}\frac{P_4}{k_x^{(1)}} \tag{16}$$

Division of both sides of Equation (16) by the common displacement $\bar{U}_{11} = \bar{U}_{22} = \bar{U}_{33} = \bar{U}_{44}$ may provide the following relation

$$\frac{U^{(2\times2)}}{\bar{U}_{11}} - \frac{P_1}{k_x^{(1)}\bar{U}_{11}} - \alpha_{12}\frac{P_2}{k_x^{(1)}\bar{U}_{22}} - \alpha_{13}\frac{P_3}{k_x^{(1)}\bar{U}_{33}} - \alpha_{14}\frac{P_4}{k_x^{(1)}\bar{U}_{44}} = 1 + \bar{\alpha}_{12} + \bar{\alpha}_{13} + \bar{\alpha}_{14} \tag{17}$$

In a similar manner, the compatibility conditions in piles 2, 3 and 4 may be expressed as

$$\frac{U^{(2\times2)}}{\bar{U}_{11}} - \alpha_{21}\frac{P_1}{k_x^{(1)}\bar{U}_{11}} - \frac{P_2}{k_x^{(1)}\bar{U}_{22}} - \alpha_{23}\frac{P_3}{k_x^{(1)}\bar{U}_{33}} - \alpha_{24}\frac{P_4}{k_x^{(1)}\bar{U}_{44}} = \bar{\alpha}_{21} + 1 + \bar{\alpha}_{23} + \bar{\alpha}_{24} \tag{18a}$$

$$\frac{U^{(2\times2)}}{\bar{U}_{11}} - \alpha_{31}\frac{P_1}{k_x^{(1)}\bar{U}_{11}} - \alpha_{32}\frac{P_2}{k_x^{(1)}\bar{U}_{22}} - \frac{P_3}{k_x^{(1)}\bar{U}_{33}} - \alpha_{34}\frac{P_4}{k_x^{(1)}\bar{U}_{44}} = \bar{\alpha}_{31} + \bar{\alpha}_{32} + 1 + \bar{\alpha}_{34} \tag{18b}$$

$$\frac{U^{(2\times2)}}{\bar{U}_{11}} - \alpha_{41}\frac{P_1}{k_x^{(1)}\bar{U}_{11}} - \alpha_{42}\frac{P_2}{k_x^{(1)}\bar{U}_{22}} - \alpha_{43}\frac{P_3}{k_x^{(1)}\bar{U}_{33}} - \frac{P_4}{k_x^{(1)}\bar{U}_{44}} = \bar{\alpha}_{41} + \bar{\alpha}_{42} + \bar{\alpha}_{43} + 1 \tag{18c}$$

Consider next the vibration in the building. Let u_{s1}, u_{s2}, u_{s3}, u_{s4}, u_{s5} denote the horizontal displacements of the floor masses in the five-story building. Because the excitation is harmonic, it may be possible to assume

$$u_{s1} = U_{s1}e^{i\omega t},\ u_{s2} = U_{s2}e^{i\omega t},\ u_{s3} = U_{s3}e^{i\omega t},\ u_{s4} = U_{s4}e^{i\omega t},\ u_{s5} = U_{s5}e^{i\omega t} \tag{19a–e}$$

For convenience of presentation, consider the vibration in the frequency domain. The equations of motion in the foundation mass and in the building may be derived as

$$\omega^2 m_0 U^{(2\times2)} + K_1(U_{s1} - U^{(2\times2)}) - P_1 - P_2 - P_3 - P_4 = 0 \tag{20a}$$

$$\omega^2 m_1 U_{s1} + K_2(U_{s2} - U_{s1}) - K_1(U_{s1} - U^{(2\times2)}) = 0 \tag{20b}$$

$$\omega^2 m_2 U_{s2} + K_3(U_{s3} - U_{s2}) - K_2(U_{s2} - U_{s1}) = 0 \tag{20c}$$

$$\vdots$$

$$\omega^2 m_5 U_{s5} - K_5(U_{s5} - U_{s4}) = 0 \tag{20d}$$

The frequency-domain expressions are derived in Equations (20a–d). Remember that $K_i = k_i + i\omega c_i$ is the dynamic stiffness. By dividing both sides of Equations (20a–d) by the common displacement $\bar{U}_{11} = \bar{U}_{22} = \bar{U}_{33} = \bar{U}_{44}$, the following expressions may be derived:

$$\frac{1}{k_x^{(1)}}(\omega^2 m_0 - K_1)\frac{U^{(2\times 2)}}{\bar{U}_{11}} - \frac{P_1}{k_x^{(1)}\bar{U}_{11}} - \frac{P_2}{k_x^{(1)}\bar{U}_{22}} - \frac{P_3}{k_x^{(1)}\bar{U}_{33}}$$

$$- \frac{P_4}{k_x^{(1)}\bar{U}_{44}} + \frac{1}{k_x^{(1)}}K_1\frac{U_{s1}}{\bar{U}_{11}} = 0$$

$$\frac{1}{k_x^{(1)}}K_1\frac{U^{(2\times 2)}}{\bar{U}_{11}} + \frac{1}{k_x^{(1)}}(\omega^2 m_1 - K_1 - K_2)\frac{U_{s1}}{\bar{U}_{11}} + \frac{1}{k_x^{(1)}}K_2\frac{U_{s2}}{\bar{U}_{11}} = 0$$

$$\frac{1}{k_x^{(1)}}K_2\frac{U_{s1}}{\bar{U}_{11}} + \frac{1}{k_x^{(1)}}(\omega^2 m_2 - K_2 - K_3)\frac{U_{s2}}{\bar{U}_{11}} + \frac{1}{k_x^{(1)}}K_3\frac{U_{s3}}{\bar{U}_{11}} = 0$$

$$\frac{1}{k_x^{(1)}}K_3\frac{U_{s2}}{\bar{U}_{11}} + \frac{1}{k_x^{(1)}}(\omega^2 m_3 - K_3 - K_4)\frac{U_{s3}}{\bar{U}_{11}} + \frac{1}{k_x^{(1)}}K_4\frac{U_{s4}}{\bar{U}_{11}} = 0$$

$$\frac{1}{k_x^{(1)}}K_4\frac{U_{s3}}{\bar{U}_{11}} + \frac{1}{k_x^{(1)}}(\omega^2 m_4 - K_4 - K_5)\frac{U_{s4}}{\bar{U}_{11}} + \frac{1}{k_x^{(1)}}K_5\frac{U_{s5}}{\bar{U}_{11}} = 0$$

$$\frac{1}{k_x^{(1)}}K_5\frac{U_{s4}}{\bar{U}_{11}} + \frac{1}{k_x^{(1)}}(\omega^2 m_5 - K_5)\frac{U_{s5}}{\bar{U}_{11}} = 0 \tag{21a–f}$$

It may be understood that four compatibility conditions Equations (17)–(18a–c) and six equilibrium equations Equations (21a–f) provide a set of ten simultaneous linear equations. This set can be used to derive the transfer functions of floor displacements, foundation displacement and pile-head shear forces. The expressions are shown in Appendix 1.

It may be convenient to introduce the displacement amplitude $2E_2$ at the outcropping bedrock surface. By using the one-dimensional wave propagation theory of Reference 13 (see Appendix 2), the pile-head displacement $\bar{U}_{11}$ of pile 1 without the pile-group effect due to the kinematic loading may be described as

$$\bar{U}_{11} = \Gamma U_{ff}(0) = \Gamma(2E_1) = \Gamma D(2E_2) \tag{22}$$

In Equation (22), D denotes the following quantity:

$$D = \frac{2}{(1+\alpha_{i1})e^{i\omega(L_1/V^*_{s1})} + (1-\alpha_{i1})e^{i\omega(-L_1/V^*_{s1})}} \tag{23}$$

The quantity α_{i1} in the denominator of Equation (23) indicates the complex impedance ratio $\alpha_{i1} = \rho_1 V^*_{s1}/\rho_2 V^*_{s2}$. Substitution of Equation (22) into Equation (A1) in Appendix 1 leads to

$$\mathbf{X}\{U^{(2\times2)}/(2E_2) \quad \ldots \quad U_{s5}/(2E_2)\}^T = \Gamma D\mathbf{Z} \tag{24}$$

Once $\{U^{(2\times2)}/(2E_2) \ldots U_{s5}/(2E_2)\}^T$ are obtained in Equation (24), the transfer functions of the interstory drifts may be expressed as

$$\Delta_1 = \frac{U_{s1}}{2E_2} - \frac{U^{(2\times2)}}{2E_2} \tag{25}$$

$$\Delta_i = \frac{U_{si}}{2E_2} - \frac{U_{si-1}}{2E_2} \quad (i = 2,\ldots,5) \tag{26}$$

5.4 Total response of pile-head bending moment

Consider the total response of the pile-head bending moment. The pile-head bending moment in pile 1 including both the kinematic and inertial effects may be obtained from the total pile-head curvature $U_1''(0)$ as

$$M_1 = E_p I_p U_1''(0) \tag{27}$$

In accordance with Reference 11, it is assumed here that the horizontal pile displacement in pile 1 as a single pile due to the inertial loading may be described by the displacement U_0 at the ground surface

$$U_{11}(z) = U_0 f(z) \tag{28}$$

In Equation (28), the coordinate z is used in place of z_1 (the coordinate in the first soil layer) and $f(z)$ is given by

$$f(z) = e^{-\lambda z}(\cos\lambda z + \sin\lambda z) \tag{29}$$

The quantity λ has been defined in Equation (4) (denoted by λ_1 there). The pile-head curvature due to the inertial loading can then be expressed by

$$U_{11}''(z) = U_0 f''(z) \tag{30}$$

Consider the ratio of the curvature at the pile head as a single pile to the pile-head displacement. Then

$$\beta_{11} = \frac{U''_{11}(0)}{U_{11}(0)} = \frac{U_0 f''(0)}{U_0} = f''(0) \tag{31}$$

After some manipulation, the relation $f''(0) = \lambda^2 - \lambda^2 - \lambda^2 - \lambda^2 = -2\lambda^2$ is derived. This result leads to

$$\beta_{11} = -2\lambda^2 \tag{32}$$

In accordance with Reference 11, it is also assumed that the additional horizontal pile displacement of pile 2 due to the inertial loading at pile 1 may be expressed by

$$U_{21}(z) = \frac{3}{4}\psi(r,\theta)\left(\frac{k_x + i\omega c_x}{k_x + i\omega c_x - m\omega^2}\right)U_0 g(z) \tag{33}$$

In Equation (33), the function $g(z)$ is given by

$$g(z) = e^{-\lambda z}\left(\cos\lambda z + \sin\lambda z + \frac{2}{3}\lambda z\sin\lambda z\right) \tag{34}$$

By differentiating Equation (33) twice with respect to z, the pile-head curvature due to the additional pile displacement of pile 2 may be described as

$$U''_{21}(z) = \frac{3}{4}\psi(r,\theta)\left(\frac{k_x + i\omega c_x}{k_x + i\omega c_x - m\omega^2}\right)U_0 g''(z) \tag{35}$$

This leads to the influence coefficient for curvature

$$\beta_{21} = \frac{U''_{21}(0)}{U_{11}(0)} = \frac{3}{4}\psi(r,\theta)\left(\frac{k_x + i\omega c_x}{k_x + i\omega c_x - m\omega^2}\right)g''(0) \tag{36}$$

After some manipulation, the following relation may be derived:

$$g''(0) = -\frac{2}{3}\lambda^2 \tag{37}$$

Equation (37) gives actually

$$\beta_{21} = -\frac{1}{2}\psi(r,\theta)\left(\frac{k_x + i\omega c_x}{k_x + i\omega c_x - m\omega^2}\right)\lambda^2 \tag{38}$$

The influence coefficient under the kinematic loading will be considered next. The horizontal displacement amplitude of a point in the free-field ground may be described by

$$U_{ff}(z) = \hat{E}_1(z) = 2E_1 \cos(\omega z / V^*_{s1}) \tag{39}$$

Then the following relation is derived:

$$\bar{U}_{11}(z) = \Gamma U_{ff}(z) = \Gamma \hat{E}_1(z) \tag{40}$$

By differentiating Equation (40) twice with respect to z, the curvature distribution in pile 1 as a single pile under the kinematic loading may be evaluated by

$$\bar{U}''_{11}(z) = \Gamma \hat{E}''_1(z) \tag{41}$$

The influence coefficient with respect to the bending moment due to the kinematic loading is then expressed as

$$\bar{\beta}_{11} = \frac{\bar{U}''_{11}(0)}{\bar{U}_{11}(0)} = \frac{\Gamma \hat{E}''_1(0)}{\Gamma \hat{E}_1(0)} = \frac{\hat{E}''_1(0)}{\hat{E}_1(0)} \tag{42}$$

After some manipulation, $\hat{E}_1(0)$ and $\hat{E}''_1(0)$ in Equation (42) may be expressed as

$$\hat{E}_1(0) = 2E_1 \tag{43a}$$

$$\hat{E}''_1(0) = -2(\omega / V^*_{s1})^2 E_1 \tag{43b}$$

Substitution of Equations (43a, b) into Equation (42) provides

$$\bar{\beta}_{11} = -(\omega / V^*_{s1})^2 \tag{44}$$

The free-field ground displacement amplitude may also be described in terms of the bedrock displacement U_g as

$$U_{ff}(z) = U_g \frac{\cos \delta z}{\cos \delta L} \tag{45}$$

The parameter L is equal to L_1 used earlier. Then

$$\hat{E}_1(z) = U_g \frac{\cos \delta z}{\cos \delta L} \tag{46}$$

In accordance with Reference 11, the additional displacement is described as

$$\bar{U}_{21}(z) \approx \psi(r, \theta)\Gamma(\Gamma - 1)U_g \frac{\cos \delta z}{\cos \delta L} = \psi(r, \theta)\Gamma(\Gamma - 1)\hat{E}_1(z) \tag{47}$$

By differentiating Equation (47) twice with respect to z, the following relation may be derived:

$$\bar{U}_{21}''(z) \approx \psi(r, \theta)\Gamma(\Gamma - 1)\hat{E}_1''(z) \tag{48}$$

The influence coefficient with respect to the bending moment due to the kinematic loading may then be described as

$$\begin{aligned}\bar{\beta}_{21} &= \frac{\bar{U}_{21}''(0)}{\bar{U}_{11}(0)} \approx \frac{\psi(r, \theta)\Gamma(\Gamma - 1)\hat{E}_1''(0)}{\Gamma \hat{E}_1(0)} \\ &= \psi(r, \theta)(\Gamma - 1)\frac{\hat{E}_1''(0)}{\hat{E}_1(0)} = -\psi(r, \theta)(\Gamma - 1)\left(\frac{\omega}{V_{s1}^*}\right)^2\end{aligned} \tag{49}$$

As a consequence, the total curvature can be evaluated as

$$\begin{aligned}U_1''(0) &= \bar{\beta}_{11}\bar{U}_{11} + \bar{\beta}_{12}\bar{U}_{22} + \bar{\beta}_{13}\bar{U}_{33} + \bar{\beta}_{14}\bar{U}_{44} \\ &\quad + \beta_{11}\frac{P_1}{k_x^{(1)}} + \beta_{12}\frac{P_2}{k_x^{(1)}} + \beta_{13}\frac{P_3}{k_x^{(1)}} + \beta_{14}\frac{P_4}{k_x^{(1)}}\end{aligned} \tag{50}$$

For convenience of expression, the following form is introduced:

$$\begin{aligned}\frac{U_1''(0)}{\bar{U}_{11}} &= \bar{\beta}_{11} + \bar{\beta}_{12} + \bar{\beta}_{13} + \bar{\beta}_{14} + \beta_{11}\frac{P_1}{k_x^{(1)}\bar{U}_{11}} \\ &\quad + \beta_{12}\frac{P_2}{k_x^{(1)}\bar{U}_{22}} + \beta_{13}\frac{P_3}{k_x^{(1)}\bar{U}_{33}} + \beta_{14}\frac{P_4}{k_x^{(1)}\bar{U}_{44}}\end{aligned} \tag{51}$$

The overall pile-head bending moment in pile 1 may then be described in terms of the influence coefficients as

$$\begin{aligned}\frac{M_1}{\rho_p d^4\omega^2(2E_2)} &= \frac{E_p I_p U_1''(0)}{\rho_p d^4\omega^2(2E_2)} = \Gamma D\left\{(\bar{\beta}_{11} + \bar{\beta}_{12} + \bar{\beta}_{13} + \bar{\beta}_{14} + \beta_{11}\frac{P_1}{k_x^{(1)}\bar{U}_{11}}\right. \\ &\quad \left. + \beta_{12}\frac{P_2}{k_x^{(1)}\bar{U}_{22}} + \beta_{13}\frac{P_3}{k_x^{(1)}\bar{U}_{33}} + \beta_{14}\frac{P_4}{k_x^{(1)}\bar{U}_{44}}\right\}\frac{E_p I_p}{\rho_p d^4\omega^2}\end{aligned} \tag{52}$$

It is noted that the quantities $P_1/(k_x^{(1)}\bar{U}_{11}), \ldots, P_4/(k_x^{(1)}\bar{U}_{44})$ have been obtained.

In order to investigate the difference in pile-head moment in the corner pile and the center pile, a five-story shear building model with 3 × 3, 5 × 5 and 9 × 9 piles is treated. The model parameters used are the same

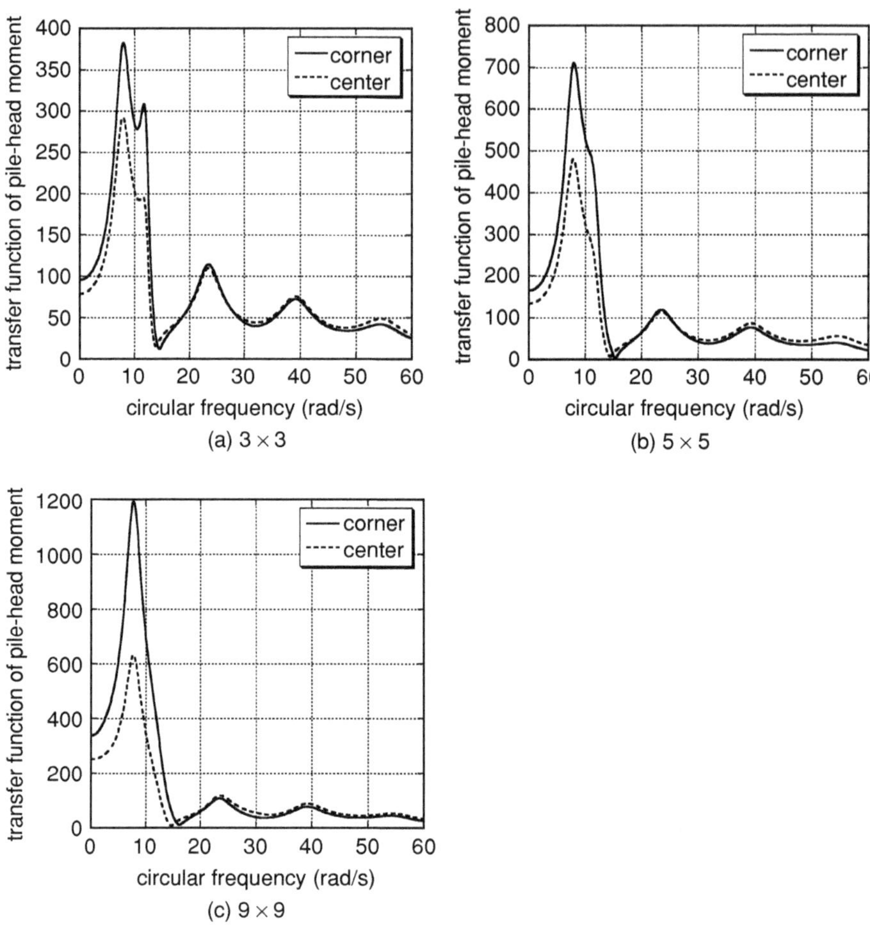

Figure 8.9 Transfer functions of pile-head moment in the corner pile and the center pile (3 × 3 piles, 5 × 5 piles, 9 × 9 piles)

as those in Section 6.2. Figure 8.9 shows the transfer functions of the pile-head moment in the corner pile and the center pile. It can be observed that the pile-head moment in the corner pile is generally larger than that in the center pile. Therefore the pile-head moment in the corner pile will be treated in the following section.

6 Numerical examples

6.1 Accuracy check

For the accuracy check, pile foundations in a uniform semi-infinite ground have been considered. The ratio S/d denotes the ratio of the pile spacing

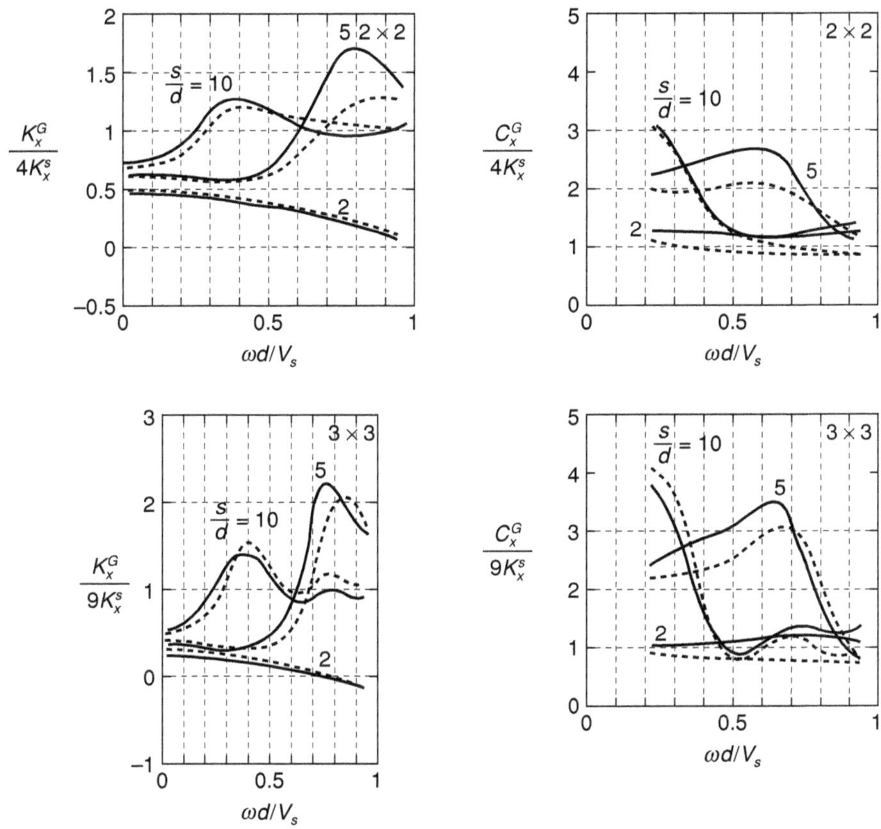

Figure 8.10 Comparison of stiffness and damping coefficients of grouped piles normalized by the sum of the stiffnesses of single piles; (solid line) three-dimensional continuum approach of Kaynia and Kausel [12] shown in Ref. 6, (dotted line) the present method. (2 × 2 piles and 3 × 3 piles)

to the pile diameter. The parameters used here are shown in Reference 6. Figure 8.10 shows the comparison of stiffness and damping coefficients of grouped piles normalized by the sum of the stiffnesses of single piles. The solid line indicates the result using the three-dimensional continuum approach of Kaynia and Kausel [12] shown in Reference 6. On the other hand, the dotted line is the result of the present method. It can be seen that the present approach has an acceptable accuracy.

6.2 *Stiffness and strength design under pile-group effect*

Three models with different pile foundations of 2 × 2 piles, 5 × 5 piles and 10 × 10 piles are considered. The piles are placed along the two sets of straight lines in the square foundation mat. The ratio of the pile spacing to the pile diameter of 1 m is specified as 5. For 2 × 2 piles, the floor masses are specified as 17.5×10^3 kg and the foundation floor mass specified as 52.5×10^3 kg. In the cases of 5 × 5 piles and 10 × 10 piles, the floor and foundation masses are modified in proportion to their areas. The depth of the surface soil is set to 20 m. The shear wave velocity (reflecting the strain-dependent property) and the mass density of the surface soil are assumed to be 100 m/s and 1.8×10^3 kg/m^3 respectively. The fundamental natural period of the surface soil is 0.8*s*. It is assumed that the hysteretic damping ratio of the surface soil is 0.05 and the Poisson's ratio is 0.45. The shear wave velocity and mass density of the bedrock are assumed to be 400 m/s and 2.0×10^3 kg/m^3. Furthermore, the hysteretic damping ratio of the bedrock is assumed to be zero. Young's modulus of the pile is set to 2.0×10^{10} N/m^2 and the mass density of the pile is assumed to be 2.4×10^3 kg/m^3. The undamped fundamental natural period of the building with a fixed base is 0.5 s, and the lowest eigenmode of the building with a fixed base is in a straight line.

Figure 8.11 demonstrates clearly the pile-group effect on the response of the building-pile system in the case of 2 × 2 piles. The solid lines in Figure 8.11(a)–(e) show the transfer functions for the 2 × 2 piles of the first-to-fifth-story interstory drifts to the bedrock surface displacement as an outcropping motion. In addition, the solid lines in Figure 8.11(f), (g) illustrate the transfer functions for the 2 × 2 piles of the foundation displacement and the pile-head bending moment to the bedrock surface displacement as an outcropping motion. On the other hand, the dotted lines in Figure 8.11(a)–(g) indicate the corresponding transfer functions without pile-group effect. These figures, without pile-group effects, have been drawn by substituting zero into all the influence coefficients in the inertial and kinematic

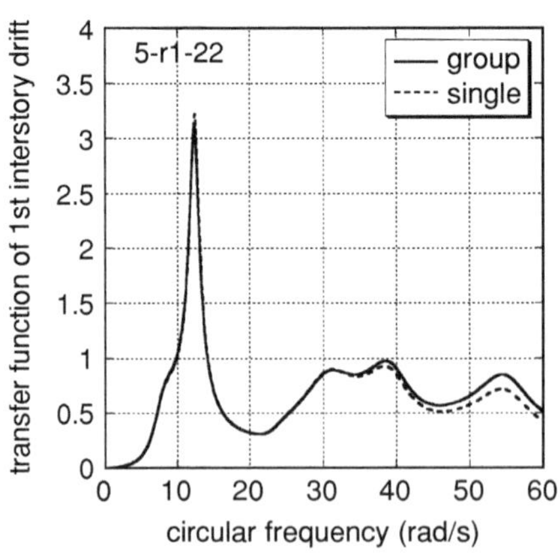

(a)

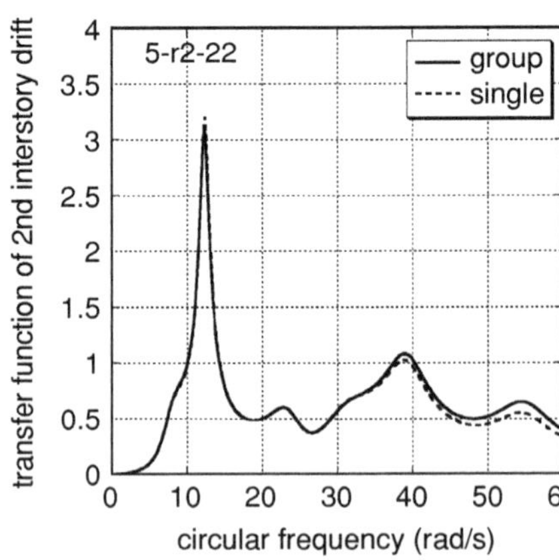

(b)

Figure 8.11 Transfer functions of the 1st, 2nd, 3rd, 4th and 5th interstory drifts, foundation displacement and pile-head moment

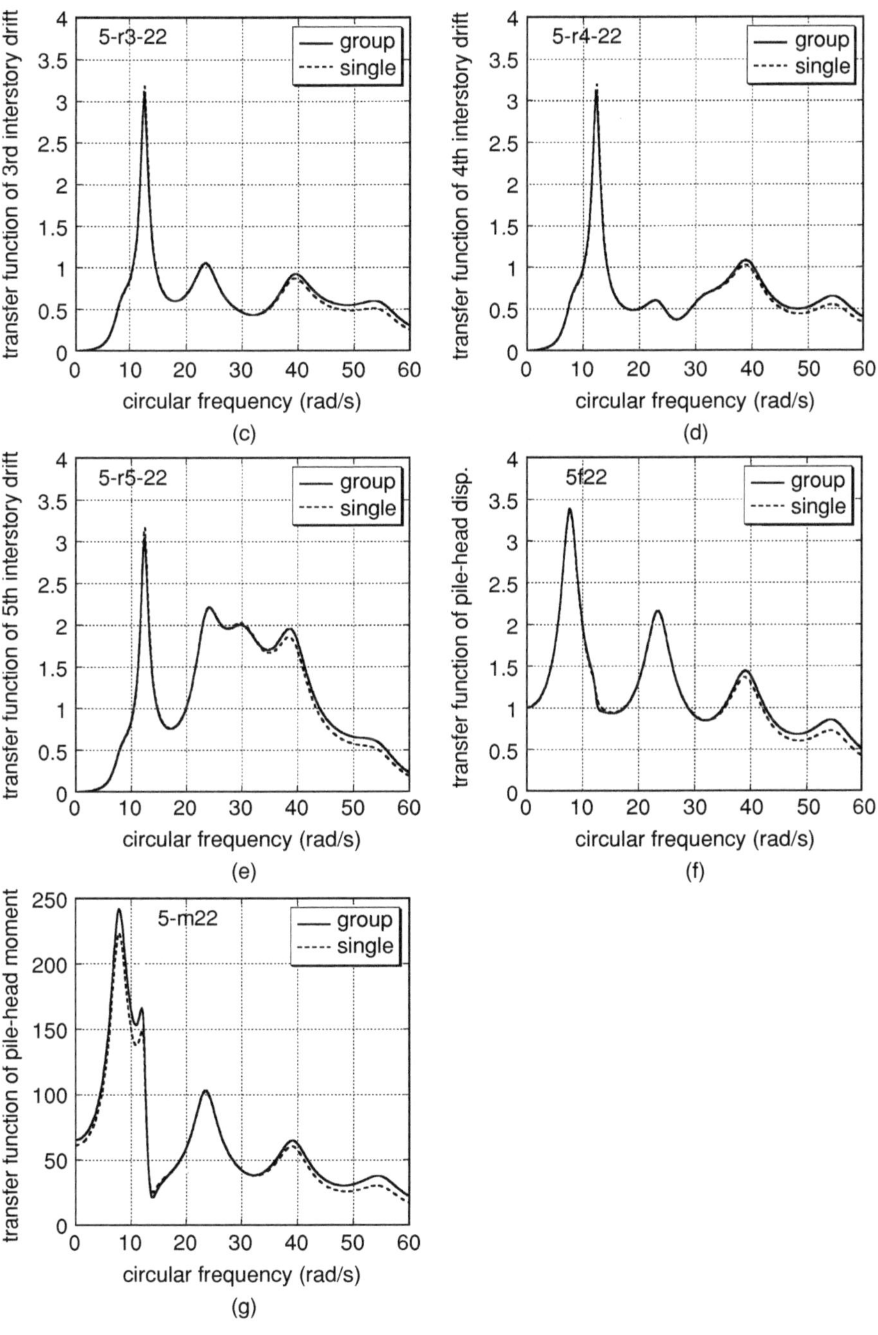

Figure 8.11 (cont'd)

interactions. It can be observed that the pile-group effect is negligible in the case of the 2 × 2 piles with the present pile spacing–pile diameter ratio. It can also be understood that the transfer functions of the interstory drifts are almost uniform at the fundamental natural circular frequency, 12.6 rad/s, of the building. This is because the lowest eigenmode of the building with a fixed base is constrained to the straight line, i.e. uniform interstory drifts in the lowest eigenmode. However, the transfer functions of the interstory drifts exhibit amplified characteristics at higher natural frequencies, and this is remarkable in the fifth story. This may result from the higher-mode effects.

Figure 8.12 clarifies the pile-group effect on the response of the building–pile system in the case of the 5 × 5 piles. The solid lines in Figure 8.12(a)–(e) show the transfer functions for the 5 × 5 piles of the first-, second-, third-,

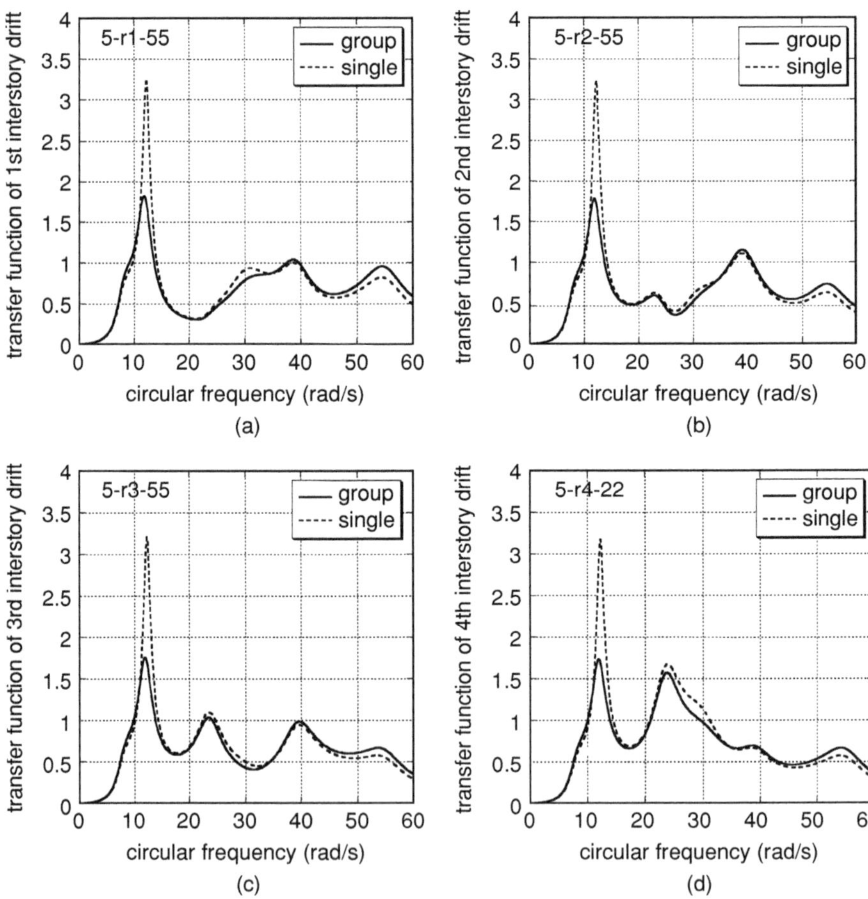

Figure 8.12 Transfer functions of the 1st, 2nd, 3rd, 4th and 5th interstory drifts, foundation displacement and pile-head moment (5 × 5 piles)

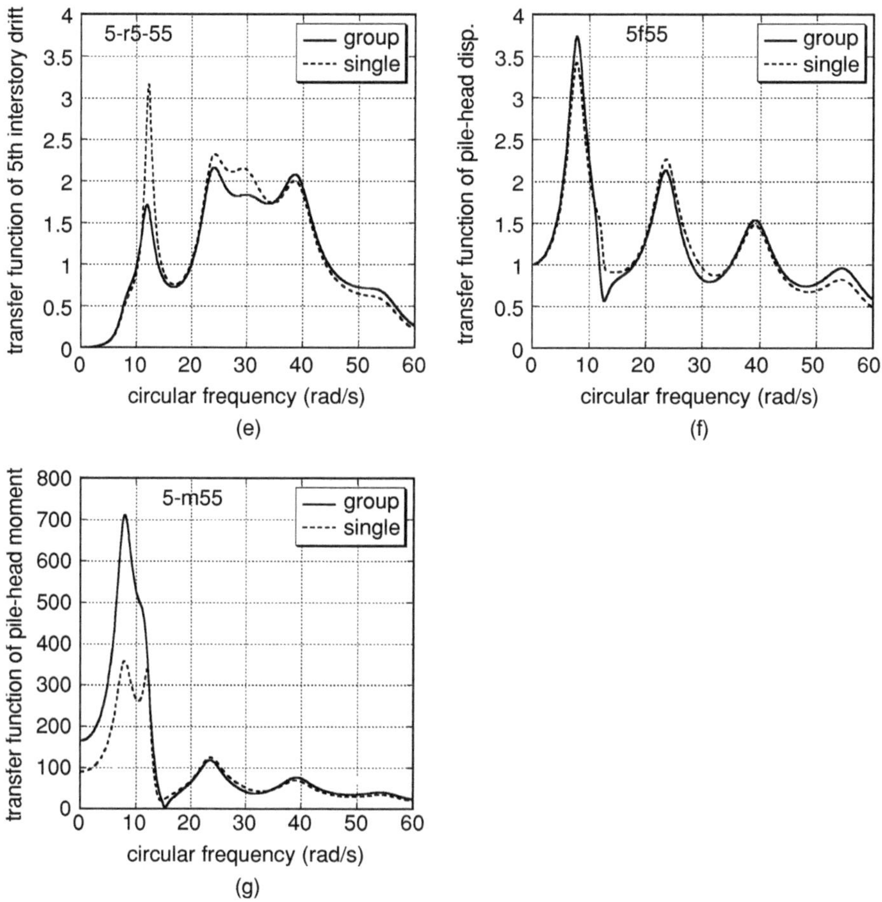

Figure 8.12 (cont'd)

fourth- and fifth-story interstory drift to the bedrock surface displacement as an outcropping motion. Furthermore the solid lines in Figure 8.12(f), (g) illustrate the transfer functions for the 5 × 5 piles of the foundation displacement and the pile-head bending moment to the bedrock surface displacement as an outcropping motion. On the other hand, the dotted lines in Figure 8.12(a)–(g) indicate the corresponding transfer functions without pile-group effect. As before, these figures have been drawn by substituting zero into all the influence coefficients in the inertial and kinematic interactions.

Figure 8.13 demonstrates the pile-group effect on the response of the building–pile system in the case of the 10 × 10 piles. Figure 8.13(a)–(e) show the transfer functions for the 10 × 10 piles of the first-, second-, third-, fourth- and fifth-story interstory drift to the bedrock surface displacement

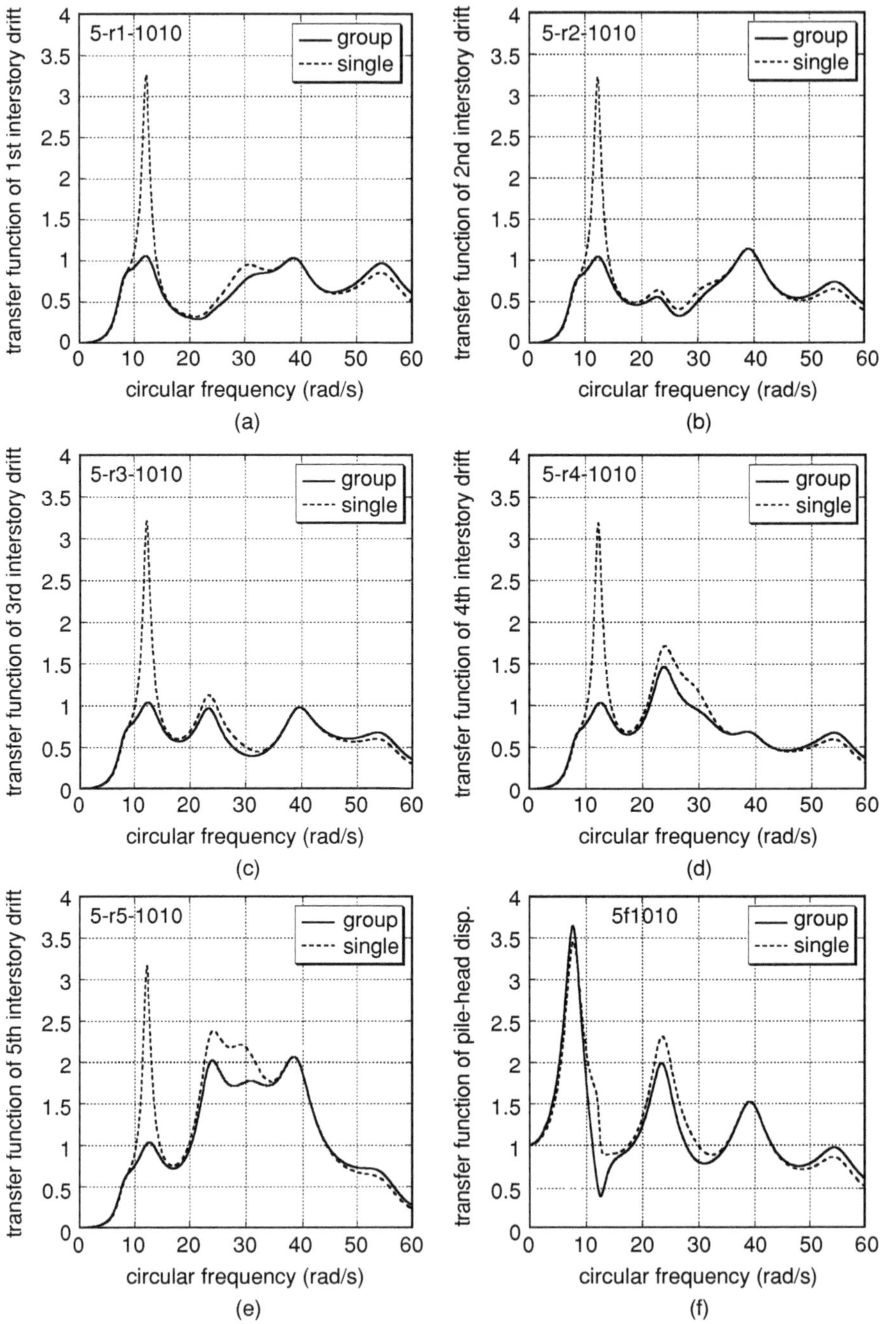

Figure 8.13 Transfer functions of the 1st, 2nd, 3rd, 4th and 5th interstory drifts, foundation displacement and pile-head moment (10 × 10 piles)

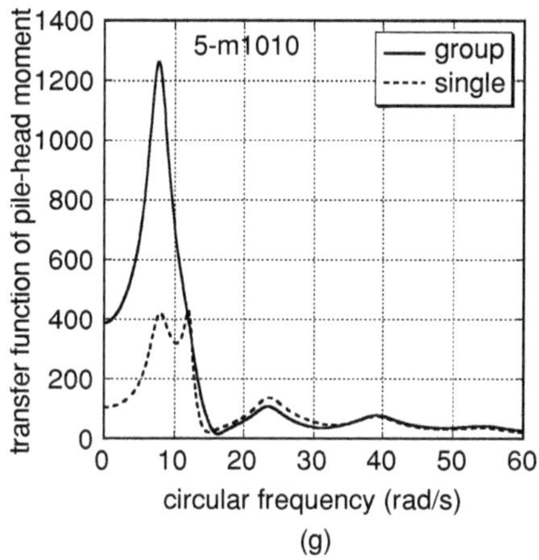

Figure 8.13 (*cont'd*)

as an outcropping motion. In addition, the solid lines in Figure 8.13(f), (g) illustrate the transfer functions for the 10 × 10 piles of the foundation displacement and the pile-head bending moment to the bedrock surface displacement as an outcropping motion. On the other hand, the dotted lines in Figure 8.13(a)–(g) indicate the corresponding transfer functions without the pile-group effect which were drawn by substituting zero into all the influence coefficients in the inertial and kinematic interactions.

It can be understood that the pile-group effect is significant in the case of the 5 × 5 piles and the 10 × 10 piles. This pile-group effect can be seen clearly in the frequency range of the building fundamental natural circular frequency, 12.6 rad/s, and of the surface-ground fundamental natural circular frequency, 7.85 rad/s.

7 Summaries

The new facts derived may be summarized as follows:

(1) An efficient method has been developed for the seismic analysis and design of buildings with closely-spaced piles. An efficient continuum model composed of a dynamic Winkler soil element and a pile has been used to describe the dynamic behavior of the pile–soil system with a single pile or a set of grouped piles.
(2) The pile-group effect has been taken into account through the influence coefficients among piles. While the influence coefficients for displacements are well known, those for curvatures or pile-head bending moments have been proposed here. This has enabled the pile-group effect on the pile-head bending moments to be evaluated efficiently.

(3) It has been shown that, while the pile-group effect decreases the inter-story drift of buildings in general, it may increase the pile-head bending moment in some cases. This means that the procedure without the pile-group effect leads to the conservative design for superstructures and requires a revised member design for piles.
(4) The pile-group effect may be significant around the frequency range of the building's fundamental natural frequency and of the surface ground's fundamental natural frequency.

While the strain dependency of soil stiffness and damping was not taken into account for the presentation of the essence of the theory in this chapter, an efficient treatment of the equivalent linear model together with the response spectrum method [14] enables this property to be included. Additional extensive computation will be necessary for further clarification of the pile-group effects on the stiffness and strength design of buildings under various design conditions.

Acknowledgment

Figures 1, 5–9, 11–13 and Equations (1)–(52), (A1)–(A5) are from *Soil Dynamics and Earthquake Engineering*, Vol. 25(5), I. Takewaki and A. Kishida, Efficient analysis of pile-group effect on seismic stiffness and strength design of buildings, pp. 355–67, 2005, with permission from Elsevier.

References

Dobry, R. and Gazetas, G., Simple method for dynamic stiffness and damping of floating pile groups, *Géotechnique*, 1988, 38 (4): 557–74.

Gazetas, G. and Dobry, R., Horizontal response of piles in layered soils, *J. Geotech. Eng.*, ASCE, 1984, 110 (1): 20–40.

Gazetas, G., Fan, K., Tazoh, T., Shimizu, K., Kavvadas, M. and Makris, N., Seismic pile-group–structure interaction, *Special Pub. of ASCE*, 'Piles under dynamic loads', ed. S. Prakash, 1992, 56–93.

Kavvadas, M. and Gazetas, G., Kinematic seismic response and bending of free-head piles in layered soil, *Géotechnique*, 1993, 43 (2): 207–22.

Kaynia, A. M. and Kausel, E., Dynamic behaviour of pile groups, *Proc. 2nd Int. Conf. Numer. Meth. Offshore Piling, Austin, Texas*, 1982, pp. 509–532.

Makris, N. and Gazetas, G., Dynamic pile–soil–pile interaction, Part II: Lateral and seismic response, *Earthquake Eng. and Struct. Dyn.*, 1992, 21: 145–62.

Nikolaou, S., Mylonakis, G., Gazetas, G. and Tazoh, T., Kinematic pile bending during earthquakes: analysis and field measurements, *Géotechnique*, 2001, 51 (5): 425–40.

Penzien, J., Scheffey, C. F. and Parmelee, R. A., Seismic analysis of bridges on long piles, *J. Eng. Mech.*, ASCE, 1964, 90 (3): 223–54.

Schnabel, P. B., Lysmer, J. and Seed, H. B., SHAKE: A computer program for earthquake response analysis of horizontally layered sites, A computer program distributed by NISEE/Computer Applications, Berkeley, Calif., 1972.

Tajimi, H. and Shimomura, Y., Dynamic analysis of soil-structure interaction by the thin layered element method, *Journal of Structural and Construction Eng.*, Archi. Inst. of Japan, 1976, 243: 41–51. In Japanese.
Takewaki, I., Doi, A., Tsuji, M. and Uetani, K., Seismic stiffness design of pile-supported building structures using dynamic Winkler-type spring models, *Journal of Structural and Construction Eng.*, Archi. Inst. of Japan, 2003, 571: 45–52. In Japanese.
Takewaki, I., Inverse stiffness design of shear-flexural building models including soil–structure interaction, *Engineering Structures*, 1999, 21 (12): 1045–54.
Takewaki, I., Response spectrum method for nonlinear surface ground analysis, *An International Journal of Advances in Structural Engineering*, 2004, 7 (6): 503–14.
Tassoulas, J. L. and Kausel, E., Elements for the numerical analysis of wave motion in layered strata, *Int. J. for Numerical Methods in Eng.*, 1983, 19 (7): 1005–32.

Appendix 1. Simultaneous linear equations for transfer functions in the five-story building model on ground with two soil layers

The transfer functions of floor displacements, foundation displacement and pile-head shear forces to the horizontal pile-head displacement are derived from the boundary conditions, continuity conditions and equilibrium equations. In view of Equations (17), (18) and (21), those equations may be expressed compactly as

$$XY = Z \tag{A1}$$

In Equation (A1)

$$X = [X_1 \cdots X_{10}]$$

$$Y = \left\{ \frac{U^{(2\times 2)}}{\bar{U}_{11}} \quad \frac{P_1}{k_x^{(1)}\bar{U}_{11}} \quad \frac{P_2}{k_x^{(1)}\bar{U}_{22}} \quad \frac{P_3}{k_x^{(1)}\bar{U}_{33}} \quad \frac{P_4}{k_x^{(1)}\bar{U}_{44}} \quad \frac{U_{s1}}{\bar{U}_{11}} \quad \frac{U_{s2}}{\bar{U}_{11}} \quad \frac{U_{s3}}{\bar{U}_{11}} \quad \frac{U_{s4}}{\bar{U}_{11}} \quad \frac{U_{s5}}{\bar{U}_{11}} \right\}^T$$

$$Z = \{1 + \bar{\alpha}_{12} + \bar{\alpha}_{13} + \bar{\alpha}_{14} \quad \bar{\alpha}_{21} + 1 + \bar{\alpha}_{23} + \bar{\alpha}_{24} \quad \bar{\alpha}_{31} + \bar{\alpha}_{32} + 1 + \bar{\alpha}_{34}$$
$$\bar{\alpha}_{41} + \bar{\alpha}_{42} + \bar{\alpha}_{43} + 1 \quad 0 \quad 0 \quad 0 \quad 0 \quad 0 \quad 0\}^T \tag{A2a–c}$$

$$X_1 = \{1 \quad 1 \quad 1 \quad 1 \quad (\omega^2 m_0 - K_1)/k_x^{(1)} \quad K_1/k_x^{(1)} \quad 0 \quad 0 \quad 0 \quad 0\}^T$$

$$X_2 = \{-1 \quad -\alpha_{21} \quad -\alpha_{31} \quad -\alpha_{41} \quad -1 \quad 0 \quad 0 \quad 0 \quad 0 \quad 0\}^T$$

$$X_3 = \{-\alpha_{12} \quad -1 \quad -\alpha_{32} \quad -\alpha_{42} \quad -1 \quad 0 \quad 0 \quad 0 \quad 0 \quad 0\}^T$$

$$X_4 = \{-\alpha_{13} \quad -\alpha_{23} \quad -1 \quad -\alpha_{43} \quad -1 \quad 0 \quad 0 \quad 0 \quad 0 \quad 0\}^T$$

$$X_5 = \{-\alpha_{14} \quad -\alpha_{24} \quad -\alpha_{34} \quad -1 \quad -1 \quad 0 \quad 0 \quad 0 \quad 0 \quad 0\}^T$$

$$X_6 = \{0 \quad 0 \quad 0 \quad 0 \quad K_1/k_x^{(1)} \quad (\omega^2 m_1 - K_1 - K_2)/k_x^{(1)} \quad K_2/k_x^{(1)} \quad 0 \quad 0 \quad 0\}^T$$

$$X_7 = \{0 \quad 0 \quad 0 \quad 0 \quad 0 \quad K_2/k_x^{(1)} \quad (\omega^2 m_1 - K_2 - K_3)/k_x^{(1)} \quad K_3/k_x^{(1)} \quad 0 \quad 0\}^T$$

$$\mathbf{X}_8 = \{0\quad 0\quad 0\quad 0\quad 0\quad 0\quad K_3/k_x^{(1)}\quad (\omega^2 m_1 - K_3 - K_4)/k_x^{(1)}\quad K_4/k_x^{(1)}\quad 0\}^T$$

$$\mathbf{X}_9 = \{0\quad 0\quad 0\quad 0\quad 0\quad 0\quad 0\quad K_4/k_x^{(1)}\quad (\omega^2 m_1 - K_4 - K_5)/k_x^{(1)}\quad K_5/k_x^{(1)}\}^T$$

$$\mathbf{X}_{10} = \{0\quad 0\quad 0\quad 0\quad 0\quad 0\quad 0\quad 0\quad K_5/k_x^{(1)}\quad (\omega^2 m_1 - K_5)/k_x^{(1)}\}^T \tag{A3a–j}$$

Appendix 2. One-dimensional wave propagation theory

Consider the one-dimensional wave propagation theory. Remember that α_{i1} is the complex impedance ratio $\rho_1 V_{s1}^* / \rho_2 V_{s2}^*$ as introduced in Section 5.3. The amplitudes E_2, F_2 of the upward and downward propagating waves respectively at the bedrock surface may be related to the amplitude E_1 of the upward propagating wave at the ground surface.

$$\begin{Bmatrix} E_2 \\ F_2 \end{Bmatrix} = [A_1] \begin{Bmatrix} 1 \\ 1 \end{Bmatrix} E_1 \tag{A4}$$

where

$$[A_1] = \begin{bmatrix} \frac{1}{2}(1 + \alpha_{i1})e^{i\omega(L_1/V_{s1}^*)} & \frac{1}{2}(1 - \alpha_{i1})e^{i\omega(-L_1/V_{s1}^*)} \\ \frac{1}{2}(1 - \alpha_{i1})e^{i\omega(L_1/V_{s1}^*)} & \frac{1}{2}(1 + \alpha_{i1})e^{i\omega(-L_1/V_{s1}^*)} \end{bmatrix} \tag{A5}$$

9 Modeling of cyclic mobility and associated lateral ground deformations for earthquake engineering applications

Ahmed Elgamal and Zhaohui Yang

1 Introduction

This chapter is concerned with the numerical modeling techniques used for the time-domain dynamic analysis of the seismic response of soil-structure systems. Attention is focused on the important aspects of soil cyclic mobility and its effects on lateral ground deformations. Incremental plasticity numerical constitutive models for cyclic loading are employed. A solid-fluid coupled finite element (FE) formulation for liquefaction scenarios is also discussed.

The chapter is divided into three sections. First, the mechanics of cyclic mobility and associated effects on the seismic response of soil systems are introduced. Second, a description of the numerical constitutive models developed to reproduce such response mechanisms is given. Third, a number of example applications using the developed soil models are presented, including studies on the following:

(1) influence of soil permeability and its spatial variation on liquefaction potential and liquefaction-induced lateral deformation,
(2) site response of a saturated dense sand profile,
(3) liquefaction of an embankment foundation and countermeasures, and
(4) ground improvement using stone columns.

2 Cyclic mobility and dilatancy

The liquefaction of soils (excess pore pressure ratio $r_u = u_e/\sigma'_v$ approaching and reaching 1.0, where u_e = excess pore pressure and σ'_v is the effective vertical stress) and associated deformations remain one the main causes of damage during earthquakes (Seed et al. 1990, Bardet et al. 1995, Sitar 1995, Japanese Geotechnical Society 1996, 1998, Ansal et al. 1999, http://peer.berkeley.edu/turkey/adapazari). Indeed, dramatic unbounded deformations (flow failure) due to liquefaction in dams and other structures (Seed et al. 1975, 1989, Davis and Bardet 1996) have highlighted the significance of this problem in earthquake engineering research.

However, liquefaction more frequently results in limited, albeit possibly high levels of deformation (Casagrande 1975, Youd et al. 1999). A survey of experimental research (triaxial and shear tests) compiled by Seed (1979) suggested that clean sands, with a relative density D_r of about 45 percent or more, appeared to exhibit the mechanism of limited strain cyclic mobility during liquefaction. A large body of more recent laboratory experiments, shaking-table and centrifuge tests, continues to corroborate the findings of earlier studies (in clean sands and non-plastic silts). In such experimental observations, uniform cohesionless soils with a reported D_r as low as 37 percent may accumulate large liquefaction-induced cyclic shear strains, but do not exhibit flow-type failures (see Elgamal et al. 1998 for an extensive literature survey).

2.1 Experimental observations

In general, at low confinement levels, dense granular soils exhibit a dilative response when subjected to shear loading conditions (Lambe and Whitman 1969). Under cyclic shear excitation, a sand mass will undergo this coupled shear-dilation process in a manner analogous to that shown in Figure 9.1 (Youd 1977). In a saturated state, the dilation-induced increase in volume is accommodated mainly owing to the migration of fluid into the created additional pore-space. The relatively low soil permeability (and/or the relatively fast rate of loading) hinders this water migration process resulting in an immediate reduction in pore-water pressures together with an associated increase in effective confinement and a resulting possible sharp slowdown of the deformation process. The slowdown occurs owing to the slowdown in dilation (and consequently in the coupled shear straining process),

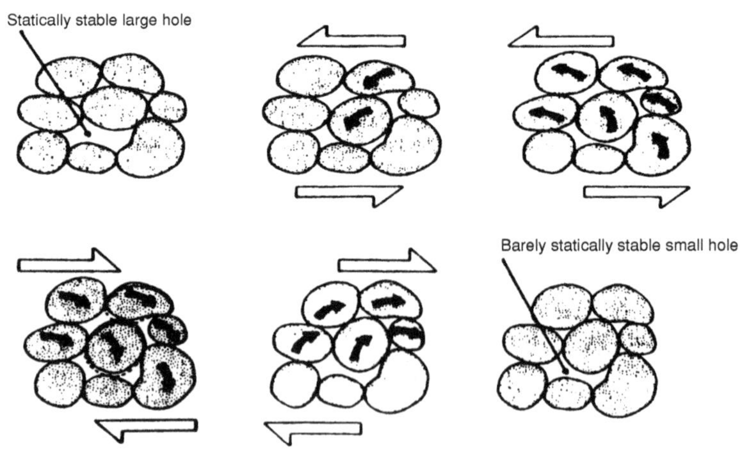

Figure 9.1 Diagrammatic cross-section of particulate group showing packing changes that occur during cyclic loading (Youd 1977)

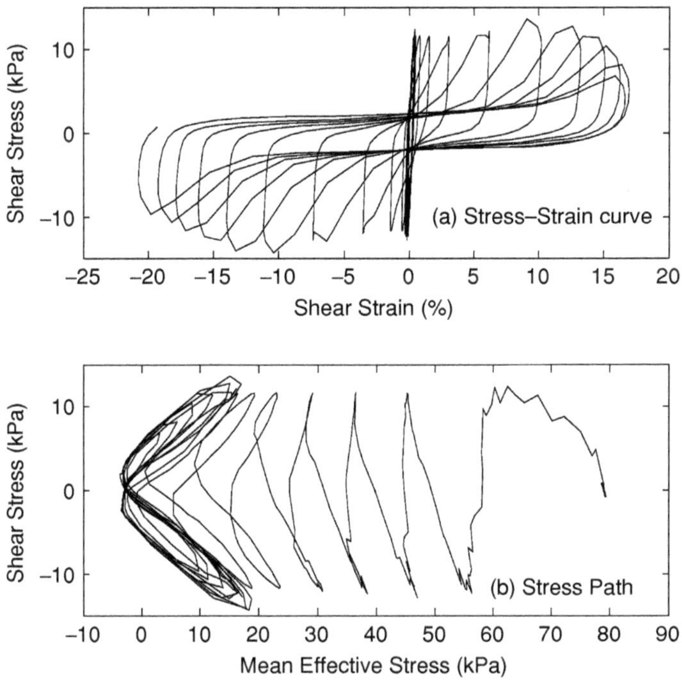

Figure 9.2 Stress–strain and stress path response for Nevada Sand (D_r = 60%) in a stress-controlled, undrained cyclic simple shear test (Arulmoli et al. 1992)

which is directly proportional to the rate of inward fluid flow and the instantaneous increase in soil stiffness resulting from the increase in effective confinement (due to pore-pressure reduction).

The mechanism of interest may be illustrated by the response shown in Figures 9.2 and 9.3 (Arulmoli et al. 1992). The simple shear test (Figure 9.2) shows:

(1) a cycle-by-cycle degradation in shear stiffness as manifested by the occurrence of increasingly larger shear strain excursions,
(2) a major portion of the large cyclic shear deformations rapidly developing at nearly constant, low shear stress and effective confinement, and
(3) a regain in shear stiffness and strength following these large shear strain excursions, along with an increase in effective confinement (due to dilatancy).

It can be seen that, as the shear strain increases, the sand skeleton is forced into dilation.

For the important situations of lateral spreading or biased strain accumulation due to a superimposed static shear stress (embankment slopes;

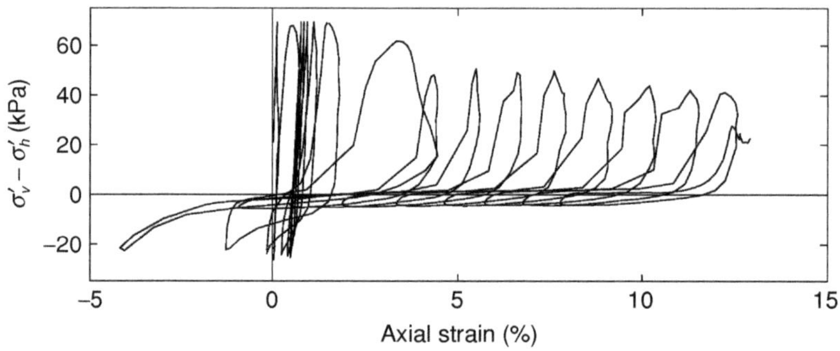

Figure 9.3 Stress–strain response during an undrained, anisotropically consolidated cyclic triaxial test of Nevada Sand at $D_r = 40\%$ (Arulmoli et al. 1992) (σ'_v and σ'_h are vertical and horizontal effective stresses respectively)

below foundations; behind retaining walls, etc.), cyclic mobility may play a dominant role. The results of an undrained triaxial test with a static shear stress bias (Figure 9.3, Arulmoli et al. 1992) shows a net increment of permanent strain accumulates in a preferred 'down-slope' direction, on a cycle-by-cycle basis. In these cycles, the dilation-induced increase in strength (shear stress) during liquefaction evolves such that a net finite increment of permanent (down-slope) shear strain occurs. The magnitude of such increments determines the total accumulated permanent deformation.

2.2 Recorded response

The cyclic mobility mechanism was evident at the Wildlife Refuge (California, USA) site. In 1987, the site was shaken by two main earthquakes (Holzer et al. 1989), including the Superstition Hills earthquake ($M_W = 6.6$), which caused a sharp increase in recorded pore-water pressure. In addition, subsequent field investigations showed evidence of site liquefaction and ground fissures. The surface records displayed peculiar acceleration spikes (Holzer et al. 1989) associated with simultaneous instants of excess pore-pressure drop (Figure 9.4).

Back-calculated stress–strain and effective-path histories (Figure 9.5, Zeghal and Elgamal 1994) based on the above acceleration and pore-pressure time-histories show that, at low effective confining pressures (high excess pore pressures), the effective stress-path clearly exhibited a reversal of behavior from contractive to dilative (Figure 9.5), as the line of phase transformation was approached (Section 3.5). This pattern of stress–strain response is similar to that shown in Figure 9.2. Such a mechanism is a consequence of soil dilation at large strain excursions, resulting in associated instantaneous pore-pressure drops.

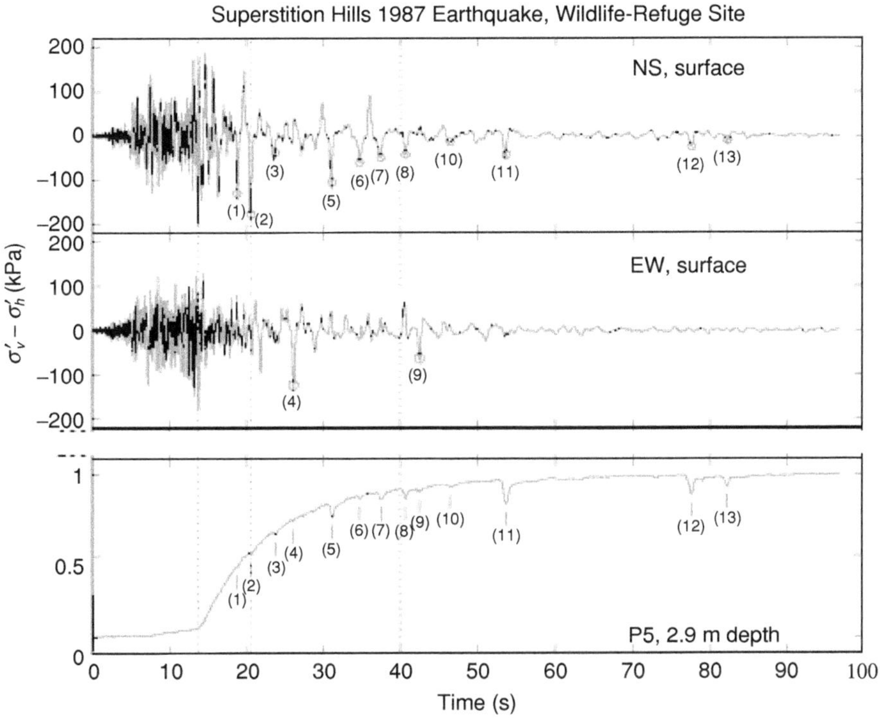

Figure 9.4 Wildlife Refuge site NS and EW surface accelerations and associated pore water pressure (at 2.9 m depth) during the Superstition Hills 1987 earthquake (Zeghal and Elgamal 1994)

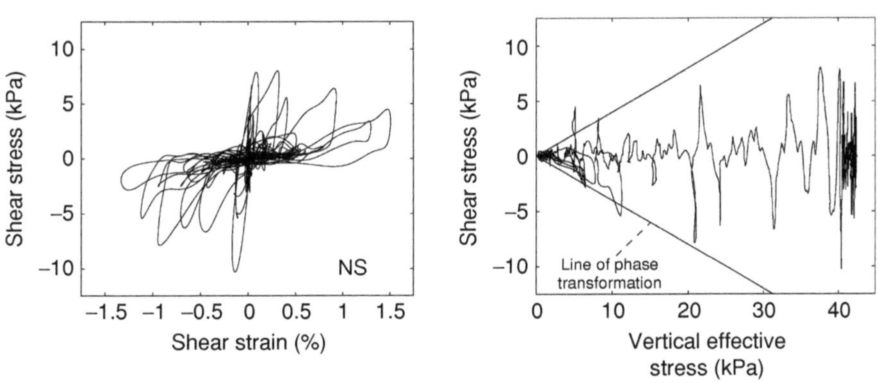

Figure 9.5 Wildlife Refuge site NS back-calculated shear stress-strain and effective stress path during the Superstition Hills 1987 earthquake (Zeghal and Elgamal 1994)

Comparing the responses of Figure 9.5 and Figures 9.2 and 9.3, it may be inferred that, in the level-ground case, spikes will be equally visible on both sides of the recorded acceleration response (Figure 9.4) and a superposed static driving shear stress (e.g., near the free face of a slope, behind a yielding retaining wall, or along a mildly inclined ground surface) may dictate a stress–strain response similar to that of Figure 9.3, with associated acceleration spikes predominantly in one direction.

3 Constitutive model

A number of computational models are available to simulate the processes associated with sand dilation during liquefaction (see Elgamal et al. 2003 and Yang et al. 2003 for a partial list). Many of the essential features of cyclic mobility have been successfully simulated in these models.

Currently, reliable computational modeling of the accumulated shear deformations associated with cyclic mobility still remains a major challenge. As indicated above, a major portion of these deformations develops at a state of low, nearly constant shear stress and effective confinement (Figures 9.2 and 9.3). This minimal change in stress state at very low confinement levels poses a significant challenge in reliably reproducing the associated shear deformations (using traditional stress-space models).

Motivated by these experimental observations, a plasticity model was developed for capturing the characteristics of cyclic mobility. In this regard, emphasis is placed on the more accurate reproduction of accumulated shear deformations in clean medium-dense cohesionless soils. In particular, an effort was made to model such deformations directly by a strain-space yield domain. The observed cyclic shear deformation patterns (Figures 9.2 and 9.3) are then accounted for by enlargement and/or translation of this domain in strain space. This model extends an existing multi-surface plasticity formulation with newly developed flow and hardening rules. The new flow rule allows for reproducing cyclic shear strain accumulation and the subsequent dilative phases observed in liquefied soil response. The new hardening rule enhances numerical robustness and efficiency.

3.1 Model description

The necessary components of classical stress-space formulation, based on the original multi-surface-plasticity framework of Prevost (1985) are initially presented. Thereafter, discussions are focused on a new non-associative flow rule and the strain-space mechanism (Parra 1996, Yang et al. 2003), the key elements in reproducing the salient cyclic mobility features of Figures 9.2 and 9.3. Herein, the adopted sign convention is such that volumetric stresses/strains are positive in compression.

The model is based on the multi-surface-plasticity framework (Iwan 1967, Mroz 1967). The constitutive equation may be written in incremental form as (Prevost 1985):

$$\dot{\sigma} = E : (\dot{\varepsilon} - \dot{\varepsilon}^p) \tag{1}$$

where $\dot{\sigma}$ is the rate of effective stress tensor, $\dot{\varepsilon}$ the rate of deformation tensor, $\dot{\varepsilon}^p$ the plastic rate of deformation tensor, and E the isotropic fourth-order tensor of elastic coefficients. The plastic rate of deformation tensor is defined by: $\dot{\varepsilon}^p = P\langle L \rangle$, where P is a symmetric second-order tensor defining the outer normal to the plastic potential surface, L the plastic loading function, and the symbol $\langle\ \rangle$ denotes the McCauley's brackets (i.e. $L = \max(L, 0)$). In the above, L is defined as: $L = (Q : \dot{\sigma})/H'$ where H' is the plastic modulus and Q a unit symmetric second-order tensor defining the outer normal to the yield surface.

3.2 Yield function

Following the standard convention, it is assumed that material elasticity is linear and isotropic, and that non-linearity and anisotropy result from plasticity (Hill 1950). The yield function (Figure 9.6) is selected as a conical surface in the principal stress space (Prevost 1985, Lacy 1986):

$$f = \frac{3}{2}(s - (p' + p'_0)\alpha) : (s - (p' + p'_0)\alpha) - M^2(p' + p'_0)^2 = 0 \tag{2}$$

in the domain $p' \geq 0$, where $s = \sigma' - p'\delta$ is the deviatoric stress tensor (σ' = effective Cauchy stress tensor, δ = second-order identity tensor), p' is the mean effective stress, p'_0 is a small positive constant such that the yield surface size remains finite at $p' = 0$ (for numerical convenience and to avoid ambiguity in defining the yield surface normal at the yield surface apex), α is a second-order deviatoric tensor defining the yield surface centre in deviatoric stress subspace, M defines the yield surface size, and ':' denotes

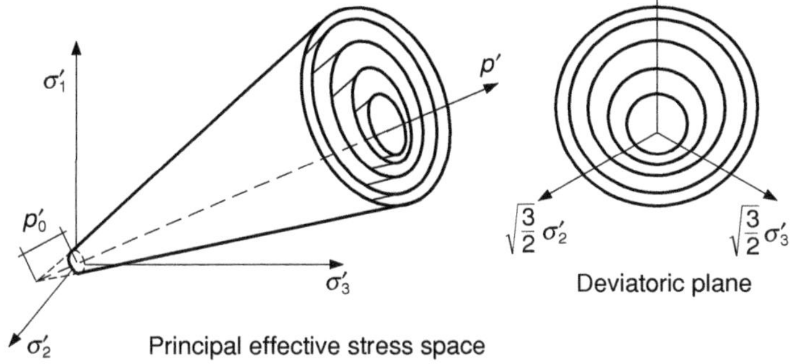

Figure 9.6 Conical yield surfaces in principal stress space and deviatoric plane (after Prevost 1985, Parra 1996, and Yang et al. 2003)

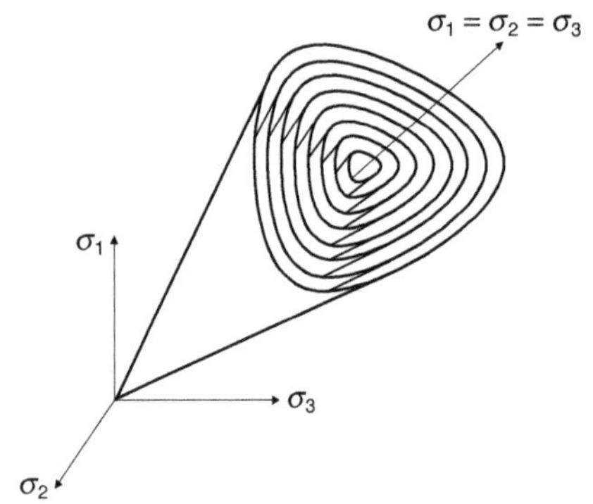

Figure 9.7 Configuration of the Lade–Duncan multi-yield surfaces in the principal stress space

the doubly contracted tensor product. In the context of multi-surface plasticity (Iwan 1967, Mroz 1967, Prevost 1985), the hardening zone is defined by a number of similar yield surfaces (Figure 9.6) with a common apex (at $-p_0'$ along the hydrostatic axis). The outmost surface is designated as the failure surface, the size of which (M_f) is related to the friction angle ϕ by $M_f = 6 \sin\phi/(3 - \sin\phi)$ (Chen and Mizuno 1990).

As described in Prevost (1985), the yield surfaces may be initially configured with a non-zero α to account for the shear strength difference between triaxial compression and extension. However, the Lode angle effect is not incorporated in this model since the yield function (Equation 2) does not include the third stress invariant. Recently, a version of this model was developed in which the yield function of Lade and Duncan (1975) is employed to address the Lode angle effect (Figure 9.7).

3.3 Shear stress–strain response

In geotechnical engineering practice, non-linear shear behavior is commonly described by a shear stress–strain backbone curve (Kramer 1996). The backbone curve at a given reference confinement p_r' can be approximated by the hyperbolic formula (Kondner 1963, Duncan and Chang 1970, see Figure 9.8):

$$\tau = G_r\gamma/(1 + \gamma/\gamma_r) \tag{3}$$

where τ and γ are the octahedral shear stress and strain respectively, G_r is the low-strain shear modulus at p_r' (Figure 9.8), and $\gamma_r = \tau_{max}/G_r$, in which τ_{max} is the maximum shear strength when γ approaches ∞. In order to reach the maximum shear strength at a finite strain, the hyperbolic curve is often capped at $\tau_f < \tau_{max}$ (Figure 9.8).

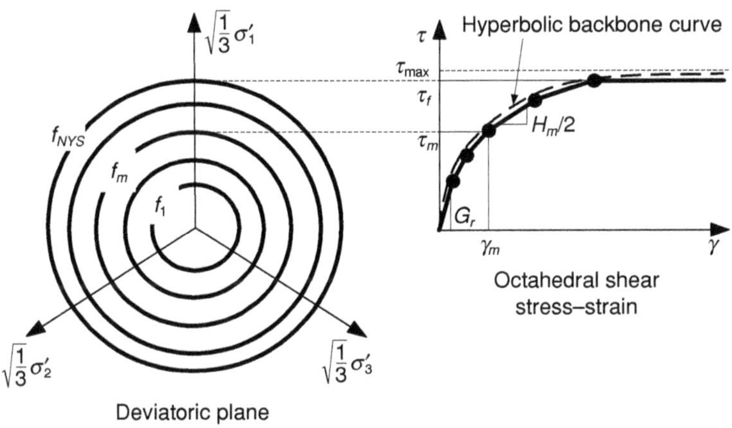

Figure 9.8 Hyperbolic backbone curve for soil non-linear shear stress–strain response and piecewise-linear representation in multi-surface plasticity (after Prevost 1985, Parra 1996)

Within the framework of multi-surface plasticity, the hyperbolic backbone curve (Equation 3) is replaced by a piecewise linear approximation (Figure 9.8). Each linear segment (Figure 9.8) represents the domain of a yield surface f_m, characterized by an elasto-plastic (tangent) shear modulus H_m and a size M_m, for $m = 1, 2, \ldots, NYS$, where NYS is the total number of yield surfaces (Prevost 1985). At the reference confinement p'_r, H_m is conveniently defined by (Figure 9.8):

$$H_m = 2(\tau_{m+1} - \tau_m)/(\gamma_{m+1} - \gamma_m) \tag{4}$$

with $H_{NYS} = 0$. Using Equation 2, the size of the surface f_m is now dictated by (Figure 9.8):

$$M_m = 3\tau_m/\sqrt{2}(p'_r + p'_0) \tag{5}$$

with $M_{NYS} = M_f$ and $\tau_{NYS} = \tau_f$. As shown in Figure 9.9, a set of yield surfaces defined using the Lade–Duncan yield function can be similarly configured by triaxial compression and triaxial extension test results simultaneously.

Finally, the low-strain shear modulus G is assumed to vary with confinement p' as follows (Prevost 1985):

$$G = G_r[(p' + p'_0)/(p'_r + p'_0)]^n \tag{6}$$

where n is a material parameter (= 0.5 typically for sand, Kramer 1996). The tangent shear moduli (Equation 4) were assumed to follow the same confinement dependence rule (Equation 6). Based on elasticity theory, the

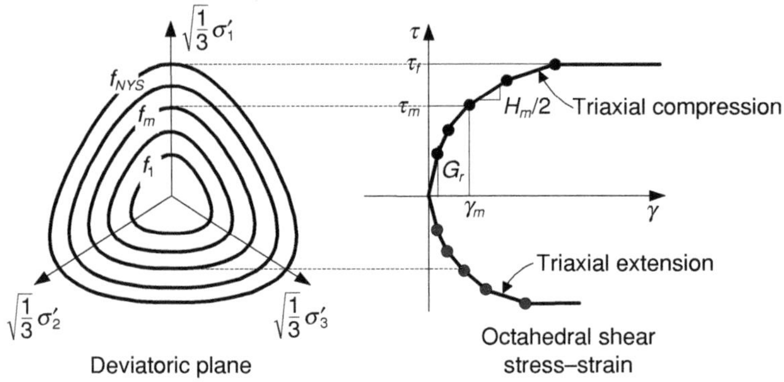

Figure 9.9 Schematic calibration of Lade–Duncan multi-yield surfaces based on triaxial compression/extension data

bulk modulus of the soil skeleton B is defined by $B = 2G(1 + \nu)/(3 - 6\nu)$, where ν is Poisson's ratio.

3.4 Hardening rule

A purely deviatoric kinematic hardening rule is employed (Prevost 1985), conveniently to generate a hysteretic cyclic response. In the context of multi-surface plasticity, translation of the yield surface is generally governed by the consideration that no overlapping is allowed between yield surfaces (Mroz 1967). Thus, contact between consecutive similar surfaces f_m and f_{m+1} (Figures 9.8 and 9.9) must occur only at conjugate points with the same direction of outward normal. A new translation rule μ (Figure 9.10) was defined as (Parra 1996):

$$\mu = [s_T - (p' + p'_0)\alpha_m] - \frac{M_m}{M_{m+1}}[s_T - (p' + p'_0)\alpha_{m+1}] \tag{7}$$

where s_T is the deviatoric stress tensor defining the position of point T (Figure 9.10) as the intersection of f_{m+1} with the vector connecting the inner surface centre $(p' + p'_0)\alpha_m$ and the updated stress state $(s + ds)$. This rule (Equation 7) is also based on the Mroz (1967) conjugate-points concept, and allows no overlapping of the yield surfaces.

3.5 Flow rule

We define Q and P as the outer normal to the yield surface and the plastic potential surface respectively. These tensors may be conveniently decomposed into deviatoric and volumetric components, giving $Q = Q' + Q''\,\delta$

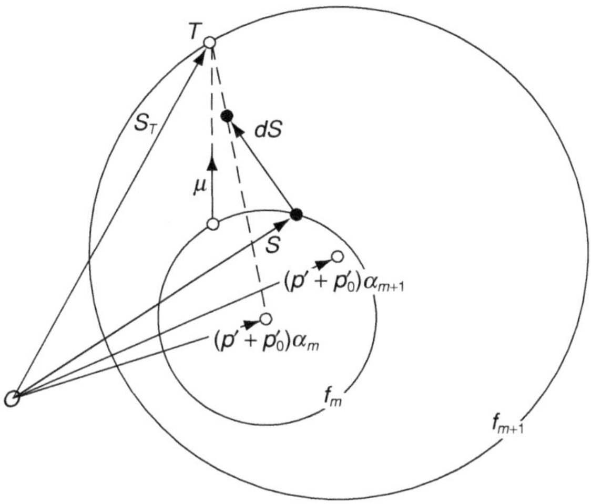

Figure 9.10 New deviatoric hardening rule (after Parra 1996)

and $P = P' + P''\ \delta$ (Prevost 1985). Non-associativity of the plastic flow is restricted to its volumetric component (Prevost 1985), i.e. $Q' = P'$ and $P'' \neq Q''$.

Under undrained conditions, shear loading inside (or outside) the phase transformation (PT) surface is accompanied by a tendency of volume contraction (or dilation), resulting in increased (or decreased) pore pressure and decreased (or increased) p' (Figure 9.11). The relative location of the stress state with respect to the PT surface may be inferred (Prevost 1985) from the stress ratio: $\eta \equiv \sqrt{3(s:s)/2}/(p' + p_0')$. Designating η_{PT} as the stress ratio along the PT surface, it follows that $\eta < \eta_{PT}$ (or $\eta > \eta_{PT}$) if the stress state is inside (or outside) the PT surface.

In our model, depending on the value of η and the sign of $\dot{\eta}$ (time rate of η), distinct contractive/dilative (dilatancy) behaviors are reproduced by specifying appropriate expressions for P''. In addition, a neutral phase ($P'' = 0$, Phase 1-2 in Figure 9.11) is proposed between the contraction ($P'' > 0$, Phase 0-1) and the dilation ($P'' < 0$, Phase 2-3) phases. This neutral phase conveniently allows for modeling the accumulation of highly yielded shear strain, as will be discussed below.

3.5.1 Contractive phase (phases 0-1, 3-4, and 4-5 in Figure 9.11)

Shear-induced contraction occurs inside the PT surface ($\eta < \eta_{PT}$) as well as outside ($\eta > \eta_{PT}$) when $\dot{\eta} < 0$. Based on experimental observations (Ishihara et al. 1975, Ladd et al. 1977) and micro-mechanical explanations (Nemat-Nasser and Tobita 1982, Papadimitriou et al. 2001), the rate of contraction

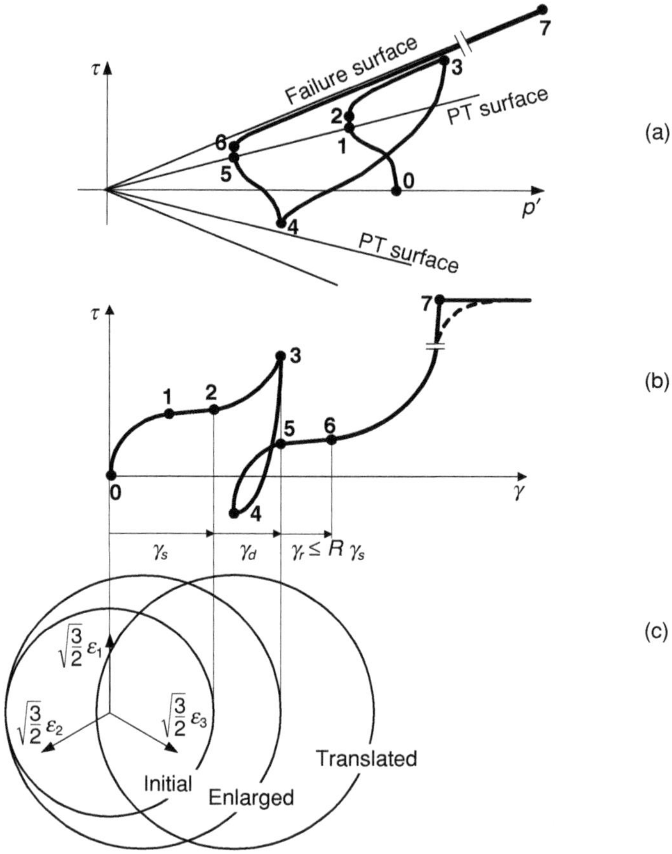

Figure 9.11 Schematic of constitutive model response showing: (*a*) octahedral stress τ – effective confinement p' response, (*b*) τ – octahedral strain γ response and (*c*) configuration of yield domain

is dictated, to a significant extent, by preceding dilation phase(s). A simple version is adopted herein by specifying P'' as a scalar function of ε_v^p. In particular, the contraction flow rule is defined by:

$$P'' = (1 - \text{sign}(\dot{\eta})\eta/\eta_{PT})(c_1 + c_2\varepsilon_c) \tag{8}$$

where c_1 and c_2 are positive calibration constants dictating the rate of contraction (or excess pore pressure increase), and ε_c is a non-negative scalar governed by the following rate equation:

$$\dot{\varepsilon}_c = \begin{cases} -\dot{\varepsilon}_v^p & (\varepsilon_c > 0 \quad \text{or} \quad -\dot{\varepsilon}_v^p > 0) \\ 0 & (\text{Otherwise}) \end{cases} \tag{9}$$

where $\dot{\varepsilon}_v^p$ is the rate of plastic volumetric strain. In other words, ε_c increases only during dilation and decreases during subsequent unloading (contraction), until it reaches zero (in phase 0-1 of Figure 9.11, since no prior dilation has taken place, ε_c remains zero). Thus, a stronger dilation phase (phase 2-3) results in a higher rate of contraction upon unloading (phase 3-4).

3.5.2 Dilative phase (phases 2-3 and 6-7 in Figure 9.11)

Dilation appears only due to shear loading outside the *PT* surface ($\eta > \eta_{PT}$ with $\dot{\eta} > 0$), and is defined herein by:

$$P'' = (1 - \eta/\eta_{PT})d_1(\gamma_d)^{d_2} \tag{10}$$

where d_1 and d_2 are positive calibration constants, and γ_d is the octahedral shear strain accumulated during this dilation phase. Equation 8 dictates a dilation tendency that increases with the accumulated strain γ_d.

The dilation rule (Figure 9.11, phase 2-3) can result in significant increases in shear stress and mean effective confining stress. This increase is limited by:

(1) *fluid cavitation* (Lambe and Whitman 1969, Casagrande 1975): If soil response is essentially undrained (fluid migration is relatively slow), the tendency for dilation can eventually drop pore pressure to the minimum value of –1.0 atmospheric pressure (cavitation). Cavitation will prevent the effective confining pressure from further increase and
(2) *critical-void-ratio soil response* (beyond stage 7 in Figure 9.11): If the soil is partially or fully drained (relatively rapid flow of pore fluid), overall volume increase is allowed. To this end, the critical-void-ratio state may be reached, whereupon further shear deformation continues to develop without additional volume/confinement change or increase in shearing resistance. In our model, a simple logic has been incorporated such that the volume remains constant ($P'' = 0$) when the critical state is reached (stage 7, Figure 9.11b). The critical state is defined based on a relationship between volumetric strain ε_v and effective confinement p'.

3.5.3 Neutral phase (phases 1-2 and 5-6 in Figure 9.11)

As the shear stress increases (Figure 9.11, phase 0-1), the stress state eventually reaches the *PT* surface ($\eta = \eta_{PT}$). At sufficiently high p' levels, dilation (phase 2-3) would follow. However, when p' is low, a significant amount of permanent shear strain may accumulate prior to dilation (Figures 9.2 and 9.3), with minimal changes in shear stress and p' (implying $P'' \approx 0$).

Such minimal change in the stress state is difficult to employ as a basis for modeling the associated extent of shear strain accumulation (during which $P'' \approx 0$). Hence, for simplicity, $P'' = 0$ is maintained during this highly yielded phase (phase 1-2) until a boundary defined in deviatoric strain space is reached (Figure 9.11c), with subsequent dilation thereafter (phase 2-3). This boundary defines an initially isotropic domain in deviatoric strain space (Figure 9.11c) as a circle of radius γ_s (expressed in terms of octahedral shear strain). This domain will enlarge or translate depending on load history, as described below.

3.5.4 Configuration of yield domain

The shear strain γ_d accumulated during dilation (phase 2-3, Figure 9.11b) may enlarge the yield domain (Figure 9.11c). Specifically, enlargement occurs when shear strain accumulated in the current dilation phase exceeds the maximum γ_d the material has ever experienced before (since phase 2-3 is the first time the material experiences dilation, the domain enlarges throughout). This logic preserves the symmetric pattern of cyclic shear deformation observed in Figure 9.2, and may be physically interpreted as a form of damage effect.

The presence of a superposed static shear stress results in biased accumulation of shear deformations, as discussed earlier (Figure 9.3). This biased accumulation is achieved through translation of the yield domain in the deviatoric strain space (strain-induced anisotropy, Bazant and Kim 1979), allowing for a yield increment γ_r to develop before the subsequent dilation (phase 5-6, Figure 9.11). According to experimentally documented accumulation patterns (Ibsen 1994, Arulmoli et al. 1992), the strain increment γ_r is proportional to the level of the previous unloading strain (phase 3-4), limited to a maximum of $R\gamma_s$ where R is a user-defined constant. Note that translation of the yield domain continues until the accumulated strain during dilation reaches the maximum γ_d previously recorded (phase 2-3 in Figure 9.11b). Thereafter, the domain enlarges again (damage effect as described above).

The initial yield domain size γ_s depends on the effective confinement p'. In the current model, this dependence is defined by the following simple linear relationship (Figure 9.12):

$$\gamma_s = \gamma_{s,\max}\left\langle \frac{p'_y - p'}{p'_y} \right\rangle \tag{11}$$

where p'_y and $\gamma_{s,\max}$ are model constants that may be easily derived from data such as that shown in Figure 9.13. In this figure, the model response under undrained monotonic loading conditions at various low confinement levels clearly indicates the influence of confinement on the extent of the accumulated shear strain. Other forms of confinement dependence may easily be prescribed, as dictated by available experimental data.

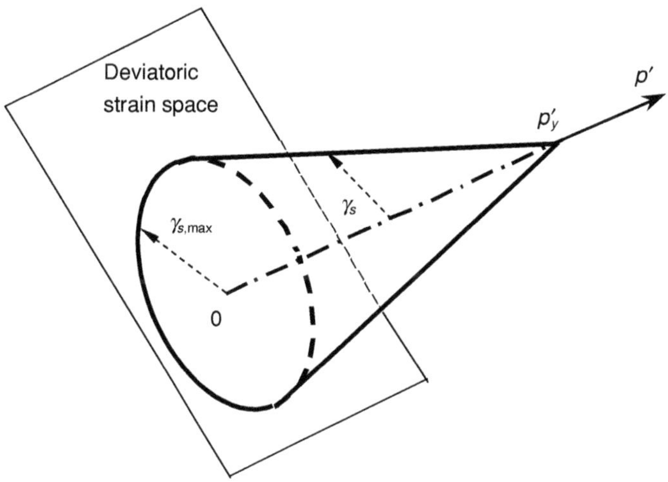

Figure 9.12 Initial yield domain at low levels of effective confinement

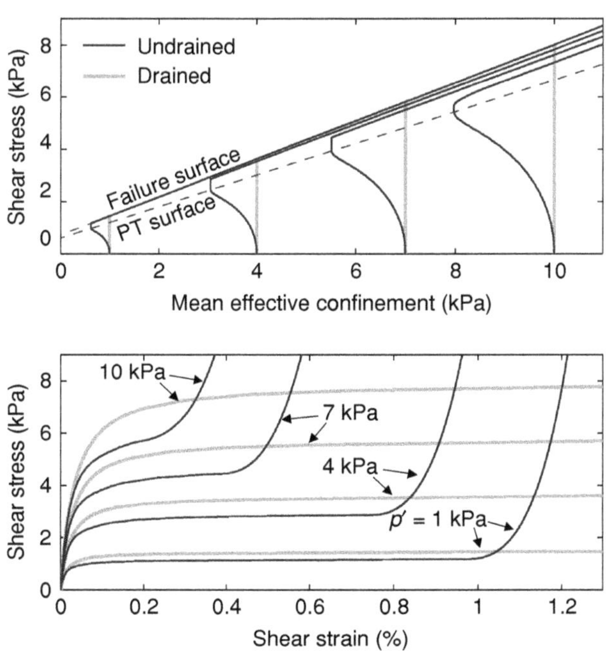

Figure 9.13 Undrained and drained monotonic simple shear stress-path and stress–strain response showing dependence of initial yield domain size on effective confinement

3.6 Model performance and calibration

In Figure 9.13, the model response under drained monotonic shear loading is also depicted for low confinements (from 1.0 to 10 kPa). However, it is emphasized that the drained volumetric response at very low confinement levels (Sture et al. 1998) is not addressed by the current formulation.

Model performance under symmetric cyclic shear loading is depicted in Figure 9.14a. This figure shows the combined effect of the gradual confinement decrease and dilation history (γ_d) on the shear stress–strain response. Finally, Figure 9.14b shows that, under biased cyclic loading, the extent of cycle-by-cycle deformation is conveniently simulated via the parameter R.

Model calibration may be carried out using results from traditional laboratory soil tests. For instance, the configuration of the yield surfaces (defining H_m and M_m) can be based on data from monotonic drained triaxial compression/extension tests, whereas the evaluation of the dilatancy parameters

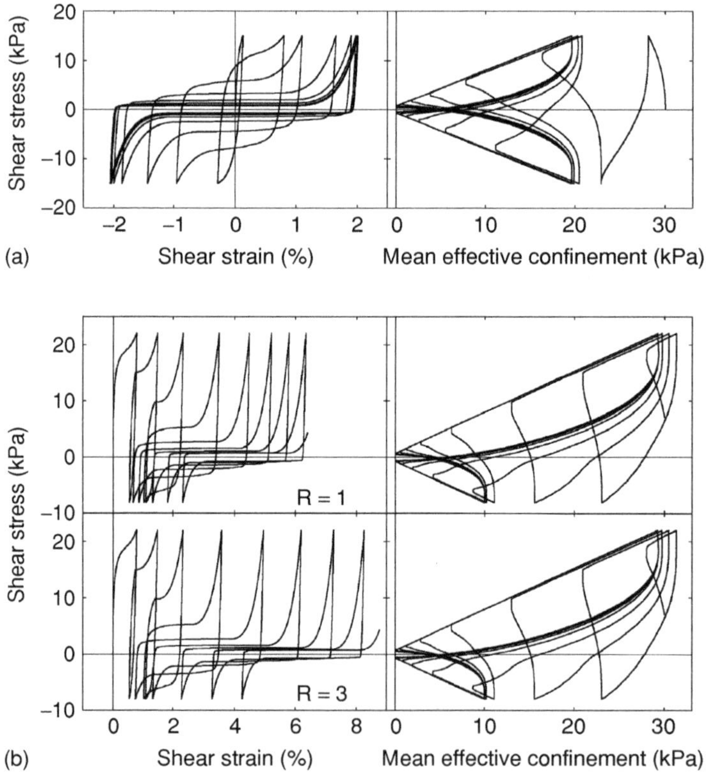

Figure 9.14 (*a*) Model simulation of undrained cyclic simple shear response (stress-controlled simulation at ±15 kPa), (*b*) effect of parameter R on undrained cyclic shear response (stress-controlled simulations at ±15 kPa with 7 kPa static shear stress bias)

and strain-space parameters are based on results from undrained cyclic triaxial compression and/or simple shear tests. More details regarding the model calibration procedures can be found in Elgamal et al. 2002a, 2003, and Yang et al. 2003.

The constitutive model described above has been incorporated into a solid–fluid fully coupled FE program to conduct one-dimensional, two-dimensional and three-dimensional dynamic analyses of soil or soil-structural systems. In the following section, this fully coupled FE formulation, which has been adopted by a class of researchers in liquefaction-related numerical simulations, is briefly presented.

4 Finite element formulation

The saturated soil system is modeled as a two-phase material based on the Biot (1962) theory for porous media. A simplified numerical formulation of this theory, known as *u-p* formulation (in which displacement of the soil skeleton u, and pore pressure p, are the primary unknowns, Chan 1988, Zienkiewicz et al. 1990), was implemented in CYCLIC (Ragheb 1994, Parra 1996, Yang 2000). The computational scheme follows the methodology of Chan (1988), based on the following assumptions: (i) small deformations and negligible rotations, (ii) densities of the solid and fluid are constant in both time and space, (iii) porosity is locally homogeneous and constant with time, (iv) incompressible soil grains, and (v) equal accelerations for the solid and fluid phases.

As described by Chan (1988), the *u-p* formulation is defined by: (i) an equation of motion for the solid–fluid mixture, and (ii) an equation of mass conservation for the mixture, incorporating the equation of motion for the fluid phase and Darcy's law:

$$\nabla \cdot (\sigma' + p\delta) - \rho(\ddot{u} - g) = 0 \qquad (12a)$$

$$\nabla \cdot \dot{u} + \frac{\dot{p}}{Q} + \nabla \cdot \left[\frac{k}{\rho_f g}(\nabla p - \rho_f \ddot{u} + \rho_f g) \right] = 0 \qquad (12b)$$

where σ' is the effective stress tensor, p the pore-fluid pressure, δ the second-order identity tensor, ρ the mass density of the mixture, u the displacement vector of the solid phase, g the gravity acceleration vector, Q the undrained mixture bulk modulus, k Darcy's permeability coefficient tensor, ρ_f the fluid mass density, g the absolute value of gravity acceleration, ∇ the gradient operator, $\nabla\cdot$ the divergence operator, and a superposed dot denotes material time derivative.

After FE spatial discretization and Galerkin approximation, the governing equations can be expressed in the following matrix form (Chan 1988):

$$M\ddot{U} + \int_{\Omega} B^T \sigma' \, d\Omega + Qp - f^s = 0 \qquad (13a)$$

$$Q^T\dot{U} + S\dot{p} + Hp - f^p = 0 \qquad (13b)$$

where M is the mass matrix, U the displacement vector, B the strain-displacement matrix, σ' the effective stress vector (determined by the soil constitutive model discussed below), Q the discrete gradient operator coupling the solid and fluid phases, p the pore pressure vector, H the permeability matrix, and S the compressibility matrix. The vectors f^s and f^p include the effects of body forces and the prescribed boundary conditions for the solid and fluid phases respectively. In Equation 13a, the first term represents the inertia force of the solid–fluid mixture, followed by the internal force due to the soil skeleton deformation and the internal force induced by pore-fluid pressure. In Equation 2b, the first two terms are the rates of volume change with time for the soil skeleton and the fluid phase respectively, followed by the seepage rate of the pore fluid (Hp). It should be noted that a high permeability k in Equation 12b results in a large-valued H matrix in Equation 13b, which behaves numerically as a penalty term that forces pore pressure p changes to be negligibly small (free drainage scenario).

Equations 13 are integrated in time using a single-step predictor multi-corrector scheme of the Newmark type (Chan 1988, Parra 1996). The solution is obtained for each time step using the modified Newton–Raphson approach (Parra 1996).

A typical element employed in the *u-p* formulation is shown in Figure 9.15, with nine nodes for the solid phase and four nodes for the fluid phase, so as to reduce the numerical difficulties associated with the nearly incompressible fluid phase (Chan 1988). Each solid node is associated with

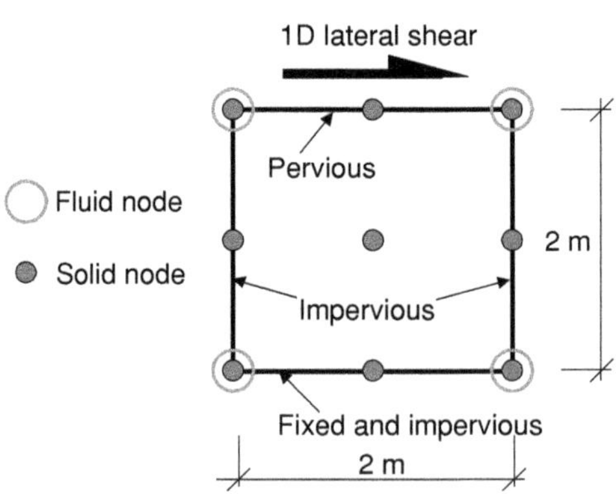

Figure 9.15 Typical 9-4-node element employed in the solid–fluid coupled formulation and boundary conditions/applied shear for the single element investigation

two-degrees-of-freedom (2 DOF) for the lateral and vertical displacements, and each fluid node is associated with 1 DOF for the pore pressure. This 9-4-node element is employed in all 2D numerical studies presented herein.

5 Applications

Scaled physical models in centrifuge tests have been increasingly employed to study static and dynamic response characteristics of soil systems. A large body of valuable experimental data from such tests is available for the calibration and verification of the numerical models. In this section, the FE program described above is used in a number of illustrative numerical simulations of centrifuge experiments. These physical/numerical studies were conducted in order to investigate:

(1) the influence of soil permeability and its spatial variation on liquefaction potential and liquefaction-induced lateral deformation (Yang and Elgamal 2002),
(2) the site response of a saturated dense sand profile (Elgamal et al. 2005),
(3) the seismic response of a liquefiable embankment foundation and the effectiveness of liquefaction countermeasures including foundation densification and sheet-pile enclosure (Elgamal et al. 2002b), and
(4) the seismic response of stone-column reinforced silty soil (Lu et al. 2005).

In each of these applications, the centrifuge experimental program is first briefly described, followed by a presentation of the numerical modeling procedure and results.

5.1 Influence of permeability on the liquefaction-induced shear deformation

The permeability of a liquefiable soil profile may affect the rate of pore-pressure buildup and subsequent dissipation during and after earthquake excitation. Consequently, effective soil confinement and available resistance to shear deformations may be significantly dependent on permeability in many practical situations. If present, spatial variation in permeability may even have a more profound impact on the available overall shear resistance. In such situations, the onset of liquefaction-induced densification may result in water or water-rich thin interlayers trapped below overlying low-permeability strata. The presence of these low-shear-strength interlayers may trigger excessive (or even unbounded) localized shear deformations (flow failure mechanism). Herein, numerical modeling is employed to investigate the influence of permeability and the spatial variation thereof on liquefaction-induced shear deformations.

5.1.1 Permeability in a uniform soil profile

In this section, the influence of permeability in a uniform soil profile composed of a single soil is investigated. A 1D model was employed to represent a 10 m thick uniform soil profile, inclined by 4 degrees (Figure 9.16) to simulate an infinite-slope response. This configuration is identical to that of the VELACS Model-2 centrifuge experiment (Dobry et al. 1995, Taboada 1995), and is modelled by specifying:

(1) for the solid phase, the two lateral sides of the mesh were tied together both horizontally and vertically to mimic a 1D shear beam effect, and the top surface was traction free, and
(2) for the fluid phase, the base and two sides were impervious, with zero prescribed pore pressure at the ground surface.

The VELACS Model-2 input excitation (harmonic, mainly 2 Hz motion, Figure 9.16) was employed in this section. Three numerical simulations were conducted, with a permeability coefficient k of 1.3×10^{-2} m/sec (gravel), 3.3×10^{-3} m/sec (VELACS Model-2 sandy gravel calibration simulation), and 6.6×10^{-5} m/sec (clean sand) respectively.

Computed lateral displacements for the three simulations are displayed in Figure 9.17a, along with the experimental response of VELACS Model 2 test (Dobry et al. 1995, Taboada 1995). It is clearly seen that: (i) computed lateral deformations with the sandy gravel k value are close to the experimental responses (part of the calibration process), and (ii) the extent of lateral deformation in this uniform profile is inversely proportional to soil permeability, i.e., a higher k results in smaller lateral deformation (near the surface, the profile with the lowest k value had a lateral translation of about 2.5 times that with the highest k value).

The relation between k and deformation is a consequence of the effect of permeability on u_e along the soil profile. Figure 9.17b shows that, in the

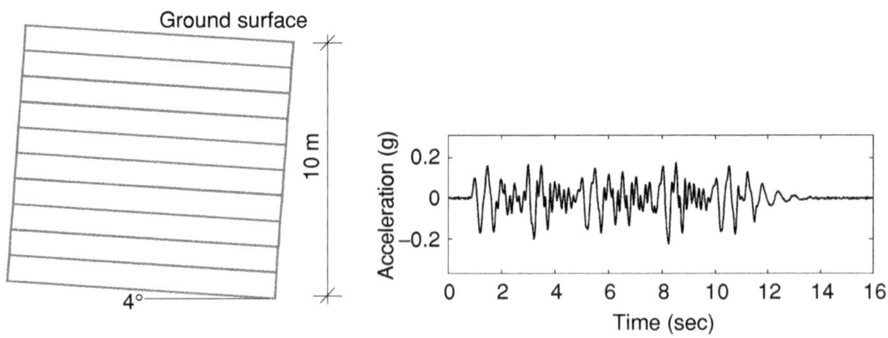

Figure 9.16 Employed soil profile configuration and input base acceleration history for uniform-profile simulations

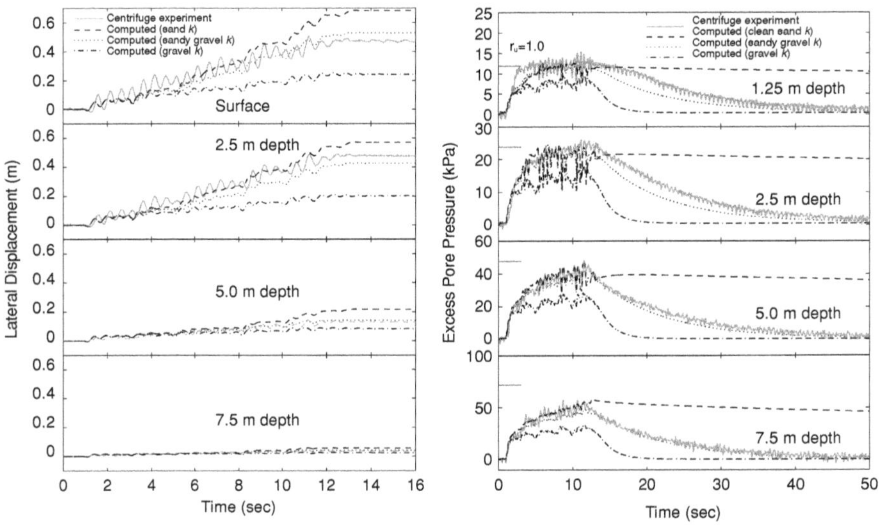

Figure 9.17 (*a*) Lateral displacement and (*b*) excess pore-pressure histories in uniform soil profile with different permeability coefficients

case of high k, fast u_e dissipation took place (even during the strong shaking phase), resulting in a low u_e profile. Consequently, high effective confinement was maintained with less reduction in shear stiffness and strength. On the contrary, in the lowest k profile, soil response was essentially undrained, with sustained high u_e long after the shaking phase.

5.1.2 Effect of a low-permeability interlayer

An inclined (4 degrees) 1D saturated soil profile was employed (Figure 9.18). Boundary conditions are identical to those of the uniform soil profile described above (Figure 9.16). Input excitation was defined in the form of 14 cycles of 1 Hz excitation, with an amplitude of 0.08 g (Figure 9.18).

Three numerical simulations were conducted, with the permeability coefficient k, profiles listed in Table 9.1. In Table 9.1, Case 1 represents a uniform sand stratum throughout (benchmark case), and Cases 2 and 3 include a 0.3 m interlayer of silt k within a uniform sandy-gravel-k stratum (Case 2) and a sand-k stratum (Case 3) respectively. The interlayer profiles were selected to allow for development of a flow failure scenario (due to the interlayer) that initiates during (Case 2) and after (Case 3) dynamic excitation as discussed below.

Figure 9.19 depicts the time histories of the computed lateral displacement at the surface (location A), immediately below the interlayer (location B),

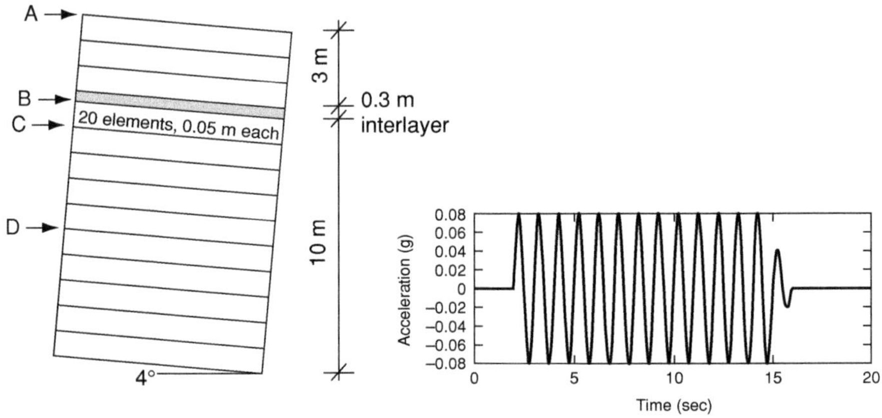

Figure 9.18 Employed soil profile and input acceleration history in low-permeability interlayer simulations

Table 9.1 Permeability coefficients employed in numerical simulations

	Permeability Coefficient k (m/sec)	
	0.3 m interlayer	*Stratum*
Case 1	3.3×10^{-4} (clean sand *k*)	3.3×10^{-4} (clean sand *k*)
Case 2	3.3×10^{-8} (silt *k*)	3.3×10^{-3} (sandy gravel *k*)
Case 3	3.3×10^{-8} (silt *k*)	3.3×10^{-4} (clean sand *k*)

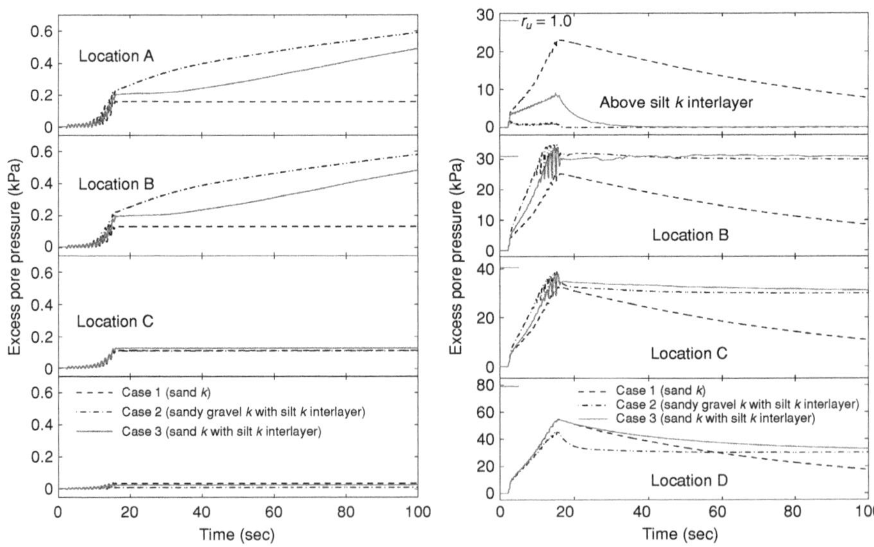

Figure 9.19 Lateral displacement and excess pore-pressure histories along soil profile

1.0 m below the interlayer (location C), and 5.0 m below the interlayer (location D). In all cases, the displacements stop in the lower sections of the soil column (location C and below) as soon as the shaking ends. The difference in response is in the upper section (location B and above). In Case 1, no further displacement is seen once the shaking stops (similar to the VELACS Model-2 experiment, Figure 9.16). In Case 2, the upper section including the silt-*k* interlayer continues to slide after the shaking ends, in the form of a flow failure. After the shaking phase, in Case 3, lateral movement appears to end in the entire column for more than 10 seconds. Thereafter, a delayed flow-failure scenario occurs.

Flow failure occurred in Case 2 owing to the loss of shear strength below the silt-*k* interlayer. As seen in Figure 9.20, this interlayer sustains r_u ($= u_e/\sigma_v'$ where σ_v' is effective vertical stress) at 1.0 (liquefied state) at location B for a long time after the shaking phase. In this illustrative computation, the critical-void-ratio state (selected here at 2 percent volume increase or about 3.3 percent void ratio increase) was reached at B before the shaking ended. Hence the lack of additional dilation and low available shear strength ($r_u = 1.0$) resulted in continued post-shaking lateral deformation (flow failure). The detrimental void ratio increase at location B was facilitated by: (1) the liquefaction-induced settlement within the sandy-gravel-*k* stratum below the thin interlayer and (2) the inability of the overlying stratum to match this rate of settlement (owing to the low permeability of the interlayer).

In Case 3, the relatively low permeability of the base layer (compared to that of Case 2) resulted in a slower liquefaction-induced settlement rate. At the end of shaking, location B had not increased in void ratio sufficiently to reach the prescribed critical-void-ratio state. The remaining tendency for dilation resulted in the availability of the shear strength to halt further lateral deformation due to the 4-degree slope inclination. Ten seconds thereafter, the continued post-liquefaction sedimentation (settlement

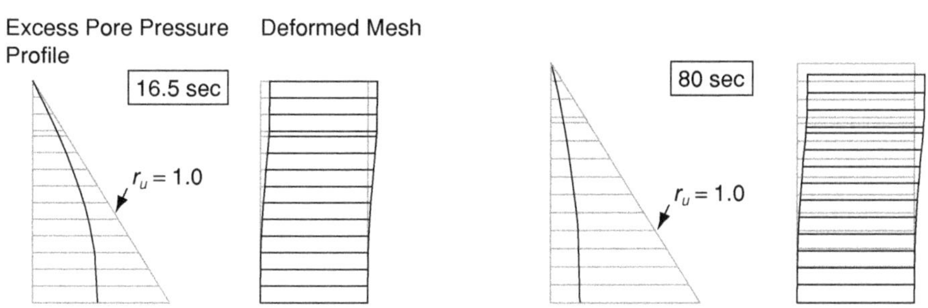

Figure 9.20 Excess pore-pressure profile and deformed mesh for case 1: uniform sand profile (deformations are not to scale and are exaggerated for clarity)

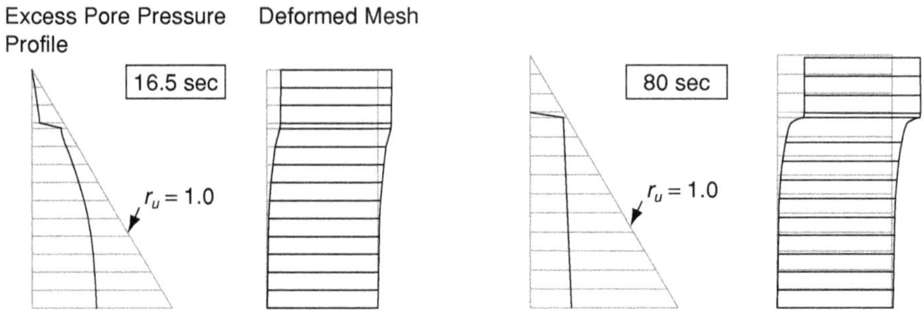

Figure 9.21 Excess pore-pressure profile and deformed mesh for case 3: clean sand profile with a silt *k* interlayer (deformations are not to scale and are exaggerated for clarity)

below the thin interlayer) eventually allowed location B to reach the critical-void-ratio state. Thereupon, the lack of dilation tendency and the liquefaction condition (r_u = 1.0) resulted in this delayed flow failure situation (Figure 9.19). Such delayed flow failure after the end of the dynamic/seismic excitation has been reported in the literature by a number of investigators (Seed 1987, Harder and Stewart 1996, Berrill et al. 1997, Bouckovalas et al. 1999, Kokusho 1999, Kokusho et al. 1999).

Finally, the influence of a low-permeability interlayer on the overall profile response may be visualized in Figures 9.20 and 9.21, in terms of u_e and deformation profiles along the soil column at 16.5 s (end of shaking) and 80.0 s. In Case 1 (Figure 9.20), lateral deformation and u_e profiles were smooth along the soil profile. With no further lateral movement observed during 16.5 s to 80 s, additional vertical settlement continued, as a result of the post-shaking dissipation and the associated re-consolidation processes. In contrast to the above, Figure 9.21 (Case 3) with the silt-*k* interlayer present shows:

(1) a very high pore-pressure gradient within the silt-*k* layer. Below this layer, the post-shaking re-consolidation process eventually results in a constant u_e distribution. This constant value is equal to the initial effective confinement (overburden pressure) imposed by the thin layer and the layers above. Dissipation of this u_e through the low-permeability interlayer may take a very long time in practical situations (if no sand boils develop).
(2) After the shaking phase, the void ratio continued to increase immediately beneath the silt-*k* layer (as discussed earlier), with large shear-strain concentration. Meanwhile, negligible additional shear strain was observed in the rest of the profile.

5.2 Site response of saturated dense sand profile

5.2.1 Centrifuge testing program

Data from one centrifuge experiment conducted at the University of California at Davis (Stevens et al. 2001) was employed to study the dynamic response of a saturated dense sand stratum. Nevada sand, at about 100% D_r, was used to represent a stiff soil formation. A pore fluid with a viscosity equal to about eight times that of water was employed, resulting in a prototype permeability coefficient within the range of medium to fine sands.

The centrifuge model was subjected to a series of earthquake-like shaking events imparted at centrifugal acceleration levels of 9.2 g, 18.1 g, 25.3 g and 37.3 g, representing a prototype stratum of 5.1 m, 10.0 m, 14.0 m and 20.6 m depth respectively. Near the model surface, recorded peak acceleration in the longitudinal direction ranged from 0.03 g to 1.73 g (in prototype scale). This wide range of peak acceleration resulted in a soil response covering linear to highly non-linear scenarios. The recorded downhole acceleration time histories at different depths along the model centerlines (where 1D response is dominant) were employed to back-calculate shear stress/strain response and to identify the stiffness, damping and dilatancy characteristics of the saturated stiff soil deposit. Numerical simulations were then performed for representative weak to strong shaking events.

5.2.2 Numerical modeling procedure and results

Figures 9.22 and 9.23 display respectively the G/G_{max} and the damping ratio data evaluated from selected shear stress–strain loops at two depths, one near the surface (1.4 m) and the other near the base (7.6 m). It may be seen from Figures 9.22 and 9.23 that dilation prevails in the saturated dense Nevada sand at shear strains higher than 0.2 percent. This dilation keeps the shear modulus from further reduction below about 20 percent of its initial value (Figure 9.22). The corresponding damping ratio does not exceed about 20–5 percent (Figure 9.23).

A 1D effective-stress shear-beam type FE model is employed. The initial low-strain shear modulus profile was defined based on the measured shear-wave velocity profile. Non-linearity and dilatancy parameters were selected based on the identified modulus reduction and damping characteristics (Figures 9.22 and 9.23), resulting in a friction angle of 42 degrees and a phase transformation angle of 22 degrees. The modulus reduction curves generated by the stress–strain model (under undrained conditions) are depicted in Figure 9.22, at the initial effective confinements corresponding to 1.4 m and 7.6 m depths respectively. These curves are in reasonable agreement with the experimental data, and show a modulus increase at large

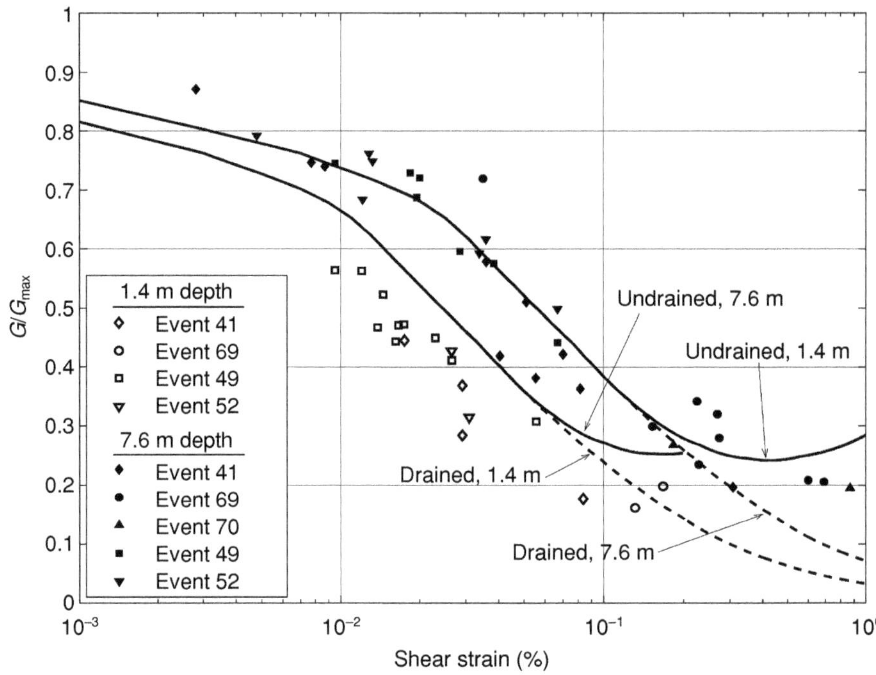

Figure 9.22 Model generated modulus reduction curves at different depths for saturated dense Nevada Sand ($D_r \approx 100\%$) and data points from centrifuge experiment

shear strain levels. This dilation effect is more pronounced at shallower depths. In addition to the hysteretic damping generated by the stress–strain model, viscous (Rayleigh) damping was employed (Figure 9.23) with an average of 3.5 percent over the frequency range of interest (1–10 Hz). Thus, the resulting overall system damping (hysteretic plus viscous) becomes comparable to the experimental data (Figure 9.23). As shown in Figure 9.23, damping levels off and tends to decrease at large shear strain levels, owing to the dilation tendency.

Representative simulation results of strong excitation (Event 41) are shown in Figure 9.24, in terms of acceleration time histories and the corresponding response spectra (5 percent damping) along the soil profile. The numerical model gives an overall satisfactory match to the experimental counterpart at all accelerometer locations, both in the time and in the frequency domains. Moreover, the dilation-induced spiky acceleration response observed in Event 41 is reproduced by the numerical model (Figure 9.24).

Figure 9.25 shows the computed shear stress–strain histories (at mid-depth) for representative weak, moderate and strong excitation, with maximum

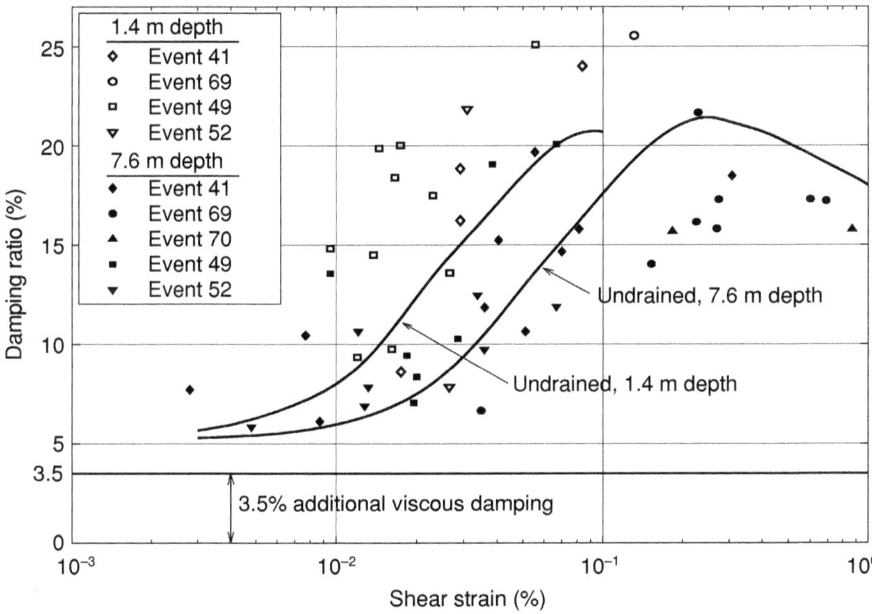

Figure 9.23 Overall model damping (including 3.5% viscous damping) at different depths for saturated dense Nevada Sand ($D_r \approx 100\%$) and data points from centrifuge experiments

shear strains of about 0.01 percent, 0.1 percent and 0.4 percent respectively. Similar to the experimental results, the computed response indicates: (1) essentially linear behavior in the low-amplitude event, (2) strong nonlinearity with minor dilatancy effects in the moderate event, and (3) significant presence of strain-stiffening effects due to dilation in the strong event.

5.3 Embankment foundation liquefaction countermeasures

5.3.1 Centrifuge testing program

Dynamic stability of a 4.5 m clayey sand embankment supported on 6 m of medium-saturated sand (Figure 9.26) was systematically tested in the centrifuge. The embankment was built at 1:1 slope, composed of a Kaolin clay and a Nevada 120 sand mixture. Nevada 120 fine sand (Arulmoli et al. 1992) was used as the liquefiable foundation soil (at $D_r \approx 40\%$). The foundation layer was saturated with a pore fluid at a prototype permeability coefficient of 5.5×10^{-4} m/s, within the range of medium sands (Lambe and Whitman 1969).

As seen in Figure 9.26, the first model constituted the benchmark case with no remedial work (Adalier 1996). In the second model, 6 m wide

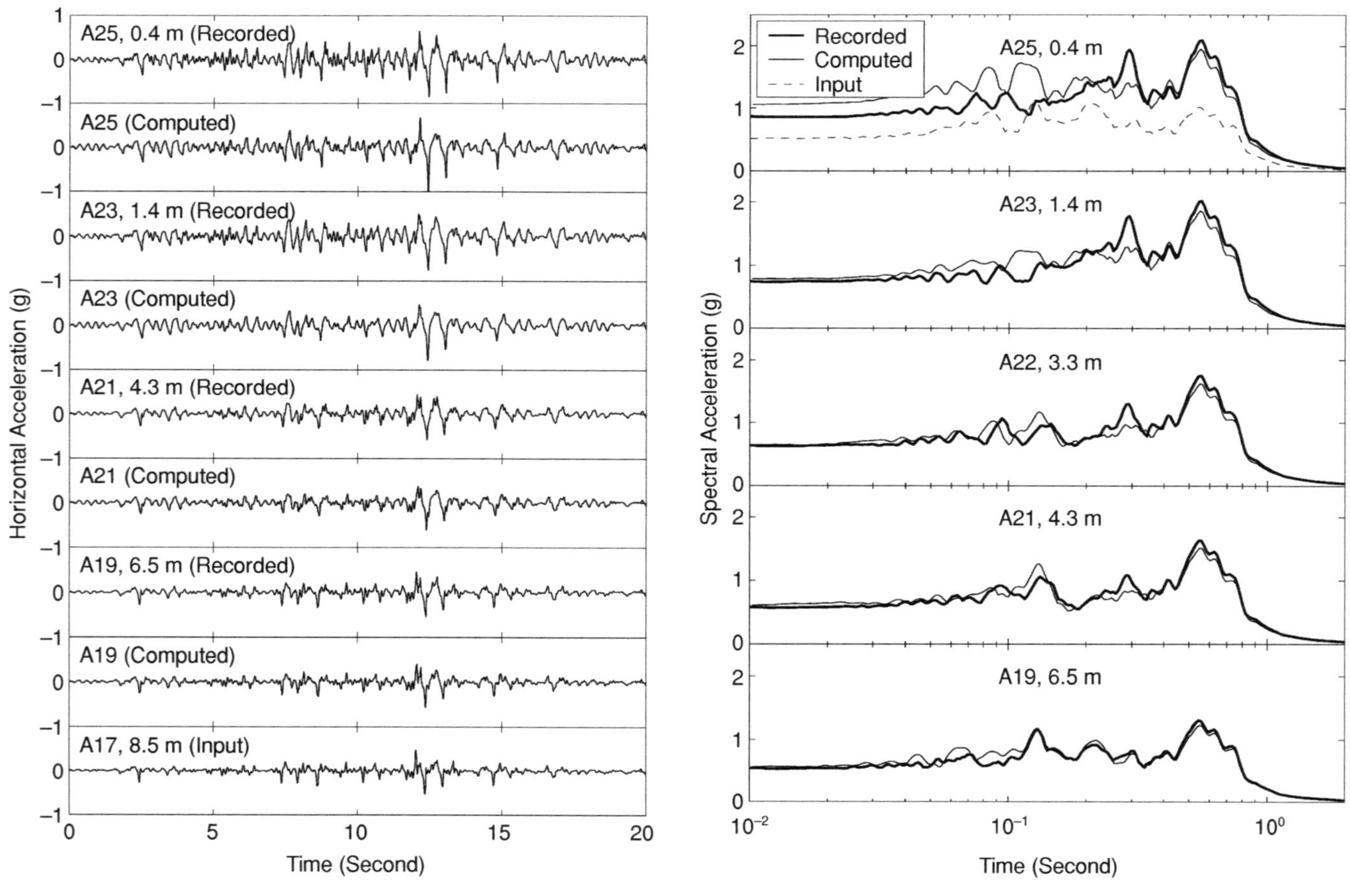

Figure 9.24 Event 41 recorded and computed acceleration histories and response spectra (5% damping)

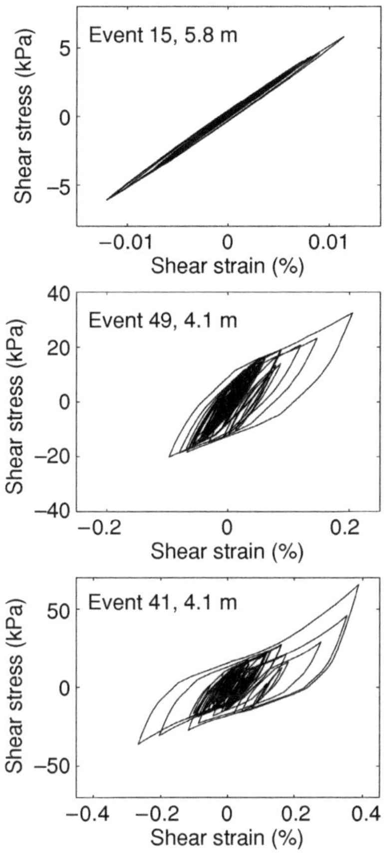

Figure 9.25 Computed shear stress–strain response at mid-depth of the sand stratum (Events 15, 49 and 41)

densified areas were placed under the embankment toes (reaching $D_r \approx 90\%$, according to the in-situ applications). Finally, in the third model, two steel sheet piles were tied together (with four steel tie-rods) and placed in the foundation layer below the embankment. One-dimensional (1D) horizontal shaking was imparted along the model's long axis. Each model was shaken at about 0.18 g peak excitation, with a uniform harmonic base input motion of 10 cycles at 1.6 Hz (Figure 9.27). The response was monitored (Figure 9.26) by accelerometers (ACC), pore pressure transducers (PPT) and Linear Variable Differential Transformers (LVDT) to measure displacement.

5.3.2 Finite element model and computation results

The FE mesh for the embankment-foundation system is shown in Figure 9.28. Boundary conditions for all simulations were:

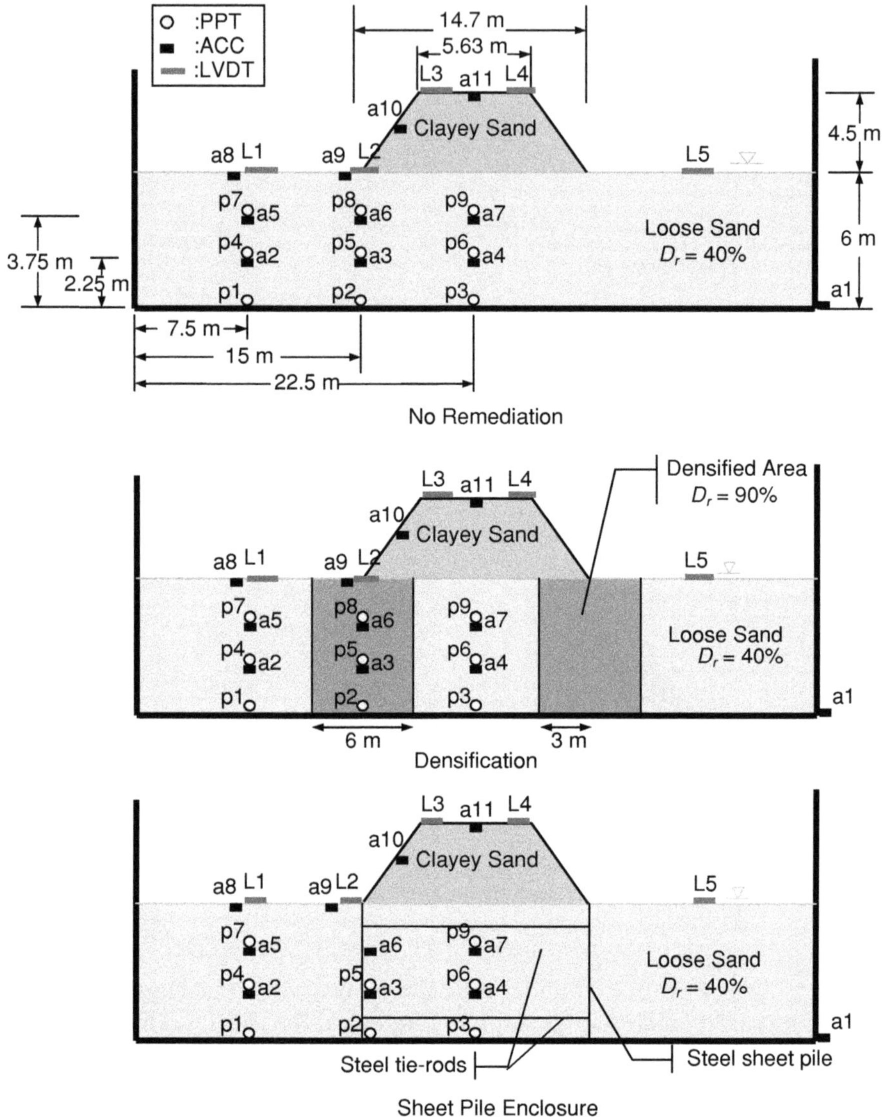

Figure 9.26 Centrifuge model setup (Adalier et al. 1998): (*a*) PPT is Pore-Pressure Transducer, (*b*) ACC is Accelerometer and (*c*) LVDT is Linear Variable Differential Transducer to measure displacement

(1) for the solid phase, horizontal input motion was specified along the base and the two lateral sides, as the recorded rigid container acceleration (Figure 9.27); all base nodes were fixed in the vertical direction; along the lateral sides, vertical motion was allowed;

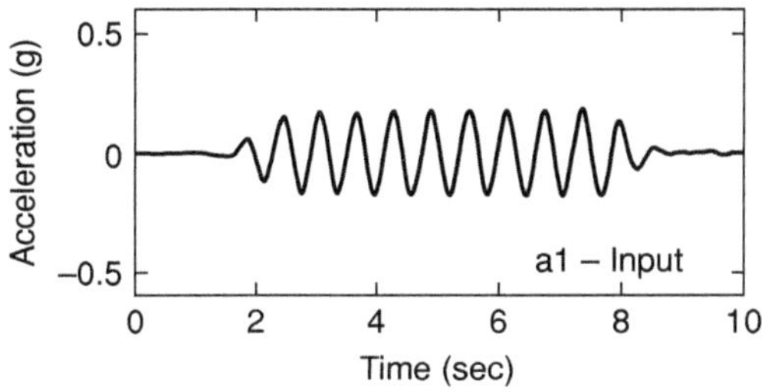

Figure 9.27 Recorded input motion (typical form)

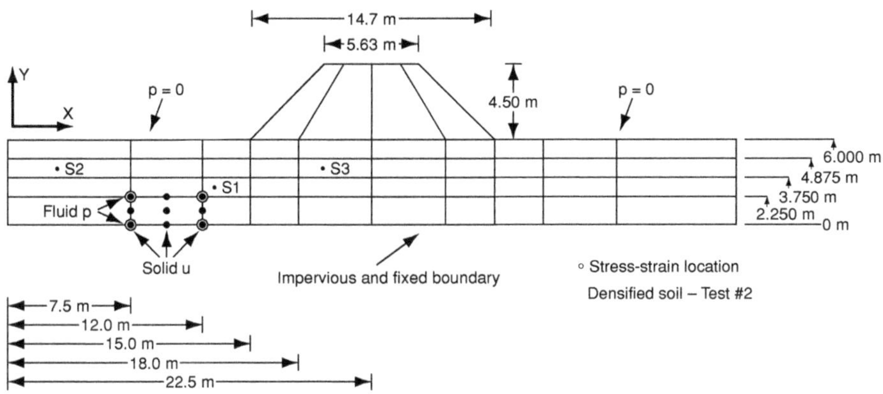

Figure 9.28 Finite element discretization and boundary conditions

(2) for the fluid phase, the base and the two sides (the container boundaries) were impervious (zero flow rate); in addition, zero pore pressure was prescribed along the foundation surface (at the water-table level) and within the entire embankment.

The employed soil stress–strain (constitutive) model was calibrated earlier (Parra, 1996) for the liquefiable embankment foundation soil (Nevada sand at a D_r of about 40 percent).

During the computational simulations, soil stress–strain responses were sampled at S1, S2 and S3 locations as shown in Figure 9.28. These locations were selected to represent zones of different response characteristics, namely: (i) S1 below the embankment toe, (ii) S3 below the embankment (near the centerline) where the vertical confinement is the highest and (iii) S2 in the free field away from the supported embankment. At these key locations, computed accelerations, pore pressures and vertical settlement will be compared to the corresponding measurements.

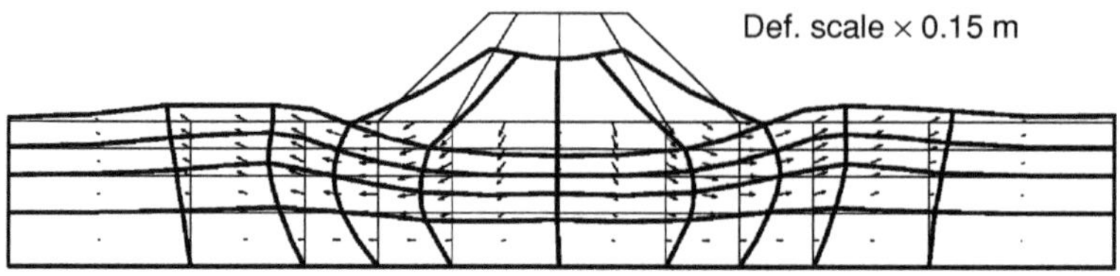

Figure 9.29 Computed deformed configuration (no remediation, deformations magnified by a factor of 6 for clarity)

5.3.2.1 EMBANKMENT ON MEDIUM NEVADA SAND (BENCHMARK TEST)

The imparted cycles of dynamic excitation (Figure 9.27) resulted in the deformed configuration of Figure 9.29, with peak vertical and lateral displacements of about 0.25 m. Primarily, the deformation pattern shows: (i) major lateral displacement and shear below the embankment toe in the foundation soil (which liquefied owing to the imparted dynamic excitation) and (ii) relatively mild lateral shear below the symmetric embankment centre. These two deformation mechanisms are discussed below in terms of soil stress/strain response.

Below embankment toe: The deformation in Figure 9.29 was associated with a large permanent lateral shear strain under the toe of 6 percent (Figure 9.30, location S1), which clearly shows the mechanisms of: (i) cycle-by-cycle shear strain accumulation and (ii) gradual loss of shear strength. The associated stress-path shows a major reduction in confinement during the first three cycles of shaking (due to excess pore pressure buildup),

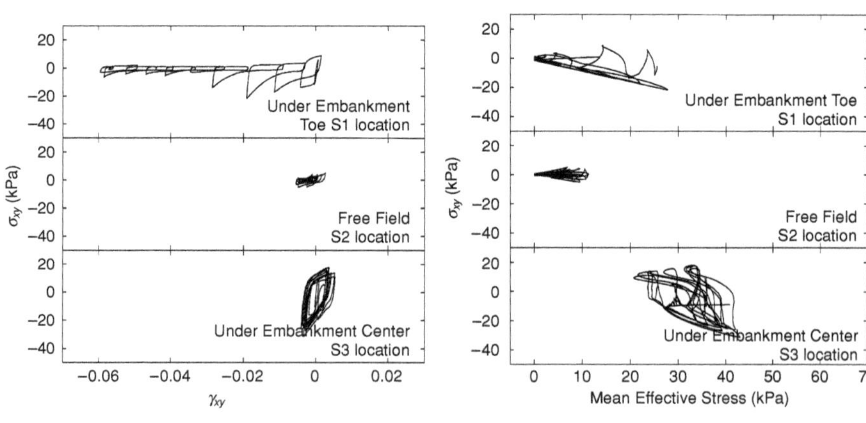

Figure 9.30 Computed shear stress–strain histories and effective stress path (no remediation)

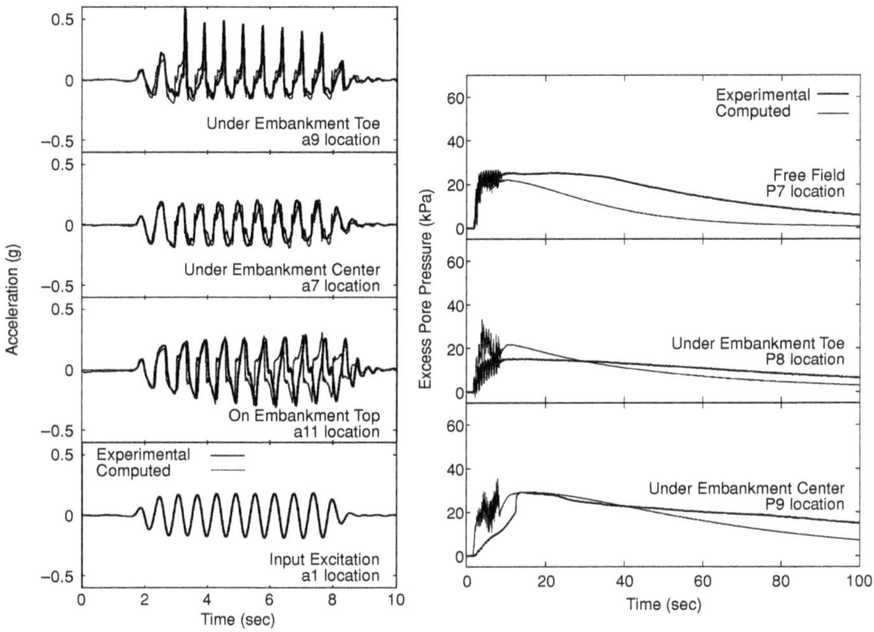

Figure 9.31 Computed vs experimental lateral acceleration and excess pore-pressure histories (no remediation)

followed by cycles of significant dilative response (stress-path along failure surface). The asymmetric phases of the dilative response (in the 'down-slope' direction) instantaneously increase the soil shearing resistance, resulting in a pattern of strong asymmetric acceleration spikes, as exhibited both computationally and experimentally (Figure 9.31, a9 location compared to a1 input). At this location, the computed and experimental pore-pressure histories were in reasonable agreement (Figure 9.31, P8 location).

Below embankment center: The computed lateral shear stress–strain response at this location shows (S3 in Figure 9.30): (i) minimal cycle-by-cycle permanent shear strain accumulation and (ii) no substantial loss in shear stiffness throughout the shaking phase. Higher effective confinement due to embankment weight appears to have sustained pore pressure well below the level of liquefaction (P9 in Figure 9.31, both computed and experimental). The corresponding stress path indicates that (S3 in Figure 9.30): (i) the dilation phase was apparent with minimal accumulated shear strain compared to S1 location under the embankment toe and (ii) the maximum loss in effective confinement was only one half of the initial value of about 40 kPa. The absence of any major 'down-slope' response at this location dictated an essentially symmetric acceleration response (a7 in Figure 9.31).

Finally, the computed shear stress and strain responses in the free field (S2 in Figures 9.30 and 9.31) show the typical cyclic loss of shear stiffness and strength owing to liquefaction. At this location, initial confinement was low, and the free-field response was characterized by low-amplitude (symmetric) cycles of shear strain. These cycles led to rapid pore-pressure buildup and liquefaction (P7 in Figure 9.31).

5.3.2.2 REMEDIATION BY DENSIFICATION

In this case, the sand columns below the embankment toe (shaded areas in Figures 9.26 and 9.28) were densified to a high D_r of about 90 percent. As no experimental data were available for Nevada sand at this high relative density, model parameters were defined based on the available data for Nevada sand at D_r of 40 percent and 60 percent (Arulmoli et al. 1992), as well as other empirical dense sand properties (Lambe and Whitman 1969). Thus, the material properties at D_r = 90% were representative of a stiff and much less liquefiable sand (soil friction angle of 38°).

In the presence of the densified zones, the computed deformed configuration (Figure 9.32) shows a pattern similar to the case of no remediation (Figure 9.29), but with smaller lateral displacements and shear under the embankment toe (only 0.15 m compared to 0.25 m without remediation). The associated stress/strain response characteristics below the embankment toe and embankment center are summarized below.

Below embankment toe: The smaller lateral deformation at this location was associated with significant cyclic dilation tendency in the dense sand, as indicated by (S1 in Figure 9.33): (i) phases of sharp increase in shear stiffness and strength, and (ii) lower cyclic shear strain accumulation in each shaking cycle. The corresponding stress path (S1 in Figure 9.33) also shows instantaneous regains in shear strength and effective confinement that are significantly more pronounced (compare S1 in Figures 9.30 and 9.33). These major dilative stress excursions within the densified material also resulted in strong asymmetric acceleration spikes manifested both experimentally and numerically (Figure 9.34, a9 compared to a1).

Below embankment center: The densified zones provided significant overall foundation strength and contained the loose sand stratum below the embankment center. At this location, contractive response is seen

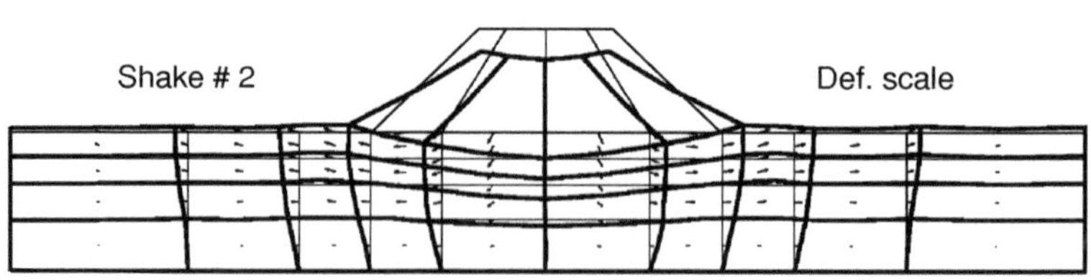

Figure 9.32 Computed deformed configuration (remediation by densification, deformations magnified by a factor of 6 for clarity)

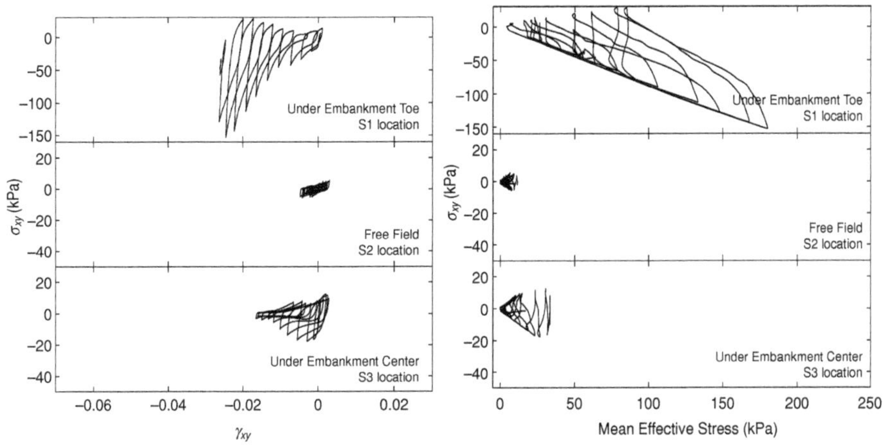

Figure 9.33 Computed shear stress–strain histories and effective stress path (remediation by densification)

to dominate (S3 in Figure 9.33), resulting in the typical pattern of cyclic loss of shear stiffness and effective confinement. The essentially symmetric computed and recorded acceleration histories are seen to be in reasonable agreement (a7 in Figure 9.34).

Finally, the free field is seen to display the typical loss of stiffness and strength due to liquefaction (S2 in Figure 9.33). At this location the response is essentially identical to the earlier no-remediation case (S2 in Figure 9.30).

5.3.2.3 REMEDIATION BY SHEET-PILE ENCLOSURE

Kimura et al. (1997) found containment sheet-piles with a drainage capability to be a most effective countermeasure. A similar conclusion was also reached by Adalier (1996) based on experimental observations. Essentially, this retrofit procedure resulted in perfect containment of the foundation soil below the embankment (Figure 9.35).

As a result of this containment remediation, soil stress–strain response remained in the relatively small strain range, with predominantly contractive behavior at all sampled locations (Figure 9.36). The reduction in shear stiffness and strength (Figure 9.36) is manifested in the computed and recorded accelerations, which show a decay in amplitude everywhere (Figure 9.37).

The computational and experimental results presented above systematically reveal a number of liquefaction-related soil response characteristics. Based on these results, the two main implications are:

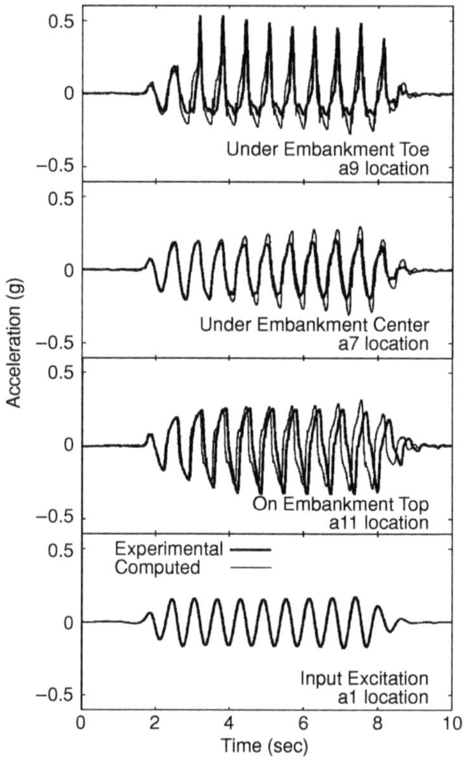

Figure 9.34 Computed and experimental lateral acceleration histories (remediation by densification)

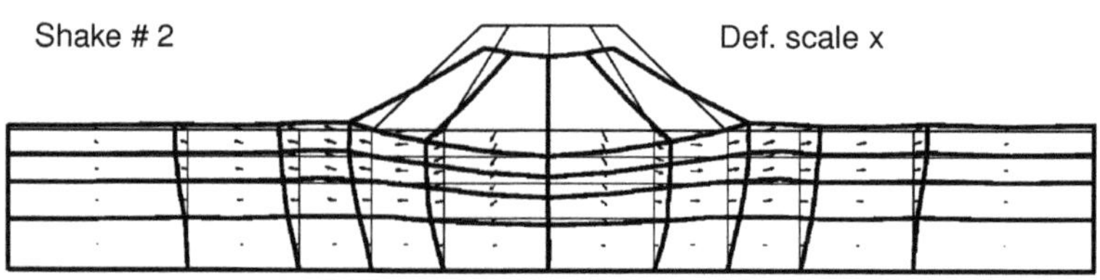

Figure 9.35 Computed deformed configuration (remediation by sheet-pile enclosure, deformations magnified by a factor of 6 for clarity)

(1) post-liquefaction dilation behavior may play a major role in dictating the soil dynamic response (shear stress-strain in general, and the extent of the permanent deformation in particular), and
(2) observed acceleration response patterns may be good indicators of the different underlying liquefaction response mechanisms.

For instance, the observed strong asymmetric acceleration spikes (in the first two models) may be associated with significant post-liquefaction dilative

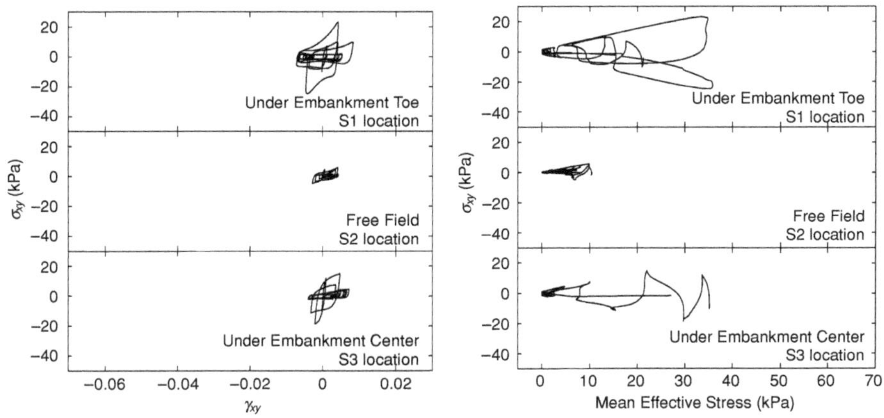

Figure 9.36 Computed shear stress–strain histories and effective stress path (remediation by sheet-pile enclosure)

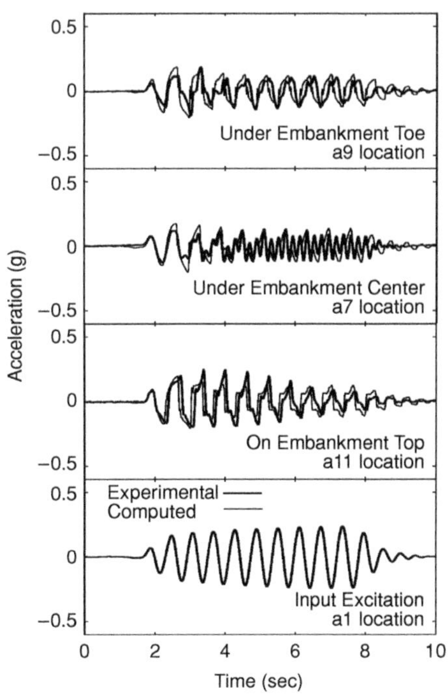

Figure 9.37 Computed and experimental lateral acceleration histories (remediation by sheet-pile enclosure)

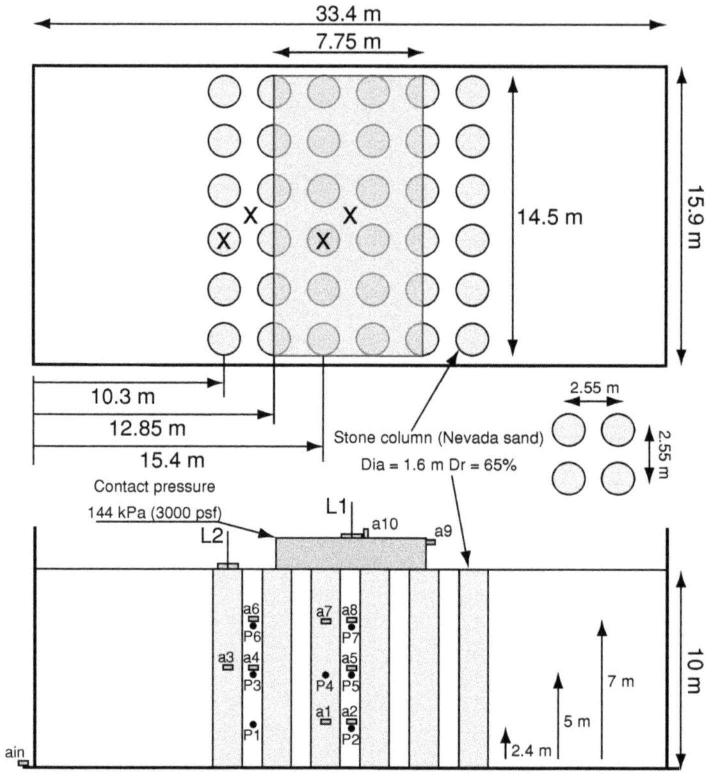

Figure 9.38 Cross-sectional and plan view of the model configuration containing 36 stone columns with an area replacement ratio of 30%

response in the 'down-slope' direction. Further, the observed steady decay in acceleration amplitudes (in the last model) indicates a gradual loss of soil stiffness and strength, with relatively small shear deformation.

5.4 Stone-column reinforced silty soil

5.4.1 Centrifuge testing program

In an attempt to verify and quantify the possible liquefaction mitigation mechanisms due to deployment of stone columns, a centrifuge testing program was conducted (Adalier et al. 2003). The centrifuge model (Figure 9.38) simulated the response of a 10 m thick saturated silt stratum with the inclusion of thirty-six 1.6 m diameter stone columns at predetermined positions (2.55 m center-to-center). This configuration provided an area replacement ratio of 30 percent within the instrumented zone below the footing. A footing surcharge was applied using a rigid steel rectangular block, with an aver-

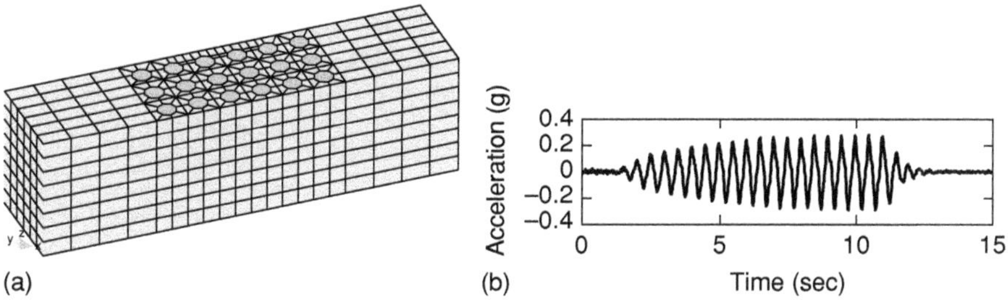

Figure 9.39 (*a*) Finite element mesh and (*b*) base input motion

age vertical contact pressure of 144 kPa. This surcharge was approximately equivalent to the vertical pressure transmitted by a 10–15-story reinforced-concrete building. The ground water table was at the soil surface.

The material representing the stone columns was Nevada No. 120 sand. Prototype permeability of the silt and the sand were 5.5×10^{-6} m/s and 3.3×10^{-3} m/s respectively. The internal friction angle is about 37 degrees for the sand and 25 degrees for the silt. The model was shaken with a uniform harmonic base input motion with a predominant 1Hz prototype frequency and a PGA of about 0.08 g in prototype scale (Figure 9.39b).

5.4.2 Numerical modeling procedure and results

As shown in Figure 9.39a, the centrifuge test model was discretized using 3D 8-node brick elements. A half mesh configuration was used owing to the geometrical symmetry (Figure 9.39a). The boundary conditions were: (1) dynamic excitation defined as the recorded base acceleration (Figure 9.39b); (2) the soil surface was traction free, with zero prescribed pore pressure except that the soil surface immediately below the footing was impervious; and (3) the base and lateral boundaries were impervious.

A static application of gravity (model own weight plus footing surcharge) was performed before imparting the seismic excitation. The resulting fluid hydrostatic pressures and soil stress states served as the initial conditions for the subsequent dynamic analysis.

Figures 9.40a and 9.40b display the computed and recorded accelerations and foundation vertical settlements. The computed foundation settlement was 0.065 m, about 40 percent less than that of a similar model with no stone-column remediation (Lu et al. 2005).

6 Summary and conclusions

A finite element numerical framework was presented for conducting non-linear seismic response analyses of soil systems. Within this framework, a constitutive model was developed for simulation of cyclic mobility response

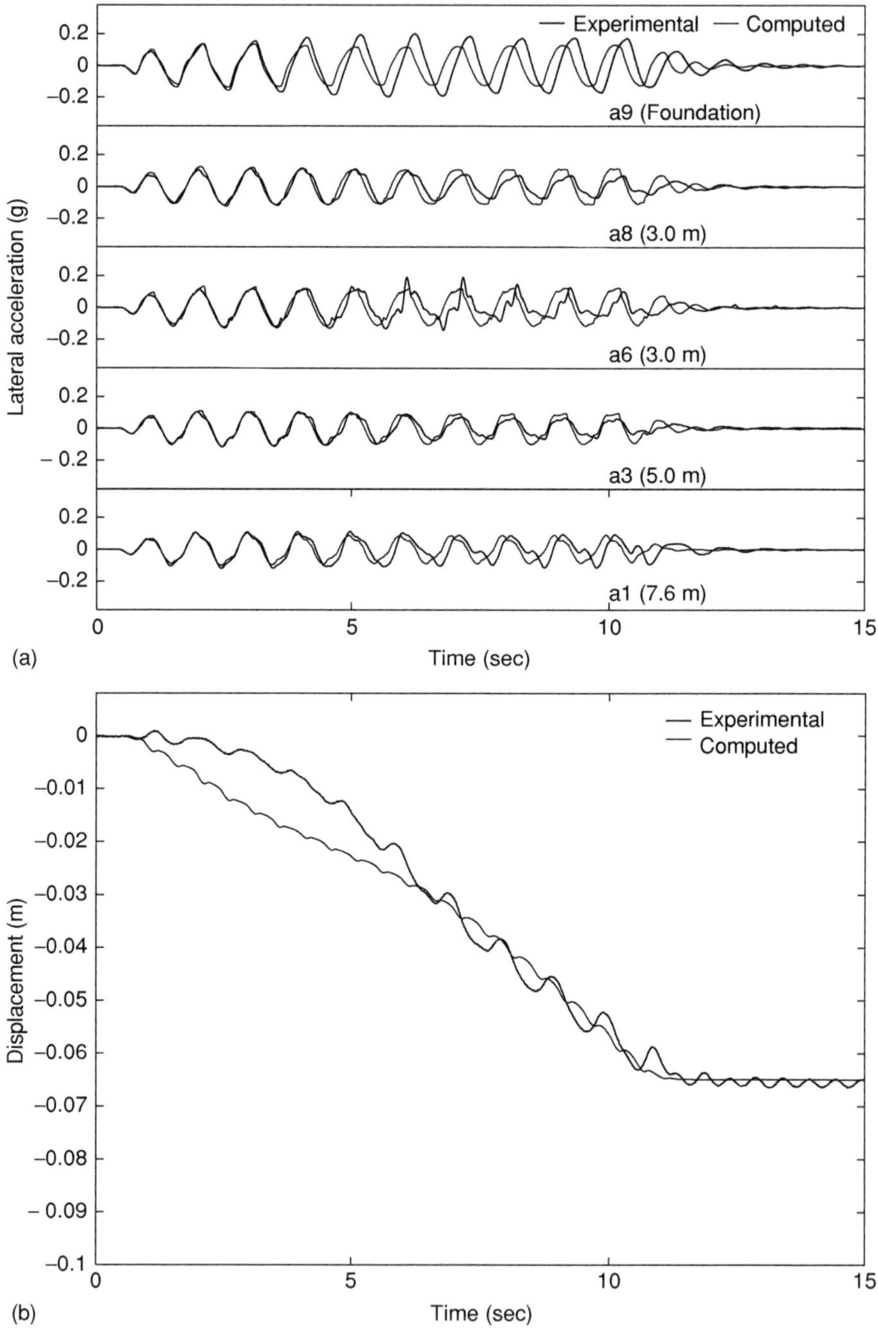

Figure 9.40 (*a*) Recorded and computed acceleration time histories; (*b*) recorded and computed foundation settlement time histories

and associated accumulation of cyclic shear deformations as observed in saturated clean sands and silts. The model, incorporated into a solid–fluid fully coupled FE formulation, allows for numerical studies of seismic response, and for assessment of associated permanent deformations.

Using the above numerical framework, a number of salient features associated with the accumulation of shear deformations in saturated cohesionless soil formations were presented. The influence of soil permeability and its spatial variation was highlighted as a primary factor in dictating the extent of such deformations.

Response of saturated dense sands was also discussed. The shear-induced high tendency for dilation was shown to affect significantly the associated dynamic response, potentially resulting in corresponding instants of large acceleration spikes.

Mitigation studies of the detrimental ground deformation effects were also addressed by the presented numerical framework. A number of scenarios for earth embankments and for shallow foundations were discussed. In this regard, numerical techniques provide an effective tool for assessment of different mitigation strategies and their impact on overall system response. Finally, more representative three-dimensional studies are becoming increasingly feasible in the current environment of prevailing PC-type desktop and laptop systems.

References

Adalier, K. (1996) Mitigation of earthquake induced liquefaction hazards, PhD thesis, Department of Civil Engineering, Rensselaer Polytechnic Institute, Troy, NY.

Adalier, K., Elgamal, A. W., Martin, G. R. (1998) Foundation liquefaction countermeasures for earth embankments, *Journal of Geotechnical and Geoenvironmental Engineering*, 124 (6): 500–17.

Adalier, K., Elgamal, A., Meneses, J. and Baez, J. I. (2003) Stone column as liquefaction countermeasure in non-plastic silty soils, *Soil Dynamics and Earthquake Engineering*, 23 (7): 571–84.

Ansal, A., Bardet, J. P., Barka, A., Baturay, M. B., Berilgen, M., Bray, J., Cetin, O., Cluff, L., Durgunoglu, T., Erten, D., Erdik, M., Idriss, I. M., Karadayilar, T., Kaya, A., Lettis, W., Olgun, G., Paige, W., Rathje, E., Roblee, C., Stewart, J. and Ural, D. (1999) Initial geotechnical observations of the November 12, 1999, Düzce earthquake, A Report of the Turkey–US Geotechnical Earthquake Engineering Reconnaissance Team, November 26.

Arulmoli, K., Muraleetharan, K. K., Hossain, M. M. and Fruth, L. S. (1992) VELACS: Verification of Liquefaction Analyses by Centrifuge Studies, Laboratory Testing Program, Soil Data Report, The Earth Technology Corporation, Project No. 90-0562, Irvine, Calif.

Bardet, J. P., Oka, F., Sugito, M. and Yashima, A. (1995) The Great Hanshin Earthquake Disaster, Preliminary Investigation Report, University of Southern California, Los Angeles, Calif.

Bazant, Z. P. and Kim, S.-S. (1979) Plastic-fracturing theory for concrete, *J. Engineering Mechanics*, ASCE, 105 (EM3): 407–28.

Berrill, J. B., Christensen, R. J., Keenan, R. J., Okada, W. and Pettinga, J. R. (1997) Lateral spreading loads on a piled bridge foundation, *Proc., Intl. Conf. on Soil Mechanics and Geotechnical Engineering*, Seismic Behaviour of Ground and Geotechnical Structures: Special Technical Session on Earthquake Geotechnical Engineering, Rotterdam: Balkema, pp. 173–83.

Biot, M. A. (1962) The mechanics of deformation and acoustic propagation in porous media, *J. Applied Physics*, 33 (4): 1482–98.

Bouckovalas, G. D., Gazetas, G. and Papadimitriou, A. G. (1999) Geotechnical aspects of the 1995 Aegion, Greece, earthquake, *Proc. 2nd Intl. Conf. on Earthquake Geotechnical Engineering*, Rotterdam: Balkema, 2: 739–48.

Casagrande, A. (1975) Liquefaction and cyclic deformation of sands – a critical review, Proceedings of the 5th Pan-American Conference on Soil Mechanics and Foundation Engineering, Buenos Aires, Argentina; also published as Harvard Soil Mechanics Series No. 88, January 1976, Cambridge, Mass.

Chan, A. H. C. (1988) A unified finite element solution to static and dynamic problems in geomechanics, PhD dissertation, University College of Swansea.

Davis, C. A. and Bardet, J. P. (1996) Performance of two reservoirs during the 1994 Northridge earthquake, *J. Geotechnical Engineering*, ASCE, 122 (8): 613–22.

Chen, W. F. and Mizuno, E. (1990) *Nonlinear Analysis in Soil Mechanics, Theory and Implementation*, New York: Elsevier.

Dobry, R., Taboada, V. and Liu, L. (1995) Centrifuge modeling of liquefaction effects during earthquakes, Keynote Lecture, *Proc. 1st Intl. Conf. on Earthquake Geotechnical Engineering, Tokyo, November 14–16*, ed. K. Ishihara, vol. 3, Rotterdam: Balkema, pp. 1291–1324.

Duncan, J. M. and Chang, C. Y. (1970) Nonlinear analysis of stress and strain in soils, *J. Soil Mechanics and Foundations Division*, ASCE, 96 (SM5): 1629–53.

Elgamal, A.-W., Dobry, R., Parra, E. and Yang, Z. (1998) Soil dilation and shear deformations during liquefaction, *Proceedings 4th Intl. Conf. on Case Histories in Geotechnical Engineering*, ed. S. Prakash, pp. 1238–59.

Elgamal, A., Parra, E., Yang, Z. and Adalier, K. (2002a) Numerical analysis of embankment foundation liquefaction countermeasures, *J. Earthquake Engineering*, 6 (4): 447–71.

Elgamal, A., Yang, Z. and Parra, E. (2002b) Computational modeling of cyclic mobility and post-liquefaction site response, *Soil Dynamics and Earthquake Engineering*, 22 (4): 259–71.

Elgamal, A., Yang, Z., Lai, T., Kutter, B. L. and Wilson, D. (2005) Dynamic response of saturated dense sand in laminated centrifuge container, *J. Geotechnical and Geoenvironmental Engineering*, ASCE, 131 (5): 598–609.

Elgamal, A., Yang, Z., Parra, E. and Ragheb, A. (2003) Modeling of cyclic mobility in saturated cohesionless soils, *Int. J. Plasticity*, 19 (6): 883–905.

Harder, L. F. and Stewart, J. P. (1996) Failure of Tapo Canyon Tailings Dam, *J. Performance of Constructed Facilities*, ASCE, 10 (3): 109–14.

Hill, R. (1950) *The Mathematical Theory of Plasticity*, London: Oxford University Press.

Holzer, T. L., Youd, T. L. and Hanks, T. C. (1989) Dynamics of liquefaction during the 1987 Superstition Hills, California, earthquake, *Science*, 244: 56–9.

Ibsen, L. B. (1994) The stable state in cyclic triaxial testing on sand, *Soil Dynamics and Earthquake Engineering*, 13: 63–72.

Ishihara, K., Tatsuoka, F. and Yasuda, S. (1975) Undrained deformation and liquefaction of sand under cyclic stresses, *Soils and Foundations*, 15 (1): 29–44.

Iwan, W. D. (1967) On a class of models for the yielding behaviour of continuous and composite systems, *J. Applied Mechanics*, ASME, 34: 612–17.

Japanese Geotechnical Society (1996) Special issue on geotechnical aspects of the January 17, 1995 Hyogoken-Nanbu earthquake, *Soils and Foundations Journal*, Tokyo.

Japanese Geotechnical Society (1998) Special issue on geotechnical aspects of the January 17, 1995 Hyogoken-Nanbu earthquake, No. 2, *Soils and Foundations Journal*, Tokyo.

Jinchi Lu, Zhaohui Yang, Korhan Adlier and Ahmed Elgamal (2004) Numerical analysis of stone column reinforced silty soil, *Proceedings of the 15th Southeast Asian Geotechnical Conference, Bangkok, Thailand, November 23–26*, vol. 1.

Kimura, T., Takemura, J., Hiro-oka, A., Okamura, M. and Matsuda, T. (1997) Countermeasures against liquefaction of sand deposits with structures, *Proceedings, 1st Intl. Conf. on Earthquake Geotechnical Engineering*, ed. K. Ishihara, Rotterdam: Balkema, 3: 1203–24.

Kokusho, T. (1999) Water film in liquefied sand and its effect on lateral spread, *J. Geotechnical and Geoenvironmental Engineering*, 125 (10): 817–26.

Kokusho, T., Kojima, T. and Nonaka, N. (1999) Emergence of water film in liquefied sand and its role in lateral flow (CD-ROM), *Proc. 12th World Conf. on Earthquake Engineering*, Auckland, New Zealand, No. 0946.

Kondner, R. L. (1963) Hyperbolic stress–strain response: cohesive soils, *J. Soil Mechanics and Foundations Division*, ASCE, 89 (SM1): 115–43.

Kramer, S. L. (1996) *Geotechnical Earthquake Engineering*, Englewood Cliffs, NJ: Prentice Hall.

Lacy, S. (1986) Numerical procedures for nonlinear transient analysis of two-phase soil system, PhD dissertation, Princeton University.

Ladd, C. C., Foott, R., Ishihara, K., Schlosser, F. and Poulos, H. G. (1977) Stress-deformation and strength characteristics, *State-of-the-art Report, Proc. 9th Intl. Conf. on Soil Mechanics and Foundation Engineering*, 2: 421–94.

Lade, P. V. and Duncan, J. M. (1975) Elastoplastic stress–strain theory for cohesionless soil, *J. Geotechnical Engineering Division*, ASCE, 101 (GT10): 1037–53.

Lambe, T. W. and Whitman, R. V. (1969) *Soil Mechanics*, New York: John Wiley.

Lu, J., Yang, Z., Adalier, K., and Elgamal, A. (2004) Numerical analysis of stone column reinforced silty soil, *Proceedings of the 15th Southeast Asian Geotechnical Conference*, Vol. 1, Bangkok, Thailand, November 23–26.

Mroz, Z. (1967) On the description of anisotropic work hardening, *J. Mechanics and Physics of Solids*, 15: 163–75.

Nemat-Nasser, S. and Tobita, Y. (1982) Influence of fabric on liquefaction and densification potential of cohesionless sand, *Mechanics of Materials*, 1: 43–62.

Papadimitriou, A. G., Bouckovalas, G. D. and Dafalias, Y. F. (2001) Plasticity model for sand under small and large cyclic strains, *J. Geotechnical and Geoenvironmental Engineering*, ASCE, 127 (11): 973–83.

Parra, E. (1996) Numerical modeling of liquefaction and lateral ground deformation including cyclic mobility and dilation response in soil system, PhD thesis, Department of Civil Engineering, Rensselaer Polytechnic Institute, Troy, NY.

Prevost, J. H. (1985) A simple plasticity theory for frictional cohesionless soils, *Soil Dynamics and Earthquake Engineering*, 4 (1): 9–17.

Ragheb, A. (1994) Numerical analysis of seismically induced deformations in saturated granular soil strata, PhD thesis, Rensselaer Polytechnic Institute, Troy, NY.

Seed, H. B. (1979) Soil liquefaction and cyclic mobility evaluation for level ground during earthquakes, *J. Geotechnical Engineering Division*, ASCE, 105 (GT2): 201–55.

Seed, H. B. (1987) Design problems in soil liquefaction, *J. Geotechnical Engineering*, 113 (8): 827–45.

Seed, H. B., Lee, K. L., Idriss, I. M. and Makdisi, F. I. (1975) The slides on the San Fernando dams during the earthquake of February 9, 1971, *J. Geotechnical Engineering Division*, ASCE, 101(7): 651–88.

Seed, H. B., Seed, R. B., Harder, L. F. and Jong, H. L. (1989) Re-evaluation of the slide in the Lower San Fernando Dam in the 1971 San Fernando earthquake, Report No. UCB/EERC-88/04, University of California, Berkeley, Calif.

Seed, R. B., Dickenson, S. E., Riemer, M. F., Bray, J. D., Sitar, N., Mitchell, J. K., Idriss, I. M., Kayen, R. E., Kropp, A., Harder, L. F. Jr and Power, M. S. (1990) Preliminary report on the principal geotechnical aspects of the October 17, 1989, Loma Prieta earthquake, Report No. UCB/EERC-90/05, Earthquake Engineering Research Center, Berkeley, Calif.

Sitar, N. (ed.) (1995) Geotechnical reconnaissance of the effects of the January 17, 1995, Hyogoken-Nanbu earthquake Japan, Report No. UCB/EERC-95/01, Earthquake Engineering Research Center, Berkeley, Calif.

Stevens, D. K., Wilson, D. W. and Kutter, B. L. (2001) Comprehensive investigation of nonlinear site response – centrifuge data report for the DKS04 Model Test, *Report UCD/CGMDR-01/03*, Department of Civil and Environmental Engineering, University of California at Davis, Calif.

Sture, S., Costes, N. C., Batiste, S. N., Lankton, M. R., AlShibli, K. A., Jeremic, B., Swanson, R. A. and Frank, M. (1998) Mechanics of granular materials at very low effective stresses, *J. Aerospace Engineering*, ASCE, 11 (3): 67–72.

Taboada, V. M. (1995) Centrifuge modeling of earthquake-induced lateral spreading in sand using a laminar box, PhD thesis, Rensselaer Polytechnic Institute, Troy, NY.

Yang, Z. (2000) Numerical modeling of earthquake site response including dilation and liquefaction, PhD dissertation, Department of Civil Engineering and Engineering Mechanics, Columbia University, New York.

Yang, Z. and Elgamal, A. (2002) Influence of permeability on liquefaction-induced shear deformation, *J. Engineering Mechanics*, ASCE, 128 (7): 720–9.

Yang, Z., Elgamal, A. and Parra, E. (2003) A computational model for cyclic mobility and associated shear deformation, *J. Geotechnical and Geoenvironmental Engineering*, ASCE, 129 (12): 1119–27.

Youd, T. L. (1977) Packing changes and liquefaction susceptibility, *Journal of the Geotechnical Engineering Division*, ASCE, 103 (GT8, August): 918–23.

Youd, T. L., Hansen, C. and Bartlett, S. (1999) Revised MLR equations for predicting lateral spread displacement, ed. T. D. O'Rourke, J.-P. Bardet and M. Hamada, Proceedings, 7th U.S.–Japan Workshop on Earthquake Resistant Design of Lifeline Facilities and Countermeasures Against Liquefaction, Technical Report MCEER-99-0019, pp. 99–114.

Zeghal, M. and Elgamal, A.-W. (1994) Analysis of site liquefaction using earthquake records, *Journal of Geotechnical Engineering*, ASCE, 120 (6): 996–1017.

Zienkiewicz, O. C., Chan, A. H. C., Pastor, M., Paul, D. K. and Shiomi, T. (1990) Static and dynamic behaviour of soils: a rational approach to quantitative solutions: I. Fully saturated problems, *Proc. Royal Society London*, A 429, pp. 285–309.

10 Bearing capacity of shallow foundations under static and seismic conditions

Deepankar Choudhury

1 Introduction

The design of shallow foundations subject to different static loadings has been an important research topic for geotechnical engineers for many decades. The devastating effects of earthquakes and other forms of damage to shallow foundations has increased the complexity of the problem. Consequently, it is necessary to obtain closed-form solutions for the earthquake-resistant design of shallow foundations, and these solutions have become an important research area. Many analytical and numerical solutions are available, and cover areas such as the limit equilibrium method, limit analysis, method of characterization, finite element analysis and other areas for the computation of the seismic bearing capacity factors required for the design of shallow foundations. In the following sections, shallow footings in various ground conditions are analyzed under both static and seismic forces.

2 Seismic bearing capacity of footings embedded in horizontal ground

A literature review reveals that the primary emphasis of researchers has been on using analytical or numerical techniques for evaluating the seismic bearing capacity of shallow footings embedded in horizontal ground.

Prandtl (1921), Terzaghi (1943), Caquot and Kerisel (1953), Meyerhof (1951), Hansen (1970), Vesic (1973), Chen (1975), Basudhar et al. (1981), Griffiths (1982), Zhu et al. (2001) and others have extensively studied the bearing capacity of shallow footings subject to static loading. Basavanna et al. (1974) obtained the dynamic bearing capacity of soils under transient loading using dynamic models. However, a limited amount of research was carried out into the seismic bearing capacity of footings. Pioneering analysis work by Meyerhof (1953, 1963) determined seismic bearing capacity factors by considering the application of pseudo-static seismic forces on the structure but only as inclined pseudo-static loads on the footing. Moore and Darragh (1956) identified the different aspects of the design of footings under earthquake conditions. Shinohara et al. (1960) also carried out pseudo-static analysis for footings on sandy soils. Selig and McKee (1961) explained the behavior of small footings under static and dynamic loading, while Shenkman and McKee (1961) reported bearing capacities of dynamically loaded footings. Sridharan (1962) reported the experimental results of the settlements of a model footing under dynamic loading, while Vesic et al. (1965) gave experimental results for the dynamic bearing capacity of footings on sand. Taylor (1968) explained the design methodology for spread footings under earthquake loadings. Prakash (1974, 1981) gave a detailed guide to the design of foundations under earthquake loading considering pseudo-static forces. The effect of seismic forces on the inertia of the supporting soil was, however, not considered. Later, correcting this omission, a small number of researchers carried out analyses of the seismic bearing capacity of shallow footings.

Sarma and Iossifelis (1990) determined the seismic bearing capacity factors for shallow strip footings by using the limit equilibrium method of slope stability analysis with inclined slices. They considered the pseudo-static effect of the inertia of the soil mass underneath a footing in terms of the horizontal seismic acceleration.

Using a Coulomb-type mechanism, i.e. with planar rupture surfaces including the pseudo-static inertia forces in the soil and on the footing, seismic bearing capacity factors were determined by Richards et al. (1993). They examined field and laboratory observations of the seismic settlements of shallow foundations on granular soils that were not attributable to changes in density in terms of seismic degradation of bearing capacity. A sliding block procedure was used in the analysis with the assumption of different mobilized wall friction angles, δ. This procedure was used for computing settlements due to the loss of bearing capacity under seismic conditions. The approach led to a design procedure for footings based on limiting the seismic settlements to a prescribed value.

Budhu and Al-Karni (1993) derived the seismic bearing capacity factors for general c-ϕ soils by considering both the horizontal and vertical seismic accelerations. Mohr–Coulomb failure theory was considered to determine the inclinations of the failure planes. The limit equilibrium solutions were described for evaluating the seismic bearing capacity factors from the similar static approach. The focus of the logspiral portion of the composite failure surface was assumed at the edge of the footing.

Seismic bearing capacity factors of a strip surface footing resting on a cohesionless soil were determined by Dormieux and Pecker (1995). The upper bound theorem of yield line design theory was used to obtain the ultimate load. Normal and tangential forces applied to the foundation and inertia forces developed within the soil volume were considered. The classical Prandtl-like mechanism was used to show that the reduction in the bearing capacity was caused mainly by load inclination.

Based on the kinematical approach of yield design theory, the seismic bearing capacity of shallow strip foundations on dry soils was analyzed by Paolucci and Pecker (1997). The seismic effects on the bearing capacity of shallow foundations on soils obeying Mohr–Coulomb failure criteria were investigated both theoretically and experimentally. The effects of load inclination, eccentricity of seismic forces and soil inertia were considered in the analysis. It had been shown that the effect of load inclination was much greater than the seismic effect on soil inertia to reduce the bearing capacity factors. Load eccentricity also played an important role in the pseudo-static approach and gave very high values of safety factors in the design of shallow foundations. Seismic bearing capacity factors of shallow strip footings were calculated by considering pseudo-static seismic forces.

Soubra (1997) used upper-bound limit analysis by considering two non-symmetrical failure mechanisms called M1 and M2. M1 consisted of a log sandwich composed of a triangular active wedge, a logspiral radial shear zone and a triangular passive wedge. M2 consisted of an arc sandwich composed of a triangular active wedge, a circular radial shear zone and a triangular passive wedge. The results were obtained using rigorous upper-bound limit analysis theory with an associated flow rule for a Coulomb material. The seismic bearing capacity factors were presented in the form of design charts.

Later, Soubra (1999) carried out the analysis for shallow strip footings using two kinematically admissible failure mechanisms M1 and M2. Here M1 was symmetrical and comprised a triangular active wedge under the footing and two radial shear zones composed of a sequence of rigid triangles. M2 was non-symmetrical and consisted of a single radial shear zone. Pseudo-static representation of earthquake forces with the use of upper-bound limit analysis theory was adopted.

Kumar and Rao (2002) carried out pseudo-static analyses for spread footings by employing the method of stress characteristics. Seismic bearing capacity factors were plotted for various horizontal earthquake acceleration coefficients and soil friction angles. Both the single-sided and double-sided failure mechanisms were considered in the analysis. The closed form solution for cohesion and surcharge components N_c and N_q respectively were found and compared with those obtained by other researchers. The N_γ values found were significantly smaller than those obtained from other theories. The effects of vertical seismic accelerations were not considered. The nature of the pressure distribution below the footing and the failure patterns were also shown together with the variation of seismic earthquake accelerations.

Kumar (2004) studied the effect of the footing–soil interface friction on the bearing capacity factor N_γ. Kumar and Ghosh (2007) had obtained the ultimate bearing capacity of two interfering rough strip footings using the method of characteristics. Again, Kumar and Kouzer (2007) had studied the effect of footing roughness on the bearing capacity factor N_γ.

In the latest research carried out by Choudhury and Subba Rao (2005), the seismic bearing capacity factors for shallow strip footings were found. The limit equilibrium method using the pseudo-static approach for seismic forces was adopted for the calculation of seismic bearing capacity factors of shallow strip footings in soil. A homogeneous isotropic c-ϕ soil with surcharge was assumed in the analysis. The soil was assumed as a rigid, perfectly plastic medium satisfying the Mohr–Coulomb failure criteria. The seismic acceleration coefficients were denoted as k_h and k_v in the horizontal and vertical directions respectively and were assumed to act both on the footing and on the inertia of the soil mass. A one-sided failure mechanism was assumed to occur in the seismic case with the formation of an asymmetrical elastic wedge with the full mobilization of the passive resistance on one side and partial mobilization on the other. A composite failure surface involving a planar and logspiral surface was considered in the analysis. The seismic bearing capacity factors with respect to cohesion, surcharge and unit weight components N_{cd}, N_{qd} and $N_{\gamma d}$ respectively were determined separately for various values of soil friction angles and seismic acceleration coefficients both in the horizontal and in the vertical directions. Unlike the assumption made by earlier researchers on the focus of the logspiral to be at the edge of the footing, a search was made in the present analysis to establish the critical focus.

2.1 *Example*

Consider Figure 10.1, which shows a horizontal shallow strip footing of base AD, embedded in horizontal ground. D_f is the embedment depth, B is the width of the footing with $D_f/B \leq 1$, and L is the footing length $L \gg B$. The base angles of the elastic wedge ADE are denoted by α_1 and α_2. In the static case $\alpha_1 = \alpha_2 = \alpha$, but for seismic case $\alpha_1 > \alpha_2$, in keeping with the direction of the horizontal seismic acceleration as shown in Figure 10.1. The central angle of the triangular wedge ADE is denoted by χ. In Figure 10.1, the forces acting on the footing are denoted as $k_h q_{ud} B$ and $(1 - k_v) q_{ud} B$, where q_{ud} is the ultimate seismic bearing capacity of the footing. In the soil below the footing within the failure zone, the forces acting are $k_h W_1$ and $(1 - k_v) W_1$ for zone I, $k_h W_2$ and $(1 - k_v) W_2$ for zone II, and $k_h W_3$ and $(1 - k_v) W_3$ for zone III where W_1, W_2 and W_3 are the weights of zones I, II and III respectively. The surcharge loads acting at the footing level are $k_h \gamma D_f$ and $(1 - k_v) \gamma D_f$. It can be noted that, in the present analysis, the critical directions of k_h and k_v which are from right to left and from bottom to top respectively are shown in Figure 10.1 which leads to the minimum seismic bearing capacity of the footing.

Considering the forces acting on the wedge ADE as shown in Figure 10.2, the forces acting on the face DE are $P_{p\gamma d1}$, P_{pqd1} and P_{pcd1}, which are the unit weight, surcharge and cohesion components of the total seismic passive resistance P_{pd1}. While $P_{p\gamma d1}$ is assumed to act at one-third of the vertical height

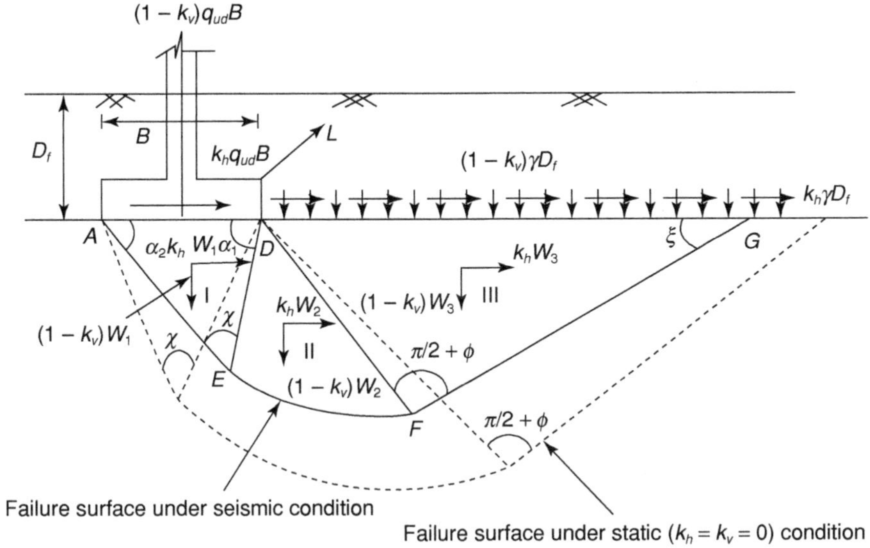

Figure 10.1 Failure surface considered by Choudhury and Subba Rao (2005) (with permission from Springer Science and Business Media and from ASCE)

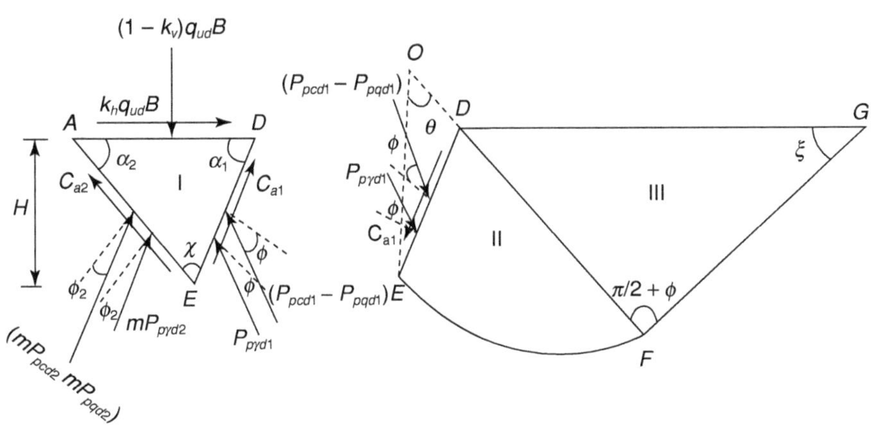

Figure 10.2 Forces considered by Choudhury and Subba Rao (2005) (with permission from Springer Science and Business Media and from ASCE)

above the base (point E), the other two components are assumed to act at the mid-height of the face DE having vertical height H. These passive forces are acting at an angle ϕ to the normal, as full mobilization of passive resistance is assumed to occur on this side of the wedge. The adhesion C_{a1} = c.DE, where c is the unit cohesion. On the face AE, with the same vertical height H, the corresponding seismic passive force components of the total seismic passive resistance mP_{pd2} will be mobilized partially as $mP_{p\gamma d1}$, mP_{pqd1} and mP_{pcd1}, where m is the mobilization factor.

The ultimate seismic bearing capacity q_{ud} is expressed in the following form:

$$q_{ud} = cN_{cd} + qN_{qd} + 0.5\gamma BN_{\gamma d} \tag{1}$$

where N_{cd}, N_{qd} and $N_{\gamma d}$ are the seismic bearing capacity factors. The closed-form design equations for each of these seismic bearing capacity factors can be obtained by considering the horizontal equilibrium of all the forces (Figure 10.2).

$$N_{cd} = \frac{1}{k_h}\left[\frac{\dfrac{K_{pcd1}}{\cos\phi}\sin(\alpha_1-\phi) - \dfrac{mK_{pcd2}}{\cos\phi_2}\sin(\alpha_2-\phi_2)}{\dfrac{1}{\tan\alpha_1}+\dfrac{1}{\tan\alpha_2}} + \frac{\sin\alpha_1\tan\phi_2\cos\alpha_2}{\sin(\alpha_1+\alpha_2)\tan\phi} - \frac{\sin\alpha_2\cos\alpha_1}{\sin(\alpha_1+\alpha_2)}\right] \tag{2}$$

$$N_{qd} = \frac{1}{k_h}\left[\frac{\dfrac{K_{pqd1}}{\cos\phi}\sin(\alpha_1-\phi) - \dfrac{mK_{pqd2}}{\cos\phi_2}\sin(\alpha_2-\phi_2)}{\dfrac{1}{\tan\alpha_1}+\dfrac{1}{\tan\alpha_2}}\right] \tag{3}$$

$$N_{\gamma d} = \frac{1}{k_h}\left[\frac{\dfrac{K_{p\gamma d1}}{\cos\phi}\sin(\alpha_1-\phi) - \dfrac{mK_{p\gamma d2}}{\cos\phi_2}\sin(\alpha_2-\phi_2)}{\left(\dfrac{1}{\tan\alpha_1}+\dfrac{1}{\tan\alpha_2}\right)^2}\right. \left. - \frac{1}{\left(\dfrac{1}{\tan\alpha_1}+\dfrac{1}{\tan\alpha_2}\right)}\right] \tag{4}$$

And from the vertical equilibrium of all the forces (Figure 10.2).

$$N_{cd} = \frac{1}{1-k_v}\left[\frac{\dfrac{K_{pcd1}}{\cos\phi}\cos(\alpha_1-\phi) + \dfrac{mK_{pcd2}}{\cos\phi_2}\cos(\alpha_2-\phi_2)}{\dfrac{1}{\tan\alpha_1}+\dfrac{1}{\tan\alpha_2}} + \frac{\sin\alpha_1\tan\phi_2\sin\alpha_2}{\sin(\alpha_1+\alpha_2)\tan\phi} + \frac{\sin\alpha_2\sin\alpha_1}{\sin(\alpha_1+\alpha_2)}\right] \tag{5}$$

$$N_{qd} = \frac{1}{1-k_v}\left[\frac{\frac{K_{pqd1}}{\cos\phi}\cos(\alpha_1-\phi)+\frac{mK_{pqd2}}{\cos\phi_2}\cos(\alpha_2-\phi_2)}{\frac{1}{\tan\alpha_1}+\frac{1}{\tan\alpha_2}}\right] \tag{6}$$

$$N_{\gamma d} = \frac{1}{1-k_v}\left[\frac{\frac{K_{p\gamma d1}}{\cos\phi}\cos(\alpha_1-\phi)+\frac{mK_{p\gamma d2}}{\cos\phi_2}\cos(\alpha_2-\phi_2)}{\left(\frac{1}{\tan\alpha_1}+\frac{1}{\tan\alpha_2}\right)^2}\right] - \frac{1}{\left(\frac{1}{\tan\alpha_1}+\frac{1}{\tan\alpha_2}\right)} \tag{7}$$

Now, by using the optimization technique, the design values of N_{cd} can be obtained by equating Equations 2 and 5, N_{qd} can be obtained by equating Equations 3 and 6, and $N_{\gamma d}$ can be obtained by equating Equations 4 and 7. The seismic passive earth pressure coefficients K_{pcd1}, K_{pqd1}, $K_{p\gamma d1}$ and K_{pcd2}, K_{pqd2}, $K_{p\gamma d2}$ can be obtained using the procedure described by Subba Rao and Choudhury (2005).

A comparative study of different methodologies to obtain the static bearing capacity factors N_{cd}, N_{qd} and $N_{\gamma d}$ are listed in Tables 10.1, 10.2 and 10.3 respectively.

Comparison of the results obtained by the analysis of Choudhury and Subba Rao (2005) in the seismic case with those obtained by Budhu and Al-Karni (1993) is given in Table 10.4. The values reported by Choudhury

Table 10.1 Comparison of bearing capacity factor N_{cd} obtained by different researchers for static case (adapted from Choudhury and Subba Rao 2005)

Different researchers	*Values of* N_{cd} *for*			
	$\phi = 10^0$	$\phi = 20^0$	$\phi = 30^0$	$\phi = 40^0$
Terzaghi (1943)	9.60	17.70	37.20	95.70
Meyerhof (1951)	8.34	14.83	30.13	75.25
Hansen (1970)	8.34	14.83	30.13	75.25
Prakash and Saran (1971)	–	17.30	36.60	94.80
Vesic (1973)	8.34	14.83	30.13	75.25
Griffiths (1982)	8.30	14.80	30.10	–
Saran and Agarwal (1991)	–	17.70	37.20	96.00
Eurocode 7 (1996)	8.34	14.83	30.13	75.25
Soubra (1999)	–	14.87	30.25	75.80
Choudhury and Subba Rao (2005)	8.31	14.74	29.86	74.28

Table 10.2 Comparison of bearing capacity factor N_{qd} obtained by different researchers for static case (adapted from Choudhury and Subba Rao 2005)

Different researchers	*Values of* N_{qd} *for*			
	$\phi = 10^0$	$\phi = 20^0$	$\phi = 30^0$	$\phi = 40^0$
Terzaghi (1943)	2.70	7.40	22.50	81.30
Meyerhof (1951)	2.50	6.40	18.40	64.10
Hansen (1970)	2.50	6.40	18.40	64.10
Prakash and Saran (1971)	–	7.40	22.40	81.30
Vesic (1973)	2.50	6.40	18.40	64.10
Griffiths (1982)	2.50	6.40	18.40	–
Saran and Agarwal (1991)	–	7.40	22.50	81.60
Eurocode 7 (1996)	2.50	6.40	18.40	64.10
Soubra (1999)	–	6.41	18.46	64.55
Choudhury and Subba Rao (2005)	2.47	6.39	18.35	63.95

Table 10.3 Comparison of bearing capacity factor $N_{\gamma d}$ obtained by different researchers for static case (adapted from Choudhury and Subba Rao 2005)

Different researchers	*Values of* $N_{\gamma d}$ *for*			
	$\phi = 10^0$	$\phi = 20^0$	$\phi = 30^0$	$\phi = 40^0$
Terzaghi (1943)	1.20	5.00	19.70	81.30
Meyerhof (1951)	0.40	2.90	15.70	93.60
Hansen (1970)	0.40	2.90	15.10	79.40
Prakash and Saran (1971)	–	3.80	19.40	115.8
Vesic (1973)	1.20	5.40	22.40	109.3
Griffiths (1982)	–	2.00	8.50	–
Saran and Agarwal (1991)	–	6.40	29.40	116.1
Bolton and Lau (1993)	0.30	1.60	7.70	–
Manoharan and Dasgupta (1995)	0.70	2.10	9.10	–
Eurocode 7 (1996)	0.53	3.93	20.09	105.9
Woodward and Griffiths (1998)	0.30	1.50	7.60	–
Soubra (1999)	0.84	4.66	21.81	120.2
Zhu et al. (2001)	0.45	3.37	17.58	97.93
Choudhury and Subba Rao (2005)	0.82	4.27	20.03	107.1

and Subba Rao (2005) are seen to be lower than those of Budhu and Al-Karni (1993) owing to the establishment of the critical position of the focus of the logspiral.

3 Seismic bearing capacity of shallow footings embedded in sloping ground

Meyerhof (1957), Siva Reddy and Mogaliah (1975), Saran et al. (1989) and Bowles (1996) conducted studies on the bearing capacity of foundations

Table 10.4 Comparison of seismic bearing capacity factors obtained by Choudhury and Subba Rao (2005) with those of Budhu and Al-Karni (1993) in seismic case for $\phi = 30^0$ and $k_v \neq 0$ (adapted from Choudhury and Subba Rao, 2005)

Seismic bearing capacity factor	*Values of* k_h	*Values of seismic bearing capacity factors obtained by*			
		Budhu and Al-Karni (1993)		*Choudhury and Subba Rao (2005)*	
		Values of k_v			
		$K_v = 0.5\ k_h$	$k_v = 1.0\ k_h$	$k_v = 0.5\ k_h$	$k_v = 1.0\ k_h$
N_{cd}	0.1	19.62	19.62	15.30	13.55
	0.2	12.76	12.76	8.80	7.09
	0.3	8.80	8.80	4.75	3.44
	0.4	5.40	–	1.96	–
N_{qd}	0.1	12.52	11.85	11.40	10.39
	0.2	7.32	6.11	6.21	5.43
	0.3	3.81	2.47	2.45	1.26
	0.4	1.76	–	0.81	–
$N_{\gamma d}$	0.1	10.21	9.46	8.40	7.76
	0.2	3.81	2.86	2.85	2.00
	0.3	1.21	0.59	0.98	0.29
	0.4	0.32	–	0.15	–

on slopes under static conditions. Hansen (1970) and Vesic (1973) proposed ground inclination factors and embedment depth factors with load inclination factors to compute the bearing capacity factors for shallow strip footings embedded in horizontal ground. There is less research on the bearing capacity of foundations embedded in slopes under seismic conditions, but some of the research is discussed later.

Pseudo-static analysis was used to determine the seismic bearing capacity of a mounded foundation near a downhill slope. Upper-bound limit analysis was adopted by Sawada et al. (1994). A logarithmic rupture surface was assumed to start at an edge of the loaded area far from the slope. Rigid body type movements of the landslide with inertia forces acting at the center of gravity were considered for a perfectly plastic medium. The kinematic energy balance principle was adopted, and the optimized results, with respect to different parameters, were compared with those obtained from other theories.

The analysis dealt with shallow strip surface footings adjacent to a slope. Pseudo-static earthquake forces both on the structure and on the soil were considered. Only a cohesionless soil was analyzed. Seismic bearing capacity factors were determined by using the limit equilibrium technique.

The analysis revealed that the bearing capacity was at a minimum when the footing was at the edge of the slope.

Sarma (1999) analyzed the bearing capacity of shallow footings placed on the surface of a slope and near-slope under seismic conditions. In the analysis, the consideration of the shear fluidization criteria originally reported by Richards et al. (1990) was modified, and the combinations of seismic horizontal and vertical acceleration coefficients with a soil friction angle and a ground inclination angle were reported.

The equivalence of limit analysis and the limit equilibrium method was proposed by Zhu (2000) to determine the one-sided critical failure mechanism for solving the least upper-bound solution of the bearing capacity factor N_γ with respect to the soil self-weight. Surface strip footings on slopes were considered. The least upper-bound values of N_γ in a variety of cases for different ground inclinations, footing base inclinations, load inclinations, horizontal and vertical seismic accelerations and soil friction angles were presented.

Askari and Farzaneh (2003) used the upper-bound method to obtain the seismic bearing capacity of shallow foundations near slopes. Kumar and Kumar (2003) gave the solution for the seismic bearing capacity of rough footings on slopes using the limit equilibrium method which was an extension of the approach of Budhu and Al-Karni (1993). Again, Kumar and Rao (2003) obtained the seismic bearing capacity factors of foundations on slopes using the method of characteristics. Kumar and Ghosh (2006) obtained the seismic bearing capacity of embedded footings on sloping ground using the method of characteristics.

In the latest research, Choudhury and Subba Rao (2006) obtained the seismic bearing capacity factors for shallow strip footings embedded in inclined ground with a general c-ϕ soil by using the limit equilibrium method. The seismic forces were considered as pseudo-static forces acting both on the footing and on the soil. A composite failure surface involving planar and logspiral surfaces was considered. Three different failure types, Type 1, Type 2 and Type 3, were considered depending on the embedment depth and ground inclinations, as will be discussed later. The seismic bearing capacity factors with respect to cohesion, surcharge and unit weight components N_{cd}, N_{qd} and $N_{\gamma d}$ respectively were determined separately for various values of soil friction angles, seismic acceleration coefficients both in the horizontal and vertical directions, ground inclinations and embedment depths. Unlike the assumption that the focus of the logspiral is at the edge of the footing, a search was made to establish the critical focus for different types of failure surfaces.

3.1 Example

Consider Figure 10.3, which shows a horizontal shallow strip footing of base AD, embedded in an inclined soil XY. D_f is the embedment depth

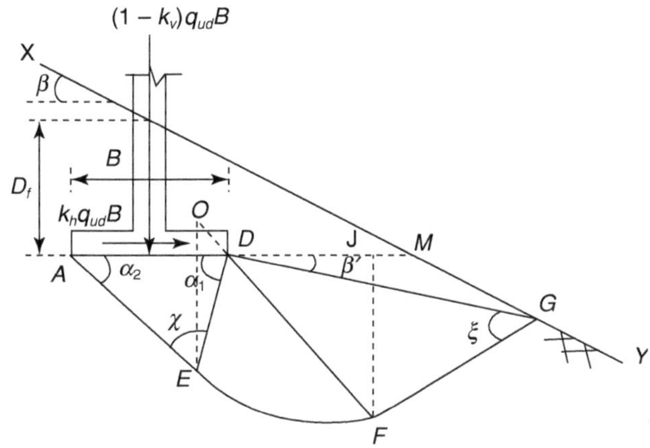

Figure 10.3 Failure surface considered by Choudhury and Subba Rao (2006) for 'Type 1' failure (with permission from ASCE)

measured along the centerline of the footing, B the width of the footing with $D_f/B \leq 1$, and L the length of the footing with $L \gg B$. The base angles of the elastic wedge ADE are denoted by α_1 and α_2. In the static case $\alpha_1 = \alpha_2 = \alpha$, but for seismic case $\alpha_1 > \alpha_2$, in keeping with the direction of horizontal seismic acceleration and facing toward the inclined soil. In Figure 10.3, the forces acting on the footing are denoted as $k_h q_{ud} B$ and $(1 - k_v) q_{ud} B$, where q_{ud} is the ultimate seismic bearing capacity of the footing. Similar to the proposed method by Choudhury and Subba Rao (2005), in the soil below the footing within the failure zone, the forces acting are $k_h W_1$ and $(1 - k_v) W_1$ for zone I, $k_h W_2$ and $(1 - k_v) W_2$ for zone II, and $k_h W_3$ and $(1 - k_v) W_3$ for zone III, where W_1, W_2 and W_3 are the weights of zones I, II and III respectively. The surcharge loads acting at the footing level are $k_h \gamma D_f$ and $(1 - k_v) \gamma D_f$. The forces acting on the wedge ADE are: on the face DE, $P_{p\gamma d1}$, P_{pqd1} and P_{pcd1}, which are the unit weight, surcharge and cohesion components respectively of the total seismic passive resistance P_{pd1}. While $P_{p\gamma d1}$ is assumed to act at one-third of the vertical height above the base (point E), the other two components are assumed to act at the mid-height of the face DE having a vertical height H. These passive forces are acting at an angle of ϕ to the normal, as full mobilization of passive resistance is assumed to occur on this side of the wedge. The adhesion $C_{a1} = c.$DE, where c is the unit cohesion. On the face AE, with the same vertical height H, the corresponding seismic passive force components of the total seismic passive resistance mP_{pd2} will be mobilized partially as $mP_{p\gamma d1}$, mP_{pqd1} and mP_{pcd1}, where m is the mobilization factor as defined by Choudhury and Subba Rao (2005).

From the geometry, depending on the values of the embedment ratio D_f/B and slope angle β, the formation of three different types of failure surfaces

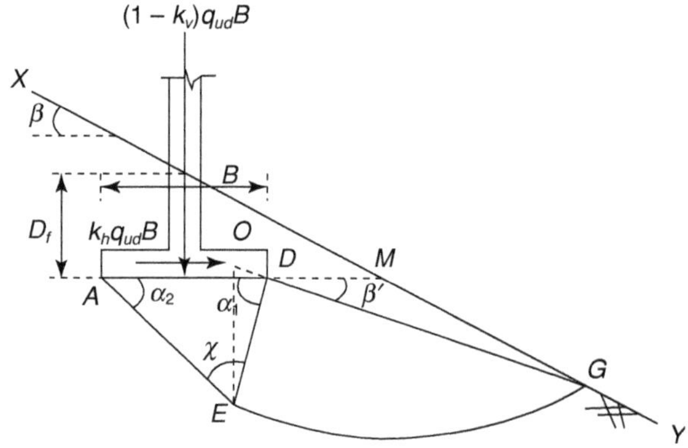

Figure 10.4 Failure surface considered by Choudhury and Subba Rao (2006) for 'Type 2' failure (with permission from ASCE)

may occur. In Figure 10.3, the failure surface is composed of the triangular wedge zone ADE, followed by a logarithmic spiral zone DFE and then followed by a partial planar zone DGF. This failure surface is termed as 'Type 1' failure surface. The line DG makes an angle β' with the horizontal line DM. Hence the problem reduces to a passive earth pressure problem with composite failure surfaces (logarithmic spiral plus planar), assuming DE as an imaginary retaining wall with ground inclination β' with the horizontal. Figure 10.4 shows the 'Type 2' failure surface, which is composed of the triangular wedge ADE and the logarithmic spiral zone DEG depending on the selection of values for the embedment ratio D_f/B and slope angle β. The line DG makes an angle of β' with the horizontal. Here, the analysis for the imaginary retaining wall DE is made considering only the logarithmic failure surface similar to the analysis of Morrison and Ebeling (1995).

The 'Type 3' failure surface is assigned to the case where the planar zone DFG in Figure 10.5 becomes full planar, i.e. the line DG merges with the horizontal line DM and the case reduces to the analysis for shallow footings embedded in horizontal ground as described by Choudhury and Subba Rao (2005) with q representing the average surcharge over the length DG.

In a similar way as described by Choudhury and Subba Rao (2005), here also, by considering both the horizontal and vertical equilibrium of all the forces acting on the triangular zone ADE, the ultimate seismic bearing capacity q_{ud} is expressed in the following form:

$$q_{ud} = cN_{cd} + qN_{qd} + 0.5\gamma BN_{\gamma d} \tag{8}$$

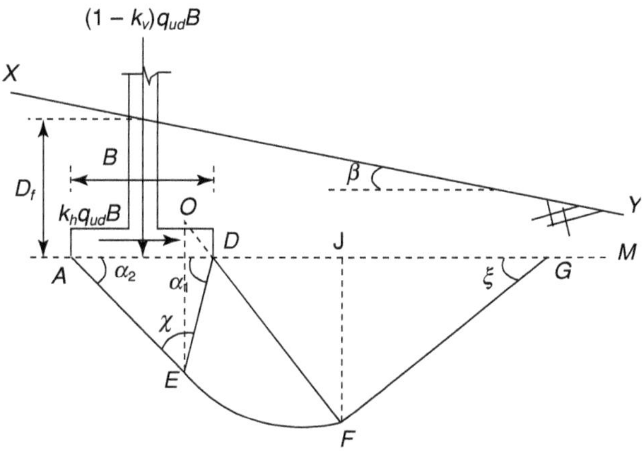

Figure 10.5 Failure surface considered by Choudhury and Subba Rao (2006) for 'Type 3' failure (with permission from ASCE)

where N_{cd}, N_{qd} and $N_{\gamma d}$ are the seismic bearing capacity factors. Typical design values of these seismic bearing capacity factors are reported in Figures 10.6, 10.7 and 10.8 respectively.

A comparative study of the results obtained by Choudhury and Subba Rao (2006) with a similar analysis for the static case is shown in Tables 10.5, 10.6 and 10.7.

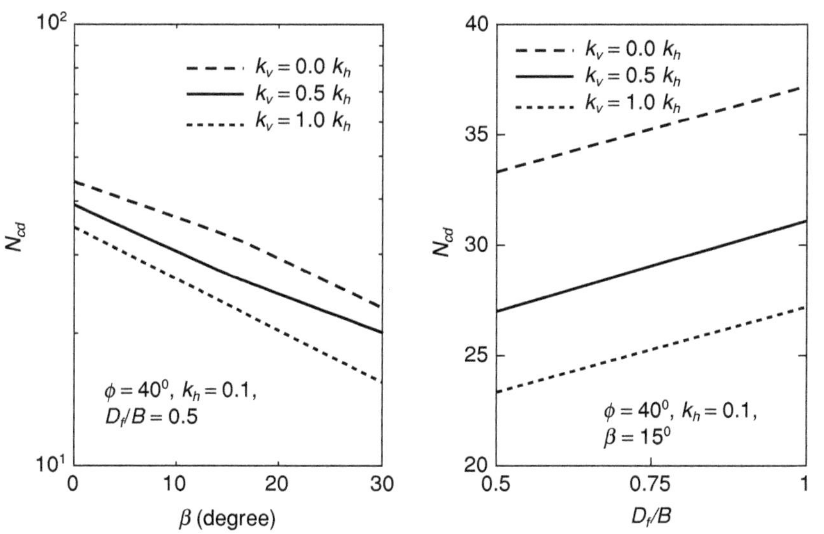

Figure 10.6 Typical design values of N_{cd} given by Choudhury and Subba Rao (2006) (with permission from ASCE)

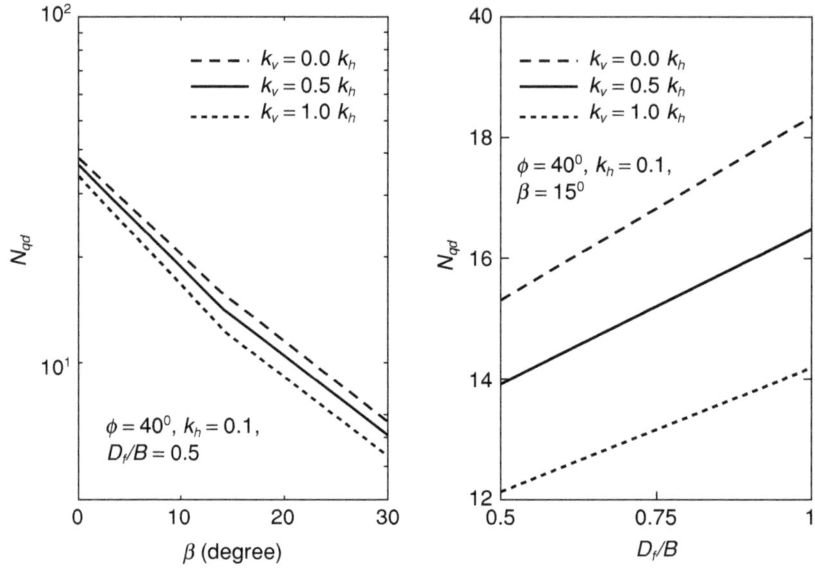

Figure 10.7 Typical design values of N_{qd} given by Choudhury and Subba Rao (2006) (with permission from ASCE)

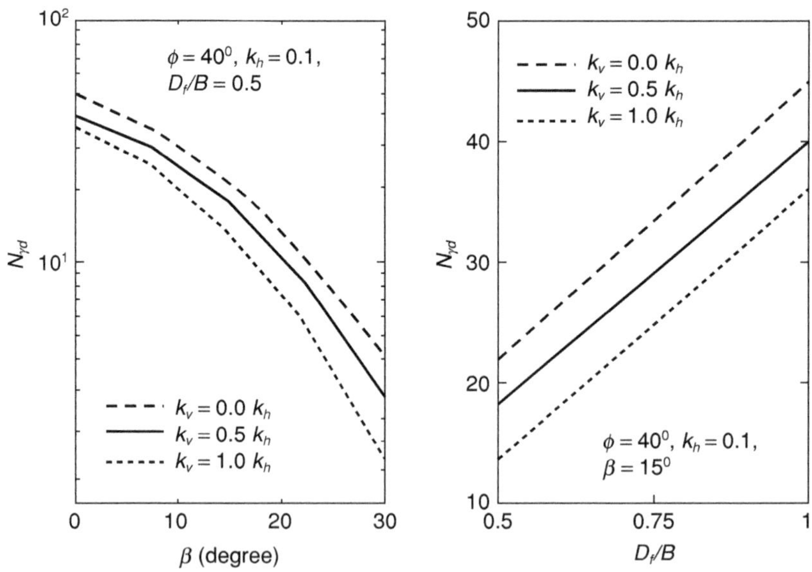

Figure 10.8 Typical design values of $N_{\gamma d}$ given by Choudhury and Subba Rao (2006) (with permission from ASCE)

Table 10.5 Comparison of bearing capacity factor N_{cd} obtained by different researchers for the static case with $\beta = 15^0$ (adapted from Choudhury and Subba Rao 2006)

Values of ϕ	*Values of* N_{cd}					
	For $D_f/B = 0.5$			*For* $D_f/B = 1.0$		
	Saran et al. (1989)	*Bowles (1996)*	*Choudhury and Subba Rao (2006)*	*Saran et al. (1989)*	*Bowles (1996)*	*Choudhury and Subba Rao (2006)*
10^0	8.20	6.40	7.96	9.00	6.40	8.31
20^0	17.36	14.45	13.29	20.04	14.83	14.83
30^0	30.31	28.56	25.60	35.46	30.14	27.89
40^0	63.45	65.42	57.74	87.52	75.31	75.77
45^0	110.53	116.26	106.33	125.96	133.73	118.84

Table 10.6 Comparison of bearing capacity factor N_{qd} obtained by different researchers for the static case with $\beta = 15^0$ (adapted from Choudhury and Subba Rao, 2006)

Values of ϕ	*Values of* N_{qd}					
	For $D_f/B = 0.5$			*For* $D_f/B = 1.0$		
	Saran et al. (1989)	*Bowles (1996)*	*Choudhury and Subba Rao (2006)*	*Saran et al. (1989)*	*Bowles (1996)*	*Choudhury and Subba Rao (2006)*
20^0	7.65	6.40	5.36	8.20	6.40	6.39
30^0	20.55	18.40	9.15	32.65	18.40	16.68
40^0	35.40	47.09	25.26	53.23	56.34	30.09
45^0	108.72	85.53	76.69	120.00	95.60	84.45

Table 10.7 Comparison of bearing capacity factor $N_{\gamma d}$ obtained by different researchers for the statc case with $\beta = 15^0$ (adapted from Choudhury and Subba Rao 2006)

Values of ϕ	*Values of* $N_{\gamma d}$			
	For $D_f/B = 0.5$		*For* $D_f/B = 1.0$	
	Saran et al. (1989)	*Choudhury and Subba Rao (2006)*	*Saran et al. (1989)*	*Choudhury and Subba Rao (2006)*
20^0	5.95	4.27	5.95	4.27
30^0	17.56	15.57	22.34	20.03
40^0	52.45	42.70	100.32	84.49
45^0	130.44	113.58	220.67	204.72

4 Summary

Recent analyses by Choudhury and Subba Rao (2005, 2006) show that, by considering the pseudo-static seismic forces, closed-form design solutions can be found for the computation of seismic bearing capacity factors for shallow foundations embedded in both horizontal and sloping soil.

The following conclusions have been drawn from their analyses:

- Seismic bearing capacity factors with respect to cohesion, surcharge and unit weight components for shallow strip footings embedded in horizontal soil have been computed for a wide range of variation in parameters such as soil friction angle (ϕ), horizontal and vertical seismic acceleration coefficients (k_h and k_v). The bearing capacity factors decrease appreciably with increases in both k_h and k_v.
- The extent of the failure zone and the depth of the failure zone decrease drastically with the increase in the seismic accelerations both in the horizontal and the vertical directions.
- Comparison with earlier results shows that the bearing capacity factors for seismic conditions determined from the present analysis are the minimum and are suitable for use in design practice.
- Seismic bearing capacity factors with respect to cohesion, surcharge and unit weight components have been computed for shallow strip footings embedded in inclined soil for a wide range of variation in parameters such as soil inclination (β), soil friction angle (ϕ), embedment ratio (D_f/B), horizontal and vertical seismic acceleration coefficients (k_h and k_v). The bearing capacity factors decrease appreciably with increases in both k_h and k_v. Bearing capacity decreases as ground inclination β increases and as the embedment depth D_f decreases.
- Closed-form design values have been proposed which can be used for the practical design of shallow footings embedded in both horizontal and sloping soil using the simple limit equilibrium method. In many cases the latest solutions reported here compare well with the previous static results for similar cases and available results for seismic cases.

Notations

α_1 = base angle on the failure side
α_2 = base angle on the side opposite to the failure side
B = width of the footing
β = soil inclination with respect to the horizontal
β' = inclination of the top of the failure surface with respect to horizontal measured from the edge of the footing
C_{a1} = adhesive force on the side of failure
C_{a2} = adhesive force on the rear side of failure
χ = central angle of the triangular wedge

D_f = depth of the footing
D_e = horizontal distance of the nearest edge of the footing from slope
ϕ_2 = partially mobilized soil friction angle on the rear side of failure
$K_{p\gamma d1}$ = seismic passive earth pressure coefficient with respect to unit weight on the side of failure
K_{pqd1} = seismic passive earth pressure coefficient with respect to surcharge on the side of failure
K_{pcd1} = seismic passive earth pressure coefficient with respect to cohesion on the side of failure
$K_{p\gamma d2}$ = seismic passive earth pressure coefficient with respect to unit weight on the rear side of failure
K_{pqd2} = seismic passive earth pressure coefficient with respect to surcharge on the rear side of failure
K_{pcd2} = seismic passive earth pressure coefficient with respect to cohesion on the rear side of failure
L = length of the footing
m = a factor used in the analysis
N_{cd} = seismic bearing capacity factor with respect to cohesion
N_{qd} = seismic bearing capacity factor with respect to surcharge
$N_{\gamma d}$ = seismic bearing capacity factor with respect to unit weight
P_{pd1} = total seismic passive earth resistance on the side of failure
P_{pd2} = total seismic passive earth resistance on the rear side of failure
$P_{p\gamma d1}$ = seismic passive earth resistance with respect to unit weight on the side of failure
P_{pqd1} = seismic passive earth resistance with respect to surcharge on the side of failure
P_{pcd1} = seismic passive earth resistance with respect to cohesion on the side of failure
$P_{p\gamma d2}$ = seismic passive earth resistance with respect to unit weight on the rear side of failure
P_{pqd2} = seismic passive earth resistance with respect to surcharge on the rear side of failure
P_{pcd2} = seismic passive earth resistance with respect to cohesion on the rear side of failure
q_{ud} = ultimate seismic bearing capacity
W_1 = weight of Zone I
W_2 = weight of Zone II
W_3 = weight of Zone III

References

Askari, F. and Farzaneh, O. (2003) Upper-bound solution for seismic bearing capacity of shallow foundations near slopes, *Géotechnique*, 53 (8): 697–702.

Basavanna, B. M., Joshi, V. H. and Prakash, S. (1974) Dynamic bearing capacity of soils under transient loading, *Bull. Indian Soc. Earthquake Tech.*, 2 (3): 67–84.

Basudhar, P. K., Valsangkar, A. J. and Madhav, M. R. (1981) Sequential unconstrained minimization in the optimal lower bound bearing capacity analysis, *Indian Geotechnical Journal*, 11: 42–55.

Bowles, J. E. (1996) *Foundation Analysis and Design*, 5th edn, New York: McGraw-Hill.

Budhu, M. and Al-Karni, A. (1993) Seismic bearing capacity of soils, *Géotechnique*, 43 (1): 181–7.

Caquot, A. and Kerisel, L. (1953) Sur le terme de surface dans le calcul des fondations en milieu pulverulent, *Proc. 3rd Int. Conf. on Soil Mechanics and Foundation Engg.*, Zurich: Icosomes, 1: 336–7.

Chen, W. F. (1975) *Limit Analysis and Soil Plasticity*, Developments in Geotechnical Engineering, Amsterdam: Elsevier.

Choudhury, D. and Subba Rao, K. S. (2005) Seismic bearing capacity of shallow strip footings, *Geotechnical and Geological Engineering*, 23 (4): 403–18.

Choudhury, D. and Subba Rao, K. S. (2006) Seismic bearing capacity of shallow strip footings embedded in slope, *International Journal of Geomechanics*, ASCE, 6 (3): 176–84.

Dormieux, L. and Pecker, A. (1995) Seismic bearing capacity of foundations on cohesionless soil, *J. Geotech. Engrg.*, ASCE, 121 (3): 300–3.

Eurocode 7 (1996) Calcul Geotechnique. AFNOR, XP, ENV 1997-1, see Sieffert and Bay-Gress 2000.

Griffiths, D. V. (1982) Computation of bearing capacity factors using finite elements, *Géotechnique*, 32 (3): 195–202.

Hansen, J. B. (1970) A revised and extended formula for bearing capacity, *Geoteknisk Inst. Bull.*, 28: 5–11.

Kumar, J. (2004) Effect of footing–soil interface friction on bearing capacity factor N_γ, *Géotechnique*, 54 (10): 677–80.

Kumar, J. and Ghosh, P. (2006) Seismic bearing capacity for embedded footings on sloping ground, *Géotechnique*, 56 (2): 133–40.

Kumar, J. and Ghosh, P. (2007) Ultimate bearing capacity of two interfering rough strip footings, *International Journal of Geomechanics*, ASCE, 7 (1): 53–62.

Kumar, J. and Kouzer, K. M. (2007) Effect of footing roughness on bearing capacity factor N_γ, *Journal of Geotechnical and Geoenvironmental Engineering*, ASCE, 133 (5): 502–11.

Kumar, J. and Kumar, N. (2003) Seismic bearing capacity of rough footings on slopes using limit equilibrium, *Géotechnique*, 53 (3): 363–9.

Kumar, J. and Rao, V. B. K. M. (2002) Seismic bearing capacity factors for spread foundations, *Géotechnique*, 52 (2): 79–88.

Kumar, J. and Rao, V. B. K. M. (2003) Seismic bearing capacity of foundations on slopes, *Géotechnique*, 53 (3): 347–61.

Meyerhof, G. G. (1951) The ultimate bearing capacity of foundations, *Géotechnique*, 2: 301–32.

Meyerhof, G. G. (1953) The bearing capacity of foundations under eccentric and inclined loads, *Proc. 3rd Int. Conf. on Soil Mechanics and Foundation Engg.*, Zurich: Icosomes, 1: 440–5.

Meyerhof, G. G. (1957) The ultimate bearing capacity of foundations on slopes, *Proc. 4th Int. Conf. on Soil Mechanics and Foundation Engg.*, 1: 334–86.

Meyerhof, G. G. (1963) Some recent research on the bearing capacity of foundations, *Canadian Geotechnical Journal*, 1 (1): 16–26.

Moore, W. W. and Darragh, R. D. (1956) Some considerations in the design of footings for earthquakes, *Proc. First World Conf. Earthquake Engrg.*, Berkeley, Calif., paper no. 28.

Morrison, E. E. and Ebeling, R. M. (1995) Limit equilibrium computation of dynamic passive earth pressure, *Canadian Geotechnical Journal*, 32: 481–7.

Paolucci, R. and Pecker, A. (1997) Seismic bearing capacity of shallow strip foundations on dry soils, *Soils and Foundations*, 37 (3): 95–105.

Prakash, S. (1974) Aseismic design of foundations, n *Earthquake Engineering*, ed. Sarita Prakashan, Meerut, pp. 333–45.

Prakash, S. (1981) *Soil Dynamics*, New York: McGraw-Hill, pp. 171–219.

Prakash, S. and Saran, S. (1971) Bearing capacity of eccentrically loaded footings, *J. Soil Mech. and Found. Div.*, ASCE, 97 (SM1): 95–111.

Prandtl, L. (1921) Uber die eindringungstestigkeit plastisher baustoffe und die festigkeit von schneiden, *Zeitschrift Fur Angewandt Mathematik Und Mechanik*, 1 (1): 15–30.

Richards, R., Elms, D. G. and Budhu, M. (1990) Dynamic fluidization of soils, *J. Geotech. Engrg.*, ASCE, 116 (5): 740–59.

Richards, R., Elms, D. G. and Budhu, M. (1993) Seismic bearing capacity and settlement of foundations, *J. Geotech. Engrg.*, ASCE, 119 (4): 662–74.

Saran, S. and Agarwal, R. K. (1991) Bearing capacity of eccentrically obliquely loaded footing, *J. Soil Mech. and Found. Div.*, ASCE, 117 (11): 1669–90.

Saran, S., Sud, V. K. and Handa, S. C. (1989) Bearing capacity of footings adjacent to slopes, *J. Geotech. Engrg.*, ASCE, 115 (4): 553–73.

Sarma, S. K. (1999) Seismic bearing capacity of shallow strip footings adjacent to a slope, *Proc. 2nd Int. Conf. Earthquake Geotech. Engrg.*, Lisbon: Balkema, pp. 309–13.

Sarma, S. K. and Iossifelis, I. S. (1990) Seismic bearing capacity factors of shallow strip footings, *Géotechnique*, 40 (2): 265–73.

Sawada, T., Nomachi, S. G. and Chen, W. F. (1994) Seismic bearing capacity of a mounded foundation near a down-hill slope by pseudo-static analysis, *Soils and Foundations*, 34 (1): 11–17.

Selig, E. T. and McKee, K. E. (1961) Static and dynamic behaviour of small footings, *J. Soil Mech. and Found. Div.*, ASCE, 87 (SM6): 29–47.

Shenkman, S. and McKee, K. E. (1961) Bearing capacities of dynamically loaded footings, *Symp. on Soil Dynamics*, ASTM Spec. Pub. No. 305, pp. 78–90.

Shinohara, T., Tateishi, T. and Kubo, K. (1960) Bearing capacity of sandy soil for eccentric and inclined load and lateral resistance of single piles embedded in sandy soil, *Proc. 2nd World Conf. on Earthquake Engrg., Tokyo*, 1: 265–80.

Sieffert, J. G. and Bay-Gress, Ch. (2000) Comparison of European bearing capacity calculation methods for shallow foundations, *Proc. Instn. Civil Engrs., Geotech. Engrg.*, 143 (2): 65–74.

Siva Reddy, A. and Mogaliah, G. (1975) Bearing capacity of shallow foundations on slopes, *Indian Geotechnical Journal*, 5 (4): 237–53.

Soubra, A. H. (1997) Seismic bearing capacity of shallow strip footings in seismic conditions, *Proc. Instn. Civil Engrs., Geotech. Engrg.*, 125 (4): 230–41.

Soubra, A. H. (1999) Upper bound solutions for bearing capacity of foundations, *Journal of Geotechnical and Geoenvironmental Engineering*, ASCE, 125 (1): 59–69.

Sridharan, A. (1962) Settlement studies of a model footing under dynamic load, *Proc. Second Symp. on Earthquake Engrg., Roorkee, India*, p. 377.

Subba Rao, K. S. and Choudhury, D. (2005) Seismic passive earth pressures in soils, *Journal of Geotechnical and Geoenvironmental Engineering*, ASCE, 131 (1): 131–5.

Taylor, P. W. (1968) Design of spread footings for earthquake loadings, *Proc. Fifth Australia–New Zealand Conf. Soil Mech. Found. Engrg.*, pp. 221–9.

Terzaghi, K. (1943) Theoretical Soil Mechanics, New York, John Wiley and Sons.

Vesic, A. S. (1973) Analysis of ultimate loads of shallow foundations, *J. Soil Mech. and Found. Div.*, ASCE, 99 (1): 45–73.

Vesic, A. S., Banks, D. C. and Woodward, J. M. (1965) An experimental study of dynamic bearing capacity of footings on sand, *Proc. Fifth Int. Conf. Soil Mech. Found. Engrg, Paris*, 2: 209–13.

Zhu, D. Y. (2000) The least upper-bound solutions for bearing capacity factor N_γ, *Soils and Foundations*, 40 (1): 123–9.

Zhu, D. Y., Lee, C. F. and Jiang, H. D. (2001) A numerical study of the bearing capacity factor N gamma, *Canadian Geotechnical Journal*, 38 (5): 1090–6.

11 Free vibrations of industrial chimneys or communications towers with flexibility of soil

Tadeusz Chmielewski

1 The structure of chimneys and communications towers

Industrial chimneys and communications (TV) towers are constructed as tall, slender reinforced-concrete structures, with industrial chimneys being constructed as single-flue or multi-flue structures. Single-flue chimneys are designed as a circular-ring cross-section having a diameter and wall thickness tapering continuously from the base to the top. Multi-flue industrial chimneys are also designed as a circular ring cross-section, usually with a constant diameter but with a reduced stepped thickness of wall as the height increases. The main structural elements of a chimney are the shaft and the foundation.

The structure of TV towers is similar to single-flue chimneys and consists mainly of four structural elements: (*a*) foundation, (*b*) shaft, (*c*) head and (*d*) antenna mast (usually steel). Figure 11.1 shows a selection of chimneys and TV towers.

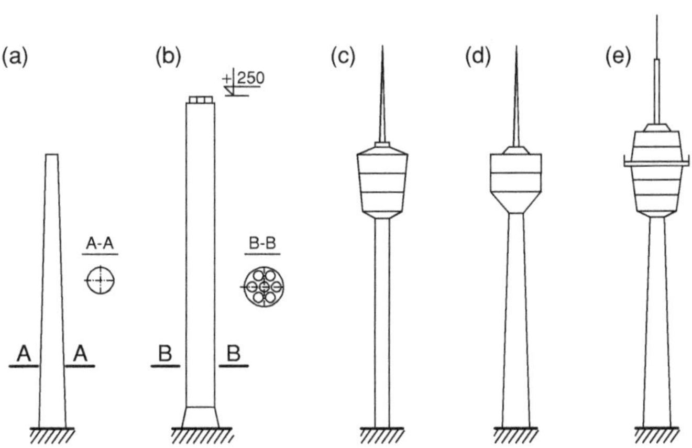

Figure 11.1 Sketches of chimneys and TV towers: (*a*) single-flue chimney; (*b*) multi-flue chimney; (*c*), (*d*) and (*e*) TV towers with different heads

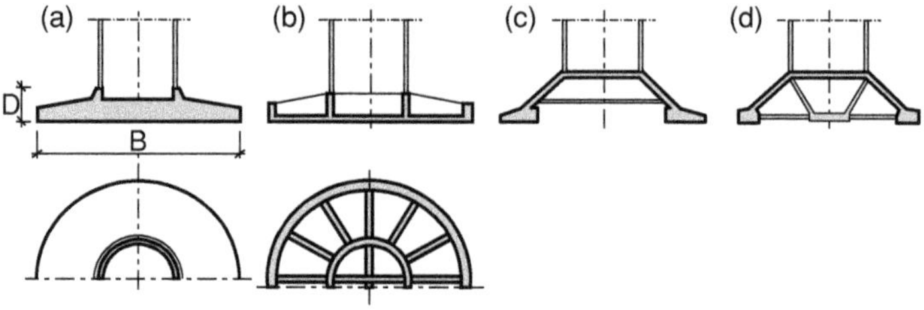

Figure 11.2 Foundations for chimneys or TV towers

2 Industrial-chimney and TV-tower foundations

The foundations for structures such as industrial chimneys or TV towers are constructed, as for other civil engineering structures (buildings, bridges, etc.), for the purpose of transmitting superstructural loads to the soil.

Typical foundations types for chimneys or TV towers are shown in Figure 11.2: (*a*) circular plate foundations, (*b*) circular plate with ribbed foundations, (*c*) conical single shell foundations and (*d*) conical double shell foundations.

The transmission of the superstructure loads, i.e. loads from the chimney or the TV tower, to the soil may be achieved by using either: (*a*) shallow foundations, where the depth D is generally less than the base diameter B, as shown in Figure 11.2, or (*b*) deep foundations – mostly piles.

3 Modeling of the soil–structure interaction

Every civil engineering structure is founded on soil and consists of two parts: (*a*) the superstructure – the upper part; and (*b*) the substructure – the foundation which interfaces the superstructure with the supporting soil. The analysis of any structure must take into account the soil–structure system. Now days there are few problems in computing the stress or displacement state of any structure, given a set of boundary conditions. Also, the progress in geomechanics, with the application of numerical methods, allows engineers to analyze three-dimensionally the soil so as to solve many of the complicated problems of soil mechanics.

From the design practice of foundation engineering, two kinds of problems may be considered. First, knowledge of the stresses and displacements in the three-dimensional domain of the soil is required. This method of determining soil stresses and displacements is expensive and requires in-situ geological investigation. This method is usually applied to prestigious structures, e.g. dams, tunnels, soil–steel bridges, etc., and is known as the soil–structure interaction problem. Second, the structural designer is interested mainly in the influence of the soil on the structure's behavior and not in the stresses, displacements, pore pressure, etc., in the soil below the foundation. In this case it is necessary to calculate both the base shear resistance and foundation settlements. This approach is one of the standard methods and procedures used for the analysis and design of foundations.

4 Theoretical and experimental free vibrations of a tall industrial chimney on a flexible soil

4.1 Introduction

The dynamic response of many structures due to wind or seismic forces can be evaluated by modal analysis. In order to represent the restraint conditions of a structure, such as a tall industrial chimney or a TV tower, natural frequencies and mode shapes are computed to take into account the dynamic phenomenon of soil–structure interaction.

Excellent examples of such studies are given by Solari and Stura [1]. In their paper the natural frequencies and mode shapes of a structure interacting with the soil and modeled as a non-prismatic cantilever beam with concentrated masses has been evaluated through the application of the Rayleigh–Ritz method. An overview of previous studies on this subject is also given.

On the one hand, the present research aims at a theoretical evaluation of the natural frequencies and mode shapes of a tall multi-flue industrial chimney interacting with the soil by using the finite element method. On the other hand, the actual free vibrations of the chimney are measured in order to confirm the results of the assumed calculation model and to obtain important data on the effect of the soil interacting with the chimney.

4.2 Description of the chimney

The six-flue, 250 m high industrial chimney considered in this work is located at the power station in Opole, Poland. The general view of the chimney with the longitudinal and cross-section and a selection of the most important dimensions is shown in Figure 11.3 and in Reference 2. This reinforced concrete chimney has two structural shells: an outer shell and an inner shell with differing thicknesses as the height increases. Both shells are joined at floors approximately 10 m apart. The floor system is formed by groups of steel beams. The chimney is placed on a circular foundation slab, 4 m deep and 50 m in diameter, lying directly on the soil.

4.3 Calculation model of the soil–chimney interaction

The chimney has been idealized as a linear elastic, homogeneous beam connected to the foundation, treated as three-dimensional, resting on a soil stratum of finite depth overlaying a rigid halfspace. The part of the chimney above the foundation was modeled as a one-dimensional structure divided into twenty-five beam elements undergoing axial deformations in order

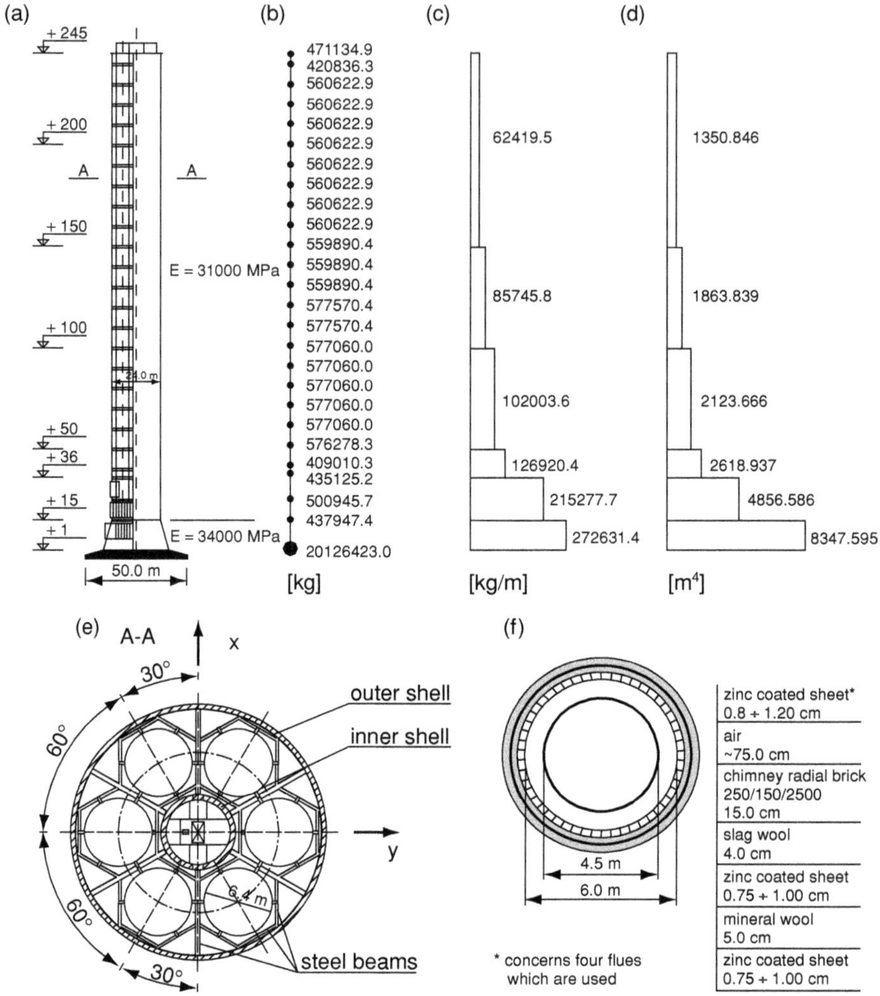

Figure 11.3 Industrial chimney at Opole power station: (*a*) view and longitudinal section, (*b*) mass distribution of the floors and the six inner flues, (*c*) mass distribution of the two concentric concrete tubes, (*d*) moment of inertia distribution of the two concentric concrete tubes, (*e*) cross-section of the chimney, (*f*) cross-section of one flue

to formulate the element stiffness matrices. For these beam elements, mass element matrices were formulated using the lumped mass approach. In the system mass matrix, the mass of reinforced-concrete shells, floors and the six inner flues were included. The foundation of the chimney was divided into twenty space elements with eight nodes for each element (SOLID

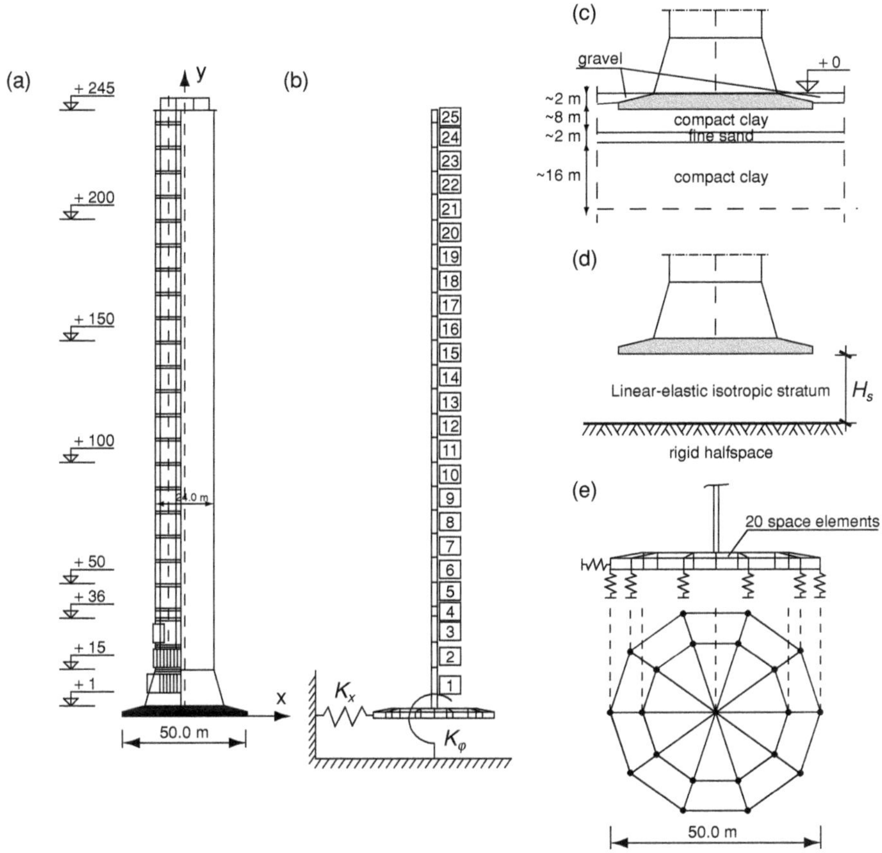

Figure 11.4 (*a*) View of chimney, (*b*) subdivision of the chimney into elements with soil springs, (*c*) description of the soil under the foundation, (*d*) model of the soil under the foundation to evaluate the soil spring, (*e*) subdivision of the foundation slab

elements of the SAP90+ computer program). Discretization of the chimney and the foundation is shown in Figure 11.4.

The analysis assumes that the cross-section of the chimney comprises two concentric concrete tubes. They are designed only to carry dead and live loads, and are supported by the common foundation. There are six internal steel cylinders; but, as the power station has only four power units, only four cylinders are used. The cross-section of a cylinder is shown in Figure 11.3f. With the two unused steel cylinders, whose diameters are 450 cm, there is no zinc-coated sheet inside them. The dead load of the all-steel cylinders is carried by the floors which are formed by groups of steel beams. This is the reason that the stiffnesses of the six flues were neglected. Their mass was included only in the system mass matrix.

Beam elements undergoing axial deformations were used, and shear deformations were neglected. The analysis assumed that the effect of the shear deformations and the torsional inertia should be negligible for the first and second modes.

The strength of the chimney concrete in compression had been previously evaluated based on samples taken from both concrete tubes. The modulus of elasticity of the concrete was also evaluated by experimental testing. The resulting data are shown in Figure 11.3a.

The description of the soil underlying the circular foundation slab of the chimney is shown in Figure 11.4c. There are, in the literature, many models of the soil and many publications describing the soil–structure interaction. In this paper the author applied the soil model shown in Figure 11.4d. For engineering purposes, for the circular foundation with a radius R_0, resting on a soil stratum of finite depth H_s over a rigid halfspace, Kausel [3] proposed the following approximate expressions for the soil spring constant (see Figure 11.4) for $H_s/R_0 \geq 2$:

$$K_x = \frac{8GR_0}{2-\nu}\left(1+\frac{R_0}{2H_s}\right),$$

$$K_\varphi = \frac{8GR_0^3}{3(1-\nu)}\left(1+\frac{R_0}{6H_s}\right) \tag{1}$$

where G is the shear modulus of soil and ν the Poisson ratio.

In the numerical analysis, the relationship between the shear modulus and the shear wave velocity of the soil ϑ_s given in Reference 4 was taken into account.

$$G = \vartheta_s^2 \rho_g \tag{2}$$

where ρ_g is the density of soil. Based on the real geotechnical conditions under the chimney foundation, the shear wave velocity of soil was estimated to be in the range $\vartheta_s \in \langle 130 - 300 \rangle$ *m/s*.

4.4 Natural vibration frequencies and modes of the chimney: numerical results

The free vibrations of the linear MDOF systems are governed by Equation (1), [5, 6]:

$$\boldsymbol{M}\ddot{q} + \boldsymbol{K}\ddot{q} = 0 \tag{3}$$

where $\boldsymbol{M}$, $\boldsymbol{K}$, q are the system mass matrix, stiffness matrix and displacement vector respectively.

Table 11.1 Computed values of the natural periods of the chimney

Shear wave velocity ϑ_s	*Natural period* T, s			
	T_1	T_2	T_3	T_4
$\vartheta_s = 150\ m/s$	4.786	0.985	0.464	0.300
$\vartheta_s = 200\ m/s$	4.627	0.940	0.400	0.278
$\vartheta_s = 250\ m/s$	4.551	0.918	0.374	0.243
$\vartheta_s = 300\ m/s$	4.508	0.906	0.391	0.216
$\vartheta_s = 2000\ m/s$	4.372	0.869	0.341	0.176

The solution of the differential equation, Equation (3), leads to the matrix eigenvalue problem of Equation (4):

$$\boldsymbol{K}\phi_i = \omega_i^2 \boldsymbol{M}\phi_i \tag{4}$$

where ϕ_i is the natural mode shape, with the natural frequency ω_i ($i = 1, 2, \ldots, N$).

The system stiffness and the system mass matrices, ***K*** and ***M*** respectively, have been evaluated based on the model of the chimney described in Section 4.3. The first n eigenvalues, n natural frequencies and n natural modes, of the chimney were calculated using SAP90+. A selection of the numerical results for the first four ($i = 1, 2, 3, 4$) natural vibration periods and mode shapes are presented in Table 11.1 and Figures 11.5 and 11.6 respectively.

4.5 Free vibration test and measuring devices

The most appropriate method to investigate the soil effects on the free vibrations of the industrial chimney was to measure the free full-scale vibration response of the actual chimney.

The vibration measurements focused on the determination of the lowest natural frequencies (highest natural periods) and the natural modes. Applying high-sensitivity geophone sensors, vibration velocities with frequencies as low as 0.2 Hz can be recorded. The basic data of the geophones are as follows: (*a*) type of geophone: Lennartz electronic LE-3D/5s; (*b*) frequency range: 0.2 – 40 Hz; (*c*) sensitivity: 400 V/m/s (10 . . . 16 V supply); and (*d*) RMS noise at 1 Hz: < 1 nm/s. The experimental setup was completed by a front-end system and with a laptop controlling the measurement, as shown in Figure 11.7. The position of the geophone sensor in the cross-section of the chimney is shown schematically in Figure 11.8. Excitation due to wind is permanently present and generally corresponds to white noise if the measuring time for each point is at least 1200 s. That means that the vibration response spectra are dominated by the natural frequencies [7].

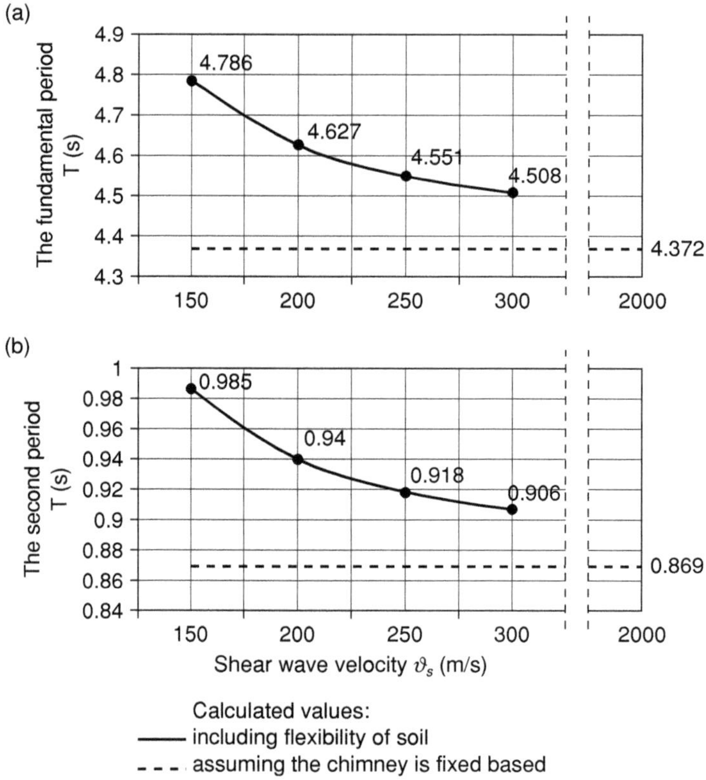

Figure 11.5 The first (*a*) and second (*b*) natural periods as functions of ϑ_s for the soil–chimney interaction

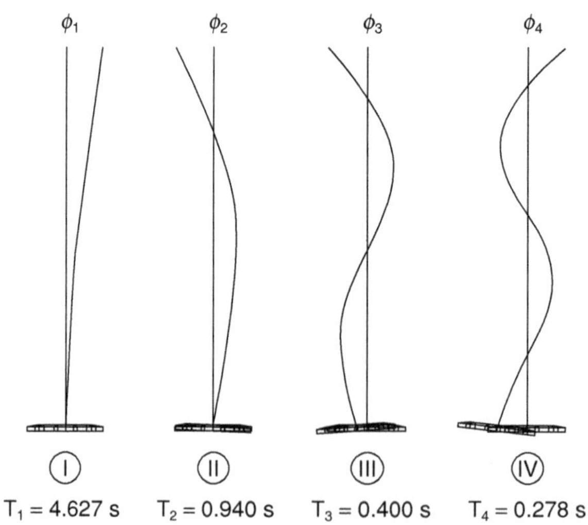

Figure 11.6 The first four natural mode shapes and natural periods of the soil–chimney interaction for ϑ_s = 200 m/s

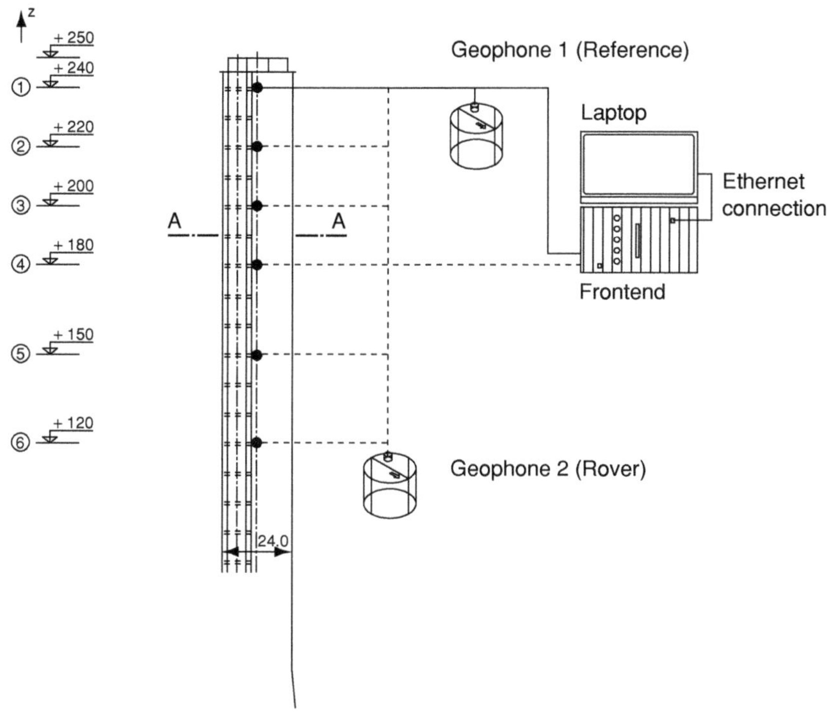

Figure 11.7 Points of measurement and the measuring devices

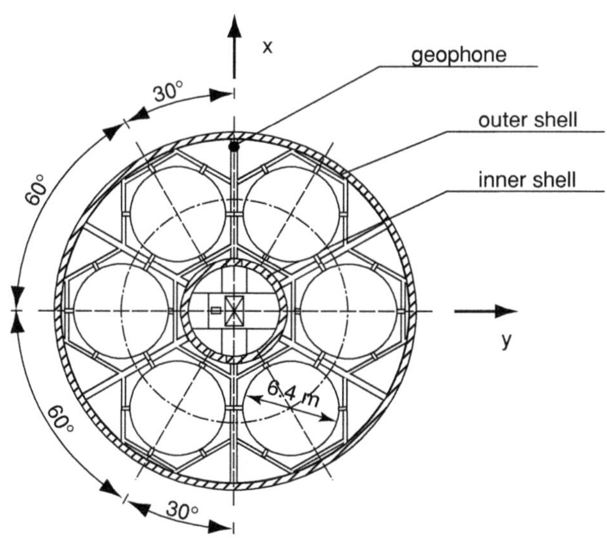

Figure 11.8 Positions of the geophone sensors in the cross-section of the chimney and the definition of coordinates

The determination of the natural modes requires the use of two geophones operating simultaneously. For this purpose, altogether six measuring points in the upper part of the chimney were chosen as shown in Figure 11.7. A geophone was fixed at reference point 1 at level 240 m near the top of the tower and a second geophone was moved, step by step, from point 1 down to the level of 120 m. From the ratio of associated peak amplitudes in the vibration response spectra of the two actual points, the jth ordinate of the corresponding i^{th} natural mode ϕ_{ij} at the actual location of the roving sensor can be determined according to [8]:

$\{\phi_{xi}\} = \{\bar{\phi}_{x,i1}, \bar{\phi}_{x,i2}, \ldots, \bar{\phi}_{x,iN}\}$ i^{th} natural mode (in the x direction)
N: number of measuring points

with

$$\phi_{x,ij} = \frac{\dot{x}_{ij}(f_i)}{\dot{x}_{ir}(f_i)}$$

j: selected measuring point (roving sensor),

f_i: i^{th} natural frequency $\left(f_i = \dfrac{\omega_1}{2\pi}\right)$,

r: reference point (fixed sensor),

$\dot{x}_{ij}(f_i)$: peak value of the ith peak in the Fourier velocity spectrum measured by the roving sensor at point j (from short-term measurements),

$\dot{x}_{ir}(f_i)$: peak value of the ith peak in the Fourier velocity spectrum measured by the fixed sensor at point r ($r = 6$ in the present case) from the short-term measurements (refer to the length of the point 1 data file which includes time data of 163.84 s). From the data the Fourier transform is calculated. This means that the mean of 8 of these spectra refers to a measuring time of 21.8453 minutes which is named as the 20-minutes average in the paper,

$\bar{\phi}_{x,ij}$: 20-minutes average of $\phi_{x,ij}$.

The signs of the mode shape ordinates results from the comparison of the phase angles in the frequency domain. Mode shapes in the y-direction are evaluated in the same way.

4.6 Comparison of theoretical and experimental results

Figure 11.9 shows the representative Fourier spectra of the vibration response due to a weak wind load, in December 2001, evaluated from measurement data at the reference point. It can be seen that the spectra are each dominated by two well-separated peaks representing the lowest two natural frequencies for the corresponding direction. The peaks are specified by the

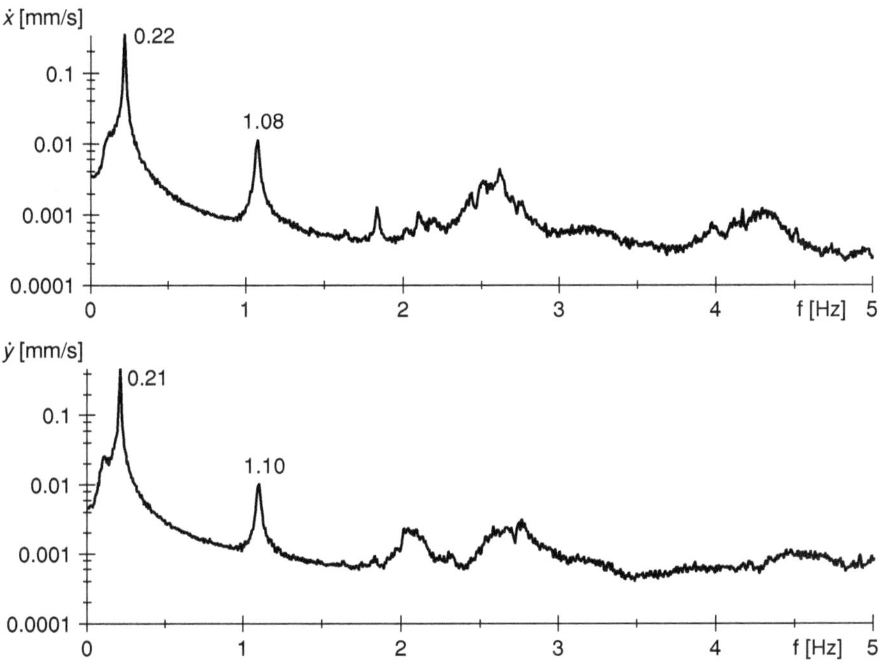

Figure 11.9 Fourier velocity spectra response at the 240 m height; point 1 (2-hour-average = 6.20-minute average)

numerical values of the frequencies. Additionally, the velocity amplitudes assigned to the first and second natural frequency for both the fixed and the roving sensor corresponding to the first and second natural mode have been determined and are shown in Figure 11.10.

The first and second natural mode in the x and the y directions determined from measurements and the natural modes calculated on the basis of the theoretical assumptions given in Section 4.3 for different values of the shear wave velocity ϑ_s are presented in Figures 11.11, 11.12 and 11.13. Tables 11.2 and 11.3 and Figure 11.14 compare the fundamental and second natural period values of the measurements and the calculations.

Aiming at a verification of the quality of the numerical model, a useful tool is given by the Modal Assurance Values (MAC), which is based on a comparison of mode shapes taking advantage of the orthogonality relation

$$MAC(N,E) = \frac{|\{\phi_N\}^T\{\phi_E\}|^2}{[\{\phi_N\}^T\{\phi_N\}][\{\phi_E\}^T\{\phi_E\}]} \tag{5}$$

where:

$\{\phi_N\}$ is the natural mode from the numerical calculations (either in the x or the y direction),

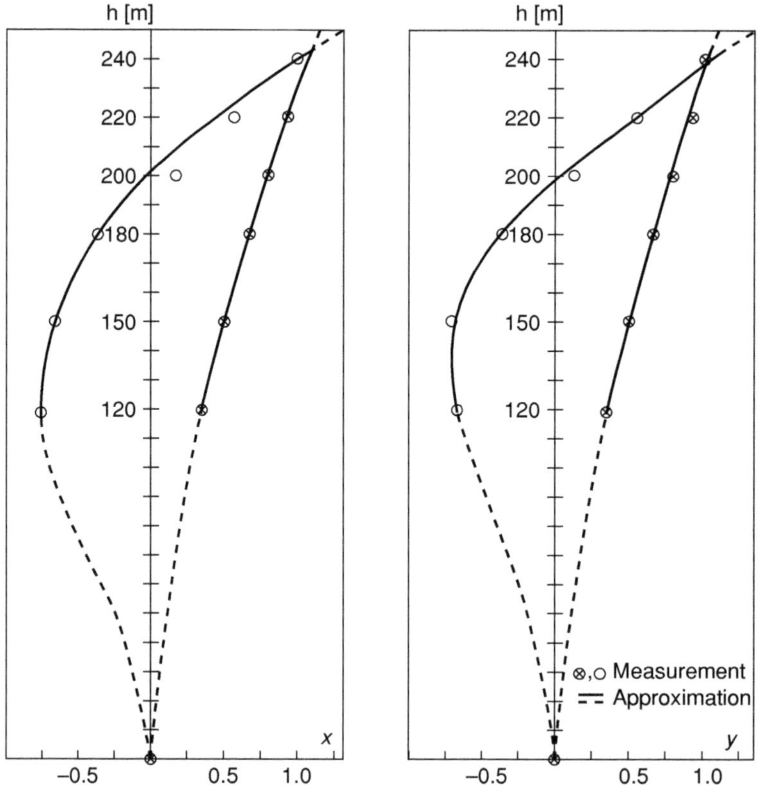

Figure 11.10 The first and the second natural mode shapes in the x and the y directions determined from measurements

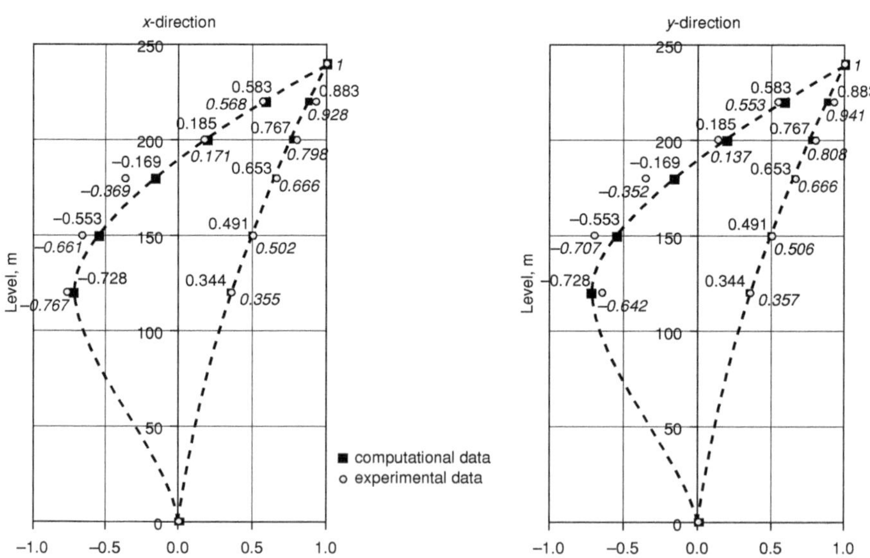

Figure 11.11 Comparison of the computed and the measured first and second natural mode shapes for $\vartheta_s = 150$ m/s

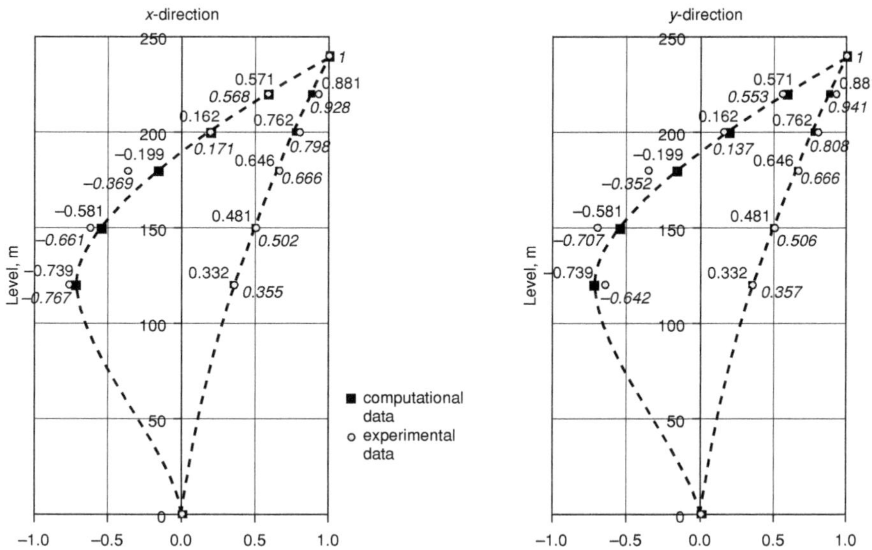

Figure 11.12 Comparison of the computed and the measured first and second natural mode shapes for $\vartheta_s = 200$ m/s

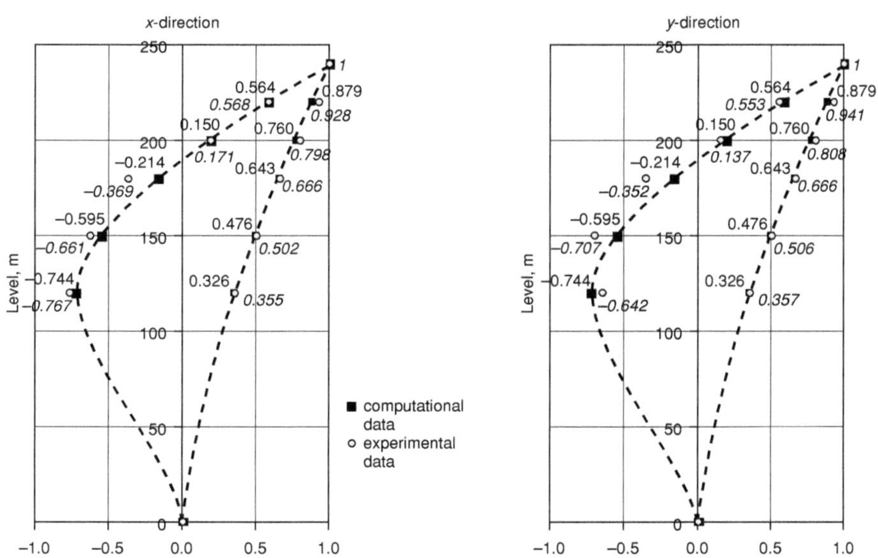

Figure 11.13 Comparison of the computed and the measured first and second natural modes shapes for $\vartheta_s = 250$ m/s

Table 11.2 Comparison of computed and experimental values of the fundamental period of the chimney

Shear wave Velocity ϑ_s	*First period* T_1, s			*The comparison values*	
	Calculation values	*Values from measurements*		$\frac{col.2 - col.3}{col.3} \cdot 100\%$	$\frac{col.2 - col.4}{col.4} \cdot 100\%$
		x-x *direction*	y-y *direction*		
1	*2*	*3*	*4*	*5*	*6*
$\vartheta_s = 150\ m/s$	4.786			5.30	0.50
$\vartheta_s = 200\ m/s$	4.627	4.545	4.762	1.80	2.83
$\vartheta_s = 250\ m/s$	4.551			0.13	4.43
$\vartheta_s = 300\ m/s$	4.508			0.81	5.33
$\vartheta_s = 2000\ m/s$	4.372			3.81	8.19

col. = column in table

Table 11.3 Comparison of computed and experimental values of the second period of the chimney

Shear wave Velocity ϑ_s	*Second period* T_2, s			*The comparison values*	
	Calculation values	*Values from measurements*		$\frac{col.2 - col.3}{col.3} \cdot 100\%$	$\frac{col.2 - col.4}{col.4} \cdot 100\%$
		x-x *direction*	y-y *direction*		
1	*2*	*3*	*4*	*5*	*6*
$\vartheta_s = 150\ m/s$	0.985			6.37	8.36
$\vartheta_s = 200\ m/s$	0.940	0.926	0.909	1.51	3.41
$\vartheta_s = 250\ m/s$	0.918			0.86	0.99
$\vartheta_s = 300\ m/s$	0.906			2.16	0.33
$\vartheta_s = 2000\ m/s$	0.869			6.16	4.40

col. = column in table

$\{\phi_E\}$ is the natural mode determined from measurement (either in the x or the y direction).

Using the MAC-values, different modes are compared. There are 2 × 2 modes available: (*a*) the first and second mode from computation, (*b*) the first and second mode from measurement. This means that comparing modes from the computation and the measurement altogether 2 × 2 = 4 MAC-values can be calculated. These 4 values are shown in Figure 11.15 (see the notations of the axes in Figure 11.15). The very good value of the MAC proves the quality of the measurement and the calculation model.

Considering the computation of MAC (Figure 11.15) for the first two natural mode shapes in the x and the y direction, good agreement of the

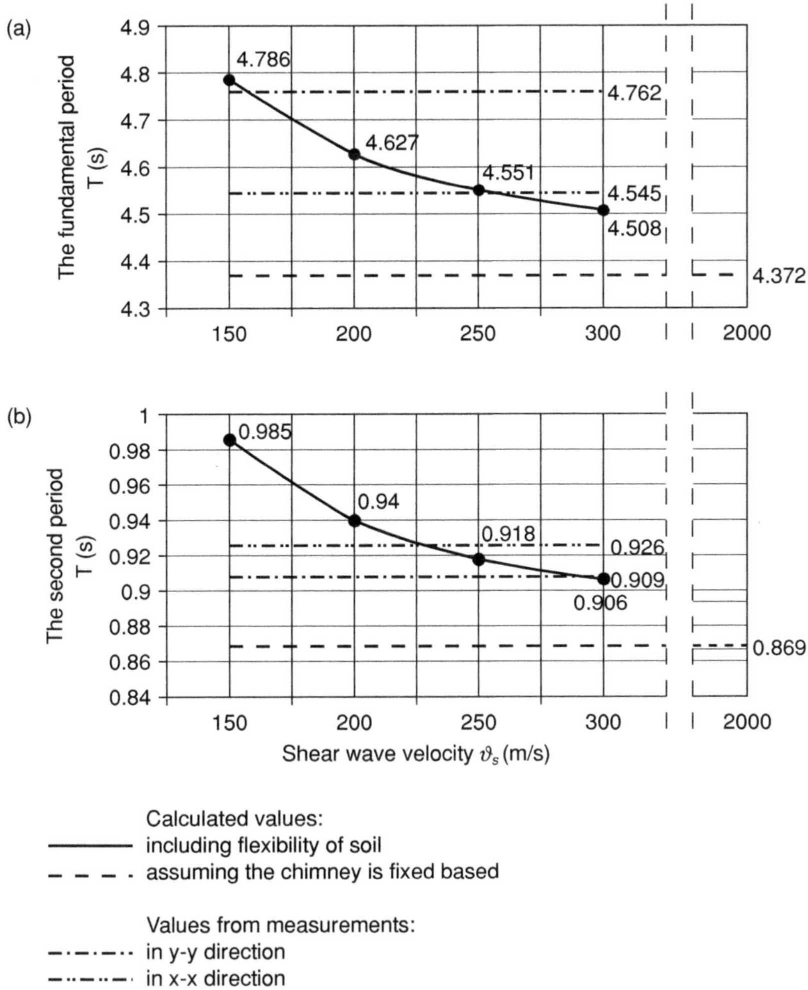

Figure 11.14 Comparison of the computed and the experimental values: (*a*) first natural period of the chimney, (*b*) second natural period of the chimney

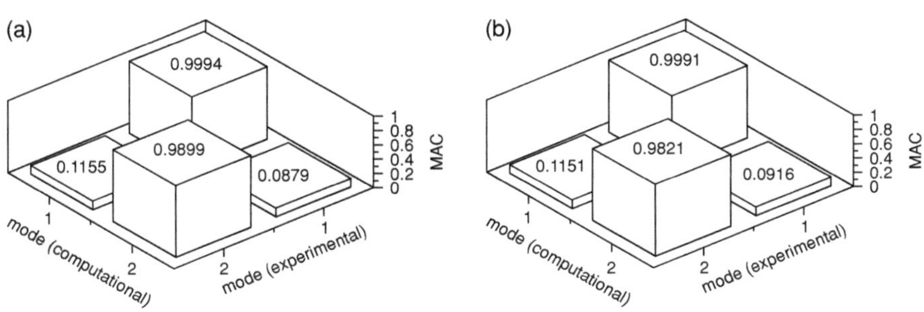

Figure 11.15 Computation of the MAC-values in the x (left) and y (right) direction for $\vartheta_s = 250$ m/s

experimental and numerical mode shapes is confirmed. The main diagonal elements approximate the ideal value of one. Off-diagonal elements move about 0.1, indicating that different mode shapes can be interpreted as being linearly independent.

5 Conclusions

1. Based on the application of two geophone sensors used in the experimental investigation of the free vibration of the 250 m high industrial chimney, the first and second natural periods and natural shape could be separated.
2. Soil flexibility under the foundation of the chimney has considerable influence on the natural modes and natural periods. The theoretical values for the first and second natural periods with and without soil influence differ as follows:

   - about 5.83% for the first natural period for $\vartheta_s = 200$ m/s,
   - about 8.17% for the second natural period for $\vartheta_s = 200$ m/s.

   This has been shown to be correct in the experiment.
3. The natural periods are influenced by the kind of soil and soil moisture content under the chimney foundation. For the soil and the low soil moisture content considered, for which $\vartheta_s > 200$ m/s (ϑ_s about 250 m/s), this influence is small for the first and second natural periods.
4. The industrial chimney of the power station in Opole has only one plane of symmetry because two flues are unused and not ready for use. This was confirmed in the tests in which the first natural periods in the x and the y directions differed by about 4.5%, but for the second natural periods the difference was about 1.8%.
5. The comparison of the first natural mode shape from the computations and the experiment leads to a difference of up to 9%. For a corresponding comparison of the second natural mode, the difference is in the range of 3% to 54%. The high value of 54% is obtained by comparing mode shape components close to the node near to the height of 200 m where amplitudes are small.

Acknowledgments

This research has been conducted as a Joint Research Project of the University of Technology at Opole, Poland, and the Brandenburg University of Technology at Cottbus, Germany. The author would like to thank the authorities of these two universities for their support of this work. Part of this work was funded by the Commission of the European Communities under the FP5, contract Number G1MA-CT-2002-04058 (CESTI)

References

Beirow, B., Schroth, G. and Osterrieder, P. Dynamic diagnostic of transmission towers, *Proc. Second International Conference 'Structural Dynamics Modelling'*, Glasgow: Bell & Bain, 1996.

Breuer, P., Chmielewski, T., Górski, P. and Konopka, E. Application of GPS technology to measurement of displacements of high-rise structures due to weak winds, *J. Wind Eng. Ind. Aerodyn.*, 2002, 90: 223–30.

Castelani, A. *Construzioni in zona sismica*, Milan: Masson Italia Editori, 1983.

Chmielewski, T., Górski, P., Beirow, B. and Kretzschmar, J. Theoretical and experimental free vibrations of tall industrial chimney with flexibility of soil, *Engineering Structures*, 2005, 27: 25–34.

Chmielewski, T. and Zembaty, Z. *Dynamics of Structures*, Warsaw: Arkady, 1998. In Polish.

Chopra, A. K. Dynamics of Structures, Upper Saddle River, NJ: Prentice Hall International, 1995.

Kausel, E. Res. Rep. R74-11, Department of Civil Engineering, MIT, Cambridge, Mass., 1974.

Luz, E. Zur experimentellen Modalanalyse von Bauwerken, *Materialprüfung*, 1986, 28: 301–6.

Solari, G. and Stura, D. An evaluation technique for vibration modes of structures interacting with soil, *Engineering Structures*, 1981, 3: 225–32.

12 Assessment of settlements of high-rise structures by numerical analysis

Rolf Katzenbach, Gregor Bachmann and Christian Gutberlet

1 Foundations of high-rise buildings

The development of foundation techniques for high-rise buildings can be shown by the examples of Frankfurt am Main. After 1950, a massive increase in the importance of the service sector started, resulting in an increasing number of high-rise buildings, typical of financial centres, e.g. Frankfurt am Main, New York, London and Hong Kong. Thus, Frankfurt am Main not only grows in size, but also in height (Figure 12.1).

The first generation of high-rise buildings in Frankfurt am Main were founded shallowly, with respect to the foundation techniques available at that time, and reached settlements of between 20 cm and 34 cm owing to

Figure 12.1 Skyline of Frankfurt am Main

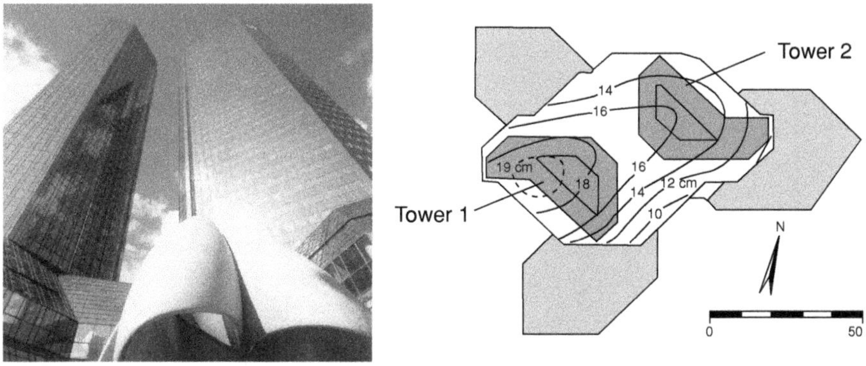

Figure 12.2 View and settlement plot of the Deutsche Bank towers, Frankfurt am Main

the settlement-sensitive Frankfurt Clay. Because of the settlements, and especially the differential settlements of these high-rise buildings, corrections to the shallow foundations had to be made during and after the construction. A prime example is the shallowly founded towers of the Deutsche Bank where settlements of between 10 cm and 22 cm were reached (Figure 12.2).

These settlements and differential settlements were handled, as far as possible, during the construction by adjusting the structure and using special construction specifications for the elevators and the glass façade. In order to control the differential settlements between the different sections of the building, hydraulic jacks were installed by which the level of the flat sections could be regulated. The system proved itself, but it required a great technical effort, turning the entire building almost into a machine. A similar approach was applied to the old building of the 166 m high Dresdner Bank (Figure 12.3 left). In this case, compression cushions made of rubber (5 m × 5 m, Figure 12.3 right) were placed beneath the northeastern corner of the building and afterwards filled with water to compensate for the tilting. After the construction process of the building was completed, the water in the compression cushions was replaced by concrete.

These correction techniques, which were used on many high-rise building projects during the 1970s and the 1980s, were laborious and did not satisfactorily solve the problem. Consequently, there was a search for a more adequate and economic foundation technique. This led to the development of the Combined Pile-Raft Foundation (CPRF) (Katzenbach and Reul 1997) and similarly to the developments in London (Hooper 1973, Hight and Green 1976, Cooke et al. 1981). The CPRF is a new design concept, rather than a new foundation type, as the foundation piles are not treated as 'stand-alone', as for a conventionally designed pile foundation, but are considered as part of a complex foundation. As the stability of the foundation in terms of base failure at a CPRF is usually guaranteed by means of

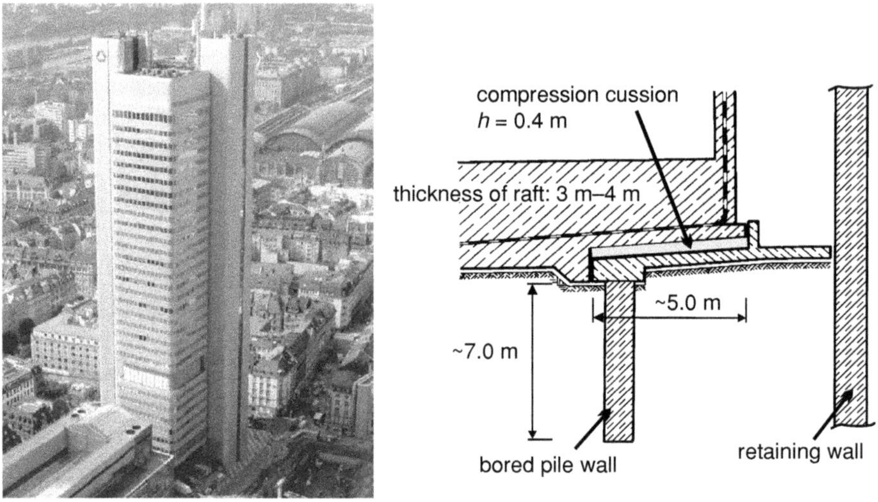

Figure 12.3 View and section with the compression cushions of the Dresdner Bank building, Frankfurt am Main

the foundation raft, the piles of a CPRF mainly have the task of reducing settlements. A positive effect of the CPRF is that the deep foundation elements can be arranged systematically at places with high loads, or eccentric loading from the superstructure, in order to harmonize the settlements, e.g. the Congress Centre Messe Frankfurt (Barth and Reul 1997) or the Skyper Building (Katzenbach et al. 2005a), and to reduce the risk of punching shear and, thus, effectively reduce the thickness of the raft (Love 2003).

The overall behavior of a CPRF is decided by the interactions occurring during loading (Figure 12.4), which were identified by Butterfield and Banerjee in 1971 as follows:

- pile–soil interaction
- pile–pile interaction
- raft–soil interaction
- pile–raft interaction

The interactions of a CPRF led to a more or less monolithic compound system of foundation and soil with a significantly increased equivalent Young's modulus compared to the undisturbed subsoil (Randolph 1994). So, for a raft foundation which is stable without piles, a small number of piles might reduce settlements and correct inclinations (Burland et al. 1977). This stabilizing effect does not proportionally increase with the number of piles, but rather tends to a limit depending on the number of piles. Adding

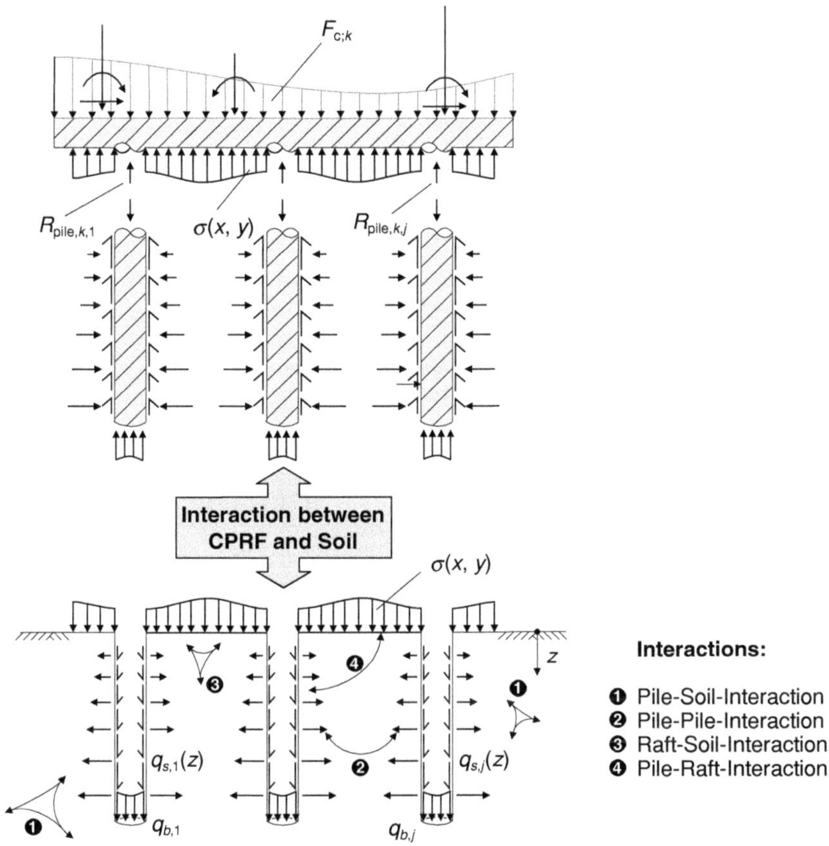

Figure 12.4 Schematic sketch of a Combined Pile Raft Foundation (CPRF) and the interactions linking to the bearing behavior

further piles, in the same area beyond this limit, will not lead to further settlement reduction (Hooper 1979, Cooke 1986).

Owing to the interactions previously mentioned, the stress states at the shaft of a CPRF pile are distinctly different from the stress states in the subsoil surrounding a conventional pile. The part of the load which is transferred, via the raft, into the subsoil increases the hydrostatic pressure on the pile shafts; this increase in the normal stresses from the stress level of a conventional pile foundation σ'_{pile} to the stress level of a CPRF σ'_{CPRF} is called $\Delta\sigma_{\text{compression}}$ (Figure 12.5) due to the raft–soil interaction. So the failure shear stress $q_{s,f}$ at the pile shaft, according to the failure criterion of Mohr–Coulomb, is computed by:

$$q_{s,f} = \sigma'_{\text{CPRF}} \cdot \tan\varphi' + c' = (\sigma'_{\text{pile}} + \Delta\sigma'_{\text{compression}}) \cdot \tan\varphi' + c' \tag{1}$$

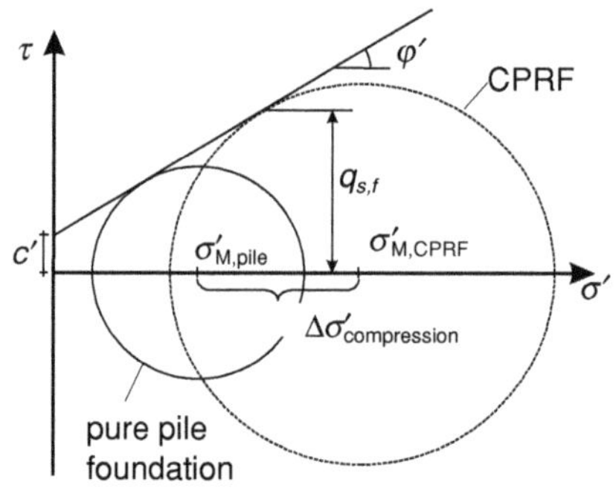

Figure 12.5 Effect of increased normal stress levels at a CPRF pile in comparison to a conventional pile foundation

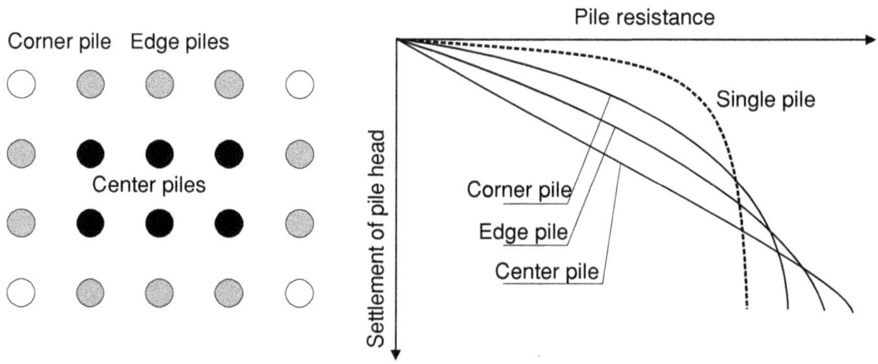

Figure 12.6 Effect of pile position on bearing behavior (Kempfert and Smoltczyk 2001)

So the maximum shear stress at the pile shaft is always larger for a CPRF than for a corresponding conventionally designed pile foundation (Katzenbach et al. 2006).

The interactions between the structural elements of a CPRF also influence the bearing behavior of the piles depending on their positions. Owing to the stronger compression, a center pile is not able to show distinctly plastic behavior; it rather remains elastic throughout the loading. Contrary to this, corner piles exhibit a bearing behavior which is more similar to that of a single pile but still influenced by the interactions (Figure 12.6).

The application of CPRFs is limited by the extent of the transferred loads and the subsoil characteristics. High-rise buildings may be too heavy to be founded on CPRFs: for example, the Commerzbank building in Frankfurt am Main (Figure 12.7 left). Initial designs for a CPRF for the Commerzbank

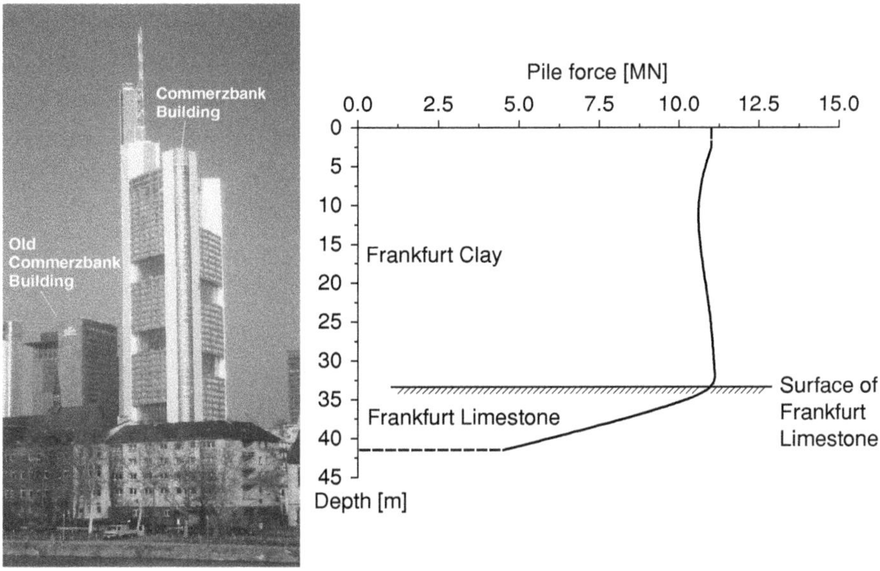

Figure 12.7 (*a*) left: View of the Commerzbank Building in Frankfurt am Main; (*b*) right: pile force distribution of the rock pile foundation (Holzhäuser 1998)

building were unsuccessful, so eventually a rock pile foundation was designed, as the piles have been extended into the massive limestone layers which are situated beneath the Frankfurt Clay. By means of the rock piles, the settlements and tilting of the adjacent buildings, mainly the old Commerzbank Building, could be minimized. As measurements in the piles show, all the forces are more or less directly transferred to the part of the pile which is situated in the compact limestone, so the shaft friction in the softer Frankfurt Clay and the participation of the raft, via base pressure, is negligible (Holzhäuser 1998). This can be shown by the distribution of the pile force in one of the central piles of the Commerzbank Building (Figure 12.7 right) exhibiting hardly any shaft friction mobilization in the Frankfurt Clay which is typical for end-bearing piles under such conditions.

For this reason, for CPRFs, it is required that the piles do not attract too much load; otherwise the interactions between raft, soil and pile shaft become too weak and the CPRF design is not effective. For example, the CPRF piles of the Main Tower (Figure 12.8) end just before the surface of the massive limestone layer, so that small settlements occur but are large enough to activate the resistances in the Frankfurt Clay.

The knowledge gained from the developments concerning the founding of high-rise buildings in Frankfurt am Main led to the German CPRF Guideline (Hanisch et al. 2002) compiling all the requirements on design

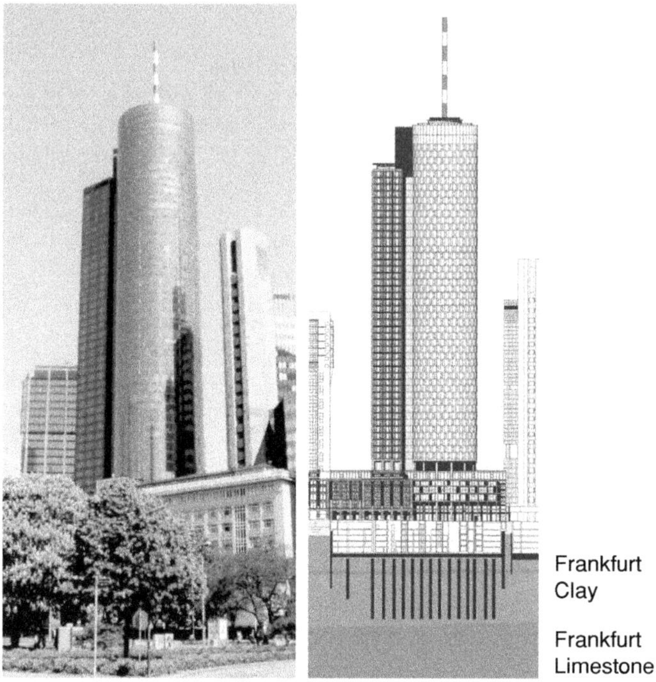

Figure 12.8 View and section of the main tower in Frankfurt am Main

and construction of a CPRF. According to this guideline, the bearing behavior and the load transfer within a CPRF have to be monitored by a geotechnical expert, qualified in this subject owing to the knowledge demands originating from the soil, the superstructure and the foundation required by the observational method (Peck 1969). The monitoring of a CPRF is an indispensable component of the safety concept and is used for the following purposes:

- verification of the computational model and computational approach,
- realtime detection of possible critical states,
- examination of the calculated settlements during the construction process, and
- quality assurance related to the recording of evidence

during the construction process and the service life of the building.

Moreover, the guideline comprises the requirement for an adequate computational model which has to be validated by back-analysis of a CPRF under comparable conditions (see section 3). This back-analysis has to be based on geotechnical measurements which provide the database for subsequent simulations. The measurements performed on current projects can be used for a new stage in an optimization cycle for the computational model. Thus, a steady updating of the computational model is achieved, which is only possible by evaluating a sufficient amount of field measurement data.

2 Measuring geotechnical field data as the basis for a computational model

The evolution of a geotechnical computational model is both dependent upon the results of laboratory experiments and linked to the collection of sufficient data from field measurements. These field measurements comprise not only data gained by subsoil exploration but also knowledge of soil, the structure's deformations and force flows. This knowledge can only be gained by a comprehensive geotechnical monitoring program. The measurements usually applied in geotechnical measurement programs comprise, in addition to the geodetical measurements, the following components:

- extensometers
- inclinometers
- load cells
- contact pressure cells
- pore pressure cells
- strain gages

More detailed information can be found in Dunnicliff (1993).

Extensometers measure the deformation of the soil in the longitudinal direction of a bore hole. The composition of several extensometers to a multi-point extensometer, as shown in Figure 12.9, measures the deformation

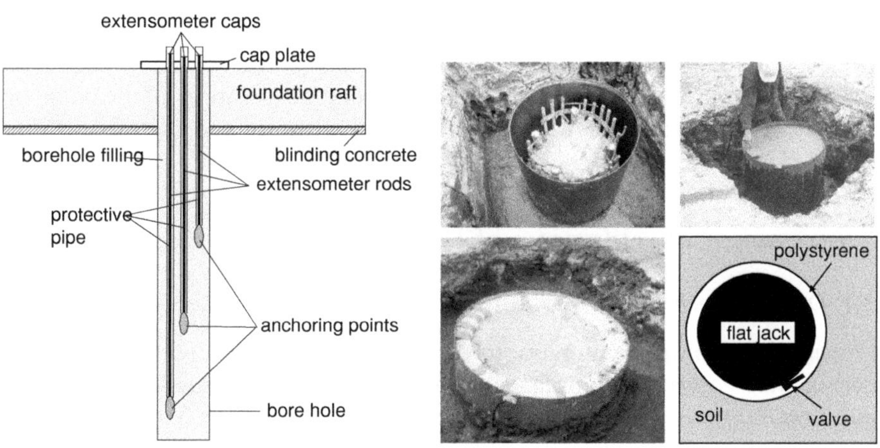

Figure 12.9 Multi-point bore-hole extensometer (left) and pile head during the preparation of a load cell installed on the pile head (right)

at different depths. Such multi-point bore-hole extensometers are used under the foundation raft as well as those adjacent to the foundation.

Inclinometers are used to measure the course of inclination of the axis of a pipe fixed in a bore hole. The lateral displacement of the observed axis is calculated from the resulting measurement.

Load cells consist of a flat jack filled with oil where the external load is measured by the pressure of the fluid. The load cell is placed on a mortar bed on the prepared pile head (Figure 12.9). The diameter of the pressure cell is smaller than the pile diameter to protect the valve, the remaining sides usually being filled with a polystyrene ring.

The total contact pressure under the raft is recorded by contact pressure cells which are usually embedded in the blinding concrete. However, it must be remembered that the ratio of the area covered, using contact pressure cells, is very small in comparison to the area covered by the raft. Therefore, relying on too small a number of contact pressure cells may lead to significant misjudgments.

Piezometers are installed under the raft to evaluate the pore water pressures. Effective (that means inducing settlements) contact pressures are calculated from the difference between the total pressures (obtained from contact pressure cells) and the neutral pressures (obtained from the piezometers). For this reason it is advantageous to place these types of measuring devices close to each other.

Axial strains in the piles are measured as a mean value over a distance from about 1 m to 3 m. With the knowledge of the stiffness of the concrete – which may not be easily determined – the pile load in the actual section can be calculated. With two or more strain gages, each measuring the strains in two directions in the cross-section, the axial strain is measured as a mean value. The measuring devices are placed in pairs on opposite sites at the inner side of the reinforcement cage.

Geodetic measurements are undertaken not only for the preservation of evidence for adjacent buildings but also for displacement monitoring used for the observational method. The permanent observation of heaving or settlement of buildings can be carried out by a motorized digital levelling device (Figure 12.10 left, Katzenbach et al. 2005a).

3 Assessment of settlements of high-rise structures

3.1 Development of a computational model

Owing to the complex interactions of CPRF, the computation for the design necessitates a similar complex computational model. Usually, such computations are carried out by means of the Finite Element Method (FEM); however, sometimes different numerical methods such as the Boundary Element Method are used (Davis and Poulos 1972, O'Neill and Ghazzaly 1977, Chow 1987, Kuwabara 1989, Viggiani 1998). Closed-form solutions

Figure 12.10 Motorized digital levelling device (left) and levelling board (right)

or empirical approaches are used only for approximate calculations or for special cases.

The computational model, in general, consists of:

- spatial discretization, i.e. the modeled section, the boundary conditions, the modeling of the constituent structural parts (subsoil, raft, piles, etc.) and the meshing,
- constitutive relationships, and
- time discretization including the initial conditions.

By numerical modeling, the bearing behavior of a pile can be correctly assessed only if the computational model is validated. This is due, mainly, to the special effects occurring in the pile shaft zone during shearing. Shearing in granular media is always linked to the formation of shear banding (Vardoulakis 1980) and large plastic strains, especially in the case of material softening, causing a loss of the well-thought-through mathematical description of the system (Hill 1962). This, again, causes numerical instabilities and, much more importantly, a general dependence of the computational results on the mesh refinement. The width of shear bands computed by the conventional approach is always oriented on the average element length in the model region concerned. This can be overcome by using higher-order continuum approaches (regularization by introducing an internal length of the material) or by calibrating the computational model with given data.

Regarding deep foundations, this phenomenon is concerned with the shear zone at a pile shaft. The discretization of the shaft zone can be carried out by different methods; the most used variant is coupling the soil and pile by shaft zone elements (Figure 12.11). These are continuum elements, with a specified width normal to the shaft surface, which exhibit the same material behavior as the remaining soil elements but fix the shear band to a certain width owing to the previously described dependence.

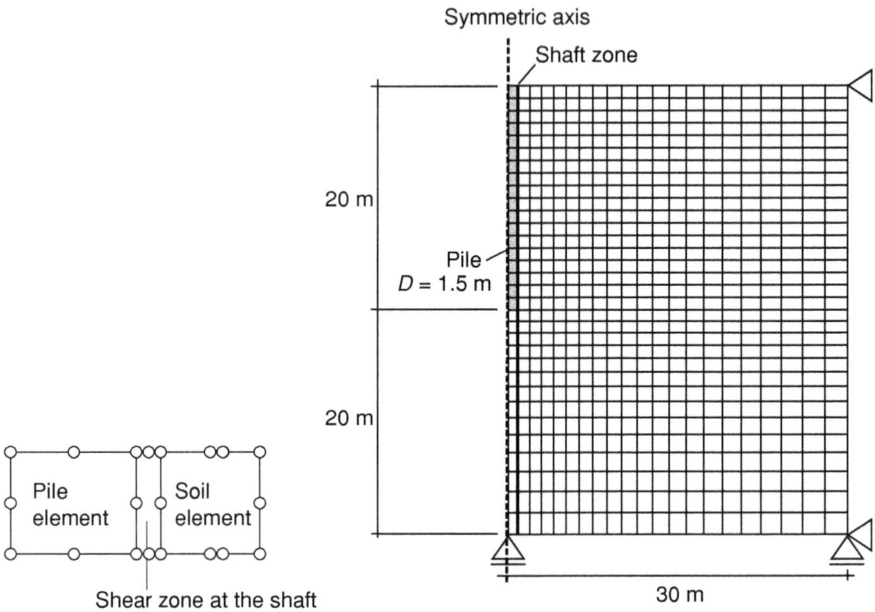

Figure 12.11 left: Discretization of the shear zone between the pile and the soil by means of continuum elements with a specified width; right: spatial discretization of the computational example

The influence of the element width on the computational results is shown in the subsequent example. A single bored pile with a length of 20 m and a diameter of 1.5 m is modeled using axial symmetry. The whole spatial discretization is shown in Figure 12.11.

The time discretization includes an initial primary stress state with a geostatic stress distribution, the excavation of the soil material in the contour of the pile, the pile installation and, lastly, the loading of the pile head by applying displacements on the upper pile nodes. The material behavior of the pile has been simulated in the analysis as linear-elastic, whereas for the simulation of the soil material behavior an elasto-plastic model was used (Figure 12.12). The constitutive model for the soil consists of two yield surfaces: the pressure-dependent, perfectly plastic shear failure surface F_S (cone) and the compression-cap yield surface F_C (cap). The equation for the cone (Equation 2), which is directly adopted from the classical Drucker–Prager failure criterion (Drucker and Prager 1952), resembles the Mohr–Coulomb failure criterion utilizing deviatoric stresses t and hydrostatic stresses p instead of the shear stress τ and the normal stress σ respectively. Corresponding to this, the friction angle φ and the cohesion c are converted into their analogies β and d respectively.

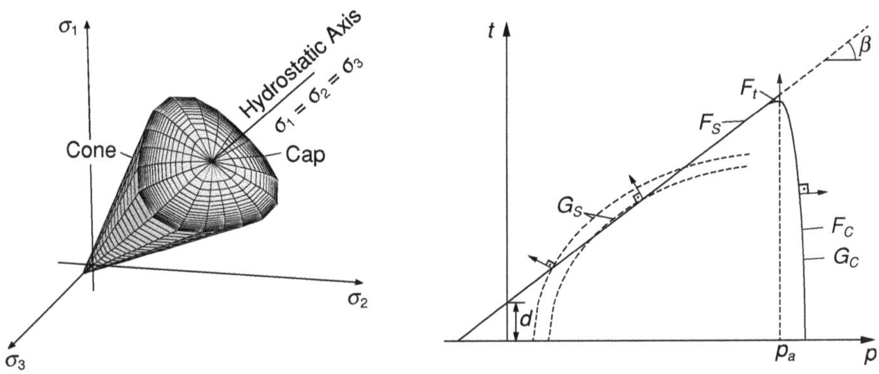

Figure 12.12 Yield surfaces in the 3D stress space (left) and in the *p*-*t*-plane (right) of the modified Drucker–Prager/cap model

$$F_S = t - d - p \tan \beta = 0 \tag{2}$$

with: t deviatoric stress with a distinction between compression and extension by

$$t = \frac{1}{2}q\left(1 + \frac{1}{K} - \left(1 - \frac{1}{K}\right)\cos(3\theta)\right)$$

q deviatoric stress in general
θ Lode's angle
d cohesion in the *p*-*t*-plane
p hydrostatic stress
β friction angle in the *p*-*t*-plane

Stresses inside the yield surfaces cause only linear elastic deformations while stresses on the yield surfaces lead to plastic deformations (Chen and Mizuno 1990). Stresses on the cap (Equation 3) involve volumetric hardening: i.e. plastic strains in addition to the elastic strains are caused, and the cap is moved by the increasing stresses. The relationship between the hydrostatic stresses and the plastic strains forms the hardening rule (Figure 12.13); an initial cap position at non-zero strains and non-zero stresses respectively represents a pre-loading and thus an over-consolidation state. Contrary to this, shearing can cause softening, i.e. the cap is taken back to the actual stress state until the critical state, with vanishing volume strains, is reached. The evolution of the plastic strains is guided by the plastic potential (Figure 12.12 right), which is formulated by Equation 4 for the cone and by Equation 5 for the cap.

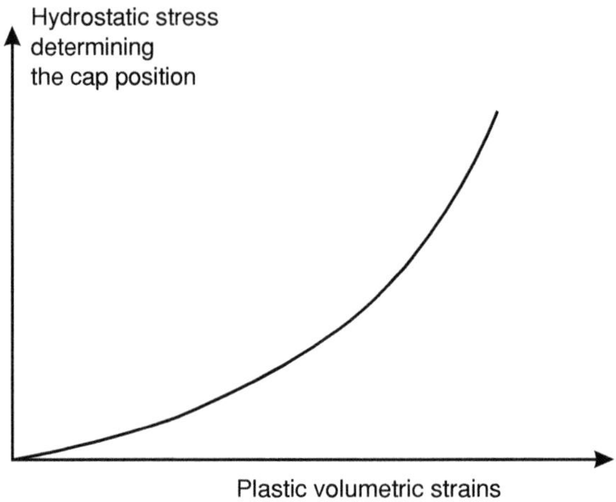

Figure 12.13 Hardening rule of the modified Drucker–Prager/cap model

$$F_C = \sqrt{(p - p_a)^2 + \left(\frac{Rt}{1 + \alpha - \dfrac{\alpha}{\cos\beta}} \right)^2} - R(d + p_a \tan\beta) = 0 \tag{3}$$

$$G_S = \sqrt{[(p - p_a)\tan\beta]^2 + \left[\frac{t}{1 + \alpha - \dfrac{\alpha}{\cos\beta}} \right]^2} \tag{4}$$

$$G_C = \sqrt{(p - p_a)^2 + \left(\frac{Rt}{1 + \alpha - \dfrac{\alpha}{\cos\beta}} \right)^2} \tag{5}$$

With this model, three variant computations with different shaft zone widths were carried out. Varying the width of the shaft zone elements provides distinctly differing relationships between the activated pile resistance, which is equal to the applied load, and the settlements occurring at the pile head. As is depicted in Figure 12.14, increasing the width of the shaft zone elements means an increase in the 'elastic range' of the overall system including the pile and the subsoil and, thus, an increase in bearing capacity. Therefore, the bearing capacity and the settlement behavior of piles cannot be computed straightforwardly, but rather the computational model has to be calibrated by back-analyzing load tests or case histories.

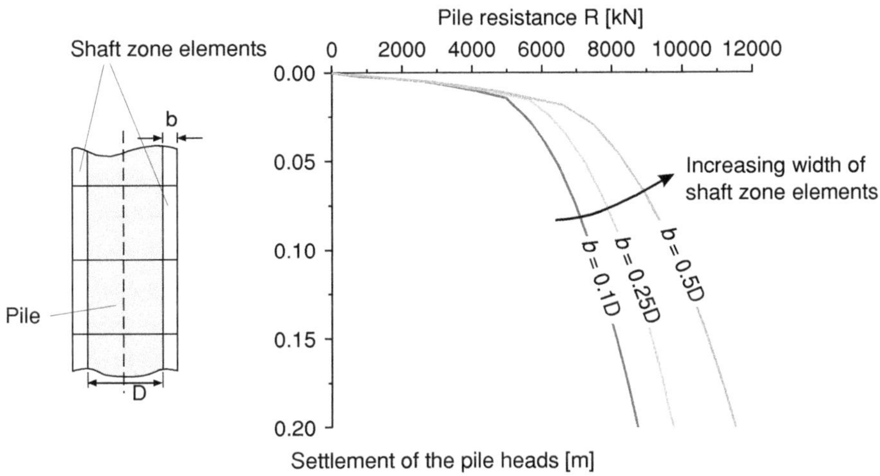

Figure 12.14 Differing pile resistance–settlement relationships for varying widths of shaft zone elements

In the previous example, a quite complex constitutive model was applied. In engineering practice, often simpler models are used, e.g. the linear-elastic Mohr–Coulomb model or the classical Drucker–Prager model. When using such models, attention has to be given to the angle of dilatancy ψ which is constant in the models previously referred to. During elastic behavior, dilatancy is related to the Poisson's ratio ν, while during plastic shearing the angle of dilatancy ψ overrules ν (Figure 12.15). The assumption of a constant angle of dilatancy is a simplification as over-consolidated cohesive soils and dense non-cohesive soils usually exhibit a change in dilatancy behavior during shearing. The estimation of the angle of dilatancy must be correct, as the influence on the bearing behavior of the simulated pile is large. Potts (2003) shows that a numerical computation of a single pile with a high angle of dilatance exhibits no distinct failure, but rather a continuous increase in the resistance forces (Figure 12.16 left); this is due to the restraints resulting from the attempts of the soil to have a large volume expansion under shearing. In the case of a zero angle of dilatancy, the volume expansion is equally zero, resulting in much more realistic computational results (Figure 12.16 right).

Generally, it is important to choose an appropriate constitutive model for each soil. Hypo-plastic models have proven to be adequate for modeling the material behavior of non-cohesive soils (Gudehus, Kolymbas) while elasto-plastic models such as the Cam Clay model (Schofield and Wroth 1968) and the modified Drucker–Prager/cap model are accepted as adequate constitutive models for cohesive soils. If the time-dependent evolution of stresses and strains or forces and displacements respectively is to be considered, a coupled consolidation analysis should be performed.

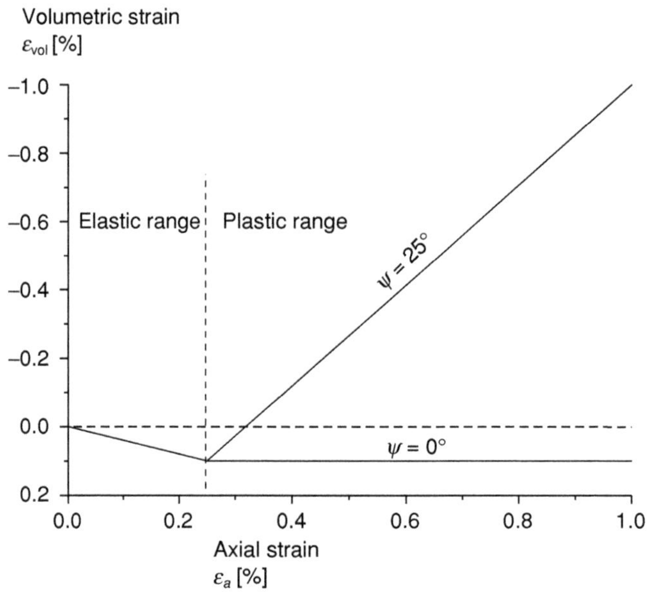

Figure 12.15 Dilation during a simulated triaxial test shown by means of the volumetric strains ε_{vol} versus the axial strains ε_a (Potts 2003)

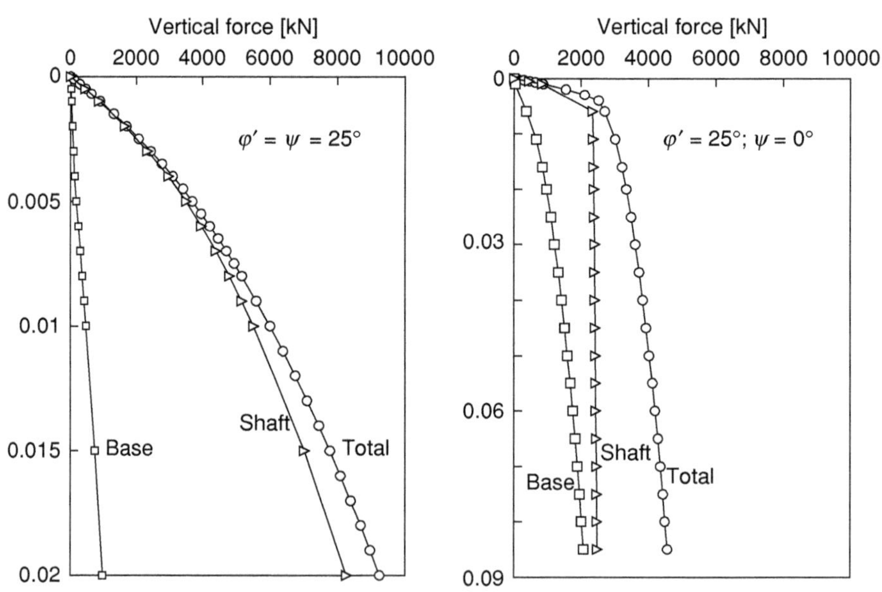

Figure 12.16 Influence of the angle of dilatancy on the computational results of a simulated pile (Potts 2003)

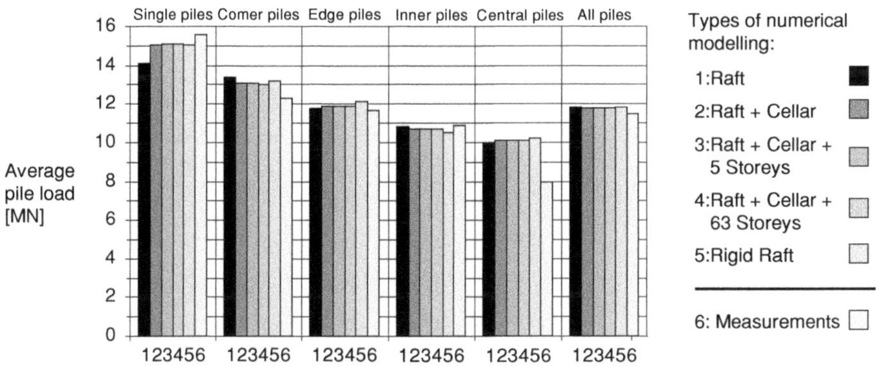

Figure 12.17 Results of the comparative investigations on the influence of the level of abstraction of the superstructure (Holzhäuser 1998)

A further question concerns the size of the mapped region and the level of abstraction of the superstructure. As the FEM, as applied in geotechnical engineering, provides approximate solutions to boundary value problems in continuum mechanics, the correct choice of the size and, thus, the position of the boundaries is highly important. In general, the boundaries should be positioned such that the changes due to the loading history do not cause significant strains in the regions close to the boundaries. Concerning the depth of the modeled region, an approach similar to soil exploration can be undertaken. Using the specifications for the soil exploration formulated in EC 7, part 2, concerning the level of abstraction of the superstructure, investigations by means of comparative back-analyses of the Commerzbank Building (Figure 12.17) indicate that modeling the raft is sufficient, so that the modeling of the superstructure of the buildings is unnecessary (Holzhäuser 1998).

Using the above information, the development, validation and successful application of a computational model for the assessment of settlements of high-rise structures is shown in the following section. The computational model was calibrated by back-analyzing the settlements of the Messeturm in Frankfurt am Main (Figure 12.18). The 256 m high Messeturm is situated in the exhibition district in Frankfurt am Main. A shallow foundation, as considered in the initial design, would have caused settlements of up to 40 cm plus large differential settlements. Consequently, the building was founded on a CPRF consisting of 64 bored piles of diameter 1.5 m arranged in three rings with staggered lengths of up to 35 m. The subsoil is composed of a surface-near quaternary stratum of mainly sand and gravel reaching down to 10 m beneath the surface, underlain by tertiary Frankfurt Clay. At the place of the Messeturm, the Frankfurt Clay reaches down to depths of 70 m to 75 m where it is underlain by compact limestone layers.

Figure 12.18 Messeturm in Frankfurt am Main

The FE model used for the analysis mapped one-eighth of the whole raft (Figure 12.19) utilizing the threefold symmetry of the construction. The analysis comprised several steps including the excavation process and the groundwater lowering and re-rising (Reul 2000).

The material behavior of the piles and the raft was simulated as linear-elastic in the FE analysis, whereas for the simulation of the material behavior of the soil the modified Drucker–Prager/cap model was used.

The calculated settlements of 19 cm differed from the measured values of 14 cm. The deviation was ascribed to the consolidation of the Frankfurt Clay which was still ongoing when the settlement measurement was carried out. The basic shape of the settlement distribution of the raft is, in both cases, nearly equal (Figure 12.20). However, the results of the numerical analysis match the measurement data very well. This was a first successful step in validating the computational model.

3.2 Assessment of settlements of high-rise buildings for geotechnical serviceability limit state (SLS) proofs

The main purpose of assessing settlements of high-rise buildings is to provide proof of the building's serviceability, including the proper functioning

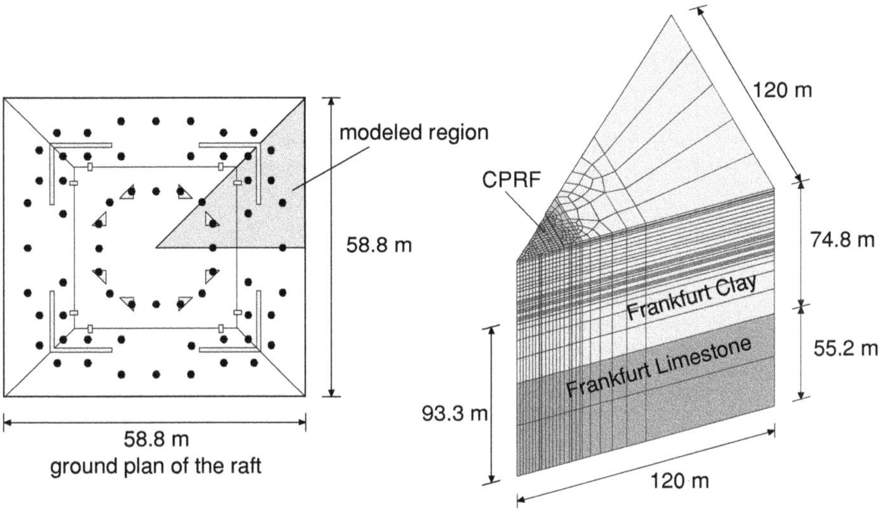

Figure 12.19 Ground plan and FE mesh of the Messeturm

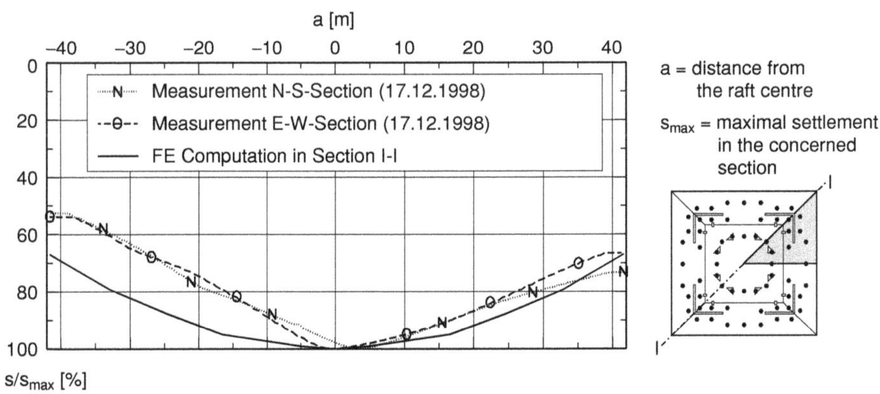

Figure 12.20 Distribution of the relative settlements of the Messeturm CPRF (Reul 2000)

of installations, elevators, etc., as well as the unequal settlements between different parts of the building. Generally, differential settlements are very important, and the resulting angular distortions which have to be limited with respect to evolving demands to safeguard the serviceability and the stability of the structure (Figure 12.21).

The computational model, which has been validated in the previous section, was applied to subsequent analyses, e.g. simulations for the Eurotheum building (Katzenbach et al. 2005b) and the Main Tower (Moormann and Katzenbach 2002). The latest example of the application of the computational

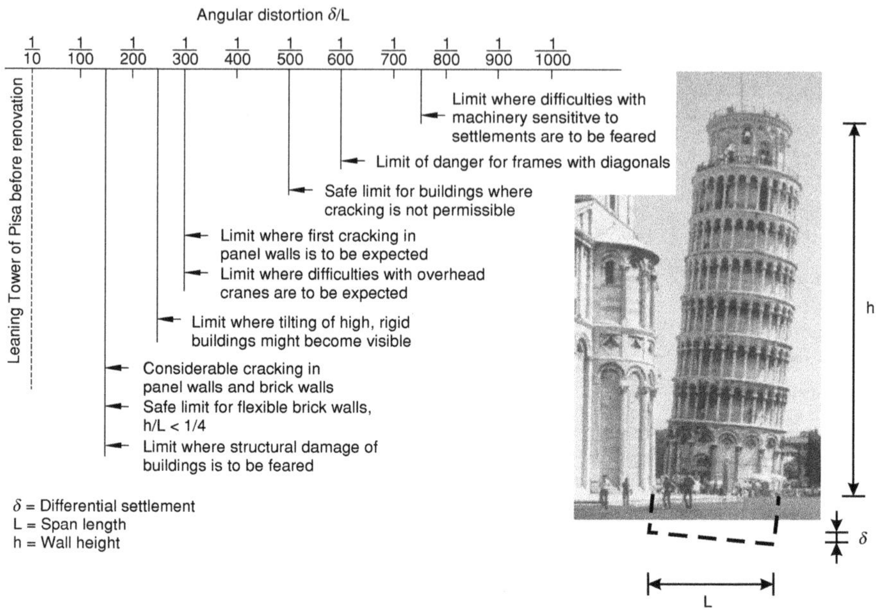

Figure 12.21 Suggested relationship between angular distortion and building performance according to Bjerrum (1963)

model is the settlement calculation for the Skyper building (Figure 12.22, Katzenbach et al. 2005a). This office complex comprises a 151 m high tower consisting of 40 stories above ground and an adjacent office and residential building consisting of 6 stories leading to a load eccentricity in the building. The whole complex has a 3-story basement beneath the surface which has a maximum excavation depth of 14 m in the area of the tower and up to 11 m in the remaining area.

Owing to the lack of symmetry within the building, the FE mesh comprises a large number of elements. The boundaries in the horizontal plane have been determined by considering the zone influenced by the Skyper building. The depth of the model was defined by the compact limestone rock layer which, owing to its high stiffness and strength, limits the region needed to be modeled.

The computed settlements ranged from 1.6 cm beneath the flat parts of the building to 7.2 cm beneath the tower where the majority of the loads are applied (Figure 12.23). The computed settlement distribution agrees very well with the measured deformations. The homogenous settlement distribution with small angular distortions indicates that the placing of the piles is effective.

SLS investigations have also to be carried out to gain knowledge about the settlements in the vicinity of the actual building project caused by its

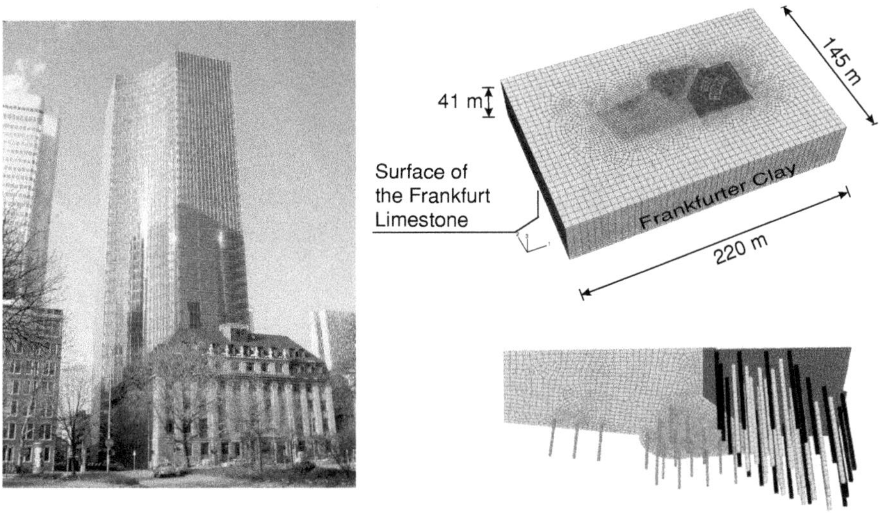

Figure 12.22 View and FE mesh of the Skyper Building in Frankfurt am Main (Katzenbach et al. 2005a)

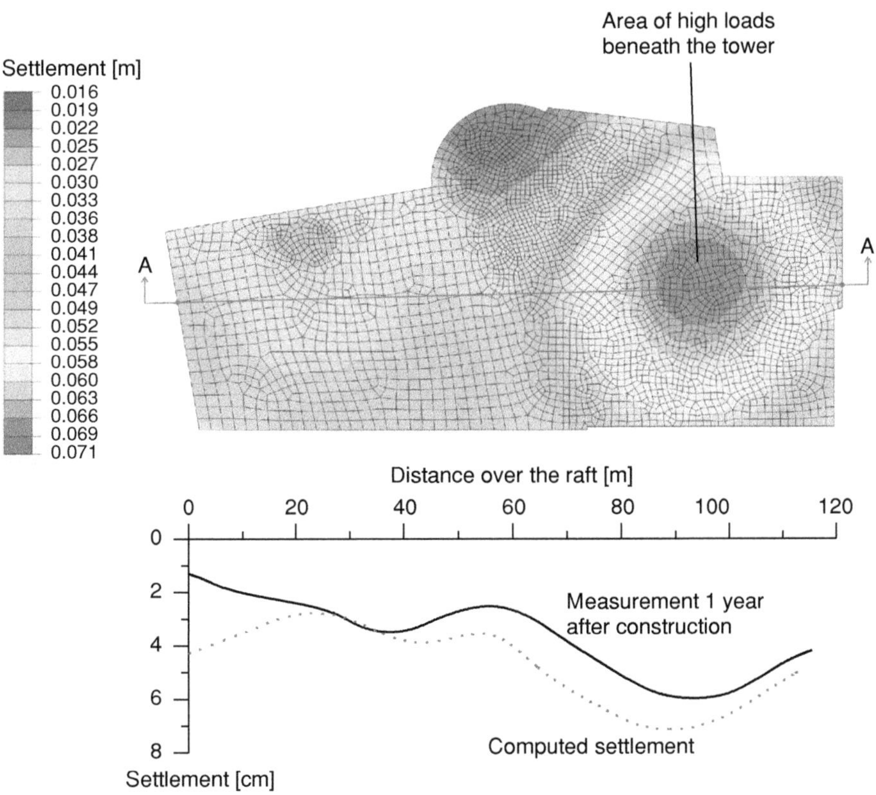

Figure 12.23 Computed settlement plot and comparison between computed and measured settlements over longitudinal section A–A (exaggerated)

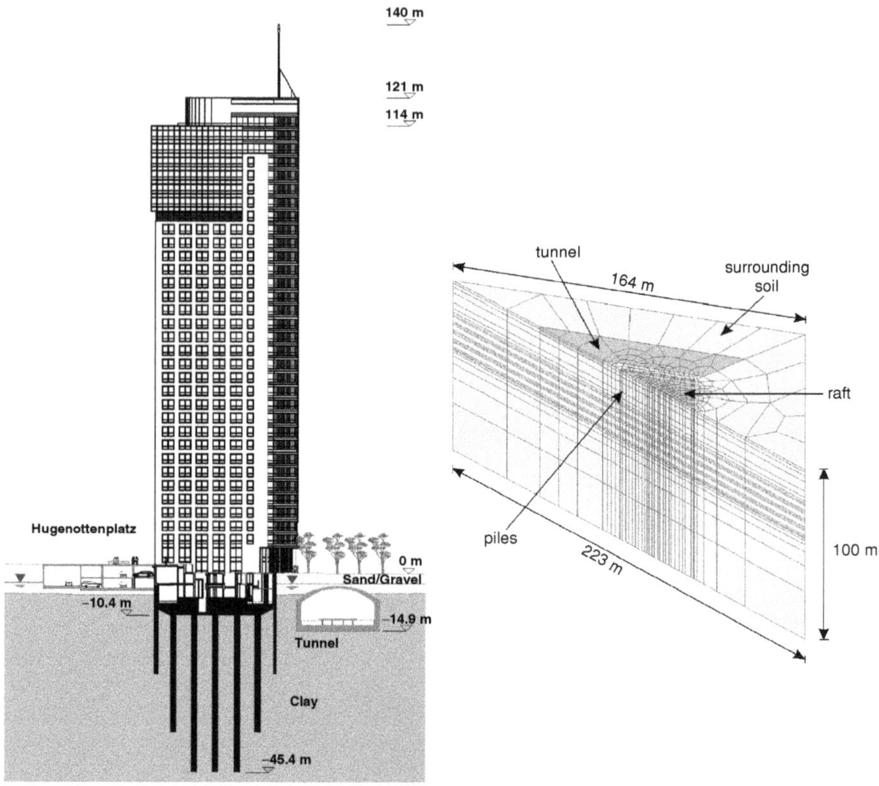

Figure 12.24 left: Cross-section of the City Tower Offenbach with closely adjoining subway tunnel; right: FE mesh of the City Tower Offenbach with modeled tunnel section

loads. Thus, attention had to be paid to buildings sensitive to deformation and tilting as well as to subterranean infrastructure elements. This problem is evident regarding the construction of the City Tower in Offenbach am Main, Germany, which had to be founded directly adjacent to a subway line (Figure 12.24). These complexities necessitated a fully three-dimensional computational model able to map all the interactions occurring during all construction stages and the in-service phase.

The total characteristic load acting on the foundation is about 600 MN. Maximum settlements related to this stress state are up to 8 cm at the centre of the raft. The differential settlements do not exceed 1 cm between the edge and the centre of the raft. The predicted values of 0.5–1.4 cm for the horizontal displacements of the adjoining tunnel do not exceed the acceptable range.

3.3 Assessment of settlements of high-rise buildings for geotechnical ultimate limit state (ULS) proofs

Recent developments in numerical geomechanics show how the proofs of stability of geotechnical systems, such as foundations or slopes, must be performed. One method is the reduction of the strength of the soil in terms of the 'φ-c-reduction' (Dawson et al. 1999, Schweiger 2005), which is commonly applied in the examples of slopes. This method does not focus on loads which must be increased to induce failure; rather, the material is artificially weakened until failure occurs. This weakening is carried out by gradually reducing the shear parameters φ and c until there is no equilibrium. Contrary to this approach, it is possible to drive a 'numerical load test', in which (similar to a field load test) the load applied to the structure is continually increased with the soil strength parameters remaining unchanged. This approach is recommended by the German CPRF guideline for the ultimate-state analyses of CPRFs (Hanisch et al. 2002). The advantage of this method is that such a computational model is closer to the physical reality than one with artificially decreased strength parameters; indeed, the numerical load test model can function not only for the ULS proofs but also for the SLS proofs. A computation using φ-c-reduction cannot provide realistic deformations and displacements.

The ultimate limit state design according to EC 7 is governed by the following inequality:

$$F_{c;d} \leq R_{c;d} \quad \text{resp.} \quad F_{c;k} \cdot \gamma_F \leq R_{c;k}/\gamma_R \tag{6}$$

This means that the design values of the applied forces $F_{c;d}$ compressing the pile must always be less than or equal to the maximum design value of the associated resistance force $R_{c;d}$ in the ULS. For a CPRF, the inequality for the proof of the ultimate limit state is formed by the sum of forces acting on the CPRF $\Sigma F_{c;d}$ and the overall resistance of the CPRF $R_{tot;d}$ in the ULS:

$$\Sigma F_{c;d} \leq R_{tot;d} \quad \text{resp.} \quad \Sigma F_{c;k} \cdot \gamma_F \leq R_{tot;k}/\gamma_R \tag{7}$$

Table 12.1 Parameters used for the simulation of the single pile

Parameter	*Symbol*	*Dimension*	*Value*
Friction angle	φ'	[°]	20
Cohesion	c'	[kN/m^2]	20
Young's modulus	E	[kN/m^2]	50000
Poisson's ratio	ν	[–]	0.2
Unit weight	γ	[kN/m^3]	19
Buoyant unit weight	γ'	[kN/m^3]	9
Shape factor	α	[–]	0
Shape factor	R	[–]	0.1
Shape factor	K	[–]	0.795

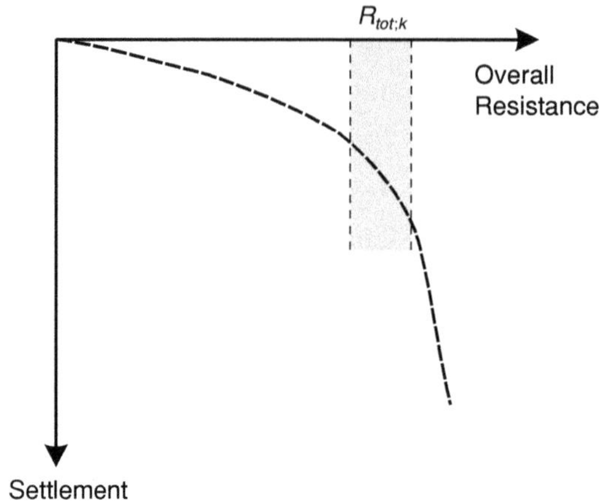

Figure 12.25 Non-linear system behavior of the CPRF and the determination of the overall resistance in the ULS

It is important to remember that the bearing capacity of single piles is not considered in this context. The overall resistance force (analogously to the pile resistance) is dependent upon the settlement. The overall resistance force of a CPRF in the ULS is defined as that point at which the increasing of the settlement becomes increasingly superproportional (Figure 12.25).

As, according to EC 7, no numeric determination of the safety level is required, it is sufficient to ensure that no failure will occur before the subsequent resistance force level – derived from the ULS condition (Equation 7) – is reached (Figure 12.26):

Figure 12.26 Non-linear system behavior of the CPRF and the determination of the overall resistance in the ULS

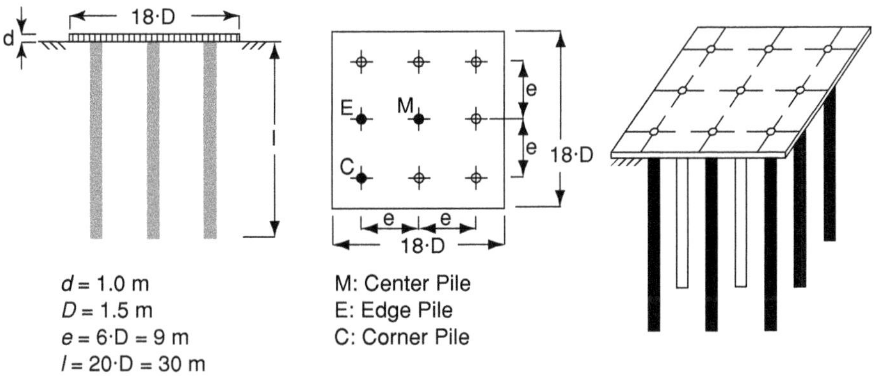

Figure 12.27 Model for the CPRF design example

$$R_{tot;k} \geq \Sigma F_{c;k} \cdot \gamma_F \cdot \gamma_R \tag{8}$$

Owing to the favorable interactions within a CPRF, a very distinct failure rarely appears; in most cases there is a smooth increase in the slope of the resistance–settlement curve as shown in Figure 12.26. This methodology is used in the subsequent example.

The design example consists of a CPRF with 9 regularly arranged piles each with a diameter of 1.5 m and a length of 30 m (Figure 12.27). The subsoil is modeled as in the pile example in section 3 by means of the modified Drucker+–Prager/cap model and with the parameters given for Frankfurt Clay.

The numerical model utilizes the double symmetry of the foundation system, thus only one quarter is modeled (Figure 12.28). The forces due to the superstructure are the steady actions $\Sigma G_{c;k}$ = 90 MN and the variable actions $\Sigma Q_{c;k}$ = 30 MN.

With this numerical model, a numerical load test was performed by steadily increasing the loads to generate the characteristic relationship between the settlement of the raft and the total load equal to the overall resistance of the CPRF. In Figure 12.29, the evolution of the overall resistance is plotted against the settlement of the middle of the raft.

The evolution of the overall resistance force shows no distinct failure state but rather a continuous decrease in the system stiffness after having left a pseudo-elastic range. This has to be ascribed to the increasing inelastic volumetric deformations which occur due to the hardening of the cap in the constitutive model. Because of this, no unique resistance force at the ultimate limit state is indicated in the graph and Equation (8) will be used.

The design value of the total load on the CPRF is computed using:

$$F_{c;d} = \Sigma G_{c;k} \cdot \gamma_G + \Sigma Q_{c;k} \cdot \gamma_Q \tag{9}$$

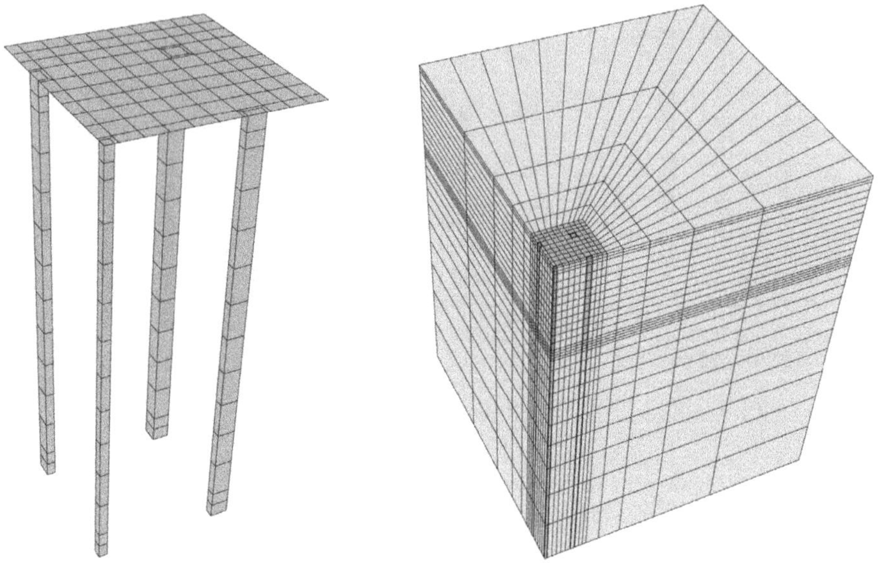

Figure 12.28 Discretization of the CPRF and the whole model

As the partial safety factors depend on the country of application, the German factors are chosen according to DIN 1054 (2005):

$$\gamma_G = 1.35, \quad \gamma_Q = 1.5, \quad \gamma_R = 1.4$$

Applying these safety factors to Equation (9) we find:

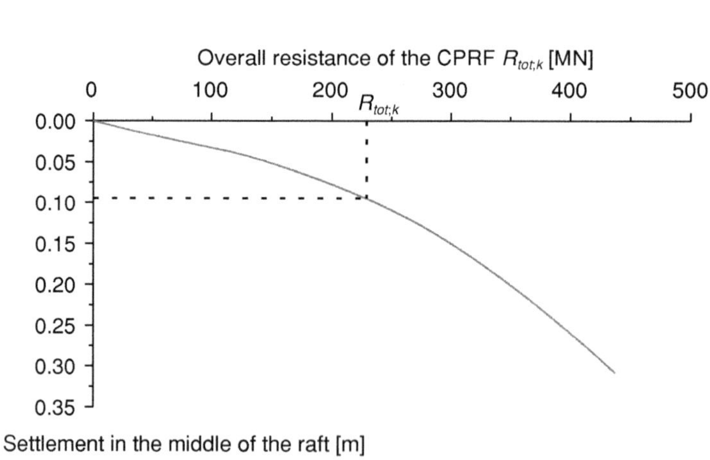

Figure 12.29 Evolution of the overall resistance of the example CPRF related to the settlement in the center of the raft

$$F_{c;d} = \Sigma G_{c;k} \cdot \gamma_G + \Sigma Q_{c;k} \cdot \gamma_Q = 90 \text{ MN} \cdot 1.35 + 30 \text{ MN} \cdot 1.5 = 167 \text{ MN} \quad (10)$$

According to Equation (8) the result of Equation (10) is multiplied by the safety factor for the overall resistance, giving:

$$F_{c;d} \cdot \gamma_R = 167 \text{ MN} \cdot 1.4 = 234 \text{ MN} \quad (11)$$

From Figure 12.29, it can be seen that up to this loading of 234 MN no significant failure has occurred. Thus, the stability of the foundation has been proved. Settlements of about 9 cm correspond to the assumed loading of 234 MN.

3.4 Assessment of settlements for structural analysis

The assessment of the settlements of the foundations of high-rise structures is not particularly relevant for the serviceability limit of the foundation with respect to the superstructure, but does provide results which can be passed to the structural engineers.

Differential settlements on monolithic structures always cause strain in the structure. This can be shown in the subsequent example. A beam is set on three supports; as it has no hinge in the middle, it is statically indeterminate. For this reason, any differential settlement δ of the centre support necessarily causes a bending moment in the beam (Figure 12.30) which has to be considered for the proof of internal stability of the structure, the raft.

Structural engineers concerned with the proof of internal stability and serviceability of the structure use programs for their calculations not based on continuum mechanics such as the programs needed by geotechnical engineers

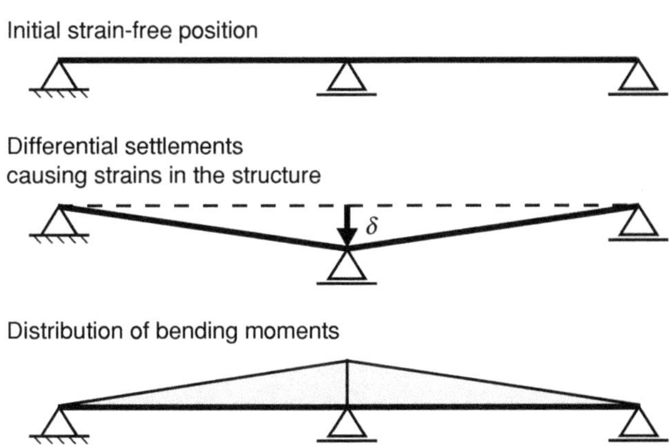

Figure 12.30 Effect of differential settlements on monolithic structures

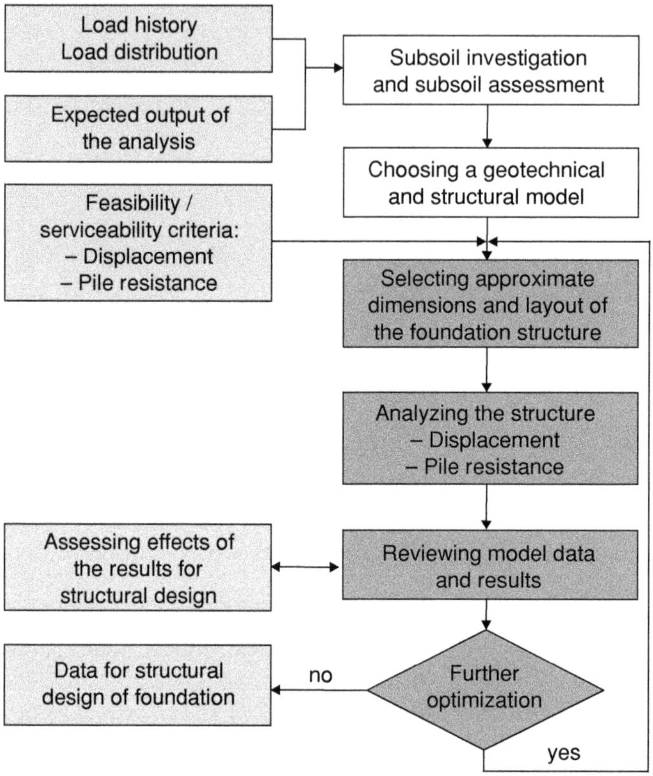

Figure 12.31 Flowchart for the design of a CPRF including the interactions between geotechnical and structural engineers

for the determination of the soil-structure interaction (Katzenbach et al. 2003, Figure 12.31).

The interaction between geotechnical engineers and structural engineers is indicated in the foundation of the City Tower in Offenbach, previously introduced. The settlements and differential settlements of the building are assessed by means of a fully three-dimensional, continuum-based FE model, while for the structural design of the raft a simpler FE model using shell and spring elements representing the raft and piles respectively was used (Figure 12.32). The spring stiffness c_i for the structural model was derived from the three-dimensional model by:

$$c_i = \frac{N_i}{s_i} \qquad (12)$$

where N_i represents the normal force and s_i the settlement at the head of the pile Number *i*. Depending on the pile position and the considered limit state, the spring stiffness is varied between 135 MN/m and 210 MN/m.

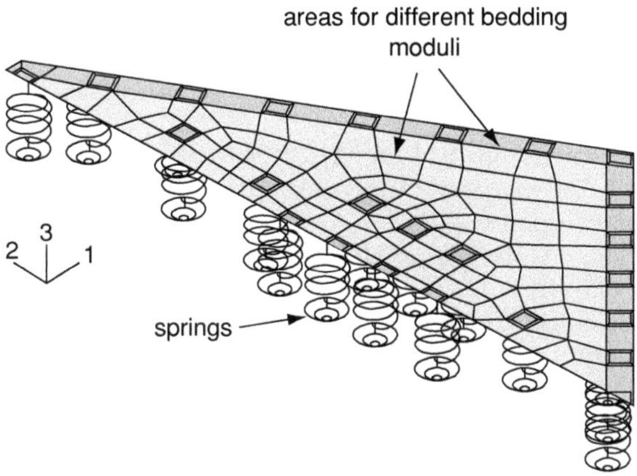

Figure 12.32 Complex continuum model for assessing the soil–structure interaction and simplified foundation model for structural design purposes

The bedding modulus for the raft was determined by back-analyzing the settlements, derived from the three-dimensional model, using the structural model (Figure 12.32). From the structural model, the reinforcement required in the raft could be easily determined.

4 Conclusion

Settlements of high-rise structures have to be assessed for both the serviceability limit state and the ultimate limit state. This includes the assessment of deformations in adjacent buildings and infrastructure units to prevent damage caused by the new building project. The assessment of the evolving deformations has to be carried out by computations involving the Finite Element Method which is widely used by geotechnical engineers. But these complex models require experience in their use for such things as mesh generation, choice of constitutive models, determination of the constitutive parameters boundary and initial conditions. In this context, it has been shown that the influence of the level of abstraction of the superstructure is not relevant. Much more important is the evaluation of the measured results as a data basis for the calibration and validation of the computational model. With such an adequate, calibrated and validated computational model, an effective foundation for the considered high-rise structures can be knowledgably designed.

References

Barth, U. and Reul, O. (1997) Congress Centre Messe Frankfurt • Piled raft for the control of large eccentricities. 4th Darmstadt Geotechnical Conf., March 1997,

Darmstadt, Darmstadt Geotechnics No. 2, Institute and Laboratory of Geotechnics, Technische Universität Darmstadt, pp. 75–85.

Bjerrum, L. (1963) Allowable settlements of structures, *Proceedings of the European Conference on Soil Mechanics and Foundation Engineering, Wiesbaden*, 2: 135–7.

Burland, J. B., Broms, B. B. and De Mello, V. F. B. (1977) Behaviour of foundations and structures, *Proc. 9th International Conference on Soil Mechanics and Foundation Engineering, Tokyo*, pp. 495–546.

Butterfield, R. and Banerjee, P. K. (1971) The problem of pile group–cap interaction, *Géotechnique*, 21 (2): 135–42.

CEN (2004) Eurocode 7: Geotechnical design

Chen, W. F. and Mizuno, E. (1990) Nonlinear analysis in soil mechanics: theory and implementation. *Development in Geotechnical Engineering 53*, Amsterdam: Elsevier.

Chow, Y. K. (1987) Iterative analysis of pile–soil–pile interaction, *Géotechnique*, 37 (3): 321–33.

Cooke, R. W. (1986) Piled raft foundation on stiff clay – a contribution to design philosophy, *Géotechnique*, 36 (2): 169–203.

Cooke, R. W., Bryde-Smith, D. W., Gooch, M. N. and Sillet, D. F. (1981) Some observations of the foundation loading and settlement of a multi-storey building on a piled raft foundation in London Clay, *Proceedings ICE 107*, London, 1: 433–60.

Davis, E. H. and Poulos, H. G. (1972) The analysis of pile raft systems, *The Australian Geomechanics Journal*, G2 (1): 21–7.

Dawson, E. M., Roth, W. H. and Drescher, A. (1999) Slope stability analysis by strength reduction, *Géotechnique*, 49 (S): 835–40.

DIN (1054) Subsoil – verifications of the safety of earthworks and foundations (German Standard), January 2005.

Drucker, D. C. and Prager, W. (1952) Soil mechanics and plastic analysis or limit design, *Quarterly of Applied Mathematics*, 10: 157–65.

Dunnicliff, J. (1993) *Geotechnical Instrumentation for Monitoring Field Performance*, New York: John Wiley.

Hanisch, J., Katzenbach, R. and König, G. (2002) *Kombinierte Pfahl-Plattengründungen*, Berlin: Ernst & Sohn.

Hight, D. W. and Green, P. A. (1976) The performance of a piled raft foundation of a tall building in London, *Proc. Vth European Conference on Soil Mechanics and Foundation Engineering, Vienna*, 1 (2): 467–72.

Hill, R. (1962) Acceleration waves in solids, *Journal of the Mechanics and the Physics of Solids*, 10: 1–16.

Holzhäuser, J. (1998) Experimentelle und numerische Untersuchungen zum Tragverhalten von Pfahlgründungen im Fels, *Mitteilungen des Institutes und der Versuchsanstalt für Geotechnik der Technischen Universität Darmstadt*, Heft 42.

Hooper, J. A. (1973) Observations on the behaviour of a piled raft foundation in London clay, *Proceeding of the Institution of Civil Engineers*, 55 (2): 855–77.

Hooper, J. A. (1979) Review of behaviour of piled raft foundation, CIRIA, report No. 83.

Katzenbach, R., Bachmann, G. and Gutberlet, C. (2006) Soil–structure interaction by high-rise buildings in megacities, *Proceedings of the Indian Geotechnical Conference*, 14–16 December, Chennai.

Katzenbach, R., Bachmann, G. and Schwarz, W. (2005a) Ongoing monitoring of deep excavation pits for damage prevention. *Darmstadt Geotechnics*, Institute and Laboratory of Geotechnics, Technische Universität Darmstadt, 17 March, 13: 199–212.

Katzenbach, R. and Reul, O. (1997) Design and performance of piled rafts, *Proc. 14th ICSMFE, Hamburg*, 4: 2253–6, Rotterdam: Balkema.

Katzenbach, R., Schmitt, A. and Turek, J. (2003) Introduction, in *Interaction between Structural and Geotechnical Engineers*, ed. R. Katzenbach and J. Turek, COST Action C7 Soil–Structure Interaction in Urban Civil Engineering, London: Thomas Telford.

Katzenbach, R., Schmitt, A. and Turek, J. (2005b) Assessing settlement of high-rise structures by 3D simulations. *Computer-Aided Civil and Infrastructure 20* (2005), pp. 221–9.

Kempfert, H.-G. and Smoltczyk, U. (2001) Pile foundations, in *Geotechnical Engineering Handbook*, d. U. Smoltczyk, 6th edn, Berlin: Ernst & Sohn.

Kuwabara, F. (1989) An elastic analysis for piled raft foundations in a homogenous soil, *Soils and Foundations*, 29 (1): 82–92.

Love, J. P. (2003) Use of settlement reducing piles to support a raft structure, Proceedings of the Institution of Civil Engineers, *Geotechnical Engineering 156*, October, GE4: 177–81.

Moormann, Chr. and Katzenbach, R. (2002) Numerical analysis of the improvement caused by foundation piles in deep excavations, *8th International Symposium on Numerical Models in Geomechanics*, NUMOG VIII, 10–12 April, Rome.

O'Neill, M. W. and Ghazzaly, O. I. (1977) Analysis of three dimensional pile groups with nonlinear soil response and pile–soil–pile interaction, *Proc. 9th Annual Offshore Technology Conference, OTC, Houston, Texas*, pp. 245–56.

Peck, R. B. (1969) Advantages and limitations of the Observational Method in applied soil mechanics, *Géotechnique*, 44 (4): 619–36.

Potts, D. M. (2003) Numerical analysis: a virtual dream or practical reality? Rankine lecture 2003, *Géotechnique*, 53 (6): 535–73.

Randolph, M. F. (1994) Design methods for pile groups and piled rafts, *XIIIth International Conference on Soil Mechanics and Foundation Engineering, New Delhi*, 4: 61–82.

Reul, O. (2000) In-situ-Messungen und numerische Studien zum Tragverhalten der Kombinierten Pfahl-Plattengründung, *Mitteilungen des Institutes und der Versuchsanstalt für Geotechnik der Technischen Universität Darmstadt*, Heft 53.

Schofield, A. N. and Wroth, C. P. (1968) *Critical State Soil Mechanics*, London: McGraw-Hill.

Schweiger, H. F. (2005) Application of FEM to ULS design (Eurocode) in surface and near surface structures, *Proc. 11th International Conference of the International Association of Computer Methods and Advances in Geomechanics (IACMAG)*, Turin, pp. 419–30.

Vardoulakis, I. (1980) Shear band inclination and shear modulus of sand in biaxial tests, *International Journal of Numerical and Analytical Methods in Geomechanics*, 4: 103–19.

Viggiani, C. (1998) Pile groups and piled rafts behaviour, *Proc. Conf. Deep Foundations on Bored and Auger Piles, Ghent*, Rotterdam: Balkema, 77–91.

13 Analysis of coupled seepage and stress fields in the rock mass around the Xiaowan arch dam

Chai Junrui, Wu Yanqing and Li Shouyi

Key words: Xiaowan arch dam, rock mass, seepage field, stress field, coupled solution, multi-level fracture network model

Abstract

The Xiaowan arch dam, constructed across the Lancangjiang river in Yunnan province, is, at 292 m in height, the highest arch dam in China. Owing to the combined significant head of water and arch action, it was necessary to analyze the interaction between the seepage and the stress fields in the rock mass around the dam. Numerical solutions for the coupled seepage and stress fields were analyzed using the multi-level fracture network model and the finite element method. The numerical results showed that the water storage in the reservoir had the four following major effects: (1) considerable change in the seepage field; (2) increase in the effective vertical stress in the rock foundation near the dam; (3) increase of tensile stress in the abutment rock mass; and (4) coupled action between the seepage and the stress fields must be considered.

1 Introduction

The Xiaowan hydropower station is located across the Lancangjiang river at the common boundary between Fengqing county and Nanjian county in the southwest of Yunnan province, the People's Republic of China (Figure 13.1). The dam is 4 km downstream of where the Hehuijiang river flows into the Lancangjiang river. In the dam area the Lancangjiang river flows from the north to the south, with the dam arc shape slightly facing the west bank (Figure 13.2). The valley at the dam site is narrow and shaped like a 'V'. The slopes of the two banks are steep, with a difference in elevation of approximately 1000 m from the valley floor to the peak. The topographic mean slopes of the two banks is between 40° and 42°, with the two banks being symmetrical. The foundation rocks at the dam site are mainly metamorphosed rocks: black micaceous and granite-metamorphosed gneiss, plagioclase-metamorphosed gneiss, schist with a small

Figure 13.1 Location of the Xiaowan arch dam in China

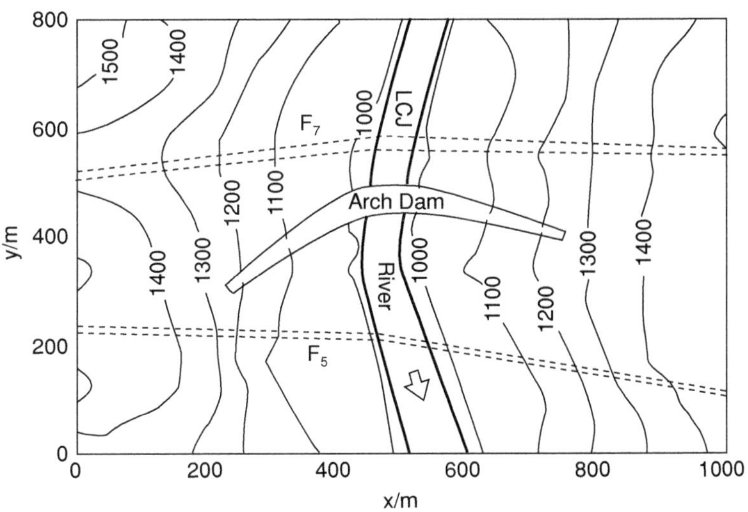

Figure 13.2 Plane of the Xiaowan arch dam and surrounding area

amount of granite. The drainage-basin area upstream of the dam site is 11,300 km^2 with an annual mean runoff of 1200 m^3/s.

The dam is a double d-arch with a maximum height of 292 m. The normal storage level of the reservoir is 1240 m, with a total reservoir volume of 145.57×10^8 m^3. The diversion tunnel and the flood-discharge tunnel are on the left bank, and the underground power plant and the transportation tunnel are on the right bank. The total capacity of the hydropower station is 4200 MW.

In the dam area, the geological formations are strongly developed. Fault F_7, perpendicular to the river, is upstream of the dam site. Its strike is near WE, dip direction NE with a dip angle of between 78° to 90°. The fault zone is 80 cm to 200 cm wide and filled with fault mud. The upper effect zone is 8.6 m to 12.7 m wide with the down effect zone 8.0 m to 10.1 m wide. The fault is a compression twist fault 5000 m long. Fault F_5 is perpendicular to the river, downstream of the dam site. Its dip direction is NE with a dip angle 70° to 80°. The fault zone is a mainly compressed fault about 20 cm wide. In addition, there are three main groups of fractures at the dam site. The strike of the first group of fractures is NNE and NNW, with a dip angle 60° to 80°, a space of 20 cm to 50 cm, a length 4 m to 7 m, and with high permeability. The second group of fractures is with a strike of NWW direction, a big dip angle, a space of 30 cm to 50 cm, a length of 3 m to 6 m and low permeability. The third group of fractures is with slow or moderate dip angles, mainly distributed at the weathering zone and slopes, and has high permeability due to the large opening of the fractures resulting from the weathering and unloading action.

There are two kinds of groundwater in the dam area: phreatic and pressure-bearing. The phreatic groundwater is stored in the weathering zone and has a sharp hydraulic gradient. In the natural condition, the Lancangjiang river is supplied by groundwater. The groundwater flows through fractures and is non-homogeneous, anisotropic and stress-influenced. Seepage through the small fractures can be regarded as continuum seepage, while seepage along the large F_7 and F_5 faults can be regarded as discontinuum seepage.

The construction of the Xiaowan hydroelectric power station started in 2001, and its construction was accelerated owing to the development of the west of China. From 2010, a water head of about 250 m will be formed between the upstream and the downstream of the dam. Owing to the combined head of water and the arch action, it was necessary to analyze the interaction between the seepage and stress fields in the rock mass foundation around the dam. Numerical solutions of the coupled seepage and stress fields were analyzed using the multi-level fracture network model and the finite element method (FEM).

2 The numerical model

As the dam lies between the F_5 and F_7 faults, the numerical modeling must include these faults to reflect their effects, the space distribution of the coupled seepage and stress fields. Consequently, the numerical model for the coupled seepage and the stress fields takes in the NS directional extension of the area between the F_5 fault and the F_7 fault (Figure 13.2). The south and north boundaries are the Gouyazi valley and the Xiushan valley respectively, which are deeply cut and suitable for the seepage and stress boundaries. The east and west boundaries are the two banks where

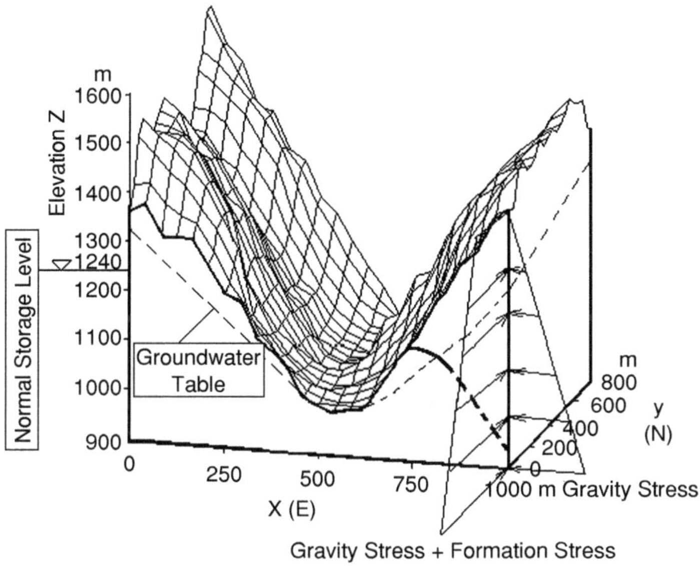

Figure 13.3 Three-dimensional numerical model

the elevation is 1500 m and where the bore holes and water-table observation holes were distributed. Thus, the numerical model is 1000 m long in the NS direction, 800 m wide in the WE direction, and 600 m in height difference (Figures 13.2 and 13.3).

According to inverse analysis for the ground stress, the residual formation stress in the dam area is about 5 MPa in the NS direction, but is released in the left bank [1]. For the stress boundary conditions, shown in Figure 13.3, the gravity stress is applied to the section $x = 1000$ m and to the left bank of the section $y = 0$; the gravity stress and the residual formation stress are applied to the right bank of the section $y = 0$; the sections $x = 0$, $y = 800$ m, and $z = 900$ m are all zero-displacement boundaries.

The multi-level fracture network model is utilized to analyze the coupled seepage and stress fields in the rock mass around the dam [2, 3]. This model is a new development of the previous rock mass hydraulics models [4–14]. According to the scale and permeability, the ruptures, faults, fractures, joints and pores in the rock mass are divided into four levels: level 1, real fracture network; level 2, random fracture network; level 3, equivalent continuum system; and level 4, continuum system. The level 1 real fracture network is composed of larger ruptures and faults, the distribution, mechanical and hydro-mechanical parameters of which can be determined from practical geological information. Based on fracture measurements and statistics, the level 2 random fracture network is simulated using the Monte-Carlo technique. The level 3 equivalent continuum system is established to reflect the overall behavior of small joints by means of hydraulic conductivity tensor theory. The level 4 continuum system is composed of pores in the rock, the behavior of which can be represented by homogeneous and isotropic parameters. These four levels of fracture networks are related owing to the water and energy balance. In the fracture network of each level, the different interactions between seepage and stress are introduced to construct the multi-level fracture network model for coupled seepage and stress fields in the rock mass. In this multi-level fracture network, the connecting fracture network (level 1 real fracture network and the connecting level 2 random fracture network) and the equivalent continuum system (generalized rock matrix) have different interaction relationships between seepage and stress as shown in Table 13.1 [2].

In Equations (2) and (5), J_f is the gradient of the hydraulic head along the flow (seepage) path.

With the Table 13.1 interaction relationships between seepage and stress, the multiple-level fracture network seepage model and the multiple-level fracture network stress model can be combined to form the multiple-level fracture network model for coupled seepage and stress fields in the rock mass. The coupled seepage and stress model can be solved numerically using the FEM and the iterative approach [2].

Table 13.1 Interaction relationships between seepage and stress in the rock mass

Type	*Item*	*Interaction relationships*	*Relationship equations*	
Generalized rock matrix	Action of seepage on stress	Seepage applies to the hydrostatic seepage pressure p and to the seepage body force f to influence the stress in the rock mass.	$p = \gamma(H - z)$ $f = \gamma J_f$	(1) (2)
	Action of stress on seepage	Stress makes the volumetric strain ε_V and the porosity n of the rock mass change to influence the hydraulic conductivity K, thus also influencing the seepage field.	$K = K(n)$ $n = n(\varepsilon_V)$, $\varepsilon_V = \varepsilon_V(\sigma_{ij})$	(3)
Connecting fracture network	Action of seepage on stress	Seepage applies to the hydrostatic seepage pressure p and to the tangent hauling force t_w on the fracture walls to influence the stress in the rock mass.	$p = n\gamma(H - z)$ $t_w = \frac{b}{2} n\gamma J_f$	(4) (5)
	Action of stress on seepage	Stress makes the apertures b of the fractures change to influence the hydraulic conductivity K_f of the fractures, thus also influencing the seepage field.	$K_f = K_f(b)$ $b = b(\sigma_{ij})$	(6)

In the numerical model shown in Figures 13.2 and 13.3, the F_5 and F_7 faults are the connecting fracture network, while other zones are the equivalent continuum system. The numerical model shown in Figure 13.3 has 1582 nodes and 1118 elements, including 964 solid elements (eight noded iso-parametric hexahedron elements with eight nodes) and 172 joint elements (Figure 13.4). The numerical parameters for all zones are listed in Table 13.2.

3 The numerical results and analysis

The coupled seepage and stress fields in the numerical model for the normal storage level of 1240 m are analyzed using the above model and the FEM.

Taking into account the curtain grouting, the total seepage discharge within the area numerically modeled is 0.3216 m^3/s. Other numerical results are shown in Figures 13.5–12, in which the normal stresses are effective stresses. Tension is considered positive.

It can be shown from the analysis of the above numerical results that:

(1) Before dam construction, the river was supplied by groundwater. After dam construction, the upstream reservoir water-storage design level of 1240 m increased the water-table height (Figure 13.5). With the significant water-head difference between the upstream and downstream

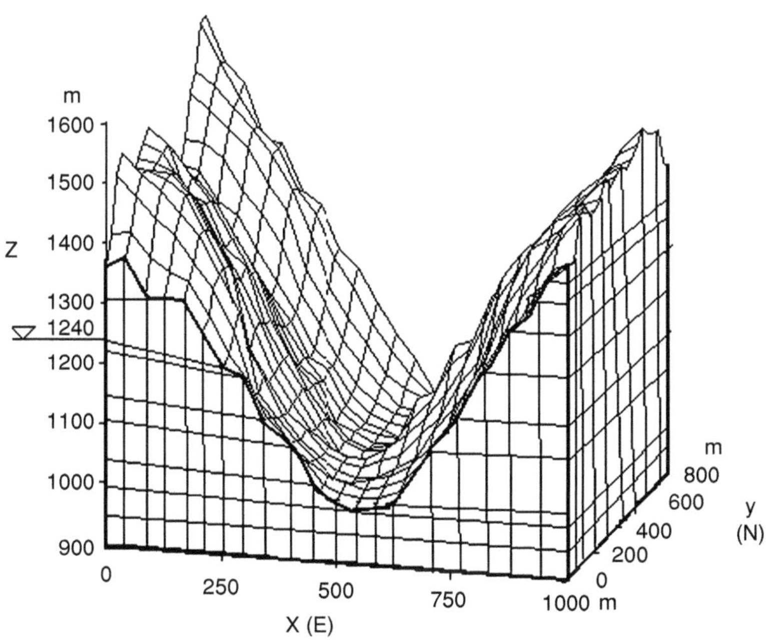

Figure 13.4 Finite element mesh

Table 13.2 Numerical parameters for all zones

Zone	*Elastic modulus (GPa)*	*Poisson's ratio*	*Specific gravity (kN/m³)*	*Hydraulic conductivity (×10⁻⁶ m/s)*			*Pore ratio*
				K_x	K_y	K_z	
F_7, F_5 fault zone	9.8	0.30	22	0.0068	0.0068	0.0068	0.202
F_7, F_5 effecting zone	9.8	0.30	22	10.88	31.48	1.11	0.198
Weathering zone	12.5	0.30	22	2.55	15.51	0.59	0.114
Granite – Metamorphosed gneiss	35	0.21	27	2.66	3.94	1.00	0.053
Plagioclase – Metamorphosed Gneiss	30	0.25	27	2.66	3.94	1.00	0.072
Curtain grouting zone	30	0.25	27	10^{-3}	10^{-3}	10^{-3}	0.050

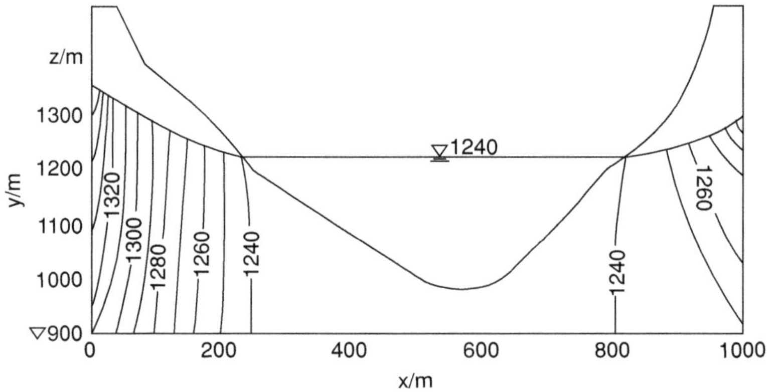

Figure 13.5 Contour map of hydraulic head (m) at cross-section y = 492 m after storage

faces of the dam, a steep hydraulic gradient was formed in the dam foundation (Figure 13.6), and the hydraulic gradient in the curtain grouting zone (upstream of the dam foundation) is clearly larger than in other zones.

(2) As the shear stress values τ_{xy}, τ_{yz}, τ_{xz} are, in most areas, relatively small (Figures 13.9 and 13.12), the normal stresses σ_x, σ_y, σ_z could be regarded as the principal stresses.

(3) Figure 13.7 shows that the effective normal stress σ_z after storage increases at the cross-section y = 492 m near to the dam owing to the action of the weight of the dam and the weight of the reservoir water, although seepage pressure trends to decrease it.

(4) Figure 13.8 shows that the normal stress σ_x after storage increases at the cross-section y = 492 m near to the dam owing to the reservoir water pressure applied to the two banks.

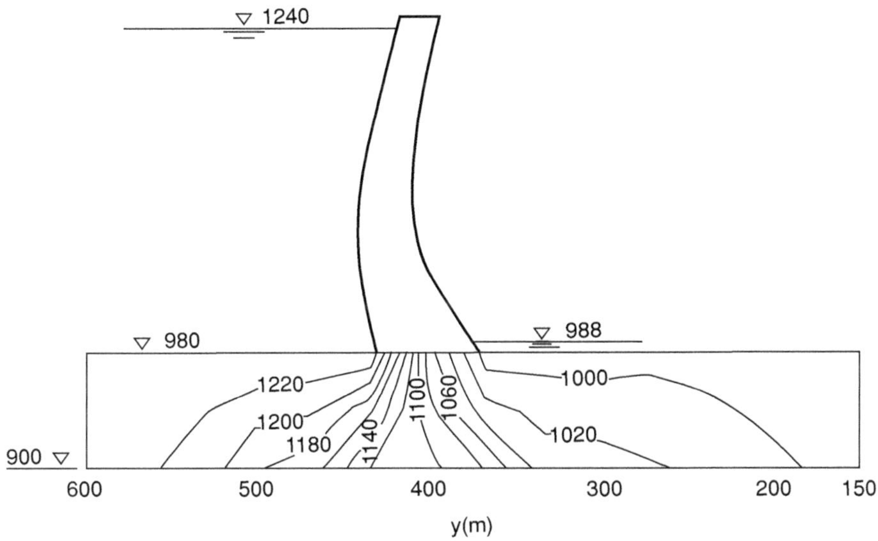

Figure 13.6 Contour map of hydraulic head (m) at longitudinal section of riverbed center after storage

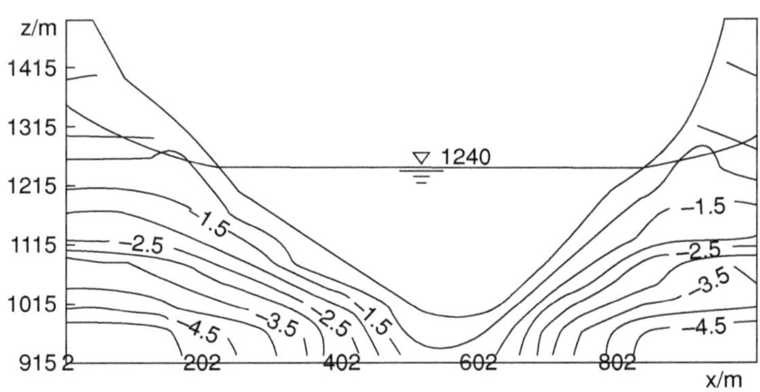

Figure 13.7 Contour map of σ_z MPa at cross-section y = 492 m after storage

(5) Figures 13.10 and 13.11 show that the tensile stresses σ_x, σ_y after storage increase at the section z = 998 m, resulting from abutment thrust due to the arch action.

4 Conclusions

The numerical results show that the coupled seepage and stress fields in the rock mass around the Xiaowan arch dam due to the reservoir considerably change the seepage field, increase the effective vertical stress in

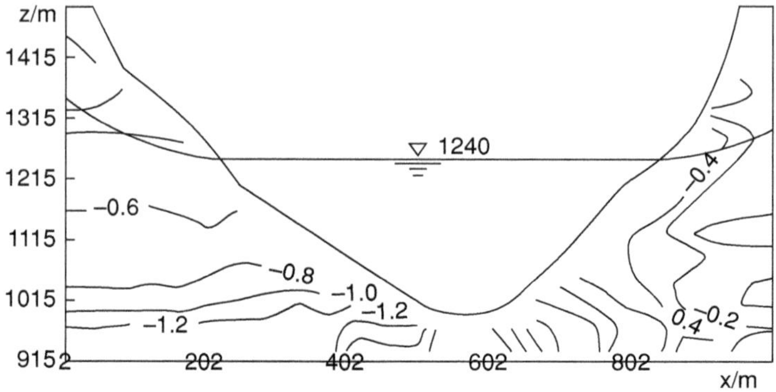

Figure 13.8 Contour map of σ_x MPa at cross-section $y = 492$ m after storage

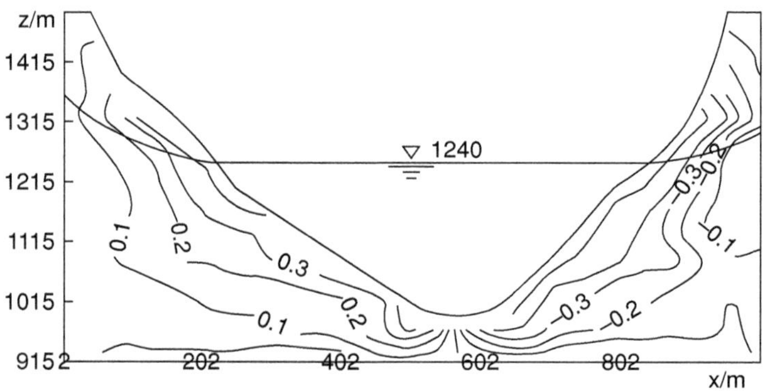

Figure 13.9 Contour map of τ_{xz} MPa at cross-section $y = 492$ m after storage

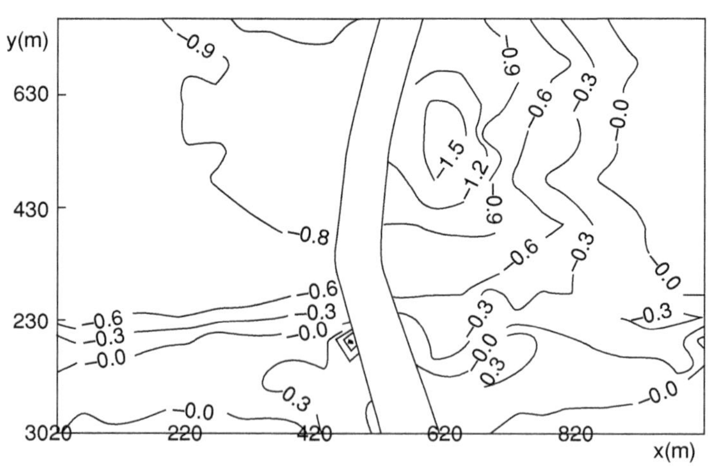

Figure 13.10 Contour map of σ_x MPa at cross-section $z = 998$ m after storage

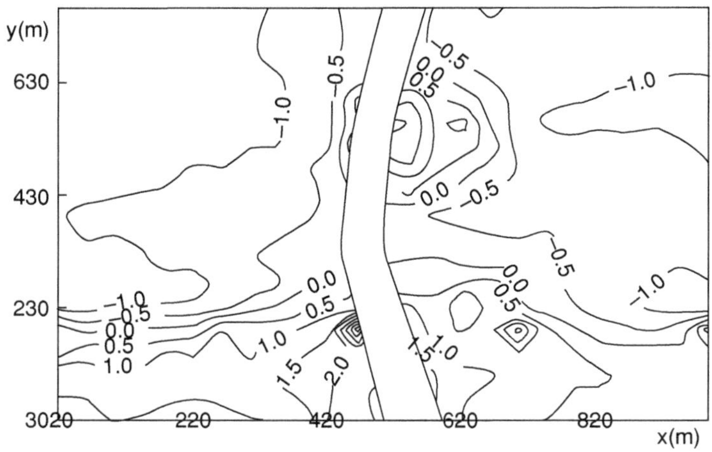

Figure 13.11 Contour map of σ_y MPa at cross-section z = 998 m after storage

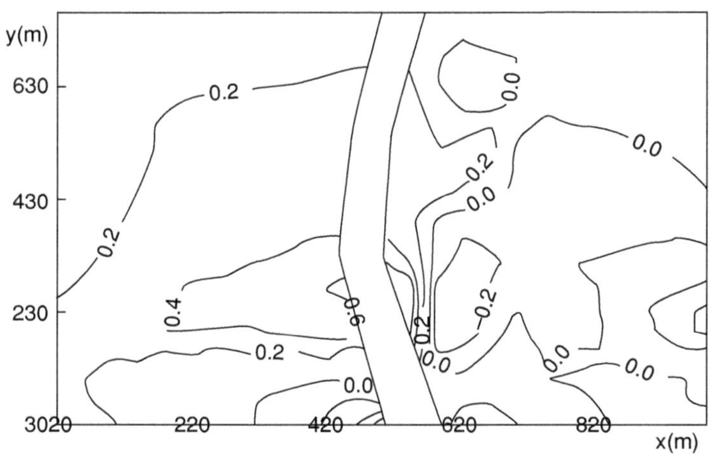

Figure 13.12 Contour map of τ_{xy} MPa at cross-section z = 998 m after storage

the rock near the dam and increase the tensile stress in the abutment rock mass. Consequently, the coupled action between the seepage and stress fields must be considered due to the significant head of water and the arch action. Further, the multi-level fracture network model and the FEM can be used to determine the coupled action between seepage and stress in the rock mass foundation.

Acknowledgements

The financial support of the following projects is gratefully acknowledged: Project 10202015 sponsored by the National Natural Science Foundation of China (NSFC)

Project 2003-106 sponsored by the Scientific Research Foundation (SRF) for the Returned Overseas Chinese Scholars (ROCS) by the State Education Ministry (SEM)

Research Project 03JK098 sponsored by the Shaanxi Provincial Education Department (SNED)

Project 2003-C1 sponsored by the China Three Gorges University (CTGU)

Scientific Innovation Project 106-210303 sponsored by Xi'an University of Technology (XAUT)

References

Barenblatt, G. I., Zheltov, I. P. and Kochina, I. N. Basic concepts in the theory of seepage of homogeneous liquids in fissured rocks, *J. Appl. Math. Mech, USSR*, 1960, 24 (5): 1286–1303.

Chai, J. R. Analysis of coupled seepage and temperature fields in concrete dam, *Commun. Numer. Meth. Engng*, 2002, 18 (6): 399–409.

Chai, J. R. *Seepage Theory for Dam Engineering*. Llasa: Tibet People Press, 2001. In Chinese.

Chai Junrui. Research on dynamical behavior of high arch dams in China, in *Proceedings of the 4th Asia-Pacific Conference on Shock and Impact Loads on Structures*, ed. Lok Tat Seng, Lim Chee Hiong, Li Xibing, Liu Deshun, Singapore: CI-Premier, 2001, pp. 129–34.

Chai Junrui. Review on Dynamic Stability Analysis of Rock Mass in High Arch Dam Abutment, in *Proceedings of Third International Conference on Dam Safety Evaluation*, ed. S. P. Kaushish, M. Gopalakrishnan, New Delhi: INCOLD and CBIP, 2001, pp. 189–94.

Chai Junrui, Li Shouyi and Wu Yanqing. Multi-level fracture network model for coupled seepage and stress fields in rock mass, *Communications in Numerical Methods in Engineering*, 2004, 20 (2).

Chai Junrui, Liu Haowu and Wu Yanqing. Hauling forces applied on fissure walls in fractured rock mass by fluid flow, in *New Development in Rock Mechanics and Rock Engineering – The Proceedings of the 2nd International Conferences of NDRM'2002*, ed. Yunmei Lin, Chun'an Tang, Xiating Feng, Shuhomg Wang, Paramus, NJ: Rinton Press, 2002, pp. 80–3.

Chai Junrui, Wu Yanqing and Yuan Jiguo. Double mechanical effects of flow through fracture network in rock mass on fracture walls, in *Frontiers of Rock Mechanics and Sustainable Development in the 21st Century – Proceedings of the 2001 ISRM International Symposium – 2nd Asia Rock Mechanics Symposium*, ed. Wang Sijing, Fu Bingjun, Li Zhongkui, Lisse: Balkema, 2001, pp. 189–91.

Huang, R. Q. *The Stability Analysis of Abutment Rock Masses and High Slopes in Area around the Xiaowan Arch Dam*. Chengdu: Chengdu University of Technology, 1995. In Chinese.

Lewis, R. W. and Ghafouri, H. R. A novel finite element double porosity model for multi-phase flow through deformable fractured porous media, *Int. J. Num. Anal. Meth. Geomec.*, 1997, 21: 789–816.

Lewis, R. W., Roberts, B. J. and Schrefler, B. A. Finite element modeling of two-phase heat and fluid flow in deforming porous media, *Transport in Porous Media*, 1989, 4: 319–34.

Lewis, R. W. and Schrefler, B. A. *The Finite Element Method in the Static and Dynamic Deformation and Consolidation of Porous Media*, 2nd edn, Chichester: John Wiley, 1998.

Oda, M. An equivalent continuum model for coupled stress and fluid flow analysis in jointed rock masses, *Water Resources Research*, 1986, 22 (13): 1854–65.

Schrefler, B. A., Zhan, X. Y. and Simoni, L. A coupled model for water flow, air flow and heat flow in deforming porous media, *Int. J. Num. Meth. Heat and Fluid Flow*, 1995, 5: 531–47.

14 Development of bucket foundation technology for operational platforms used in offshore oilfields

Shihua Zhang, Quanan Zheng, Haiying Xin and Xinan Liu

Keywords: bucket foundation, finite element method, offshore oilfield, Shengli oilfield

1 Introduction

Bucket foundations are one of the foundation technologies used in the construction of offshore oil and gas platforms as illustrated in Figure 14.1. The technology is especially suitable for shallow water areas where there are thick sediment layers on the sea floor. The bucket foundation platform has high stability, mobility and re-usability while its relatively simple structure saves construction materials and costs. It is installed by using negative pressure suction methods, which reduce the on-site construction

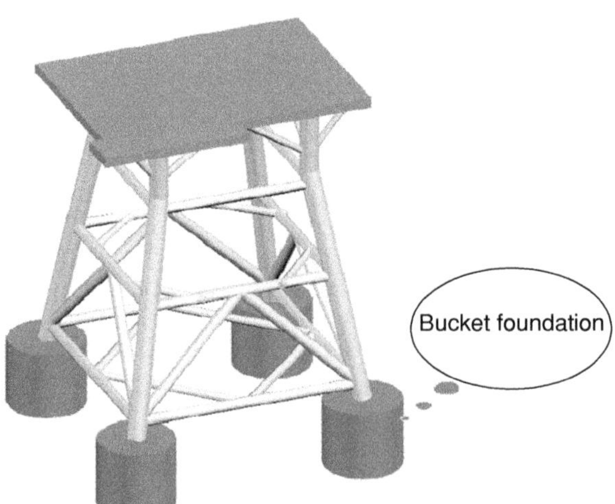

Figure 14.1 Schematic drawing of a mobile bucket foundation platform

times. The installation process minimally disturbs the marine environment and is thus an environment-friendly technology, widely used since the early 1990s (Baerheim 1995; Erbich et al. 1995; Rusaas 1995; Tjelta 1995a, 1995b). During platform installation, a key technology is how to press a bucket foundation efficiently so that it penetrates the sea floor's sediment layers. In this process, water pumps are used to produce a negative pressure inside the bottom-up bucket. As a result, the relatively high pressure of the sea water and the atmosphere outside the bucket forces interstitial water in the surrounding sediment to flow into the bucket. This water flow forms a seepage field in the sea-floor sediments both outside and inside the bucket. Field operations show that this seepage field can cause favorable and unfavorable effects on the suction penetration of the bucket foundation. For instance, the penetration resistance with seepage can be reduced to half of that without seepage. On the other hand, the increase in effective stress between the surrounding sediment and the bucket wall may hinder penetration. Therefore, it is important to analyze the dynamics of the seepage field during the penetration process to increase the successful installation of the bucket foundation platform.

During the development process of bucket foundation technology, at the Drilling Technology Research Institute, for the Shengli oilfield of China, the finite element (FE) method was used to simulate the dynamics of the seepage field generated by the penetration of the bucket foundation into the sea floor under a variety of environmental and working conditions. In particular, heterogeneous sediment layers were taken into account. The simulation results were compared with a series of physical model tests both in the laboratory and at sea. Following the successful simulation, the FE method was used to investigate all penetration conditions before an operational bucket foundation platform was installed and to guide the construction of the bucket foundation in the Chengbei offshore oilfield, a branch of the Shengli oilfield of China. The successful erection of the operational bucket foundation platform indicated that the technology worked well under the engineering geological conditions of the Yellow River submersed delta (Sun 2000; Zhang and Chu 2000; Zhang et al. 2004).

2 Theories and the finite element model results

According to the Darcy's law (Greenkorn 1983; Cedergren 1997), the steady-state seepage velocity $\vec{v}$ in the field of sea-floor sediments can be expressed as follows:

$$\vec{v} = -\left(k_x \frac{\partial h}{\partial x} \vec{i} + k_y \frac{\partial h}{\partial x} \vec{j} + k_z \frac{\partial h}{\partial x} \vec{k} \right), \tag{1}$$

where x, y and z constitute a Cartesian coordinate system with z positive upward, k_x, k_y and k_z are the permeability coefficients along the x, y and

z coordinates respectively, h is the head of water, and $\frac{\partial h}{\partial x}, \frac{\partial h}{\partial y}, \frac{\partial h}{\partial z}$ are the head-of-water gradients along the x, y and z coordinates respectively. In terms of the mass conservation, i.e. $\nabla \cdot \vec{v} = 0$, the governing equation for the seepage field, derived from Equation (1), can be written as:

$$\frac{\partial}{\partial x}\left(k_x \frac{\partial h}{\partial x}\right) + \frac{\partial}{\partial y}\left(k_y \frac{\partial h}{\partial y}\right) + \frac{\partial}{\partial z}\left(k_z \frac{\partial h}{\partial z}\right) = 0. \tag{2}$$

The boundary conditions for this problem are given as follows: on the boundary of the sea-floor surface, the general head of water is given by:

$$h = h^*, \tag{3}$$

where h^* is a known head of water. On the boundaries of non-permeable borders including the waterproof layer within the sea-floor sediments and the bucket wall, the seepage velocity satisfies:

$$\vec{v} \cdot \vec{n} = 0, \tag{4}$$

where $\vec{n}$ is a normal unit vector of the boundary surface. Also, on the vertical outside boundary of the sediment border far away from the bucket, the seepage velocity is given as:

$$\vec{v} = 0. \tag{5}$$

Equations (1) and (2) with the boundary conditions of Equations (3) to (5) were solved using the commercial FE software ANSYS. In the model, the bucket body is simulated by a very thin walled cylinder, the thickness of the wall being modeled by one finite element. The computational domain of the sediment surrounding the bucket is 15 to 20 times the bucket diameter, with the penetration of the bucket being vertically downward. The finite elements are 3D axisymmetric and vertical from the sea-floor surface to the waterproof layer, as shown in Figures 14.2 and 14.3.

Results from the FE simulations are shown in Figures 14.4–7. Figure 14.4 shows the head of water on both sides of the bucket wall. Figure 14.5 shows the head-of-water gradient on both sides of the bucket wall. Figure 14.6 shows the head-of-water gradient around the lower edge of the bucket wall. Figure 14.7 shows the seepage vector field near the lower edge of the bucket wall.

From the FE results shown in Figures 14.4–7, it can be seen that, driven by the suction pressure inside the bucket, strong head-of-water gradients are generated near to the lower edge of the bucket. Then the strong head-of-water gradients drive a seepage flow into the bucket with the maximum inflow vector occurring near to the lower edge of the bucket. The seepage

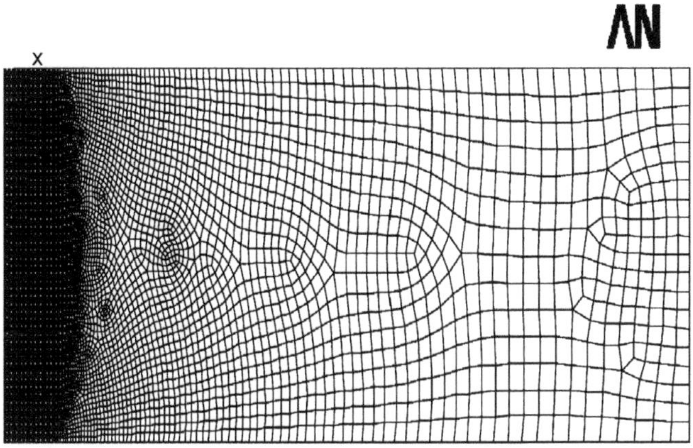

Figure 14.2 FE mesh on the *x–z* plane

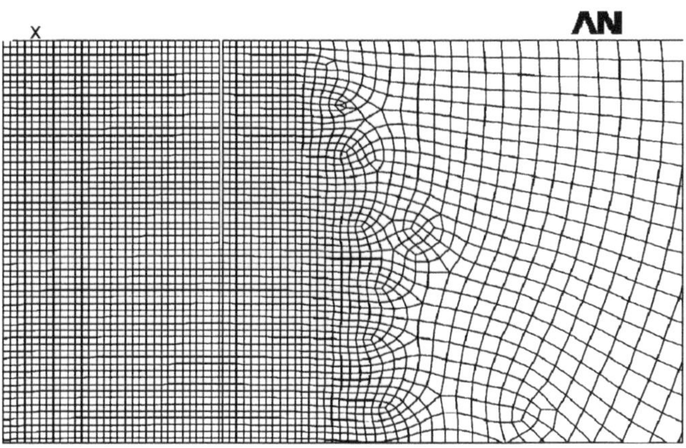

Figure 14.3 Enlarged FE subdomain near the bucket wall in Figure 14.2. The long, narrow vertical bar represents the bucket wall. The inside of the bucket is on the right-hand side of the vertical bar

inflow field is controlled by the boundary conditions and parameters used in the model, so that the variation of inflow field during the bucket penetration process can be simulated before the on-site operations. The effects of seepage flow on the suction penetration of the bucket foundation are double-edged. The favorable effect reduces the resistance to penetration. The unfavorable effect increases the effective stress between the surrounding sediment and the bucket wall, which hinders the penetration of the bucket into the layers of sediment. Consequently, the most appropriate suction pressure must be determined before the on-site erection of the platform by using laboratory model tests and FE simulations.

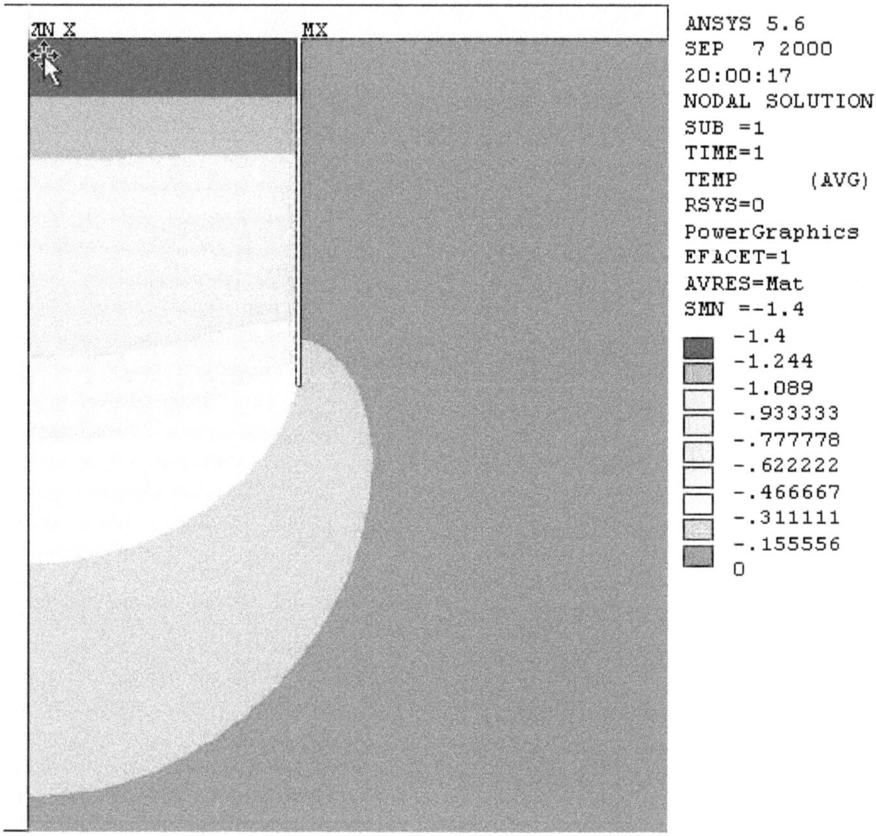

Figure 14.4 Head of water on both sides of the bucket wall generated by FE. The inside of the bucket is on the right-hand side of the vertical bar

3 Laboratory model tests

The objectives of the laboratory model tests were as follows:

(1) To verify the fundamental concepts. This was the first adoption, in China, of the bucket foundation platform for offshore oil and gas development. Laboratory model tests would help engineers and scientists to understand the concept and to test its feasibility under on-site conditions.
(2) To validate results of the numerical model. The FE model was to be used to simulate the bucket foundation platform penetration process and to determine the engineering parameters. The feasibility of the method and the accuracies of the derived parameters were determined and compared with those from laboratory model tests.

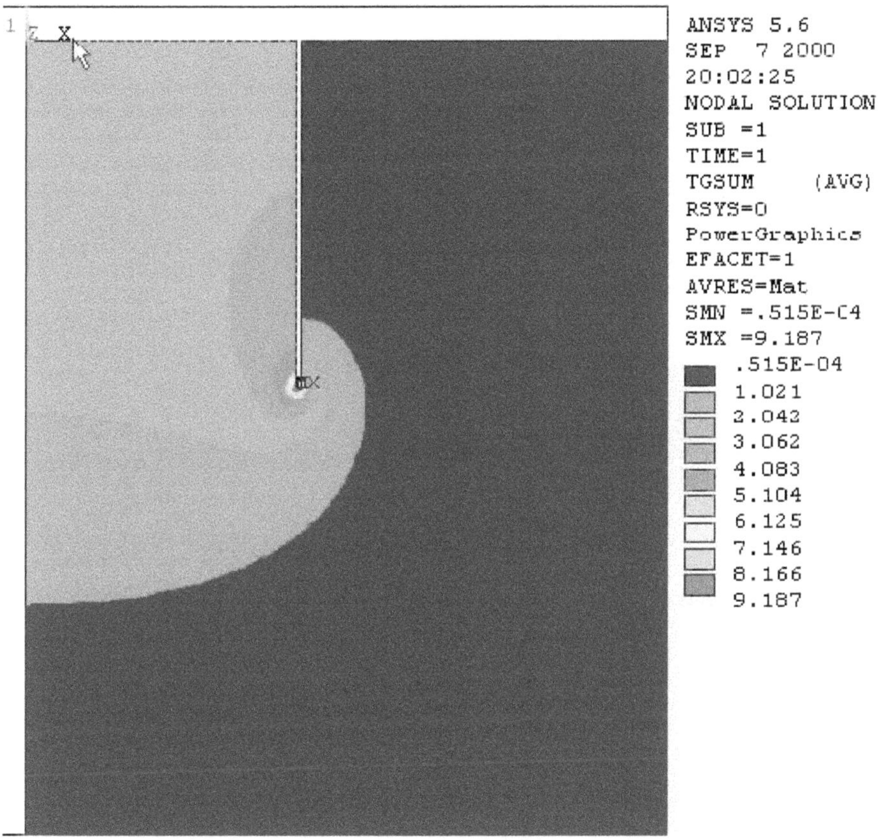

Figure 14.5 The head of water gradient on both sides of the bucket wall generated by FE. The inside of the bucket is on the right-hand side of the vertical bar

(3) To measure the engineering parameters. Laboratory model tests produced considerable data to answer questions raised in the engineering design and during on-site operations.
(4) To predict on-site problems and seek their solutions.

3.1 Model ocean and model sea-floor base

The physical model tests were carried out in the Offshore Engineering Simulation Laboratory of the Shengli Petroleum Administrative Bureau, located in Dongying City, Shandong Province, People's Republic of China. The laboratory has a test water-pool facility 52 m long and 24 m wide, with an adjustable water depth of up to 0.70 m. A model ocean for the bucket foundation model tests was constructed within the pool utilizing

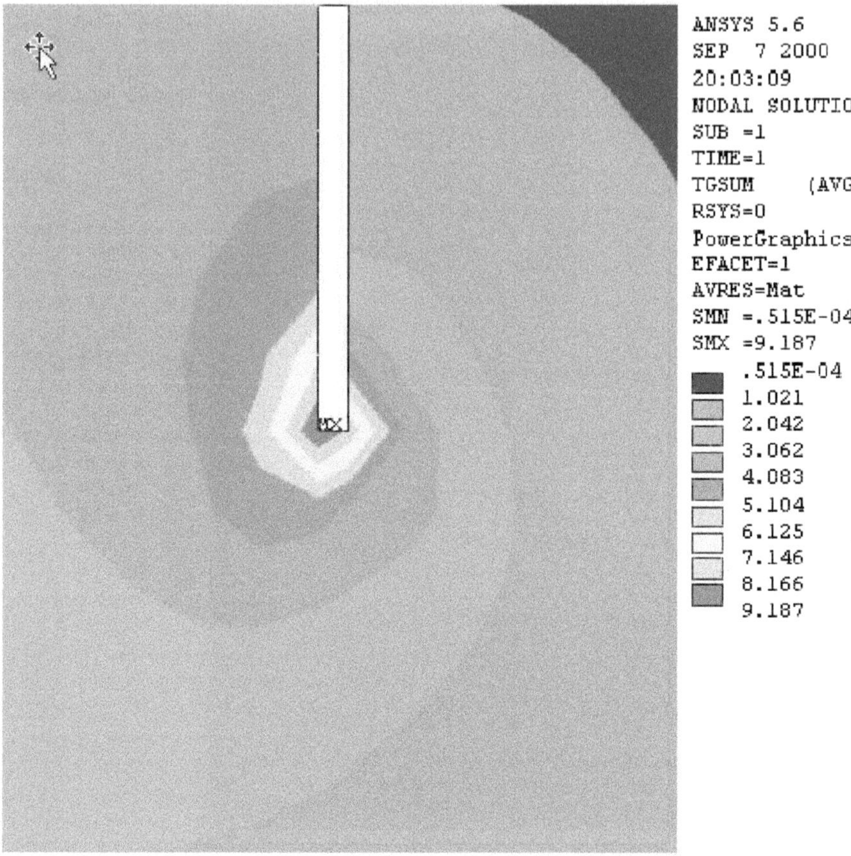

Figure 14.6 Close-up view of the head of water gradient around the lower edge of the bucket wall. The inside of the bucket is on the right-hand side of the vertical bar

equipment such as the mobile measurement bridge and the overhead traveling crane. The area of the model ocean was 15 m by 15 m with a water depth of 0.14 m. The model sea-floor base was a flat and solid sediment layer 1.6 m thick. The sediment material was silt from the Yellow River Delta, the same as at the site where the operational bucket foundation platform was to be erected. The permeability coefficient of the silt was 1.56×10^{-5} cm s^{-1}.

3.2 Model bucket penetration tests

A series of single model bucket and 4-bucket model platform penetration tests were carried out in the model ocean. For the single-bucket tests, the diameters of model buckets are from 180 to 820 mm, with heights

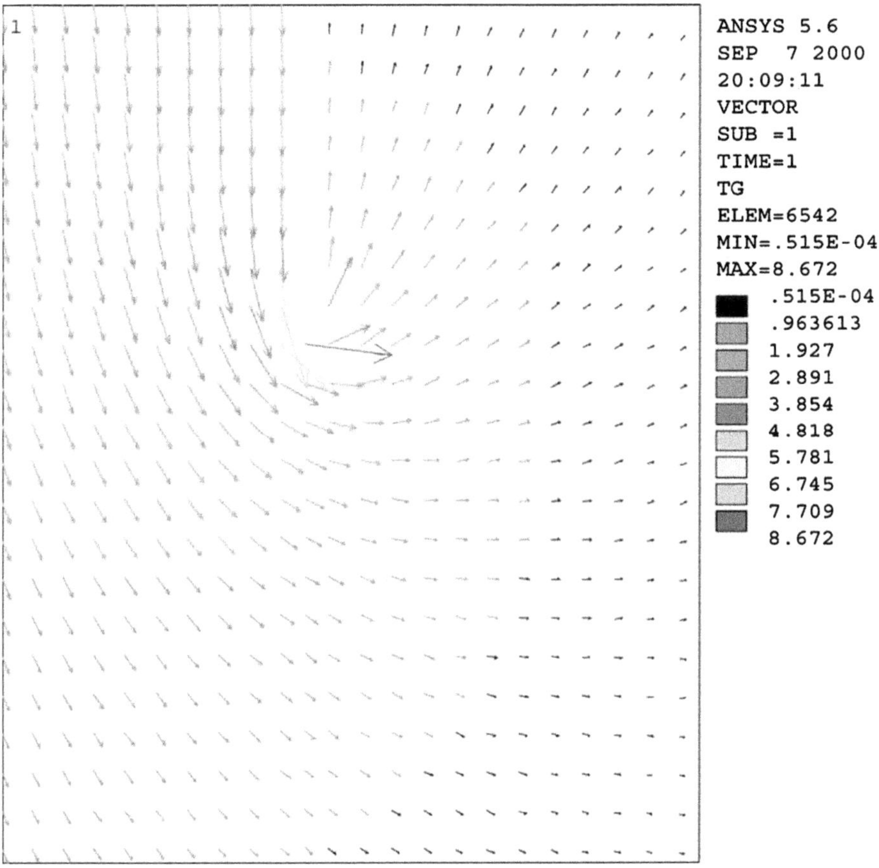

Figure 14.7 Seepage vector field around the lower edge of the bucket wall. The inside of the bucket is on the right-hand side of the vertical bar

between 370 to 1200 mm, as shown in Figure 14.8. The following tests were carried out:

(1) Static-pressure penetration test. The purpose of this test was to determine the curves of the penetration resistance against depth in the case of a single-bucket penetration into the silt by static pressure. The curves were to be compared with those of the negative-pressure penetration.
(2) Negative-pressure penetration tests. The purpose of these tests was to determine the mechanical processes that occur during the negative-pressure penetration of the bucket, to observe the variations of silt properties, the way the silt body collapses, and to determine the maximum negative pressure required to keep the silt body stable.

Figure 14.8 Model buckets used in the laboratory single bucket penetration tests. The buckets are constructed with 5 mm-thick steel sheet

Negative-pressure penetration tests for a 4-bucket foundation model platform were carried out in the model ocean. The purpose of the tests was to determine the mechanical processes occurring during the negative penetration of the 4-bucket model platform and to test the methods used to adjust the platform to be vertical and level. Figure 14.9 shows the model platform ready for the penetration tests.

3.3 Major results from the laboratory model tests

The results of laboratory model tests verified that the bucket foundation platform technology is feasible under the conditions of the engineering geology in the Yellow River submersed delta. Local engineers and scientists gained direct experience from the tests, and were ready to carry out field model tests before an operational platform was erected.

In the laboratory model tests, the piezometers, which were inserted at different depths into the sea-floor base and at different locations around the center of the bucket, were used to measure the head of water. Comparisons of the head-of-water values in the seepage field derived from the numerical simulation and those measured from the model tests indicated that the maximum relative deviation between the two data sets was 24.8 percent, the minimum was 0.5 percent, and the mean 9.5 percent. These

Figure 14.9 A 4-bucket foundation model platform prepared for a penetration test into the model ocean

results indicate that the FE model could be used as a calculation tool for the bucket foundation platform technology.

The laboratory model tests were used to produce the data needed for engineering design and field installation of the operational platform. The data were confirmed by field model tests and then passed to the users. In order to avoid silt-body collapse during the bucket foundation penetration, the maximum allowable negative pressure was determined by model tests. In the 4-bucket tests, platform tilt was considered. To solve this problem, 4-compartment water-tank technology was developed. The tank was loaded on to the top of the platform during penetration. By adjusting the water level in each compartment of the tank, platform tilt could be controlled.

4 Field model tests

After the successful laboratory model tests, further model tests took place in the ocean environment. The ocean model tests were designed as an intermediate phase between the laboratory model and the operational platform. Their purposes were to confirm the laboratory results and to gain direct experience of field penetration and installation of the bucket foundations. If the test models could withstand all ocean loading conditions, such as winds, tides, currents and waves, then the technology could be applied, with confidence, in constructing the operational platform.

Figure 14.10 Single bucket foundation in the sea before the penetration test. On the top is the water tank (painted blue and red) and on the bottom is the bucket (painted red)

The test site was near the site for the operational platform CB20B in water 6 m deep and with a 20 m thick top layer of the sea-floor silt sediment. The single-bucket and the 4-bucket foundation penetration tests were carried out at this site. Figure 14.10 shows the single-bucket foundation in the sea prior to the penetration test. The model is the same size as that on the operational platform and consists of two parts. On the top is a water tank 4000 mm high (painted blue and red) and on the bottom is the 4400 mm high bucket (painted red). Their diameters are 4000 mm.

The FE model was used to simulate the seepage field. The computation domain of the sediment surrounding the buckets was 80 m in diameter with a depth of 26.7 m, which was the total thickness of the sea-floor sediments. For a given penetration depth, the head of water and its gradient were calculated, and then the penetration resistance determined (Johnson and DeGraff 1988). Further, the penetration resistance was measured during the single-bucket penetration test. The results from the two methods are shown in Figure 14.11. It can be seen that the results from the two methods are similar, i.e. that the penetration resistance increases with penetration depth, increasing from 2000 kN at 1 m depth to 10,000 kN at 4 m depth.

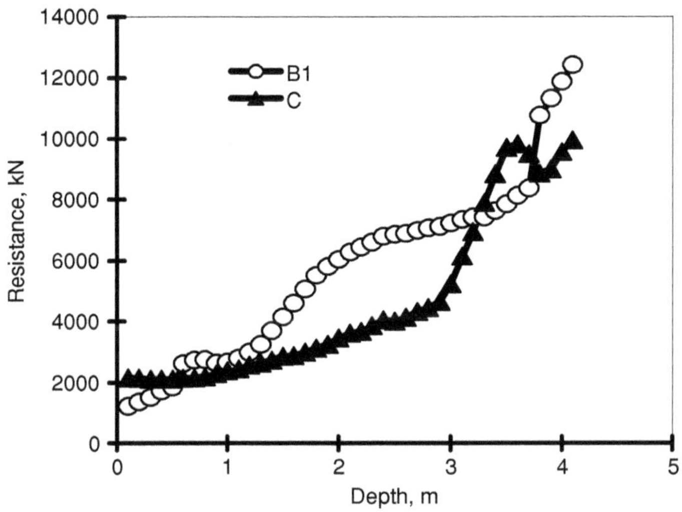

Figure 14.11 Suction penetration resistance vs. the penetration depth. Curve B1 represents the results derived from the FE model. Curve C represents the data measured from the single bucket penetration field test

Figure 14.12 shows a 4-bucket foundation model platform in the sea prior to the penetration test. This test was carried out at the same site as the single-bucket test. The results from the laboratory model tests were also validated through this test. The experience obtained from this 4-bucket foundation model platform penetration test served as an important reference for the erection of the operational platform.

5 CB20B production platform

The Shengli oilfield is the second-largest oil-production base in China. Its main body lies in the Yellow River Delta, with its working area covering the land and offshore shallow water. Long-term monitoring indicates that the sea floor in the offshore working area is relatively stable. The materials of the sea-floor sediment layer, which may reach 20 to 30 m thick, are silt and clay originating from the river deposits. Consequently, the geological conditions are favorable for constructing the bucket foundation platform.

The CB20B platform is the first production platform with a bucket foundation in the Shengli oilfield. The water depth at the platform site is 8.9 m. The platform has three parts: a deck module, deck frames and a 4-bucket foundation. The buckets have a diameter of 4000 mm, a height of 4400 mm and are made of 20 mm steel sheet. The designed penetration depth of the bucket is 4 m.

Figure 14.12 A 4-bucket foundation model platform in the sea before the penetration test

5.1 Geological conditions

A marine geological survey was carried out prior to the installation of the CB20B platform to determine the material parameters of the sediments in the region of the platform for use in the numerical analysis. The survey showed that the sea-floor sediments were composed of four layers of silt and clay with a total thickness of 26.7 m. Table 14.1 lists the thickness and permeability coefficients for the four sediment layers.

Figure 14.13 shows the curves of the side friction resistance and end resistance against the penetration depth obtained from the core pressure

Table 14.1 Thickness and permeability coefficients of sea-floor sediment layers at the CB20B platform site

Layer no.	*Sediment material*	*Thickness of layer (m)*	*Coef. of horizontal permeability (cm s^{-1})*	*Coef. of vertical permeability (cm s^{-1})*
1	silt	5.2	4.28×10^{-6}	2.35×10^{-6}
2	clay	3.0	2.75×10^{-7}	5.60×10^{-8}
3	silt	2.5	3.05×10^{-6}	5.43×10^{-7}
4	silt	16.0	3.40×10^{-8}	4.30×10^{-7}

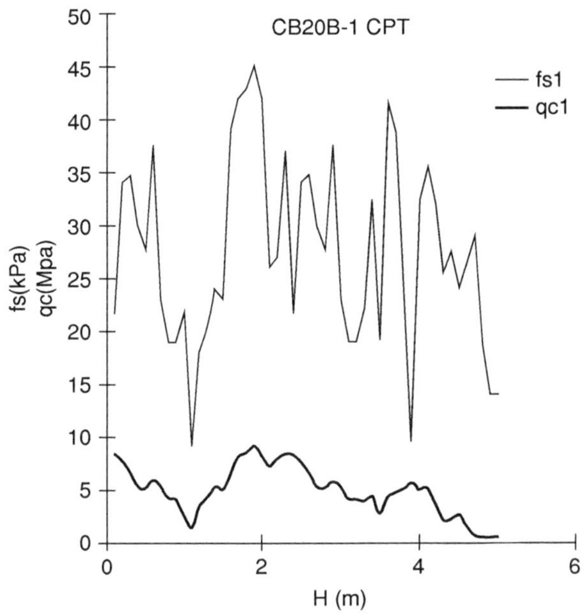

Figure 14.13 Friction resistances vs. the penetration depth for the sea-floor sediments taken at the site of the CB20B platform. Symbols 'fs' represents the side friction resistance per unit area and 'qc' the end resistance per unit area and H the penetration depth

test (CPT). The critical hydraulic gradient of the first (top) layer is 18.13. If the factor of safety (FOS) for the engineering operation is 3.0, then the allowable critical hydraulic gradient is 6.04.

5.2 Determination of the allowable suction pressure

The FE model results show that during the penetration process the seepage gradient at the lower end of the bucket wall increases considerably (see Figure 14.7). In terms of the allowable critical hydraulic gradient of the first layer, the allowable suction pressures are determined. The results are listed in Table 14.2, and these data are used to guide the installation operation for the CB20B platform.

Table 14.2 Allowable suction pressure against penetration depth determined by the FE method

H (m)	0.8	1.2	1.6	2.0	2.4	2.8	3.2	3.6	4.0
$P^{-\kappa}$	87.8	117.4	144.6	170.9	196.7	222.6	248.9	276.0	305.0

H: Penetration depth, $P^{-\kappa}$: Allowable suction pressure.

Figure 14.14 Bucket foundation with deck frames of the CB20B platform erected on the dock. The T-shaped structure on the top is the 4-compartment water tank

5.3 Field penetration and installation

The CB20B platform was constructed on land before being installed in the sea. The platform consisted of two major structural parts: part 1, the bucket foundation with deck frames; and part 2, the upper deck module. Figure 14.14 shows the bucket foundation with deck frames erected in the dock ready to be moved to the well site. The T-shaped structure on the top is the 4-compartment water tank of total volume 340 m^3, used to produce extra pressure for penetration and to adjust the level of the platform. During the in-sea erection, part 1 was hoisted up and placed vertically on the sea floor by a floating crane. Then water was pumped into the 4-compartment water tank. Forced by its own weight and the load water-tank weight, the bucket foundation penetrated the sea floor. Then the suction pumps were turned on to generate negative pressure in the bucket foundation. The foundation was gradually penetrated to the designed depth. After that, the 4-compartment water tank and other work equipment were removed and part 2, the upper deck module, was installed on top of part 1 and the fieldwork thus completed. The CB20B platform was successfully erected

Figure 14.15 CB20B bucket foundation production platform developed for the Shengli Oilfield in China. It is the first one of its kind in China and went into production in October 1999

on the well site, and production started in October 1999. Figure 14.15 shows the platform in November 1999.

During the bucket foundation penetration process, instruments and sensors were used to detect the negative pressure, penetration resistance, penetration rate and platform angle. As an example, Figure 14.16 shows the negative-pressure set-up process during the bucket foundation penetrating the sea floor. The four curves (shown in different colours) were measured for each of the four bucket foundations. An automatic monitoring system showed that the bucket foundation penetration depth was 4.2 m and the platform tilt was less than 0.5°. All technological specifications reach the designed standards.

6 Summary

Bucket foundation technology is a reliable, low-cost and environment-friendly technology. The experience of the Shengli oilfield of China indicates that this technology is especially suitable for the construction of oil and gas development platforms in the shallow-water areas which have thick sediment layers on the sea floor such as the Yellow River submersed delta. During

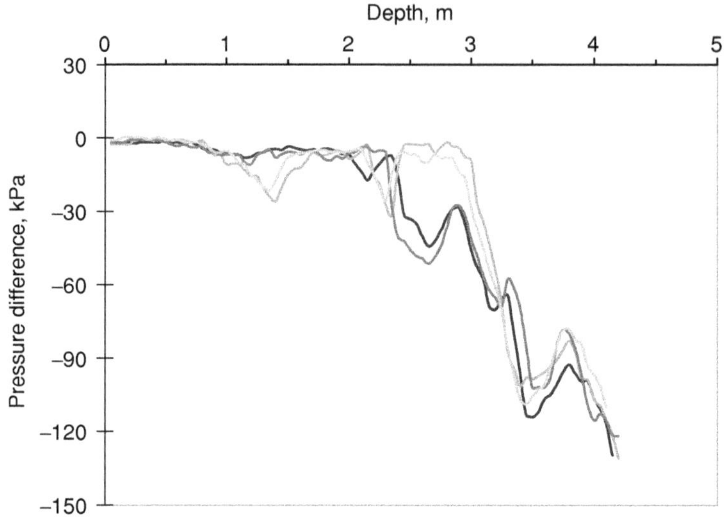

Figure 14.16 Negative pressure set up process during bucket foundation penetrating into the sea floor. The four curves (each in different colours) were measured for each of the four bucket foundations

the technological development, the FE method was used in the simulation of the seepage field generated by the suction penetration of the bucket foundation into the sea floor. Laboratory model tests and field model tests were used to validate the simulation results and to provide technical parameters and data for the engineering design, field penetration and installation of the platform. Using known and locally developed technologies, an operational bucket foundation platform, CB20B, was successfully erected at the designed site in the Chengbei offshore oilfield, Shengli oilfield of China, and has been put into production since late 1999.

Acknowledgments

This work was supported by the Chinese State 863 High Tech Development Program and partially supported by NASA (NAG513636 for QZ). The authors express their special thanks to Dongchang Sun, Songsen Xu and Xinjie Chu for their help and assistance in the laboratory experiments and in the field collection of data.

References

Baerheim, M. (1995) Development and structural design of the bucket foundations for the Europipe jacket, *Proceedings of Offshore Technology Conference 1995*, Report No. OTC7792, pp. 859–68.

Cedergren, H. R. (1997) *Seepage, Drainage and Flow Nets*, New York: John Wiley.

Erbich, C., Rognlien, B. and Tjelta, T. L. (1995) Geotechnical design of bucket foundations, *Proceedings of Offshore Technology Conference 1995*, Report No. OTC7793, pp. 869–83.

Greenkorn, R. A. (1983) *Flow Phenomena in Porous Media*, New York: Marcel Dekker.

Johnson, R. B. and DeGraff, J. V. (1988) *Principles of Engineering Geology*, New York: John Wiley.

Rusaas, P. (1995) Design, operations planning and experience from the marine operations for the Europipe jacket with bucket foundations, *Proceedings of Offshore Technology Conference 1995*, Report No. OTC7794, pp. 885–95.

Sun, D. (2000) Analysis of the suction penetration characteristics of bucket foundation platform, *J. Oceanogr. Huanghai Bohai Seas*, 18: 01–05.

Tjelta, T. L. (1995a) Geotechnical aspects of bucket foundations replacing piles for the Europipe 16/11E jacket, *Proceedings of Offshore Technology Conference 1995*, Report No. OTC7379, pp. 73–82.

Tjelta, T. L. (1995b) Geotechnical experience from the installation of the Europipe jacket with bucket foundations, *Proceedings of Offshore Technology Conference 1995*, Report No. OTC7795, pp. 897–908.

Zhang, S. and Chu, X. (2000) Field test and study of bucket foundation suction penetration and its application, *J. Oceanogr. Huanghai Bohai Seas*, 18: 51–5.

Zhang, S., Zheng, Q. and Liu, X. (2004) Finite element analysis of suction penetration seepage field of bucket foundation platform with application to offshore oilfield development, *Ocean Engineering*, 31: 1591–9.

Index

For Product Safety Concerns and Information please contact our EU representative GPSR@taylorandfrancis.com
Taylor & Francis Verlag GmbH, Kaufingerstraße 24, 80331 München, Germany

www.ingramcontent.com/pod-product-compliance
Ingram Content Group UK Ltd.
Pitfield, Milton Keynes, MK11 3LW, UK
UKHW020024200926
475702UK00014B/2046

9780367385880